Dynamic Vision for Perception and Control of Motion

Ernst D. Dickmanns

Dynamic Vision for Perception and Control of Motion

Ernst D. Dickmanns, Dr.-Ing.
Institut für Systemdynamik und Flugmechanik
Fakultät für Luft- und Raumfahrttechnik
Universität der Bundeswehr München
Werner-Heisenberg-Weg 39
85579 Neubiberg
Germany

British Library Cataloguing in Publication Data
Dickmanns, Ernst Dieter
Dynamic vision for perception and control of motion
1. Computer vision - Industrial applications 2. Optical detectors 3. Motor vehicles - Automatic control 4. Adaptive control systems
I. Title
629'.046
ISBN-13: 9781846286377

Library of Congress Control Number: 2007922344

ISBN 978-1-84628-637-7 e-ISBN 978-1-84628-638-4 Printed on acid-free paper

9 8 7 6 5 4 3 2 1

Springer Science+Business Media
springer.com

Preface

During and after World War II, the principle of feedback control became well understood in biological systems and was applied in many technical disciplines to relieve humans from boring workloads in systems control. N. Wiener considered it universally applicable as a basis for building intelligent systems and called the new discipline "Cybernetics" (the science of systems control) [Wiener 1948]. Following many early successes, these arguments soon were oversold by enthusiastic followers; at that time, many people realized that high-level decision–making could hardly be achieved only on this basis. As a consequence, with the advent of sufficient digital computing power, computer scientists turned to quasi-steady descriptions of abstract knowledge and created the field of "Artificial Intelligence" (AI) [McCarthy 1955; Selfridge 1959; Miller *et al.* 1960; Newell, Simon 1963; Fikes, Nilsson 1971]. With respect to achievements promised and what could be realized, a similar situation developed in the last quarter of the 20th century.

In the context of AI also, the problem of computer vision has been tackled (see, *e.g.*, [Selfridge, Neisser 1960; Rosenfeld, Kak 1976; Marr 1982]. The main paradigm initially was to recover a 3-D object shape and orientation from single images (snapshots) or from a few viewpoints. On the contrary, in aerial or satellite remote sensing, another application of image evaluation, the task was to classify areas on the ground and to detect special objects. For these purposes, snapshot images, taken under carefully controlled conditions, sufficed. "Computer vision" was a proper name for these activities since humans took care of accommodating all side constraints to be observed by the vehicle carrying the cameras.

When technical vision was first applied to vehicle guidance [Nilsson 1969], separate viewing and motion phases with static image evaluation (lasting for minutes on remote stationary computers in the laboratory) had been adopted initially. Even stereo effects with a single camera moving laterally on the vehicle between two shots from the same vehicle position were investigated [Moravec 1983]. In the early 1980s, digital microprocessors became sufficiently small and powerful, so that onboard image evaluation in near real time became possible. DARPA started its program "On strategic computing" in which vision architectures and image sequence interpretation for ground vehicle guidance were to be developed ('Autonomous Land Vehicle' ALV) [Roland, Shiman 2002]. These activities were also subsumed under the title "computer vision", and this term became generally accepted for a broad spectrum of applications. This makes sense, as long as dynamic aspects do not play an important role in sensor signal interpretation.

For autonomous vehicles moving under unconstrained natural conditions at higher speeds on nonflat ground or in turbulent air, it is no longer the computer which "sees" on its own. The entire body motion due to control actuation and to perturbations from the environment has to be analyzed based on information coming from many different types of sensors. Fast reactions to perturbations have to be derived from inertial measurements of accelerations and the onset of rotational rates, since vision has a rather long delay time (a few tenths of a second) until the enormous amounts of data in the image stream have been digested and interpreted sufficiently well. This is a well-proven concept in biological systems also operating under similar conditions, such as the vestibular apparatus of vertebrates with many cross-connections to ocular control.

This object-oriented sensor fusion task, quite naturally, introduces the notion of an extended presence since data from different times (and from different sensors) have to be interpreted in conjunction, taking additional delay times for control application into account. Under these conditions, it does no longer make sense to talk about "computer vision". It is the overall vehicle with an integrated sensor and control system, which achieves a new level of performance and becomes able "to see", also during dynamic maneuvering. The computer is the hardware substrate used for data and knowledge processing.

In this book, an introduction is given to an integrated approach to dynamic visual perception in which all these aspects are taken into account right from the beginning. It is based on two decades of experience of the author and his team at UniBw Munich with several autonomous vehicles on the ground (both indoors and especially outdoors) and in the air. The book deviates from usual texts on computer vision in that an integration of methods from "control engineering/systems dynamics" and "artificial intelligence" is given. Outstanding real-world performance has been demonstrated over two decades. Some samples may be found in the accompanying DVD. Publications on the methods developed have been distributed over many contributions to conferences and journals as well as in Ph.D. dissertations (marked "Diss." in the references). This book is the first survey touching all aspects in sufficient detail for understanding the reasons for successes achieved with real-world systems.

With gratitude, I acknowledge the contributions of the Ph.D. students S. Baten, R. Behringer, C. Brüdigam, S. Fürst, R. Gregor, C. Hock, U. Hofmann, W. Kinzel, M. Lützeler, M. Maurer, H.-G. Meissner, N. Mueller, B. Mysliwetz, M. Pellkofer, A. Rieder, J. Schick, K.-H. Siedersberger, J. Schiehlen, M. Schmid, F. Thomanek, V. von Holt, S. Werner, H.-J. Wünsche, and A. Zapp as well as those of my colleague V. Graefe and his Ph.D. students. When there were no fitting multi-microprocessor systems on the market in the 1980s, they realized the window-oriented concept developed for dynamic vision, and together we have been able to compete with "Strategic Computing". I thank my son Dirk for generalizing and porting the solution for efficient edge feature extraction in "Occam" to "Transputers" in the 1990s, and for his essential contributions to the general framework of the third-generation system *EMS* vision. The general support of our work in "control theory and application" by K.-D. Otto over three decades is appreciated as well as the infrastructure provided at the institute ISF by Madeleine Gabler.

Ernst D. Dickmanns

Acknowledgments

Support of the underlying research by the Deutsche Forschungs-Gemeinschaft (DFG), by the German Federal Ministry of Research and Technology (BMFT), by the German Federal Ministry of Defense (BMVg), by the Research branch of the European Union, and by the industrial firms Daimler-Benz AG (now DaimlerChrysler), Dornier GmbH (now EADS Friedrichshafen), and VDO (Frankfurt, now part of Siemens Automotive) through funding is appreciated.

Through the German Federal Ministry of Defense, of which UniBw Munich is a part, cooperation in the European and the Trans-Atlantic framework has been supported; the project "AutoNav" as part of an American-German Memorandum of Understanding has contributed to developing "expectation-based, multifocal, saccadic" (EMS) vision by fruitful exchanges of methods and hardware with the National Institute of Standards and Technology (NIST), Gaithersburgh, and with Sarnoff Research of SRI, Princeton.

The experimental platforms have been developed and maintained over several generations of electronic hardware by Ingenieurbüro Zinkl (VaMoRs), Daimler-Benz AG (VaMP), and by the staff of our electromechanical shop, especially J. Hollmayer, E. Oestereicher, and T. Hildebrandt. The first-generation vision systems have been provided by the Institut für Messtechnik of UniBwM/LRT. Smooth operation of the general PC-infrastructure is owed to H. Lex of the Institut für Systemdynamik und Flugmechanik (UniBwM /LRT/ ISF).

Contents

1 Introduction

The field of "vision" is so diverse and there are so many different approaches to the widespread realms of application that it seems reasonable first to inspect it and to specify the area to which the book intends to contribute. Many approaches to machine vision have started with the paradigm that easy things should be tackled first, like single snapshot image interpretation in unlimited time; an extension to more complex applications may later on build on the experience gained. Our approach on the contrary was to separate the field of dynamic vision from its (quasi-) static counterpart right from the beginning and to derive adequate methods for this specific domain. To prepare the ground for success, sufficiently capable methods and knowledge representations have to be introduced from the beginning.

1.1 Different Types of Vision Tasks and Systems

Figure 1.1 shows juxtapositions of several vision tasks occurring in everyday life. For humans, snapshot interpretation seems easy, in general, when the domain is well known in which the image has been taken. We tend to imagine the temporal context and the time when the image has been shot. From motion smear and unusual poses, the embedding of the snapshot in a well-known maneuver is concluded. So in general, even single images require background knowledge on motion processes in space for more in-depth understanding; this is often overlooked in machine or computer vision. The approach discussed in this book (bold italic letters in Figure 1.1) takes motion processes in "3-D space and time" as basic knowledge required for understanding image sequences in an approach similar to our own way of image interpretation. This yields a natural framework for using language and terms in the common sense.

Another big difference in methods and approaches required stems from the fact that the camera yielding the video stream is either stationary or moving itself. If moving, linear or/and rotational motion also may require special treatment. Surveillance is done, usually, from a stationary position while the camera may pan (rotation around a vertical axis, often also called yaw) and tilt (rotation around the horizontal axis, also called pitch) to increase its total field of view. In this case, motion is introduced purposely and is well controlled, so that it can be taken into account during image evaluation. If egomotion is to be controlled based on vision, the body carrying the camera(s) may be subject to strong perturbations, which cannot be predicted, in general.

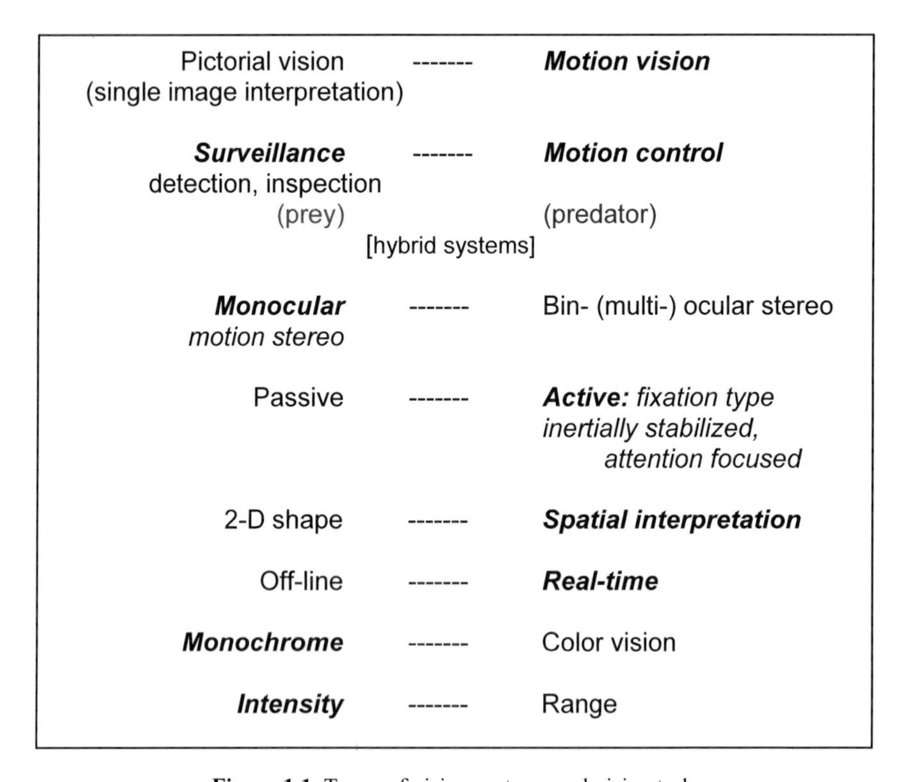

Figure 1.1. Types of vision systems and vision tasks

In cases with large rotational rates, motion blur may prevent image evaluation at all; also, due to the delay time introduced by handling and interpreting the large data rates in vision, stable control of the vehicle may no longer be possible.

Biological systems have developed close cooperation between inertial and optical sensor data evaluation for handling this case; this will be discussed to some detail and applied to technical vision systems in several chapters of the book. Also from biologists stems the differentiation of vision systems into "prey" and "predator" systems. The former strive to cover a large simultaneous field of view for detecting predators sufficiently early and approaching from any direction possible. Predators move to find prey, and during the final approach as well as in pursuit they have to estimate their position and speed relative to the dynamically moving prey quite accurate to succeed in a catch. Stereovision and high resolution in the direction of motion provides advantages, and nature succeeded in developing this combination in the vertebrate eye.

Once active gaze control is available, feedback of rotational rates measured by inertial sensors allows compensating for rotational disturbances on the own body just by moving the eyes (reducing motion blur), thereby improving their range of applicability. Fast moving targets may be tracked in smooth pursuit, also reducing motion blur for this special object of interest; the deterioration of recognition and tracking of other objects of less interest are accepted.

Since images are only in two dimensions, the 2-D framework looks most natural for image interpretation. This may be true for almost planar objects viewed approximately normal to their plane of appearance, like a landscape in a bird's-eye view. On the other hand, when a planar surface is viewed with the optical axis almost parallel to it from an elevation slightly above the ground, the situation is quite different. In this case, each line in the image corresponds to a different distance on the ground, and the same 3-D object on the surface looks quite different in size according to where it appears in the image. This is the reason why homogeneously distributed image processing by vector machines, for example, does have a hard time in showing its efficiency; locally adapted methods in image regions seem much more promising in this case and have proven their superiority. Interpreting image sequences in 3-D space with corresponding knowledge bases right from the beginning allows easy adaptation to range differences for single objects. Of course, the analysis of situations encompassing several objects at various distances now has to be done on a separate level, building on the results of all previous steps. This has been one of the driving factors in designing the architecture for the Third-generation "expectation-based, multi-focal saccadic" (EMS) vision system described in this book. This corresponds to recent findings in well-developed biological systems where for image processing and action planning based on the results of visual perception, different areas light up in magnetic resonance images [Talati, Hirsch 2005].

Understanding motion processes of 3-D objects in 3-D space while the body carrying the cameras also moves in 3-D space, seems to be one of the most difficult tasks in real-time vision. Without the help of inertial sensing for separating egomotion from relative motion, this can hardly be done successfully, at least in dynamic situations.

Direct range measurement by special sensors such as radar or laser range finders (LRF) would alleviate the vision task. Because of their relative simplicity and low demand of computing power, these systems have found relatively widespread application in the automotive field. However, with respect to resolution and flexibility of data exploitation as well as hardware cost and installation volume required, they have much less potential than passive cameras in the long run with computing power available in abundance. For this reason, these systems are not included in this book.

1.2 Why Perception and Action?

For technical systems which are intended to find their way on their own in an ever changing world, it is impossible to foresee every possible event and to program all required capabilities for appropriate reactions into its software from the beginning. To be flexible in dealing with situations actually encountered, the system should have perceptual and behavioral capabilities which it may expand on its own in response to new requirements. This means that the system should be capable of judging the value of control outputs in response to measured data; however, since outputs of control affect state variables over a certain amount of time, ensuing time histories have to be observed and a temporally deeper understanding has to be developed. This is exactly what is captured in the "dynamic models" of systems theory (and what biological systems may store in neuronal delay lines).

Also, through these time histories, the ground is prepared for more compact "frequency domain" (integral) representations. In the large volume of literature on linear systems theory, time constants T as the inverse of eigenvalues of first-order system components, as well as frequency, damping ratio, and relative phase as characteristic properties of second-order components are well known terms for describing temporal characteristics of processes, *e.g.*, [Kailath 1980]. In the physiological literature, the term "temporal Gestalt" may even be found [Ruhnau 1994a, b], indicating that temporal shape may be as important and characteristic as the well known spatial shape.

Usually, control is considered an output resulting from data analysis to achieve some goal. In a closed-loop system, where one of its goals is to adapt to new situations and to act autonomously, control outputs may be interpreted as questions asked with respect to real-world behavior. Dynamic reactions are now interpreted to better understand the behavior of a body in various states and under various environmental conditions. This opens up a new avenue for signal interpretation: beside its use for state control, it is now also interpreted for system identification and modeling, that is, learning about its temporal behavioral characteristics.

In an intelligent autonomous system, this capability of adaptation to new situations has to be available to reduce dependence on maintenance and adaptation by human intervention. While this is not yet state of the art in present systems, with the computing power becoming available in the future, it clearly is within range. The methods required have been developed in the fields of system identification and adaptive control.

The sense of vision should yield sufficient information about the near and farther environment to decide when state control is not so important and when more emphasis may be put on system identification by using special control inputs for this purpose. This approach also will play a role when it comes to defining the notion of a "self" for the autonomous vehicle.

1.3 Why Perception and Not Just Vision?

Vision does not allow making a well-founded decision on absolute inertial motion when another object is moving close to the ego-vehicle and no background can be seen in the field of view (known to be stationary). Inertial sensors like accelerometers and angular rate sensors, on the contrary, yield the corresponding signals for the body they are mounted on; they do this practically without any delay time and at high signal rates (up to the kHz range).

Vision needs time for the integration of light intensity in the sensor elements (33 1/3, respectively, 40 ms corresponding to the United States or European standard), for frame grabbing and communication of the (huge amount of) image data, as well as for feature extraction, hypothesis generation, and state estimation. Usually, three to five video cycles, that are 100 to 200 ms, will have passed until a control output derived from vision will hit the real world. For precise control of highly dynamic systems, this time delay has to be taken into account.

Since perturbations should be counteracted as soon as possible, and since visually measurable results of perturbations are the second integral of accelerations with corresponding delay times, it is advisable to have inertial sensors in the system for early pickup of perturbations. Because long-term stabilization may be achieved using vision, it is not necessary to resort to expensive inertial sensors; on the contrary, when jointly used with vision, inexpensive inertial sensors with good properties for the medium- to high-frequency part are sufficient as demonstrated by the vestibular systems in vertebrates.

Accelerometers are able to measure rather directly the effects of most control outputs; this alleviates system identification and finding the control outputs for reflex-like counteraction of perturbations. Cross-correlation of inertial signals with visually determined signals allows temporally deeper understanding of what in the natural sciences is called "time integrals" of input functions.

For all these reasons, the joint use of visual and inertial signals is considered mandatory for achieving efficient autonomously mobile platforms. Similarly, if special velocity components can be measured easily by conventional devices, it does not make sense to try to recover these from vision in a "purist" approach. These conventional signals may alleviate perception of the environment considerably since the corresponding sensors are mounted onto the body in a fixed way, while in vision the measured feature values have to be assigned to some object in the environment according to just visual evidence. *There is no constantly established link for each measurement value in vision as is the case for conventional sensors.*

1.4 What are Appropriate Interpretation Spaces?

Images are two-dimensional arrays of data; the usual array size today is from about 64 × 64 for special "vision" chips to about 770 × 580 for video cameras (special larger sizes are available but only at much higher cost, *e.g.,* for space or military applications). A digitized video data stream is a fast sequence of these images with data rates up to ~ 11 MB/s for black and white and up to three times this amount for color.

Frequently, only fields of 320 × 240 pixels (either only the odd or the even lines with corresponding reduction of the resolution within the lines) are being evaluated because of computing power missing. This results in a data stream per camera of about 2 MB/s. Even at this reduced data rate, the processing power of a single microprocessor available today is not yet sufficient for interpreting several video signals in parallel in real time. High-definition TV signals of the future may have up to 1080 lines and 1920 pixels in each line at frame rates of up to 75 Hz; this corresponds to data rates of more than 155 MB/s. Machine vision with this type of resolution is way out in the future.

Maybe, uniform processing of entire images is not desirable at all, since different objects will be seen in different parts of the images, requiring specific image processing algorithms for efficient evaluation, usually. Very often, lines of discontinuity are encountered in images, which should be treated with special methods differing essentially from those used in homogeneous parts. Object- and situation-dependent methods and parameters should be used, controlled from higher evaluation levels.

The question thus is, whether any basic feature extraction should be applied uniformly over the entire image region. In biological vision systems, this seems to be the case, for example, in the striate cortex (V1) of vertebrates where oriented edge elements are detected with the help of corresponding receptive fields. However, vertebrate vision has nonhomogeneous resolution over the entire field of view. Foveal vision with high resolution at the center of the retina is surrounded by receptive fields of increasing spread and a lower density of receptors per unit of area in the radial direction.

Vision of highly developed biological systems seems to ask three questions, each of which is treated by a specific subsystem:

1. Is there something of special interest in a wide field of view?
2. What is it precisely, that attracted interest in question one? Can the individual object be characterized and classified using background knowledge? What is its relative state "here and now"?
3. What is the situation around me and how does it affect optimal decisions in behavior for achieving my goals? For this purpose, a relevant collection of objects should be recognized and tracked, and the likely future behavior should be predicted.

To initialize the vision process at the beginning and to detect new objects later on, it is certainly an advantage to have a bottom-up detection component available all over the wide field of view. Maybe, just a few algorithms based on coarse resolution for detecting interesting groups of features will be sufficient to achieve this goal. The question is, how much computing effort should be devoted to this bottom-up component compared to more elaborate, model based top-down components for objects already detected and being tracked. Usually, single objects cover only a small area in an image of coarse resolution.

To answer question 2 above, biological vision systems direct the foveal area of high resolution by so-called saccades, which are very fast gaze direction changes with angular rates up to several hundred degrees per second, to the group of features arousing most interest. Humans are able to perform up to five saccades per second with intermediate phases of smooth pursuit (tracking) of these features, indicating a very dynamic mode of perception (time-sliced parallel processing). Tracking can be achieved much more efficiently with algorithms controlled by prediction according to some model. Satisfactory solutions may be possible only in special task domains for which experience is available from previous encounters.

Since prediction is a very powerful tool in a world with continuous processes, the question arises: What is the proper framework for formulating the continuity conditions? Is the image plane readily available as plane of reference? However, it is known that the depth dimension in perspective mapping has been lost completely: All points on a ray have been mapped into a single point in the image plane, irrespective of their distance, which has been lost. Would it be better to formulate all continuity conditions in 3-D physical space and time? The corresponding models are available from the natural sciences since Newton and Leibnitz have found that differential equations are the proper tools for representing these continuity conditions in generic form; over the last decades, simulation technology has provided the methods for dealing with these representations on digital computers.

In communication technology and in the field of pattern recognition, video processing in the image plane may be the best way to go since no understanding of the content of the scene is required. However, for orienting oneself in the real world through image sequence analysis, early transition to the physical interpretation space is considered highly advantageous because it is in this space that occlusions become easily understandable and motion continuity persists. Also, it is in this space that inertial signals have to be interpreted and that integrals of accelerations yield 3-D velocity components; integrals of these velocities yield the corresponding positions and angular orientations for the rotational degrees of freedom. Therefore, for visual dynamic scene understanding, images are considered intermediate carriers of data containing information about the spatiotemporal environment. To recover this information most efficiently, all internal modeling in the interpretation process is done in 3-D space and time, and the transition to this representation should take place as early as possible. Knowledge for achieving this goal is specific to single objects and the generic classes to which they belong. Therefore, to answer question 2 above, specialist processes geared to classes of objects and individuals of these classes observed in the image sequence should be designed for direct interpretation in 3-D space and time.

Only these spatiotemporal representations then allow answering question 3 by looking at these data of all relevant objects in the near environment for a more extended period of time. To be able to understand motion processes of objects more deeply in our everyday environment, a distinction has to be made between classes of objects. Those obeying simple laws of motion from physics are the ones most easily handled (*e.g.*, by some version of Newton's law). Light objects, easily moved by stochastically appearing (even light) winds become difficult to grasp because of the variable properties of wind fields and gusts.

Another large class of objects – with many different subclasses – is formed by those able to sense properties of their environment and to initiate movements on their own, based on a combination of the data sensed and background knowledge internally stored. These special objects will be called **subjects**; all animals including humans belong to this (super-) class as well as autonomous agents created by technical means (like robots or autonomous vehicles). The corresponding subclasses are formed by combinations of perceptual and behavioral capabilities and, of course, their shapes. Beside their shapes, individuals of subclasses may be recognized also by stereotypical motion patterns (like a hopping kangaroo or a winding snake).

Road vehicles (independent of control by a human driver or a technical subsystem) exhibit typical behaviors depending on the situation encountered. For example, they follow lanes and do convoy driving, perform lane changes, pass other vehicles, turn off onto a crossroad or slow down for parking. All of the maneuvers mentioned are well known to human drivers, and they recognize the intention of performing one of those by its typical onset of motion over a short period of time. For example, a car leaving the center of its lane and moving consistently toward the neighboring lane is assumed to initiate a lane change. If this occurs within the safety margin in front, egomotion should be adjusted to this (improper) behavior of other traffic participants. This shows that recognition of the intention of other subjects is important for a defensive style of driving. This cannot be recognized without knowledge of temporally extended maneuvers and without observing behavioral patterns of subjects in the environment. Question 3 above, thus, is not answered by interpreting image patterns directly but by observing symbolic representations resulting as answers to question 2 for a number of individual objects/subjects over an extended period of time.

Simultaneous interpretation of image sequences on multiple scales in 3-D space and time is the way to satisfy all requirements for safe and goal-oriented behavior.

1.4.1 Differential Models for Perception "Here and Now"

Experience has shown that the simultaneous use of differential and integral models on different scales yields the most efficient way of data fusion and joint data interpretation. Figure 1.2 shows in a systematic fashion the interpretation scheme developed. Each of the axes is subdivided into four scale ranges. In the upper left corner the point "here and now" is shown as the point where all interaction with the real world takes place. The second scale range encompasses the local (as opposed to global) environment which allows introducing new differential concepts compared to the pointwise state. Local embedding, with characteristic properties

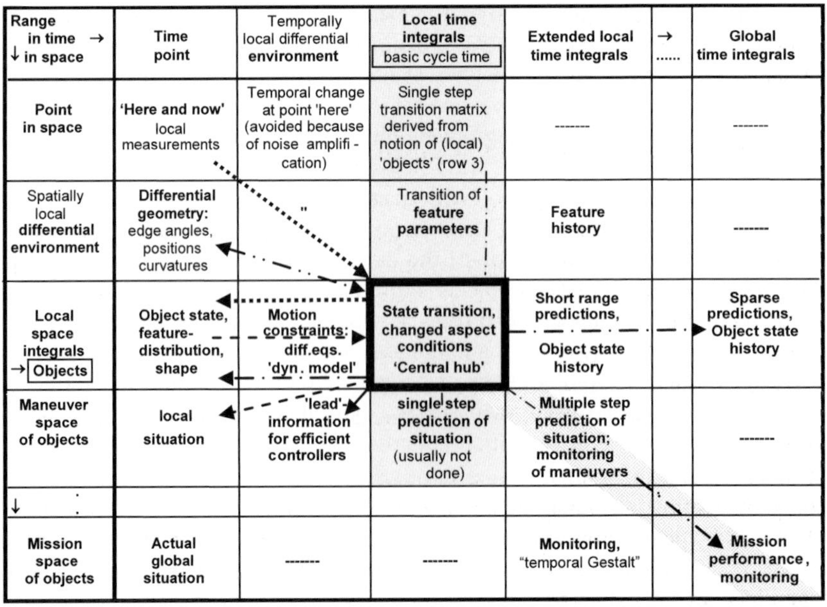

Range in time → ↓ in space	Time point	Temporally local differential environment	Local time integrals (basic cycle time)	Extended local time integrals	→	Global time integrals
Point in space	'Here and now' local measurements	Temporal change at point 'here' (avoided because of noise amplification)	Single step transition matrix derived from notion of (local) 'objects' (row 3)	-------		-------
Spatially local differential environment	Differential geometry: edge angles, positions curvatures	"	Transition of feature parameters	Feature history		-------
Local space integrals → Objects	Object state, feature-distribution, shape	Motion constraints: diff.eqs. 'dyn. model'	State transition, changed aspect conditions 'Central hub'	Short range predictions, Object state history		Sparse predictions, Object state history
Maneuver space of objects	local situation	'lead' information for efficient controllers	single step prediction of situation (usually not done)	Multiple step prediction of situation; monitoring of maneuvers		-------
↓ :						
Mission space of objects	Actual global situation	-------	-------	Monitoring, "temporal Gestalt"		Mission performance, monitoring

Figure 1.2. Multiple interpretation scales in space and time for dynamic perception. Vertical axis: 3-D space; horizontal axis: time

such as spatial or temporal change rates, spatial gradients, or directions of extreme values such as intensity gradients are typical examples.

These differentials have shown to be powerful concepts for representing knowledge about physical properties of classes of objects. Differential equations represent the natural mathematical element for coding knowledge about motion processes in the real world. With the advent of the Kalman filter [Kalman 1960], they have become the key element for obtaining the best state estimate of the variables describing the system, based on recursive methods implementing a least-squares model fit. Real-time visual perception of moving objects is hardly possible without this very efficient approach.

1.4.2 Local Integrals as Central Elements for Perception

Note that the precise definition of what is local depends on the problem domain investigated and may vary in a wide range. The third column and row in Figure 1.2 are devoted to "local integrals"; this term again is rather fuzzy and will be defined more precisely in the task context. On the timescale, it means the transition from analog (continuous, differential) to digital (sampled, discrete) representations. In the spatial domain, typical local integrals are rigid bodies, which may move as a unit without changing their 3-D shape.

These elements are defined such that the intersection in field (3, 3) in Figure 1.2 becomes the central hub for data interpretation and data fusion: it contains the individual objects as units to which humans attach most of their knowledge about the real world. Abstraction of properties has lead to generic classes which allow subsuming a large variety of single cases into one generic concept, thereby leading to representational efficiency.

1.4.2.1 Where is the Information in an Image?

It is well known that information in an image is contained in local intensity changes: A uniformly gray image has only a few bits of information, namely, (1) the gray value and (2) uniform distribution of this value over the entire image. The image may be completely described by three bytes, even though the amount of data may be about 400 000 bytes in a TV frame or even 4 MB (2k × 2k pixels). If there are certain areas of uniform gray values, the boundary lines of these areas plus the internal gray values contain all the information in the image. This object in the image plane may be described with much less data than the pixel values it encompasses.

In a more general form, image areas defined by a set of properties (shape, texture, color, joint motion, *etc.*) may be considered image objects, which originated from 3-D objects by perspective mapping. Due to the numerous aspect conditions, which such an object may adopt relative to the camera, its potential appearances in the image plane are very diverse. Their representation will require orders of magnitude more data for an exhaustive description than its representation in 3-D space plus the laws of perspective mapping, which are the same for all objects. Therefore, an object is defined by its 3-D shape, which may be considered a local spatial integral of its differential geometry description in curvature terms. Depending on the task at hand, both the differential and the integral representation, or a combination of both may be used for visual recognition. As will be shown for the example of road vehicle guidance, the parallel use of these models in different parts of the overall recognition process and control system may be most efficient.

1.4.2.2 To Which Units Do Humans Affix Knowledge?

Objects and object classes play an important role in human language and in learning to understand "the world". This is true for their appearance at one time, and also for their motion behavior over time.

On the temporal axis, the combined use of differential and integral models may allow us to refrain from computing optical flow or displacement vector fields, which are very compute-intensive and susceptible to noise. Because of the huge amount of data in a single image, this is not considered the best way to go, since an early transition to the notion of physical objects or subjects with continuity conditions in 3-D space and time has several advantages: (1) it helps cut the amount of data required for adequate description, and (2) it yields the proper framework for applying knowledge derived from previous encounters (dynamic models, stereotypical control maneuvers, *etc.*). For this reason, the second column in Figure 1.2 is avoided intentionally in the 4-D approach. This step is replaced by the well-known observer techniques in systems dynamics (Kalman filter and derivatives, Luenberger observers). These recursive methods reconstruct the time derivatives of state variables by prediction error feedback and knowledge about the dynamic behavior of the object and (for the Kalman filter) of the statistical properties of the system (dubbed "plant" in systems dynamics) and of the measurement processes. The stereotypical behavioral capabilities of subjects in different situations form an important part of the knowledge base.

Two distinctly different types of "local temporal integrals" are used widely: Single step integrals for video sampling and multiple step (local) integrals for maneuver understanding. Through the imaging process, the analog motion process in the real world is made discrete along the time axis. By forming the (approximate, since linearized) integrals, the time span of the analog video cycle time (33 1/3 ms in the United States and 40 ms in Europe, respectively, half these values for the fields) is bridged by discrete transition matrices from kT to $(k + 1)T$, k = running index.

Even though the intensity values of each pixel are integrals over the full range or part of this period, they are interpreted as the actually sampled intensity value at the time of camera readout. Since all basic interpretations of the situation rest on these data, control output is computed newly only after this period; thus, it is constant over the basic cycle time. This allows the analytical computation of the corresponding state transitions, which are evaluated numerically for each cycle in the recursive estimation process (Chapter 6); these are used for state prediction and intelligent control of image feature extraction.

1.4.3 Global Integrals for Situation Assessment

More complex situations encompassing many objects or missions consisting of sequences of mission elements are represented in the lower right corner of Figure 1.2. Again, how to best choose the subdivisions and the absolute scales on the time axis or in space depends very much on the problem area under study. This will be completely different for a task in manufacturing of micro-systems compared to one in space flight. The basic principle of subdividing the overall task, however, may be according to the same scheme given in Figure 1.2, even though the technical elements used may be completely different.

On a much larger timescale, the effect of entire feed-forward control time histories may be predicted which have the goal of achieving some special state changes or transitions. For example, lane change of a road vehicle on a freeway, which may take 2 to 10 seconds in total, may be described as a well-structured sequence of control outputs resulting in a certain trajectory of the vehicle. At the end of the maneuver, the vehicle should be in the neighboring lane with the same state variables otherwise (velocity, lateral position in the lane, heading). The symbol "lane change", thus, stands for a relatively complex maneuver element which may be triggered from the higher levels on demand by just using this symbol (maybe together with some parameters specifying the maneuver time and, thereby, the maximal lateral acceleration to be encountered). Details are discussed in Section 3.4.

These "maneuver elements", defined properly, allow us to decompose complex maneuvers into stereotypical elements which may be pieced together according to the actual needs; large sections of these missions may be performed by exploiting feedback control, such as lane following and distance keeping for road vehicles. Thereby, scales of distances for entire missions depend on the process to be controlled; these will be completely different for "autonomously guided vehicles" (AGVs) on the factory floor (hundreds of meters) compared to road vehicles (tens of km) or even aircraft (hundreds or thousands of km).

The design of the vision system should be selected depending on the task at hand (see next section).

1.5 What Type of Vision System Is Most Adequate?

For motion control, due to inertia of a body, the actual velocity vector determines where to look to avoid collisions with other objects. Since lateral control may be applied to some extent and since other objects and subjects may have a velocity vector of their own, the viewing range should be sufficiently large for detecting all possible collision courses with other objects. Therefore, the simultaneous field of view is most critical nearby.

On the other hand, if driving at high speed is required, the look-ahead range should be sufficiently large for reliably detecting objects at distances which allow safe braking. At a speed of 30 m/s (108 km/h or about 65 mph), the distance for braking [with a deceleration level of 0.4 Earth gravity g (9.81 m/s^2, that is $a_x \approx -4$

m/s²) and with 0.5 seconds reaction time] is 15 + 113 = 128 m. For half the magnitude in deceleration (– 2 m/s^2, *e.g.*, under unfavorable road conditions) the braking distance would be 240 m.

Reliable distance estimation for road vehicles occurs under mapping conditions with at least about 20 pixels on the width of the vehicle (typically of about 2 m in dimension). The total field of view of a single camera at a distance of 130 m, where this condition is satisfied, will be about 76 m (for ~ 760 pixel per line). This corresponds to an aperture angle of ~ 34°. This is certainly not enough to cover an adequate field of view in the near range. Therefore, at least a bifocal camera arrangement is required with two different focal lengths (see Figure 1.3).

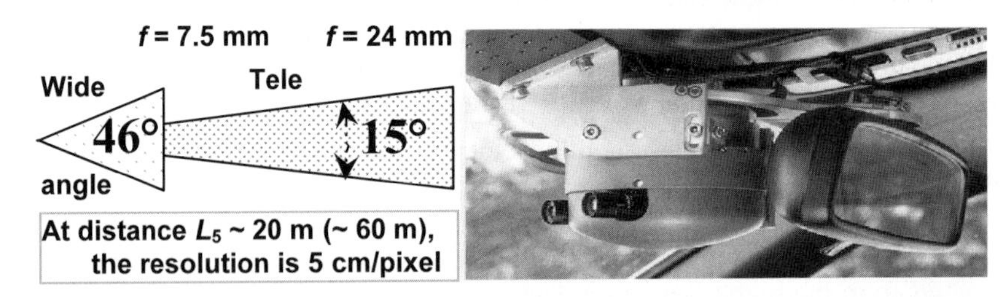

Figure 1.3. Bifocal arrangement of miniature TV–cameras on a pan platform in front of the rear view mirror of test vehicles **VaMP** and **VITA 2**, Prometheus, 1994. Left: Fields of view and ranges (schematically), right: System realized in VaMP

For a rather flexible high performance "technical eye" a trifocal camera arrangement as shown in Figure 1.4 is recommended. The two wide-angle CCD-cameras with focal length of 4 to 6 mm and with divergent optical axes do have a central range of overlapping image areas, which allows stereo–interpretation nearby. In total, a field of view of about 100 to 130 degrees can be covered; this allows surveying about one–third of the entire panorama.

The mild telecamera with three to four times the focal length of the wide-angle one should be a three–chip color camera for more precise object recognition. Its field of view is contained in the stereo field of view of the wide-angle cameras such that trinocular stereointerpretation becomes possible [Rieder 1996].

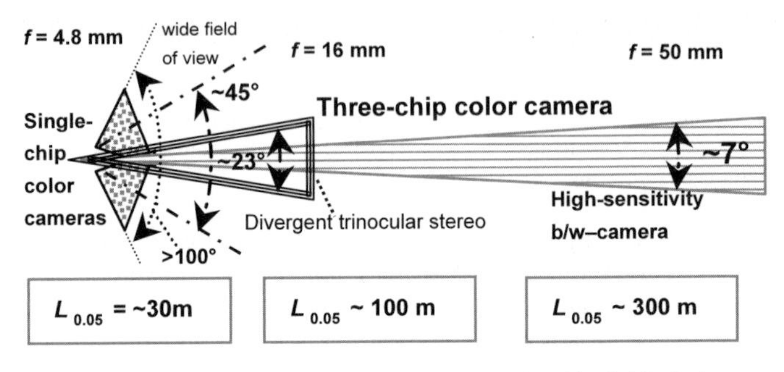

Figure 1.4. Trifocal camera arrangement with wide field of view

To detect objects in special areas of interest far away, a camera with a third focal length (again with a factor of 3 to 4 relative to the mild telelens), and the field of view within that of the mild telecamera should be added (see Figure 1.4). This camera may be chosen to be especially light-sensitive; black-and-white images may be sufficient to limit the data rate. The focal length ratio of 4 does have the advantage that the coarser image represents the same scene at a resolution corresponding to the second pyramidal stage of the finer one.

This type of sensor combination is ideally suited for active viewing direction control: the coarse resolution, large simultaneous field of view allows discovering objects of possible interest in a wide area, and a viewing direction change will bring this object into the center of the images with higher resolution. Compared to a camera arrangement with maximal resolution in the same entire field of view, the solution shown has only 2 to 4 % the data rate. It achieves this in exchange for the need of fast viewing direction control and at the expense of delay times required to perform these gaze changes. Figure 1.5 gives an impression of the fields of view of this trifocal camera arrangement.

Figure 1.5. Fields of view of trifocal camera arrangement. Bottom: Two divergent wide angle cameras; top left: mild tele camera, top right: strong tele-camera. Dashed white lines show enlarged sections

The lower two wide-angle images have a central region of overlap marked by vertical white lines. To the left, the full road junction is imaged with one car coming out of the crossroad and another one just turning into the crossroad; the rear of this vehicle and the vehicle directly in front can be seen in the upper left image of the mild telecamera. This even allows trinocular stereo interpretation. The region marked in white in this mild teleimage is shown in the upper right as a full image of the strong telecamera. Here, letters on the license plate can be read, and it can be seen from the clearly visible second rearview mirror on the left-hand side that there is a second car immediately in front of the car ahead. The number of pixels per area on the same object in this image is one hundred times that of the wide-angle images.

For inertial stabilization of the viewing direction when riding over a nonsmooth surface or for aircraft flying in a turbulent air, an active camera suspension is needed anyway. The simultaneous use of almost delay-free inertial measurements (time derivatives such as angular rates and linear accelerations) and of images, whose interpretation introduces several tenths of a second delay time, requires extended representations along the time axis. There is no single time for which it is possible to make consistent sense of all data available. Only the notion of an "extended presence" allows arriving at an efficient invariant interpretation (in 4-D!). For this reason, the multifocal, saccadic vision system is considered to be the preferable solution for autonomous vehicles in general.

1.6 Influence of the Material Substrate on System Design: Technical vs. Biological Systems

Biological vision systems have evolved over millions of generations with the selection of the fittest for the ecological environment encountered. The basic neural substrate developed (carbon-based) may be characterized by a few numbers. The electrochemical units do have switching times in the millisecond (ms) range; the traveling speed of signals is in the 10 to 100 m/s range. Cross-connections between units exist in abundance (1000 to 10 000 per neuron). A single brain consists of up to 10^{11} of these units. The main processing step is summation of the weighted input signals which contain up to now unknown (multiple?) feedback loops [Handbook of Physiology 1984, 1987].

These systems need long learning times and adapt to new situations only slowly. In contrast, technical substrates for sensors and microprocessors (silicon-based) have switching times in the nanosecond range (a factor of 10^6 compared to biological systems). They are easily programmable and have various computational modes between which they can switch almost instantaneously; however, the direct cross-connections to other units are limited in number (one to six, usually) but may have very high bandwidth (in the hundreds of MB/s range).

While a biological eye is a very complex unit containing several types and sizes of sensors and computing elements, technical imaging sensors are rather simple up to now and mostly homogeneous over the entire array area. However, from television and computer graphics, it is well known that humans can interpret the images thus generated without problems in a natural way if certain standards are maintained.

In developing dynamic machine vision, two groups of thinking have formed: One tries to mimic biological vision systems on the silicon substrate available, and the other continues to build on the engineering platform developed in systems– and computer science.

A few years ago, many systems were investigated with single processors devoted to single pixels (Connection Machine [Hillis 1985, 1992], Content-Addressable Associative Parallel Processors (CAAPP) [Scudder, Weems 1990] and others). The trend now clearly is toward more coarsely granulated parallel architectures. Since a single microprocessor on the market at the turn of the century is capable of performing about 10^9 instructions per second, this means in excess of 2000 instructions per pixel of a 770 × 525 pixel image. Of course, this should not be confused with information processing operations. For the year 2010, general-purpose PC processors are expected to have a performance level of about 10^{11} instructions per second.

On the other hand, the communication bandwidths of single channels will be so high, that several image matrices may be transferred at a sufficiently high rate to allow smooth recognition and control of motion processes. (One should refrain from video norms, presently dominating the discussion, once imaging sensors with digital output are in wide use.) Therefore, there is no need for more elaborate data processing on the imaging chip except for ensuring sufficiently high intensity dynamics. Technical systems do not have the bandwidth problems, which may have forced biological systems to do extensive data preprocessing near the retina (from 120 million light sensitive elements in the retina to 1.2 million nerves leading to the lateral geniculate nucleus in humans).

Interesting studies have been made at several research institutions which tried to exploit analog data processing on silicon chips [Koch 1995]; future comparisons of results will have to show whether the space needed on the chip for this purpose can be justified by the advantages claimed.

The mainstream development today is driven by commercial TV for the sensors and by personal computers and games for the processors. With an expected increase in computing power of one order of magnitude every 4 to 5 years over the next decade, real-time machine vision will be ready for a wide range of applications using conventional engineering methods as represented by the 4-D approach.

A few (maybe a dozen) of these processors will be sufficient for solving even rather complex tasks like ground and air vehicle guidance; dual processors on a single chip are just entering the market. It is the goal of this monograph to make the basic methods needed available to a wide public for efficient information extraction from huge data streams.

1.7 What Is Intelligence? A Practical (Ecological) Definition

The sensors of complex autonomous biological or technical systems yield an enormous data rate containing information about both the state of the vehicle body relative to the environment and about other objects or subjects in the environment. It is the task of an intelligent information extraction (data interpretation) system to quickly get rid of as many data as possible, however simultaneously, to retain all of the essential information for the task to be solved. Essential information is geared to task domains; however, complex systems like animals and autonomous vehicles do not have just one single task to perform. Depending on their circumstances, quite different tasks may predominate.

Systems will be labeled intelligent if they are able to:

- recognize situations readily that require certain behavioral capabilities and
- trigger this behavior early and correctly, so that the overall effort to deal with the situation is lower than for direct reaction to some combination of values measured but occurring later (tactical – strategic differentiation).

This "insight" into processes in the real world is indicative of an internal temporal model for this process in the interpretation system. It is interesting to note that the word *"intelligent"* is derived from the Latin stem "inter-legere": To read in between the lines. This means to understand what is not explicitly written down but what can be inferred from the text, given sufficient background knowledge and the capability of associative thinking. Therefore, intelligence understood in this sense requires background knowledge about the processes to be perceived and the capability to recognize similar or slightly different situations in order to be able to extend the knowledge base for correct use.

Since the same intelligent system will have to deal with many different situations, those individuals will be superior which can extract information from actual experience not just for the case at hand but also for proper use in other situations. This type of "knowledge transfer" is characteristic of truly intelligent systems. From this point of view, intelligence is not the capability of handling some abstract symbols in isolation but to have symbolic representations available that allow adequate or favorable decisions for action in different situations which have to be recognized early and reliably.

These actions may be feedback control laws with very fast implementations gearing control output directly to measured quantities (reflex-like behavior), or stereotypical feed-forward control time histories invoked after some event, known to achieve the result desired (rule-based instantiation). To deal robustly with perturbations common in the real world, expectations of state variable time histories corresponding to some feed-forward control output may be determined. Differences between expected and observed states are used in a superimposed feedback loop to modify the total control output so that the expected states are achieved at least approximately despite unpredictable disturbances.

Monitoring these control components and the resulting state variable time histories, the triggering "knowledge-level" does have all the information available for checking the internal models on which it based its predictions and its decisions. In a distributed processing system, this knowledge level need not be involved in any of the fast control implementation and state estimation loops. If there are systematic prediction errors, these may be used to modify the models. Therefore, prediction error minimization may be used not just for state estimation according to some model but also for adapting the model itself, thereby learning to better understand behavioral characteristics of a body or the perturbation environment in the actual situation. Both of these may be used in the future to advantage. The knowledge thus stored is condensed information about the (material) world including the body of the vehicle carrying the sensors and data processing equipment (its "own" body, one might say); if it can be invoked in corresponding situations in the future, it will help to better control one's behavior in similar cases (see Chapter 3).

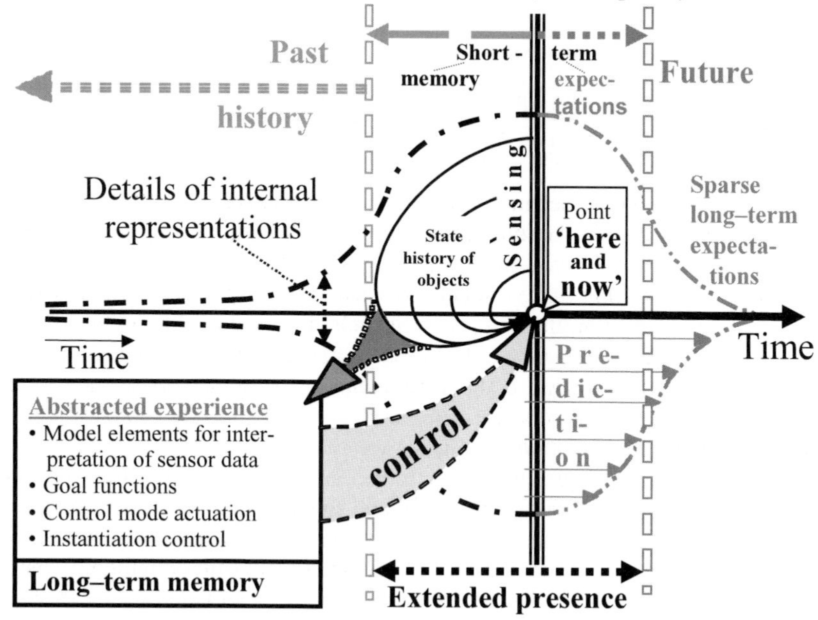

Figure 1.6. Symbolic representation of the interactions between the 'mental-' and the 'real world' (point 'here and now') in closed-loop form

Intelligence, thus, is defined as allowing deep understanding of processes and the way the "own" body may take advantage of this. Since proper reactions depend on the situation encountered, recognizing situations early and correctly and knowing what to do in these cases (decision-making) is at the core of intelligence. In the sense of steady learning, all resulting actions are monitored and exploited to improve the internal representations for better use in the future. Figure 1.6 shows a symbolic representation of the overall interaction between the (individual) "mental world" as data manipulation activity in a prediction-error feedback loop. It spans part of the time axis (horizontal line) and the "real world" represented by the spatial point "here" (where the sensors are). The spatial point "here", with its local environment, and the temporal point "now", where the interaction of the subject with the real world takes place, is the only 4-D point for the autonomous system to make real-world experience. All interactions with the world take place "here and now" (see central box). The rest of the world, its extensions in space and time, are individual constructs in the "mental world" to "make sense" of the sensor data stream and its invariance properties observed individually, and as a social endeavor between agents capable of proper information exchange.

The widely varying interpretations of similar events in different human cultures are an indication of the wide variety of relatively stable interpretation systems possible. Biological systems had to start from scratch; social groups were content with interpretations, which allowed them to adjust their lives correspondingly. Inconsistencies were accepted, in general, if satisfying explanations could be found. Progress toward more consistent overall models of "the world" was slow and took millennia for humankind.

The natural sciences as a specific endeavor of individuals in different cultural communities looking for a consistent description of "the world" and trying to avoid biases imposed by their specific cultures have come up with a set of "world models", which yield very good predictions. Especially over the last three centuries after the discovery of differential calculus by Leibnitz and Newton and most prominently over the last five decades after electronic computers became available for solving the resulting sets of equations in their most general form, these prediction capabilities soared.

In front of this background, it seems reasonable to equip complex technical systems with a similarly advanced sensor suite as humans have, with an interpretation background on the latest state of development in the natural sciences and in engineering. It should encompass a (for all practical purposes) correct description of the phenomena directly observable with its sensor systems. This includes the lighting conditions through sun and moon, the weather conditions as encountered over time and over different locations on the globe, and basic physical effects dominating locomotion such as Earth–gravity, dry and fluid friction, as well as sources for power and information. With respect to the latter ones, technical systems do have the advantage of being able to directly measure their position on the globe through the "Global Positioning System" (GPS). This is a late achievement of human technology only less than two decades of age, which is based on a collection of human-made Earth satellites revolving in properly selected orbits.

With this information and with digital maps of the continents, technical autonomous systems will have global navigation capabilities far exceeding those of biological systems. Adding all-weather capable imaging sensors in the millimeter wave range will make these systems truly global with respect to space and time in the future.

1.8 Structuring of Material Covered

Chapters 1 to 4 give a general introduction to dynamic vision and provide the basic knowledge representation schemes underlying the approach developed. Active subjects with capabilities for perception and control of behaviors are at the core of this unconventional approach.

Chapter 2 will deal with methods for describing models of objects and processes in the real world. Homogeneous coordinates as the basic tool for representing 3-D space and perspective mapping will be discussed first. Perspective mapping and its inversion are discussed next. Then, spatiotemporal embedding for circumnavigation of the inversion problems is treated. Dynamic models and integration of information over time are discussed as a general tool for representing the evolution of processes observed. A distinction between objects and *subjects* is made for forming (super-) classes. The former (treated in Chapter 2) are stationary, or obey relatively simple motion laws, in general. *Subjects* (treated in Chapter 3) have the capability of sensing information about the environment and of initiating motion on their own by associating data from sensors with background knowledge stored internally.

Chapter 4 displays several different kinds of knowledge components useful for mission performance and for behavioral decisions in the context of a complex world with many different objects and subjects. This is way beyond actual visual interpretation and takes more extended scales in space and time into account, for which the foundation has been laid in Chapters 2 and 3. Chapter 4 is an outlook into future developments.

Chapters 5 and 6 encompass procedural knowledge enabling real-time visual interpretation and scene understanding. Chapter 5 deals with extraction methods for visual features as the basic operations in image sequence processing; especially the bottom-up mode of robust feature detection is treated here. Separate sections deal with efficient feature extraction for oriented edges (an "orientation-selective" method) and a new orientation-sensitive method which exploits local gradient information for a collection of features: "2-D nonplanarity" of a 2-D intensity function approximating local shading properties in the image is introduced as a new feature separating homogeneous regions with approximately planar shading from nonplanar intensity regions. Via the planar shading model, beside homogeneous regions with linear 2-D shading, oriented edges are detected including their precise direction from the gradient components [Hofmann 2004].

Intensity corners can be found only in nonplanar regions; since the planarity check is very efficient computationally and since nonplanar image regions (with residues $\geq 3\%$ in typical road scenes) are found in $< 5\%$ of all mask locations, computer–intensive corner detection can be confined to these promising regions. In addition, most of the basic image data needed have already been determined and are used in multiple ways.

This bottom-up image feature extraction approach is complemented in Chapter 6 by specification of algorithms using predicted features, in which knowledge about object classes and object motion is exploited for recognizing and intelligent tracking of objects and subjects over time. These recursive estimation schemes from the field of systems dynamics and their extension to perspective mapping as measurement processes constitute the core of Chapter 6. They are based on dynamic models for object motion and provide the link between image features and object description in 3-D space and time; at the same time, they are the major means for data fusion. This chapter builds on the foundations laid in the previous ones. Recursive estimation is done for n single objects in parallel, each one with specific parameter sets depending on the object class and the aspect conditions. All these results are collected in the dynamic object data base (DOB).

Chapters 7 to 14 encompass system integration for recognition of roads, lanes, other vehicles, and corresponding experimental results. Chapter 7 as a historic review shows the early beginnings. In Chapter 8, the special challenge of initialization in dynamic road scene understanding is discussed, whereas Chapter 9 gives a detailed description of various application aspects for recursive road parameter and ego-state estimation while cruising. Chapter 10 is devoted to the perception of crossroads and to performing autonomous turnoffs with active vision. Detection and tracking of other vehicles is treated in Chapter 11.

Based on experience gained in these areas, Chapter 12 discusses sensor requirements for advanced vision systems in automotive applications and shows an early result of saccadic perception of a traffic sign while passing. Chapters 13 and 14 give an outlook on the concept of such an expectation-based, multifocal, saccadic (EMS) vision system and discuss some experimental results. Chapter 13 presents the concept for a dynamic knowledge representation (DKR) serving as an isolation layer between the lower levels of the system, working mainly with methods from systems dynamics/engineering, and higher ones leaning mainly on "artificial intelligence" methods. The DOB as one part of DKR is the main memory for all objects and subjects detected and tracked in the environment. Recent time histories of state variables may be stored as well; they alleviate selecting the most relevant objects/subjects to be observed more closely for safe mission performance. Chapter 14 deals with a few aspects of "real-world" situation assessment and behavior–decisions based on these data. Some experimental results with this system are given: Mode transition from unrestricted roadrunning to convoy driving, multi–sensor adaptive cruise control by radar and vision, autonomous visual lane changes, and turnoffs onto crossroads as well as onto grass-covered surfaces; detecting and avoiding negative obstacles such as ditches is one task solved in cross-country driving in a joint project with U.S. partners.

Chapter 15 gives some conclusions on the overall approach and an outlook on chances for future developments.

2 Basic Relations: Image Sequences – "the World"

Vision is a process in which temporally changing intensity and color values in the image plane have to be interpreted as processes in the real world that happen in 3-D space over time. Each image of today's TV cameras contains about half a million pixels. Twenty five (or thirty) of these images are taken per second. This high image frame rate has been chosen to induce the impression of steady and continuous motion in human observers. If each image were completely different from the others, as in a slide show with snapshots from scenes taken far apart in time and space, and were displayed at normal video rate as a film, nobody would understand what is being shown. The continuous development of action that makes films understandable is missing.

This should make clear that it is not the content of each single image, which constitutes the information conveyed to the observer, but the relatively slow development of motion and of action over time. The common unit of 1 second defines the temporal resolution most adequate for human understanding. Thus, relatively slow moving objects and slow acting subjects are the essential carriers of information in this framework. A bullet flying through the scene can be perceived only by the effect it has on other objects or subjects. Therefore, the capability of visual perception is based on the ability to generate internal representations of temporal processes in 3-D space and time with objects and subjects (synthesis), which are supported by feature flows from image sequences (analysis). This is an animation process with generically known elements; both parameters defining the actual 3-D shape and the time history of the state variables of objects observed have to be determined from vision.

In this "analysis by synthesis" procedure chosen in the 4-D approach to dynamic vision, the internal representations in the interpretation process have four independent variables: three orthogonal space components (3-D space) and time. For common tasks in our natural (mesoscale, that is not too small and not too large) environment, these variables are known to be sufficiently representative in the classical nonrelativistic sense.

As mentioned in the introduction, fast image sequences contain quite a bit of redundancy, since only small changes occur from one frame to the next, in general; massive bodies show continuity in their motion. The characteristic frequencies of human and most animal motion are less than a few oscillations per second (Hz), so that at video rate, at least a dozen image frames are taken per oscillation period. According to sampled data theory, this allows good recognition of the dynamic parameters in frequency space (time constants, eigenfrequencies, and damping). So, the task of visual dynamic scene understanding can be described as follows:

> Looking at 2-D data arrays generated by several hundred thousands of sensor elements, come up with a distribution of objects in the real world and of their relative motion. The sensor elements are arranged in a uniform array on the chip, usually. Onboard vehicles, it cannot be assumed that the sensor orientation is known beforehand or even stationary. However, inertial sensors for linear acceleration components and rotational rates are available for sensing ego-motion.

It is immediately clear that knowledge about object classes and the way their visible features are mapped into the image plane is of great importance for image sequence understanding. These objects may be grouped in classes with similar functionality and/or appearance. The body of the vehicle carrying the sensors and providing the means for locomotion is, of course, of utmost importance. The lengthy description of the previous sentence will be abbreviated by the term: the "own" body. To understand its motion directly and independently of vision, signals from other sensors such as odometers, inertial angular rate sensors and linear accelerometers as well as GPS (from the "Global Positioning System" providing geographic coordinates) are widely used.

Image data points carry no direct information on the distance at which their light sources, which have stimulated the sensor signal are in the real world; the third dimension (range) is completely lost in a single image (except maybe for intensity attenuation over longer distances). In addition, since perturbations may invalidate the information content of a single pixel almost completely, useful image features consist of signals from groups of sensor elements where local perturbations tend to be leveled out. In biological systems, these are the receptive fields; in technical systems, these are evaluation masks of various sizes. This now allows a more precise statement of the *vision task*:

> By looking at the responses of feature extraction algorithms, try to find objects and subjects in the real world and their relative state to the own body. When knowledge about motion characteristics or typical behaviors is available, exploit this in order to achieve better results and deeper understanding by filtering the measurement data over time.

For simple massive objects (*e.g.*, a stone, our sun and moon) and man-made vehicles, good "dynamic models" describing motion constraints are known very often. To describe relative or absolute motion of objects precisely, suitable reference coordinate systems have to be introduced. According to the wide scale of space accessible by vision, certain scales of representation are advantageous:

- Sensor elements have dimensions in the micrometer range (µm).
- Humans operate directly in the meter (m) range: reaching space, single step (body size).
- For projectiles and fast vehicles, the range of immediate reactions extends to several hundred meters or kilometers (km).
- Missions may span several hundred to thousands of kilometers, even one-third to one-half around the globe in direct flight.

- Space flight and lighting from our sun and moon extend up to 150 million km as a characteristic range (radius of Earth orbit).
- Visible stars are far beyond these distances (not of interest here).

Is it possible to find one single type of representation covering the entire range? This is certainly not achievable by methods using grids of different scales as often done in "artificial intelligence"- approaches. Rather, the approach developed in computer graphics with normalized shape descriptions and overall scaling factors is the prime candidate. Homogeneous coordinates as introduced by [Roberts 1965, Blinn 1977] also allow, besides scaling, incorporating the perspective mapping process in the same framework. This yields a unified approach for computer vision and computer graphics; however, in computer vision, many of the variables entering the homogeneous transformation matrices are the unknowns of the problem. A direct application of the methods from computer graphics is thus impossible, since the inversion of perspective projection is a strongly nonlinear problem with the need to recover one space component completely lost in mapping (range).

Introducing strong constraints to the temporal evolution of (3-D) spatial trajectories, however, allows recovering part of the information lost by exploiting first-order derivatives. This is the big advantage of spatiotemporal models and recursive least-squares estimation over direct perspective inversion (computational vision). The Jacobian matrix of this approach to be discussed throughout the text plays a vital role in the 4-D approach to image sequence understanding.

Before this can be fully appreciated, the chain of coordinate transformations from an object-centered feature distribution for each object in 3-D space to the storage of the 2-D image in computer memory has to be understood.

2.1 Three-dimensional (3-D) Space and Time

Each point in space may be specified fully by giving three coordinates in a well-defined frame of reference. This reference frame may be a "Cartesian" system with three orthonormal directions (Figure 2.1a), a spherical (polar) system with one (radial) distance and two angles (Figure 2.1b), or a cylindrical system as a mixture of both, with two orthonormal axes and one angle (Figure 2.1c).

The basic plane of reference is usually chosen to yield the most simple description of the problem: In orbital mechanics, the plane of revolution is selected for reference. To describe the shape of objects, planes of symmetry are preferred; for example, Figure 2.2 shows a rectangular box with length L, width B and height H. The total center of gravity S_t is given by the intersection of two space diagonals. It may be considered the box encasing a road vehicle; then, typically, L is largest and its direction determines the standard direction of travel. Therefore, the centerline of the lower surface is selected

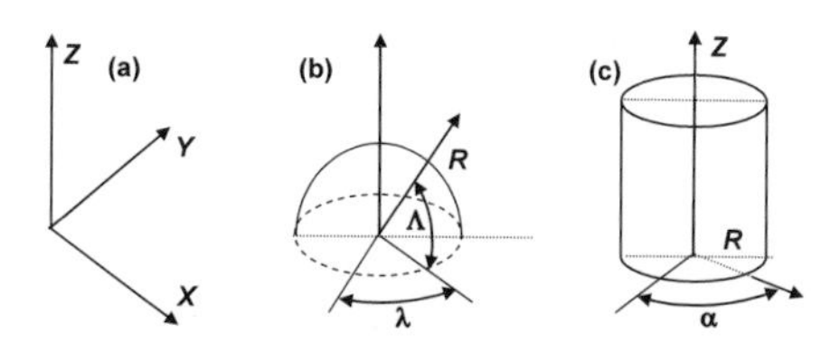

Figure 2.1. Basic coordinate systems (CS): (a) Cartesian CS, (b) spherical CS, (c) cylindrical CS

as the x-direction of a body-fixed coordinate system (x_b, y_b, z_b) with its origin 0_b at the projection S_b of S_t onto the ground plane.

To describe motion in an all-dominating field of gravity, the plane of reference may contain both the gravity and the velocity vector with the origin at the center of gravity of the moving object. The "horizontal" plane normal to the gravity vector also has some advantages, especially for vehicle dynamics since no gravity component affects motion in it.

If a rigid object moves in 3-D space, it is most convenient to describe the shape of the object in an object-oriented frame of reference with its origin at the center (possibly even the center of gravity) or some other convenient, easily definable point (probably at its surface). In Figure 2.2, the shape of the rectangular box is defined by the lengths of its sides L, B, and H. The origin is selected at the center of the ground plane S_b. If the position and orientation of this box has to be described relative to another object, the frame of reference given in the figure has to be related to the independently defined one of the other object by three translations and three rotations, in general.

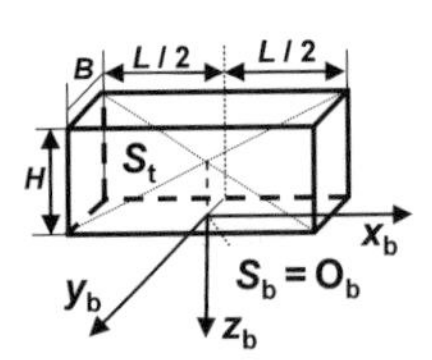

Figure 2.2. Object-oriented coordinate system for a rectangular box

To describe the object (box) shape in the new frame of reference, a coordinate transformation for all significant points defining the shape has to be performed. For the rectangular box, these are its eight corner points located at ± $L/2$ and ± $B/2$ for z_b= 0 and $-H$. The straight edges of the box remain linear connections between these points. [The selection of the coordinate axes has been performed according to the international standard for aero-space vehicles. X is in the standard direction of motion, x and z are in the plane of vehicle symmetry, and y completes a right-handed set of coordinates. The origin at the lower outside of the body alleviates measurements and is especially suited for ground vehicles, where the encasing box touches the ground due to gravity, in the normal case. Measuring altitude (elevation) positively upward requires a sign change from the positive z-direction (direction of the gravity vector in normal level flight). For this reason, some national standards for ground vehicles rotate the coordinate system by 180° around the x-axis (z upward and y to the left).]

In general, the coordinate transformations between two systems in 3-D space have three translational and three rotational components. In the 1970s, when these types of operations became commonplace in computer graphics, together with perspective mapping as the final stage of visualization for human observers, so-called *"homogeneous coordinates"* were introduced [Roberts 1965, Blinn 1977]. They allow the representation of all transformations required by transformation matrices of size 4 by 4 with different entries. Special microprocessors have been developed in the 1970s allowing us to handle these operations efficiently. Extended concatenations of several sequential transformations turn out to be products of these matrices; to achieve real-time performance for realistic simulations with visual feedback and human operators in the loop, these operations have shaped computer graphics hardware design (computer generated images, CGI [Foley et al. 1990]).

2.1.1 Homogeneous Coordinate Transformations in 3-D Space

Instead of the Cartesian vector $\underline{\mathbf{r}}_C = (x, y, z)$, the homogeneous vector

$$r_h = (p \cdot x, p \cdot y, p \cdot z, p) \tag{2.1}$$

is used with p as a scaling parameter. The specification of a point in one coordinate system can be "transformed" into a description in a second coordinate system by three translations along the axes and three rotations around reference axes, some of which may not belong to any of the two (initial and final) coordinate systems.

2.1.1.1 Translations

This allows writing translations along all three axes by the amount $\Delta\underline{r} = (\Delta x, \Delta y, \Delta z)$ in the form of a matrix · vector multiplication with the *homogeneous transformation matrix (HTM)* for translation:

$$r_1 = \begin{pmatrix} 1 & 0 & 0 & \Delta x \\ 0 & 1 & 0 & \Delta y \\ 0 & 0 & 1 & \Delta z \\ 0 & 0 & 0 & 1 \end{pmatrix} \cdot r_0 . \tag{2.2}$$

The three translation components shift the reference point for the rotated original coordinate system.

2.1.1.2 Rotations

Rotations around all axes may be described with the shorthand notation c = cos(angle) and s = sin(angle) by the corresponding HTMs:

$$R_x = \begin{pmatrix} 1 & 0 & 0 & 0 \\ 0 & c & s & 0 \\ 0 & -s & c & 0 \\ 0 & 0 & 0 & 1 \end{pmatrix}; \; R_y = \begin{pmatrix} c & 0 & -s & 0 \\ 0 & 1 & 0 & 0 \\ s & 0 & c & 0 \\ 0 & 0 & 0 & 1 \end{pmatrix}; \; R_z = \begin{pmatrix} c & s & 0 & 0 \\ -s & c & 0 & 0 \\ 0 & 0 & 1 & 0 \\ 0 & 0 & 0 & 1 \end{pmatrix}. \tag{2.3}$$

The position of the 1 on the main diagonal indicates the axis around which the rotation takes place.

The sequence of the rotations is of importance in 3-D space because the final result depends on it. Because of the dominant importance of gravity on Earth, the usual nomenclature for Euler angles (internationally standardized in mechanical engineering disciplines) requires the first rotation be around the gravity vector, defined as "heading angle" ψ (or pan angle for cameras). This reference system is dubbed the "geodetic coordinate system"; the x- and y-axes then are in the horizontal plane. The x-direction of this coordinate system (CS) may be selected as the main direction of motion or as the reference direction on a global scale (*e.g.*, magnetic North). The magnitude of rotation ψ is selected such that the x-axis of the rotated system comes to lie vertically underneath the x-axis of the new CS [*e.g.*, the body-fixed x-axis of the vehicle (x_O in Figure 2.3, upper right corner)]. As the second rotation, the turn angle of the vehicle's x-axis perpendicular to the horizontal plane has proven to be convenient. It is called "pitch angle" θ for vehicles (or tilt

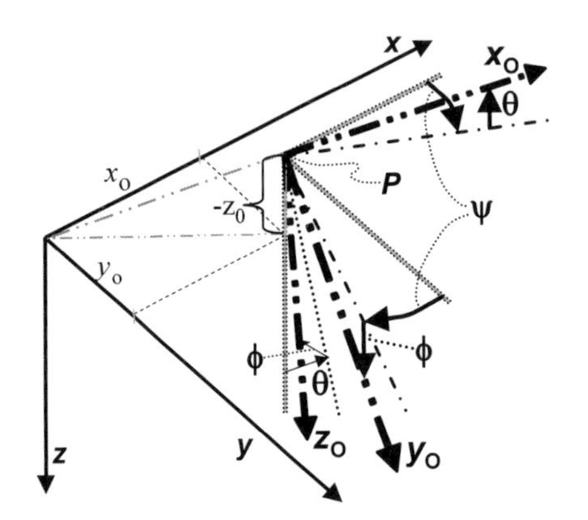

Figure 2.3. Transformation of a coordinate system

angle for cameras); the rotation takes place around an intermediate y-axis, called node axis k_y. This already yields the new x-direction x_O, around which the final rotation takes place: The roll– or bank angle ϕ indicates the angle around this axis between the plane of symmetry of the vehicle and the vertical plane. All of these angles are shown twice in the figure for easier identification of the individual axis of rotation.

2.1.1.3 Scaling

Due to Equation 2.1 scaling can be achieved simply by setting the last element in the HTM [lower right element p (4, 4)] different from 1. All components are then interpreted as scaled by the same factor p. This scaling is conveniently exploited by application to perspective mapping.

2.1.1.4 Perspective Mapping

Figure 2.4 shows some properties of perspective projection by a pinhole model. All points on a ray through the projection center P_p are mapped into a single point in the image plane at a distance f (the focal length) behind the plane $x_p = 0$. For example, the points Q_1, Q_2, and Q_3 are all mapped into the single point Q_i. This is to say that the 3-D depth to the point in the real world mapped is lost in the image. This is the major challenge for monocular vision. Therefore, the rectangle in the image plane Re_i may correspond both to the two rectangles Re_1 and Re_2 and to the trapezoids $Trap_1$ and $Trap_2$ at different ranges and with different orientations in the real world. Any four-sided polygon in space (also nonplanar ones) with the corner points on the four rays through the corners given will show up as the same (planar, rectangular) shape in the image.

To get rid of the sign changes in the image plane incurred by the projection center P_p (pinhole), the position of this plane is mirrored

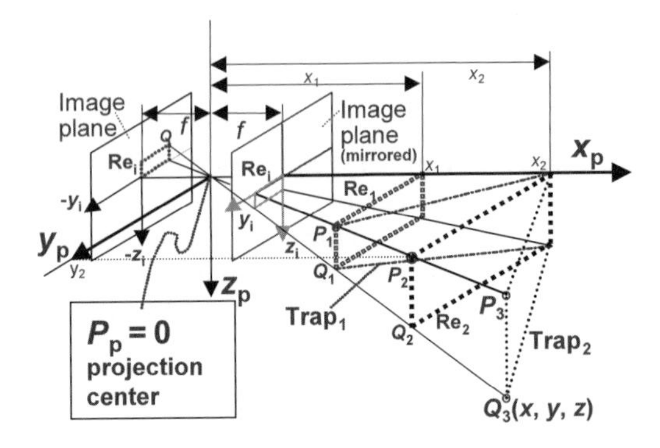

Figure 2.4. Perspective projection by a pinhole camera model

at $x_p = 0$ (as shown). This allows us to easily write down the mathematical relationship of perspective mapping when the x_p-axis is selected perpendicularly to the image plane and the origin is at the pin hole (projection center):

$$\begin{aligned} y_i / f &= y / x \\ z_i / f &= z / x, \\ \text{or} \quad y_i &= y / (x \cdot (1/f)) \\ z_i &= z / (x \cdot (1/f)). \end{aligned} \tag{2.4}$$

This perspective projection equation may be included in the HTM–scheme by the projection matrix P (P_p for pixel coordinates). It has to be applied as the last matrix multiplication and yields (for each point-vector $\mathbf{x}_{Fk}$ in the real world) the "homogeneous" feature vector "e". The image coordinates y_i and z_i of a feature point are then obtained from the "homogeneous" feature vector e by dividing the second and third component resulting from Equation 2.4a by the fourth one (see Figure 2.5 for the coordinates):

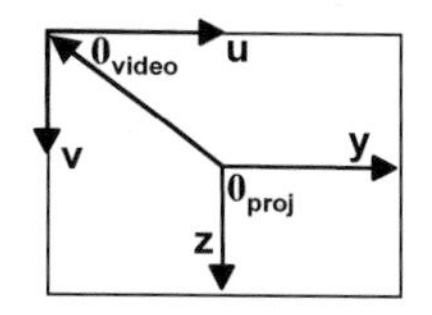

Figure 2.5. Image coordinates

$$P = \begin{pmatrix} 0 & 0 & 0 & 0 \\ 0 & 1 & 0 & 0 \\ 0 & 0 & 1 & 0 \\ 1/f & 0 & 0 & 0 \end{pmatrix}; \quad P_p = \begin{pmatrix} 0 & 0 & 0 & 0 \\ 0 & k_y & 0 & 0 \\ 0 & 0 & k_z & 0 \\ 1/f & 0 & 0 & 0 \end{pmatrix}. \tag{2.4a}$$

The image coordinates y_i and z_i of a feature point are then obtained from the "homogeneous" feature vector e by dividing the second and third component resulting from Equation 2.4a by the fourth one (see Figure 2.5 for the coordinates):

$$\begin{aligned} y_i &= (\text{second component } e_2)/(\text{fourth component } e_4) \\ z_i &= (\text{third component } e_3)/(\text{fourth component } e_4). \end{aligned} \tag{2.5}$$

The left matrix in Equation 2.4a leaves the y-component (within each image line) and the z-component (for each image line) as metric pixel coordinates.

The x-component has lost its meaning during projection and is used for scaling as indicated by Equation 2.4 and the last row of Equation 2.4a. If coordinates are to be given in pixel values in the image plane, the "1"s on the diagonal are replaced by k_y (element 2, 2) and k_z (element 3, 3) representing the number of pixels per unit length (Equation 2.4a right). Typical pixel sizes at present are 5 to 20 micrometer (approximately square), or k_y, $k_z \sim 200$ to 50 pixels per mm length.

After a shift of the origin from the center to the upper left corner (see Figure 2.5) this y-z sequence (now dubbed u-v) is convenient for the way pixels are digitized line by line by frame grabbers from the video signal. (For real-world applications, it has to be taken into account that frame-grabbing may introduce offsets in y- and z-directions, which lead to additive terms in the corresponding fourth column.)

2.1.1.5 Transformation of a Planar Road Scene into an Image

The set of HCTs needed for linking a simple road scene with corresponding features in an image is shown in Figure 2.6. The road is assumed to be planar, level and straight. The camera is somewhere above the road at an elevation H_c (usually

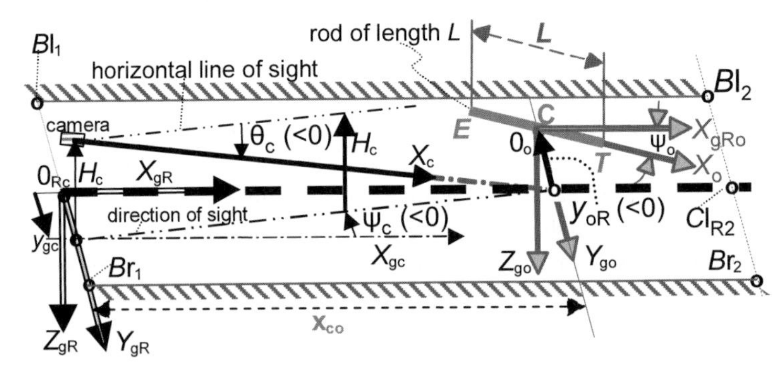

Figure 2.6. Coordinate systems for transforming a simple road scene into an image

known in vision tasks) and at a lateral offset y_{gc} (usually unknown) from the road centerline. The width B of the road is assumed to be constant in the look-ahead range; the marked centerline of the road partitions it into two equal parts. Some distance down the road there is a rod of length L as an obstacle to be detected. To simplify the task, it is assumed to be a one-dimensional object, easily describable in an object centered coordinate system to extend from $x_o = -L/2$ to $+L/2$. (The real-world object does have a cross section of some extension and shape that warrants treating it as an obstacle not to be driven over.) Relative to the road the rod does have a lateral position y_{oR} of its center point C from the centerline of the road and an orientation ψ_o between the object-fixed x-axis X_o and the tangent direction of the road X_{gRo}. Figure 2.6 shows the situation with the following CS:

1. X_o object-oriented and body-fixed in rod-direction (only one component).
2. Geodetic CS of the object (rod); geodetic CSs are defined with their X-Y-plane in the horizontal plane and their origin at the center of gravity of the object. The orientation of the x-axis in the horizontal plane is left open for a convenient choice in connection with the actual task. In this case, there is only one road direction, and therefore, X_{gRo} is selected as the reference for object orientation. There is only one rotation-angle ψ_o between the two CS 1 and 2 because gravity keeps the rod on the road surface (X_g - Y_g plane). The corresponding HTM is $R_{\psi o}$ [with a 1 in (3, 3) for rotation around the z-axis].
3. The road-centered geodetic CS (indexed "gR") at the longitudinal location of the camera has its origin at 0_{Rc}, and X_{gR} is directed along the road. Between the CS 2 and 3 there are (in the special case given here) the two translations x_{co} and y_{oR}. The corresponding HTM is T_{Ro} with entries x_{co} and y_{oR} in the last column of the first two rows.
4. The geodetic CS is at the projection center of the camera (not shown in Figure 2.6 for clarity); between the CS 3 and 4 there are again (in the special case given here) the two translations y_{gc} and H_c . The latter one is negative since z_g is posi-

tive downward. The corresponding HTM is T_{Rc} with entries y_{gc} and H_c in the last column of the second and third rows.

5. The camera-oriented CS (indexed "c"): The gaze direction of the camera is assumed to be fixed on the center of the road at the look-ahead distance of the object center. The elevation H_c of the camera above the ground and its lateral offset y_{gc} yield the camera pan (yaw) and tilt (pitch) angles ψ_c and Θ_c. The camera CS is obtained from CS 4 by two rotations: First, rotating around the Z_{gc} -axis (vertical line through camera projection center, not directly shown but indicated by the vector H_c in the opposite direction) by the amount ψ_c yields the horizontal direction of sight. The corresponding HCT–matrix is $R_{\psi c}$. Now the X-Z-plane cuts the axis X_{gR} at distance x_{co}. Within this X-Z-plane now the intermediate x-axis has to be rotated until it also cuts the axis X_{gR} at distance x_{co} (pitch angle $-\Theta_c$) and becomes X_c. The corresponding HTM is $R_{\Theta c}$ [with a 1 in position (2, 2) for rotation around the intermediate y-axis].
6. The image CS into which the scene is now mapped by perspective projection.
7. The CS for the image matrix of pixel points in computer memory. During data acquisition and transfer, shifts may occur: By misalignments of a frame grabber, unintentional shifts of the image center may occur. Intentionally, the origin of the image CS may be shifted to the upper left corner of the image (see coordinates u, v in Figure 2.5).

Since all these transformations can be applied to only one point at a time, objects consisting of sequences of straight lines (so-called "polygons") have to be described by the ensemble of their corner points. This will be treated in Section 2.2. Note here that each object described in a certain CS has to be given by the ensemble of its corner points. For example, the rod is given by the two endpoints E and T on the X_o -axis at $-L/2$ and $+L/2$. The straight road is given by its left and right boundary lines Bl and Br, and by the centerline Cl_R in the road-CS. All three lines are realized by a straight line connection between two end points with indices 1 (left side of Figure 2.6) and 2 (right); all end points of lines defining the road lie at $z = 0$. The points on the left-hand side of the road are at $y = -B/2$, and those on the right-hand side are at $+ B/2$. The centerline Cl_R is at $y = 0$.

Let us consider the transformation of the endpoint T of the rod into the image taken by the camera according to Figure 2.6. In the 3-D homogeneous object CS of the rod, this point has the coordinate description $\mathbf{x_o}^T = (L/2, 0, 0, 1)$. After transition to the geodetic CS X_{go} , the state vector according to point 2 in the list above changes to

$$x_{go} = R_{\psi o} \cdot x_o . \tag{2.6}$$

To describe the point T in the road-oriented CS, the second HTM T_{Ro} for translation from C to 0_{Rc} according to point 3 above has to be applied:

$$x_{gR} = T_{Ro} \cdot R_{\psi o} \cdot x_o . \tag{2.7}$$

For the transition to the geodetic CS of the camera, multiplication by the HTM T_{Rc} with entries y_{gc} and H_c has to be performed according to point 4 above:

$$x_{gc} = T_{Rc} \cdot T_{Ro} \cdot R_{\psi o} \cdot x_o . \tag{2.8}$$

The two translations may be combined into a single matrix containing in the upper three rows of the fourth column the sum of the elements in the corresponding rows of the HTMs in Equation 2.8. It has not been done here because the direction of the local road tangents would not be the same for a curved road. Therefore, before performing the second translation to the origin of the camera CS, a rotation around the vertical axis by the difference χ in local road direction would have to be inserted (yielding another rotation matrix $R_{\chi co}$ between T_{Rc} and T_{Ro}.

Now the two rotation-angles ψ_c and Θ_c for the optical axis of the camera have to be applied. Since ψ_c according to the definition has to be applied first, $R_{\psi c}$ has to stand to the right because this matrix will be encountered first when the column-vector $\mathbf{x_o}$ is multiplied from the right. This finally yields the state vector for the point T in camera coordinates:

$$x_{co} = R_{\psi c} \cdot R_{\theta c} \cdot T_{Rc} \cdot T_{Ro} \cdot R_{\psi o} \cdot x_o. \tag{2.9}$$

Applying perspective projection (Equation (2.4) to this 3-D-point $\mathbf{x_{co}}$ yields the "homogeneous" feature data for the image coordinates **e** according to Equation 2.5; note that the five HTMs contain the unknown variables of the vision task written below each matrix:

$$\begin{aligned} e &= [P_p \cdot R_{\theta c} \cdot R_{\psi c} \cdot T_{Rc} \cdot T_{Ro} \cdot R_{\psi o}] \cdot x_o = T_{tot} \cdot x_o. \\ &\text{unknowns: } \theta_c, \psi_c, y_{gc} (x_{co}, y_{oR})\ \psi_o! \end{aligned} \tag{2.10}$$

The explicitly written down (transposed) form $(\mathbf{e})^T = (e_1, e_2, e_3, e_4)$ then yields the image coordinates

$$y_i = e_2 / e_4 ; \quad z_i = e_3 / e_4. \tag{2.11}$$

The expression in square brackets is the same for any point to be transformed from object- into camera- coordinates. Therefore, in computer graphics, where all elements entering the HTMs are known beforehand, the so-called concatenated transformation matrix T_{tot} is computed as the product of all single HTMs once for each object and aspect condition. A single matrix-vector multiplication $T_{tot} \cdot \mathbf{x}_o$ then yields the position in homogeneous feature coordinates for the image of a point on the object at $\underline{\mathbf{x}}_o$ in object-centered coordinates. Equation 2.11 finally gives the image coordinates.

Note that these coordinates are real numbers, which means that the positions of the points mapped into the image are known to subpixel accuracy. If measurements of image features (Chapter 5) can be done to subpixel accuracy, too, the methods applied in recursive estimation (see Chapter 6) yield improved results by not rounding off feature coordinates to integer numbers (as is often done in a naive approach).

2.1.1.6 General Concatenations of HCTs; the Scene Tree

While the use of these transformation matrices in computer graphics is commonplace as a flexible tool for adaptation to new or modified tasks, in machine vision, until very recently, they have been exploited only during the formulation phase of the problem. Then, to be numerically more efficient on general-purpose processors, the resulting expressions of the matrix product T_{tot} have been hand-coded, initially. With the processing power available now, more easily adaptable codes become preferable; this is achieved by keeping each HTMs separate until numerical evaluation because in each matrix variables to be iterated may appear.

The challenge in machine vision as opposed to computer graphics is that some of the transformation parameters entering the matrices are not known beforehand but are the unknowns of the vision process, which have to be determined from image sequence analysis. Therefore, in each transformation, its sensitivity to small parameter changes has to be determined to compute the corresponding overall "Jacobian" matrices (JM, the first-order approximation for the nonlinear functional relationship describing the mapping of features on objects in the real world to those measured in the images). This rather compute-intensive operation and an efficient implementation will be discussed in Section 2.1.2.

The tendency toward separation of application-oriented aspects from those geared to the general methods of dynamic vision required a major change from the initial approach with respect to handling homogeneous coordinates. Concatenation is shifted to the evaluation of the scene model at runtime; then, both the nominal total HTM and the partial-derivative matrices for all unknown parameters and state variables are computed in conjunction (maybe numerically). This allows efficient use of intermediate results and makes the setup of new problems much easier for the user. The corresponding representation scheme for all objects and CSs in a so-called "scene tree" has been developed by D. Dickmanns (1997) and will be discussed in the following paragraphs.

Figure 2.7 without the shaded areas gives an example of a scene tree for describing the geometrical relations among several objects of relevance for the vision task shown in Figure 2.6 a single camera on a straight road. The nodes and edges in the shaded areas on the right-hand side and on top will be needed for the more general case of a camera onboard a vehicle moving on a curved road. In the straight road scene, the "object" represents the rod on the road at some look-ahead distance x_{co}; its lateral position on the road can be recovered in the image from the road boundaries nearby and from the vanishing point at the horizon (see Figure 2.8).The figure shows the resulting image, into which some labels for later image interpretation have been inserted.

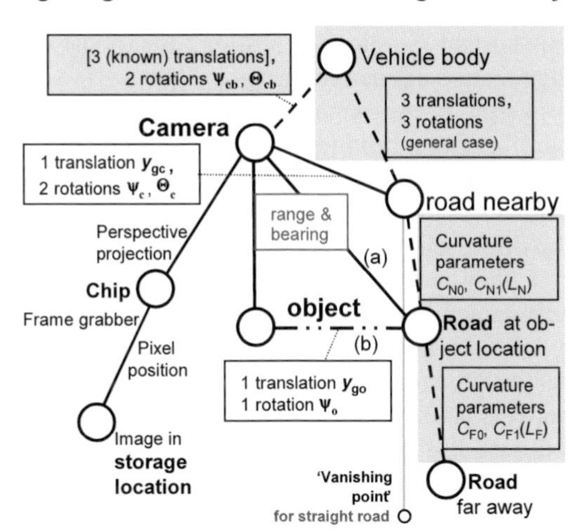

Figure 2.7. Scene tree for representing spatial relationships between objects seen and their image in perspective projection

For a horizontal straight road with parallel lines, the vanishing point, at which all parallel lines intersect, lies on the horizon line. Its distance components to the image center yield the direction of the optical axis: $-\psi_c$ to the direction of the road and $-\theta_c$ to the horizon line. The center of gravity (cg) of the rod has the averaged coordinates of the end points E and T in 3-D space.

The location of this point relative to the road centerline gives the lateral offset y_{oR} at the look-ahead distance x_{co} . The yaw angle ψ_o of the rod relative to the road can be determined by computing the difference in the positions of the end points E and T; however, the distortions from perspective mapping have to be taken into account. All these interpretations are handled automatically by the 4-D approach to dynamic vision (see Chapter 6). Since the camera looks from above almost tangentially to the plane containing the road, the distance in the real world increases with decreasing row index. If a certain geometric resolution for each pixel is required, there is a limit to the look-ahead range usable (shown on the left-hand side in Figure 2.8).

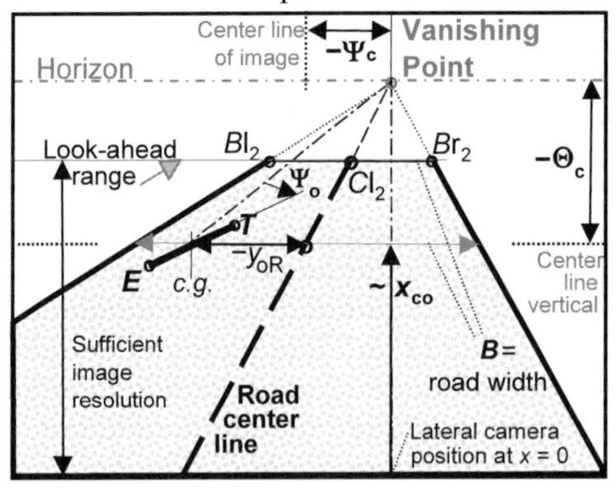

Figure 2.8. Image resulting from the scene given in Figure 2.6 after perspective mapping

For example, if each pixel is not allowed to cover more than 5 cm normal to the optical axis, the look-ahead range $L_{0.05}$ (in meters) or simply L_5 (in cm) is thus defined. This makes sense in road–scene analysis since lane markings, usually, are 10 to 50 cm wide and at least two pixels normal to a line are required for robust recognition under perturbations with edge feature extractors (Chapter 5).

Looking at sequences of images like these, the camera motion and the relative state of all objects of relevance for performing a driving mission have to be recognized sufficiently well and early with minimal time delay. The approach given in this book has proven to solve this problem reliably. Before the overall solution for precise and robust recognition can be discussed, all components needed have to be introduced first. Starting in Chapter 7, they will be applied together; the performance and complexity level will be open-ended for future growth.

Back to the scene tree: In Figure 2.7 each node represents an object in the real world (including virtual ones such as the CS at certain locations). The edges represent HCTs, *i.e.*, encodings of geometric relations. In combination with knowledge about the effects of these transformations, this allows a very compact description of all objects of relevance in the visual scene. Only the components of spatial state vectors and a few parameters of generic models for the objects are needed to represent the scene. The rest of the knowledge is coded in the object classes from which the objects hypothesized are generated.

The edges (a) and (b) at the center of Figure 2.7 (from the camera, respectively, from the "object" to the "road at the object location") are two alternative ways to determine where the object is. Edge (a) represents the case, where the bearing angles to some features of the road and to the object are interpreted separately; the road features need not necessarily be exactly at the location of the object. From these results, the location of the road and the lateral position of the object on the road can be derived indirectly in a second step. The difference in bearing angle to the road center at the range of the object yields the lateral position relative to the road. In the case of edge (b), first only the range and bearing to the object are determined. Then at the position of the object, the features of the road are searched and measured, yielding directly the explicit lateral position of the object relative to the road. This latter procedure has yielded more stable results in recursive estimation under perturbations in vehicle pitch and yaw angles (see Chapter 6).

The sequence of edges in Figure 2.7 specifies the individual transformation steps; each node represents a coordinate system (frequently attached to a physical body) and each edge represents HCTs, generally implying several HTMs. The unknown parameters entering the HCTs are displayed in the boxes attached to the edge. At the bottom of each branch, the relevant object is represented in an object-centered coordinate system; this will be discussed in Section 2.2. A set of cameras (instead of a single one) may be included in the set of nodes making their handling schematic and rather easy. This will be discussed in connection with EMS vision later.

The additional nodes and edges in the shaded areas show how easily more detailed models may be introduced in the interpretation process. Figure 2.9 gives a sketch of the type of road scene represented by the full scene tree of Figure 2.7.

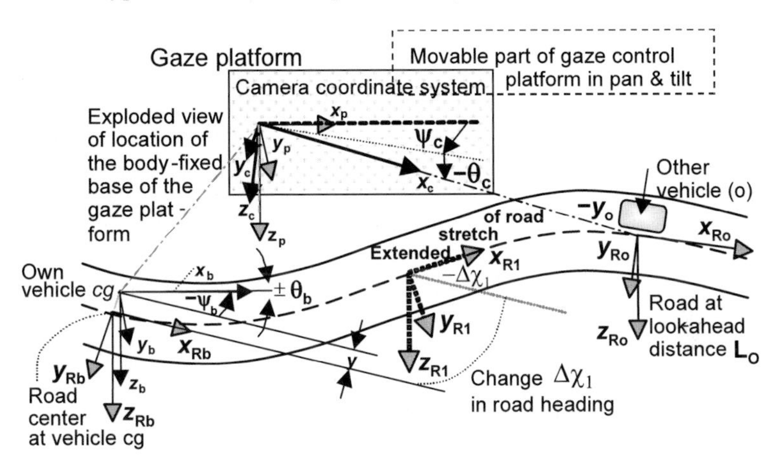

Figure 2.9. Coordinate systems for a general scene with own vehicle (index b) and one other vehicle (index o) on a curved road

Now, the position of the own vehicle relative to the road has to be determined. In the general case, these are three translational and three rotational components. Neglecting movements in bank angle (they average around 0) and in heave (vertical translation) and taking the longitudinal position as the moving origin of the vehicle CS, the same components as in the previous case have to be determined.

However, now the camera is located somewhere in the vehicle. The three translational components are usually fixed and do not change; the two rotational components from gaze control can be measured conventionally on the platform and are assumed known, error-free. So, there is no new unknown variable for active gaze control; however, the transformations corresponding to the known variables from mounting the platform on the vehicle have to be applied.

For the more general case of a curved road (shaded area to the right in Figure 2.7), the road models to be discussed in later sections have to be applied. They introduce several more unknowns into the vision process. However, using differential-geometry models minimizes the number of these terms; for planar roads, two sets of additional CSs allow large look-ahead ranges even with up to two inflection points of the road (changes of the sign of curvature; Figure 2.9 has just one).

General scheme of the scene tree: The example of a scene tree given above can be generalized for perspective mapping of many objects in the real world into images by several cameras. For practical reasons, one CS will be selected as the main reference; in vehicle guidance, this may be the geodetic CS linked to the center of gravity of the vehicle (or some easily definable one with similar advantages). This is called the "root node" and is drawn as the topmost node in standard notation.

The letter T shall designate all transformations for uniformity (both translations and rotations). The standard way of describing these transformations is from the leaves (bottom) to the root node. Therefore, when forming the total chain of transformations T_{tot} from features on objects in the real world into features in an image, denoted by K in Figure 2.10, the inverse transformation matrices T_{kj}^{-1} have to be used from the root to the leaves (left-hand side). A total transformation T_{tot} exists for each object-sensor pair, of which the object can be visually observed from the sensor. Once the scene tree has been defined for m cameras and n objects, the evaluation of the (at most $n \cdot m$) total transformation matrices is independent of the special task and can be coded as part of the general method [D. Dickmanns 1997].

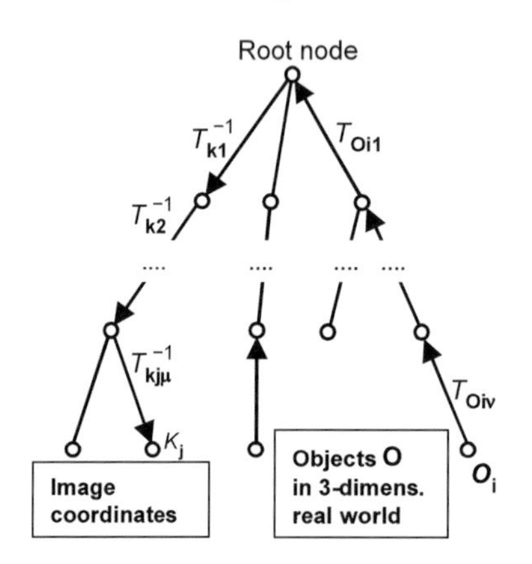

Figure 2.10. General scheme for object mapping in the scene graph

Since objects may appear and disappear during a mission, the perception system has to have the capability of autonomously inserting and deleting object branches in the scene tree. This object hypothesis generation and deletion capability is a crucial part of intelligent visual perception. Detailed discussions of various task domains will be given in later sections after the elements necessary for a flexible overall system have been introduced. Let the computation of T_{tot} be called the "traverse" of the scene graph. The recursive estimation method presented in Chapter 6 requires that this traverse is done not just once for each object-sensor pair but $(q + 1)$ times, if there are q unknown state variables and parameters entering the HTMs in T_{tot}. This model-based approach yields a first-order approximation (so-called "Jacobian matrices" or in short "Jacobians" of perspective mapping) describing the relationship between all model parameters and state components in the mentally represented world on the one hand, and feature positions in the images, on the other hand. Note that for 3-D models, there is also spatial information available in the Jacobians, allowing depth perception even with monocular vision (motion stereo). Because of this heavy workload in computation, efficient evaluation of all these traverses is very important. This is the subject of the next section.

Again, all of this is independent of the special application, once the scene tree for the set of problems has been defined.

2.1.2 Jacobian Matrices for Concatenations of HTMs

To be flexible in coding and to reduce setup time for new task domains, the elements of Jacobian matrices may be determined by numerical differencing instead of fully analytical derivations; this is rather computer-intensive. An analytical derivation based on the factored matrices may be most computer-efficient. Figure 2.11 shows in the top row the total HTM for the homogeneous feature vector e of the example Equation 2.10 from road vehicle guidance with scene perception through an active camera. The two translations T have two and one unknown components here (maximally six are possible in total); all three rotation-angles have to be determined from vision as well. So there are six unknowns in the problem, for which the entries in the Jacobian matrix have to be computed.

To obtain these elements, nominal value and the systematically perturbed values due to changes in each unknown state variable or parameter $d\underline{\mathbf{x}}_i$ have to be computed for each feature point, just one at a time to obtain partial derivatives. If there are n state variables and q parameters to be iterated during recursive estimation, $(n + q + 1)$ total transformations have to be computed for each feature point. To be efficient, the nominal values should be exploited as much as possible also for computing the $n + q$ perturbed values. This procedure for the unknown state components is sketched in the sequel; adaptable parameters will be discussed in Section 2.2 and Chapter 6.

2.1.2.1 Partial Derivatives of Homogeneous Transformation Matrices

The overall transformation matrix for each feature point $\underline{x}_{Fk}$ on the object (the rod in our case) in 3-D space was given by Equation 2.10:

$$\begin{aligned} &\mathrm{e} = [\,P_\mathrm{p} \cdot R_{\theta\mathrm{c}} \cdot R_{\psi\mathrm{c}} \cdot T_\mathrm{Rc} \cdot T_\mathrm{Ro} \cdot R_{\psi\mathrm{o}}] \cdot \mathrm{x}_\mathrm{o} = T_\mathrm{tot} \cdot \mathrm{x}_\mathrm{o}. \\ &\text{unknowns: } \theta_\mathrm{c}, \psi_\mathrm{c}, y_\mathrm{gc} (x_\mathrm{co}, y_\mathrm{oR})\ \psi_\mathrm{o}\ ! \end{aligned} \tag{2.10}$$

To describe the state of the rod relative to the camera, first its state relative to the road CS has to be specified by (ψ_o, x_{co}, y_{oR}). Then the state of the camera relative to the road is given by $[y_{gc}$, (the known elevation H_c), ψ_c, $\Theta_c]$. Under the assumption of a planar straight road, the geometric arrangement of the objects “camera” and “rod” is fully described by specifying these “state components”. The unknown state vector x_S of this problem with six components thus is

$$\mathrm{x}_\mathrm{S}^\mathrm{T} = (\psi_\mathrm{o}, x_\mathrm{co}, y_\mathrm{oR}, y_\mathrm{gc}, \psi_\mathrm{c}, \theta_\mathrm{c}). \tag{2.12}$$

These components have to be iterated during the vision process so that the underlying models yield a good fit to the visual features observed. In Equation 2.10 each R represents a single rotational and each T up to three translational components, the unknowns of which are written below them. The total HTM including perspective mapping, designated by T_t is written

$$T_{tot} = P \cdot R_{\theta c} \cdot R_{\psi c} \cdot T_{Rc} \cdot T_{Ro} \cdot R_{\psi o}. \tag{2.13}$$

The partial derivative of T_t with respect to any component of the state vector $\underline{\mathbf{x}}_S$ (Equation 2.12) is characterized by the fact that each component enters just one of the HTMs and not the other ones. Applying the chain rule for derivatives of products such as $\partial T_t/\partial(x)$ yields zeros for all the other matrices, so that the overall derivative matrix, for example, with respect to y_{gc} (abbreviated here for simplicity by just x) entering the central HTM T_{Rc} is given by

$$\begin{aligned} \partial T_t / \partial(x) = T_{tx} &= R_{\theta c} \cdot R_{\psi c} \cdot \underbrace{\partial T_{Rc} / \partial(x)} \cdot T_{Ro} \cdot R_{\psi o} \\ &= R_{\theta c} \cdot R_{\psi c} \cdot \quad T_{Rcx} \quad \cdot T_{Ro} \cdot R_{\psi o}. \end{aligned} \tag{2.14}$$

Two cases have to be distinguished for the general partial derivatives of the HTMs containing the variables to be iterated: Translations and rotations.

Translations: These components are represented by the first three values in column 4 of HTMs. In addition to the nominal value for the transformation matrix T_N as given in Equation 2.2, also the partial derivative matrices for the unknown variables are computed in parallel. The full set of these matrices is given by

$$r_1 = \begin{pmatrix} 1 & 0 & 0 & \Delta r_x \\ 0 & 1 & 0 & \Delta r_y \\ 0 & 0 & 1 & \Delta r_z \\ 0 & 0 & 0 & 1 \end{pmatrix} \cdot r_0 = T_{rN} \cdot r_0; \quad \frac{\partial T_{Rc}}{\partial(\Delta r_x)} = \begin{pmatrix} 0 & 0 & 0 & 1 \\ 0 & 0 & 0 & 0 \\ 0 & 0 & 0 & 0 \\ 0 & 0 & 0 & 0 \end{pmatrix} = T_{rx}, \tag{2.15 (a)}$$

$$\frac{\partial T_{Rc}}{\partial(\Delta r_y)} = \begin{pmatrix} 0 & 0 & 0 & 0 \\ 0 & 0 & 0 & 1 \\ 0 & 0 & 0 & 0 \\ 0 & 0 & 0 & 0 \end{pmatrix} = T_{ry} \quad ; \quad \frac{\partial T_{Rc}}{\partial(\Delta r_z)} = \begin{pmatrix} 0 & 0 & 0 & 0 \\ 0 & 0 & 0 & 0 \\ 0 & 0 & 0 & 1 \\ 0 & 0 & 0 & 0 \end{pmatrix} = T_{rz}. \tag{2.15 (b)}$$

Rotations: According to Equation 2.3 the nominal HTMs for rotation (around the x-, y- and z- axis respectively) are repeated below: s stands for the sine and c for the cosine of the corresponding angle of rotation, say α.

$$R_x = \begin{pmatrix} 1 & 0 & 0 & 0 \\ 0 & c & s & 0 \\ 0 & -s & c & 0 \\ 0 & 0 & 0 & 1 \end{pmatrix}; R_y = \begin{pmatrix} c & 0 & -s & 0 \\ 0 & 1 & 0 & 0 \\ s & 0 & c & 0 \\ 0 & 0 & 0 & 1 \end{pmatrix}; R_z = \begin{pmatrix} c & s & 0 & 0 \\ -s & c & 0 & 0 \\ 0 & 0 & 1 & 0 \\ 0 & 0 & 0 & 1 \end{pmatrix}. \tag{2.3}$$

The partial derivatives of the transformation matrices $\partial R/\partial\alpha$, may be obtained from

$$d(\sin\alpha)/d\alpha = \cos\alpha \;=\; c; \quad d(\cos\alpha)/d\alpha = -\sin\alpha \;= -s. \tag{2.16}$$

This leads to the derivative matrices for rotation

$$R'_{xx} = \begin{pmatrix} 0 & 0 & 0 & 0 \\ 0 & -s & c & 0 \\ 0 & -c & -s & 0 \\ 0 & 0 & 0 & 0 \end{pmatrix}; R'_{yy} = \begin{pmatrix} -s & 0 & -c & 0 \\ 0 & 0 & 0 & 0 \\ c & 0 & -s & 0 \\ 0 & 0 & 0 & 0 \end{pmatrix}; R'_{zz} = \begin{pmatrix} -s & c & 0 & 0 \\ -c & -s & 0 & 0 \\ 0 & 0 & 0 & 0 \\ 0 & 0 & 0 & 0 \end{pmatrix}. \tag{2.17}$$

It is seen that exactly the same entries s and c as in the nominal case are required, but at different locations and some with different signs. The constant values 1 have disappeared, of course.

2.1.2.2 Concatenations and Efficient Computation Schemes

The overall transformation matrix for each point (see Equation 2.10) together with the concatenated derivative matrices for computing the Jacobian matrices are shown in Figure 2.11. It can be seen that many multiplications of matrices are the

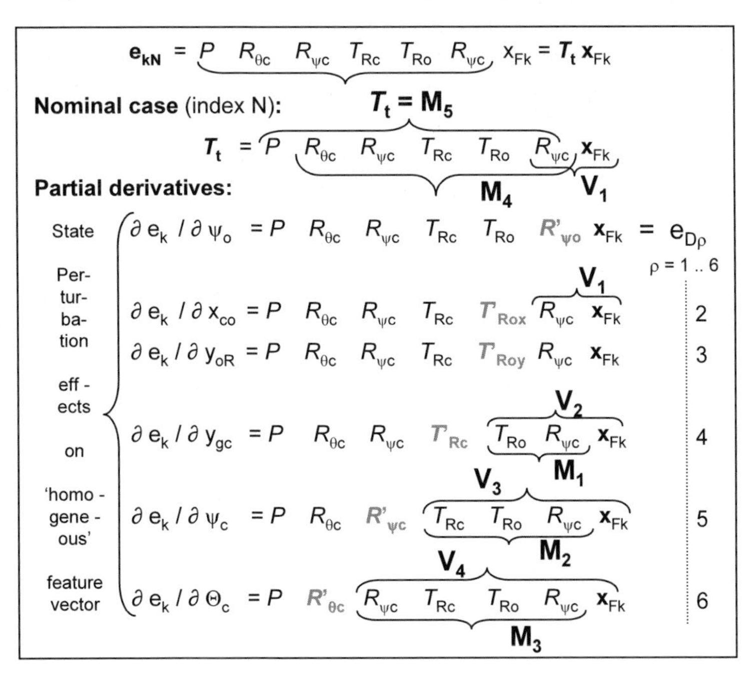

Figure 2.11. Scheme of matrix multiplication for efficient computation of concatenated homogeneous coordinate transformations (top) and elements of the Jacobian matrices

same for the nominal case and for the concatenated derivative matrices. Since the elements of the overall derivative matrix (Equation 2.14) are sparsely filled (Equation 2.15 and 2.17), let us first have a look at their matrix products for efficient coding.

The derivatives of the translation matrices all have a single "1" in the upper three rows of the last columns, the positions depend on the variable for partial derivation; the rest of the elements are zero, allowing efficient computation of the products. If such a matrix is multiplied from the right by another matrix, the multiplication just copies the last row of this matrix into the corresponding row of the product matrix where the 1 is in the derivative matrix (row 1 for x, 2 for y, and 3 for z). If such a matrix is multiplied to the left by another matrix, the multiplication just copies the ith column of this matrix into the last column of the product matrix. The index i designates the row, in which the 1 is in the derivative matrix (row 1 for x, 2 for y, and 3 for z). Note that the zeros in the first three columns lead to the effect that in all further matrix multiplications to the left, these three columns remain zero and need not be computed any longer. The significant column of the matrix product is the last one, filled in each row by the inner product of the row-vector with the last column of the old (intermediate) product matrix. This analytical insight can save 75% of the computational steps for all matrices to the left if column vectors are used and matrix multiplication starts from the right-hand side.

The derivative matrices for rotational variables have four nonzero elements. These trigonometric functions have already been evaluated for the nominal case. The nonzero elements appear in such a pattern that working with sets of row– or column–vectors for matrices cuts in half the number of multiplications necessary for the elements of the product matrix.

As a final step toward image coordinates, the leftmost matrix P for perspective projection has to be applied. Note that this matrix product and the following scaling operations with element e_4 (Equation 2.11) yield two feature positions y and z in the image for each real-world feature. Thus, two Jacobian elements are also obtained for each image feature.

To get from the partial derivative of the total homogeneous transformation matrix $T'_{t\rho}$ to the correspondingly varied feature position in the image, this matrix has to be multiplied from the right-hand side by the 3-D feature vector $\mathbf{x}_{Fk}$ for the feature point (see Figure 2.11). This yields n vectors $e_{Dk\rho}$, $\rho = 1$ to n (6 in our case), each of which has four components. This is shown in the lower part of Figure 2.11. Multiplying these expressions by a finite variation in the state component δxs_ρ results in the corresponding changes in the homogeneous feature vector:

$$\delta e_\rho = e_{D\rho} \cdot \delta x_{S\rho}. \tag{2.18}$$

The virtually displaced "homogeneous" feature position vector e (index p) around the nominal point (designated by index N) is computed from the "homogeneous" feature vectors e_N for the nominal case and δe_ρ from Equation 2.18.

$$e_{p\rho 2} = e_{N2} + e_{D\rho 2} \cdot \delta x_{S\rho}; \quad e_{p\rho 3} = e_{N3} + e_{D\rho 3} \cdot \delta x_{S\rho}; \\ e_{p\rho 4} = e_{N4} + e_{D\rho 4} \cdot \delta x_{S\rho}. \tag{2.19}$$

Now the perturbed image feature positions after Equation 2.5 are

$$y_{p\rho} = e_{p\rho 2} / e_{p\rho 4} \ ; \quad z_{p\rho} = e_{p\rho 3} / e_{p\rho 4}. \tag{2.20}$$

Inserting the proper expressions from Equation 2.19 yields, with $e_{N2} / e_{N4} = y_{pN}$

$$\begin{aligned} y_{p\rho} &= (e_{N2} + e_{D\rho 2} \cdot \delta x_{S\rho})/(e_{N4} + e_{D\rho 4} \cdot \delta x_{S\rho}) \\ &= y_{pN} \cdot [1 + e_{D\rho 2} \cdot \delta x_{S\rho} / e_{N2}]/[1 + e_{D\rho 4} \cdot \delta x_{S\rho} / e_{N4}]. \end{aligned} \tag{2.21}$$

Since the components of e_D contain unknown variations δx_ρ, a linear relationship between these unknowns and small variations in feature positions are sought. If $e_{D4} / e_{N4} << 1$, the ratio in Equation 2.21 can be approximated by

$$\begin{aligned} y_{p\rho} &= y_{pN} \cdot (1 + e_{D\rho 2} / e_{N2} \cdot \delta x_{S\rho}) \cdot (1 - e_{D\rho 4} / e_{N4} \cdot \delta x_{S\rho}) \\ &= y_{pN} \cdot [1 + (e_{D\rho 2} / e_{N2} - e_{D\rho 4}/e_{N4}) \cdot \delta x_{S\rho} \ (-) \\ &\quad - (e_{D\rho 2} \cdot e_{D\rho 4})/(e_{N2} \cdot e_{N4}) \cdot \delta x_{S\rho}^2]. \end{aligned} \tag{2.22}$$

Neglecting the last term with $\delta x^2 s_\rho$ as being at least one order of magnitude smaller than the linear term with δxs_ρ, a linear relationship between changes in y due to δxs_ρ has been found

$$\delta y_{p\rho} = y_{p\rho} - y_{pN} \approx y_{pN} \cdot (e_{D\rho 2} / e_{N2} - e_{D\rho 4} / e_{N4}) \cdot \delta x_{S\rho}. \tag{2.23}$$

The element of the Jacobian matrix linked to the horizontal (y_i) feature at point x_{Fk} in the real world and to the unknown state variable xs_ρ now becomes

$$J_{k\rho y} = \partial y_k / \partial x_{S\rho} = \delta y_{p\rho} / \delta x_{S\rho} = y_{pN} \cdot (e_{D\rho 2} / e_{N2} - e_{D\rho 4} / e_{N4}). \quad (2.24)$$

The corresponding relation for the vertical feature position in the image is obtained in a similar way as

$$J_{k\rho z} = \partial z_k / \partial x_{S\rho} = \delta z_{p\rho} / \delta x_{S\rho} = z_{pN} \cdot (e_{D\rho 3} / e_{N3} - e_{D\rho 4} / e_{N4}). \quad (2.25)$$

This approach is a very flexible scheme for obtaining the entries into the Jacobian matrix efficiently. Adaptations to changing scene trees, due to new objects appearing with knew unknown states to be determined visually, can thus be made in an easy way.

The general approach discussed leaves two variants open to be selected for the actual case at hand:

1. *Very few feature points for an object:* In this case, it may be more economic with respect to computational load to multiply the sequence of transformations in Figure 2.11 from the left by the homogeneous 3-D feature point x_{Fk} (four components). This always requires only four inner vector products (= 25% of a matrix product). So, in total, for 6 matrix vector products, 24 inner products are needed; for the 7 expressions in Figure 2.11, a total of 168 such products result.
2. *Many feature points on an object:* Multiplying (concatenating) the elemental transformation matrices for the seven expressions in Figure 2.11 from right to left, in a naive approach requires at most $16 \cdot 5 \cdot 7 = 560$ inner vector products. For each feature point in the real world on a single object, $7 \cdot 4 = 28$ inner vector products have to be added to obtain the e-vector and its six partial derivatives. Asking for the number of features m on an object for which this approach is more economic as the one above, the relation $m \cdot 168 = 560 + m \cdot 28$ has to be solved for m as the break-even point, yielding $m = 560/140 = 4$.

So for more than four features on a single object, in our case with six unknowns in five transformation matrices plus perspective projection, the concatenation of transformation matrices first, and the multiplication with the coordinates of the feature points x_{Fk} afterward, is more computer-efficient.

Considering the fact that the derivative matrices are sparsely filled, as discussed above, and that many matrix products can be reused, frequently more than once, concatenation, performed as standard method in computer graphics, also becomes of interest in computer vision. However, as Figure 2.11 shows, much larger memory space has to be allotted for the iteration of transformation variables (the partial derivative matrices and their products). Note that to the left of derivative matrices of translations, also just a vector results for all further products, as in method 1 above. Taking advantage of all these points, method 2 is usually more efficient for more that two to three feature points on an object.

2.1.3 Time Representation

Time is considered an independent variable, monotonically increasing at a constant rate (as a good approximation to experience in the spatiotemporal domain of interest here). The temporal resolution required of measurement and control processes depends on the application area; with humans as the main partner in dealing with the real world, their characteristic timescale will also predominate for the technical systems under investigation here.

Due to the fact that humans need at least 30 ms between two signals sensed, to be able to tell their correct sequence (independent of the sensory modality: tactile, auditory, or visual) [Pöppel et al. 1991; Pöppel, Schill 1995], this time-window of 30 ms is considered the "window of simultaneity". It is the basic temporal unit within which all signals are treated as simultaneous [Ruhnau 1994a, b]. This fact has also been the decisive factor in fixing the video frame rate. (More precisely, the subdivision into fields of interleaved odd and even lines and the reduced field rate by a factor of 2 was introduced to cheat human perception because of missing technological performance levels at the time of definition in the 1930s). This was done to achieve the impression of smooth analog motion for the observer, even though the fields are discrete and do represent jumps. When looking at field sequences of video signals from a static scene, taken at a large angular rate of the camera in the direction of image lines, a noticeable shift between frames can be observed. For precise interpretation and early detection of an onset of motion, therefore, the alternating fields at twice the frame rate (frequency of 50, respectively, 60 Hz) should be analyzed.

Since today's machine vision very often relies on the old standard video equipment, the basic cycle time for full images is adopted for dynamic machine vision and for control output. The sampling periods are 16 2/3 ms in the US (33 1/3 ms for full images or for each odd or even field) and 20 ms (40 ms) in Europe. The decision is justified by the fact that the corner frequency of human extremities for control actuation is about 2 Hz (arms and legs). In sampled control theory, a dozen samplings per period are considered sufficient to achieve analogue-like overall behavior. Therefore, constant control outputs over one video period are acceptable from this point of view. Note that the transition to fully digital image sensors in the near future will allow more freedom in the choice of frame rates.

Processes in the real world are described most compactly by relating temporal change rates of state variables to the values of the state variables, to the control variables involved, and to additional perturbations, which can hardly be modeled; these relations are called differential equations.

They can be transformed into difference equations according to sampled data theory with constant control output over the sampling period by numerical (or analytical) integration; perturbations will show up as added accumulated values with similar statistical properties, as in the analog case. The standard forms for (linearized) state transitions over a time period T are the state transition matrix $A(T)$ and the control effect matrix $B(T)$. $A(T)$ multiplied by the old state vector yields the homogeneous part of the new state vector; $B(T)$ describes the effect of constant unit control inputs onto the new state; multiplying $B(T)$ with the actual control output and adding this to the homogeneous part yields the new state.

Using this knowledge about motion processes of 3-D objects in 3-D space for image sequence interpretation is the core of the 4-D approach to dynamic vision developed by [Dickmanns, Wuensche 1987, 1999]. Combining temporal prediction with the first-order derivative matrix of perspective projection (the "Jacobian matrix" of spatial vision discussed in previous sections) allows bypassing perspective inversion. Since each row of the Jacobian matrix contains the first-order sensitivity elements of the relation, how the feature measured depends on each state variable, spatial interpretation may become (at least partially) possible even with monocular vision, if either the object or the observer is moving. This will be discussed further down. Temporal embedding thus alleviates image interpretation despite the higher data rates. Temporal continuity conditions and attention control can counteract these higher data rates.

In addition, the eigenvalues of the transition matrix A represent characteristic time scales of the process. These and the frequency content of control inputs and of perturbations determine the temporal characteristics of the motion process.

In the framework of mission performance, other timescales may have special importance. The time needed for stabilizing image sequence interpretation is crucial for arriving at meaningful decisions based on this perception process. About a half second to one second are typical values for generating object hypotheses and having the transients settle down from poor initialization. Taking limited rates of change of state variables into account, preview (and thus prediction) times of several seconds seem to be reasonable in many cases. Total missions may last for several hours.

With respect to flawless functioning and maintenance of the vehicle's body, special timescales have to be observed, which an autonomous system should be aware of. All these aspects will be briefly discussed in the next section together with similar multiple scale problems in the spatial domain. To be flexible, an autonomous visual perception system should be capable of easy adjustment to temporal and spatial scales according to the task at hand.

2.1.4 Multiple Scales

The range of scales in the temporal and spatial domains needed to understand processes in the real world is very large. They are defined by typical sensor and mission dimensions as well as by the environmental conditions affecting both the sensors and the mission to be performed.

2.1.4.1 Multiple Space Scales

In the spatial domain, the size of the light sensitive elements in the sensor array may be considered the lower limit of immediate interest here. Typically, 5 to 20 micrometer (μm) is common today. Alternatively, as an underlying characteristic dimension, the typical width of the electronic circuitry may be chosen. This is about 0.1 to 2 μm, and this dimension characterizes the state of the art of microprocessors. Taking the 1-meter (m) scale as the standard, since this is the order of magnitude of typical body dimension of interest, the lower bound for spatial scales then is 10^{-7} m.

As the upper limit, the distance of the main light source on Earth, the orbital radius of the planet Earth circling the Sun (about 150 million km, that is $1.5 \cdot 10^{11}$ m), may be chosen with all other stars at infinity. The scale range of practical interest thus is about 18 orders of magnitude. However, the different scales are not of equal and simultaneous interest.

Looking at the mapping conditions for perspective imaging, the 10^{-5} m range has immediate importance as the basic grid size of the sensor. For remote sensing of the environment, when the characteristic speed is in the order of magnitude of tens of m/s, several hundred meters may be considered a reasonable viewing range yielding about 3 to 10 seconds reaction time until the vehicle may reach the location inspected. If objects with a characteristic dimension of about 5 cm = 0.05 m should just fill a single pixel in the image when seen at the maximum distance of, say 200 m, the focal length f required is $f/10^{-5} = 200/0.05$ or $f = 0.04$ m or 40 mm. If one would like to have a 1 cm wide line mapped onto 2 pixel at 10 m distance (*e.g.*, to be able to read the letters on a license plate), the focal length needed is $f =$ 20 mm. To recognize lane markings 12 cm wide at 6m distance with 6 pixels on this width, a focal length of 3 mm would be sufficient. This shows that for practical purposes, focal lengths in the millimeter to decimeter range are adequate. This also happens to be the physical dimension of modern CCD-TV cameras.

Because the wheel diameters of typical vehicles are of the order of magnitude of about 1 m, objects become serious obstacles if their height exceeds about 0.1 m. Therefore, the 0.1 to 100 m range (typical look-ahead distance) is the most important and most frequently used one for ground vehicles. Entire missions, usually, measure 1 to 100 km in range. For air vehicles, several thousand km are typical travel distances since the Earth radius is about 6 371 km.

It may be interesting to note that the basic scale "1 m" was defined initially as 10^{-7} of one quarter of the circumference around the globe via both poles about 2 centuries ago.

Inverse use of multiple space scales in vision: In a visual scene, the same object may be a few meters or a few hundred meters away; the system should be able to recognize the object as the same unit independent of the distance viewed. To achieve this more easily, multiple focal lengths for a set of cameras will help since a larger focal length directly counteracts the downscaling of the image size due to increased range. This is the main reason for using multifocal camera arrangements in EMS vision. With a spacing of focal lengths by a factor of 4 (corresponding to the second pyramid stage each time), a numerical range of 16 may be bridged with three cameras.

In practical applications, a new object is most likely picked up in the wide field of view having least resolution. As few as four pixels (2×2) may be sufficient for detecting a new object reliably without being able to classify it. Performing a saccade to bring the object into the field of view of a camera with higher resolution, would result in suddenly having many pixels available, additionally. In a bifocal system with a focal length ratio of 4, the resolution would increase to 8×8 (64 pixels); in a trifocal system it would even go up to 32×32 (*i.e.*, 1K pixels) on the same area in the real world. Now the object may be analyzed on these scales in parallel. The coarse space scale may be sufficient for tracking the object with high temporal resolution up to video rate. On the high-resolution space scale, the object may then be analyzed with respect to its detailed shape, possibly on a lower timescale if computing power is limited. This approach has proven to be efficient and robust.

Even with this approach, to cover a range of possible object distances of two to three orders of magnitude, the size of objects in the images still varies over more than one order of magnitude; this fact has to be dealt with. Pyramid techniques [Burt *et al.* 1981] and multiple scales of feature extraction operators are used to achieve this. This requires that the same object be represented on different scales (with different spatial resolution and corresponding shape descriptors). Homogeneous coordinates allow representing different scales by just one parameter, the scaling factor. In the 4×4 transformation matrices, it enters at position (4, 4).

2.1.4.2 Multiple Timescales

On the time axis, the lower limit of resolution is considered the cycle time of electronic devices such as sensors and processors; it is presently in the 10^{-8} to 10^{-10} second (s) range. Typical process elements in digital computing such as message overheads for communication with other processing elements last of the order of magnitude of 0.1 to 1 ms; this also is a typical range of cycle times for conventional sensing as with inertial sensors.

The video cycle times mentioned above are the next characteristic timescale. Human reaction times are characterized by and may well be the reason for the 1–second basic timescale. Therefore, the 0.1 to 10 s scale ranges are the most important and most frequently used. "Maneuvers" as typical time histories of control outputs for achieving desired transitions from one regime of steady behavior to another last up to several minutes. Quasi-steady behaviors such as road-running on a highway or flying across an ocean may last several hours. Beyond this, the astronomically based scales of *days* and *years* predominate. One day means one revolution with respect to the sun around the Earth's axis; it has subdivisions into 24 hours of 60 minutes of 60 seconds each that is 86 400 seconds in total. One "year" means one revolution of Earth around the Sun and includes about 365 days. The corresponding lighting and climatic conditions (seasonal effects) affect the operation of vehicles in natural and man-made environments to a large degree.

The lifetimes of typical man-made objects are the order of 10 years (vehicles, sensors); human life expectancy is 5 to 10 decades. Objects encountered in the environment may be hundreds (trees, buildings, *etc.*) or many thousands of years of age (geological formations). Therefore, also in the temporal domain the range of scales of interest spans from 10^{-9} to about 10^{10} seconds or 19 orders of magnitude. Autonomous systems to be developed should have the capability of handling this range of scales as educated humans are able to do. In a single practical application, the actual range of interest is much lower, usually.

2.2 Objects

Beside background knowledge of the environmental conditions at some point or region on the globe and the variations over the seasons, most of our knowledge about "the world" is affixed to object and subject classes. Of course, the ones of most importance are those one has to deal with most frequently in everyday life. Therefore, developing the sense of vision for road or air vehicles requires knowledge about objects and subjects encountered in these contexts most frequently; but also critical events which may put the achievement of mission goals at risk have to be known, even when their appearance is rather rare.

2.2.1 Generic 4-D Object Classes

The efficiency of the 4-D approach to dynamic vision is achieved by associating background knowledge about classes of objects and their behavioral capabilities with measurement data input. This knowledge is available in generic form, that is, structural information typical for object classes is fixed while specific parameters in the models have to be adapted to the special case at hand. Motion descriptions for the *c*enter of *g*ravity (the translational trajectory of the cg in space) forming the so-called "where"-problem, are separated from shape descriptions, called the "what"-problem. Typically, summing and averaging of feature positions is needed to solve the "where"-problem while differencing of feature positions contributes to solving the "what"-problem. In the approach chosen, the *"where"*-problem consists of finding the translational transformation parameters in the homogeneous transformations involved. The *"what"*-problem consists of finding an appropriate generic shape model and the best fitting shape and photometric parameters for this model after perspective projection (possibly including rotational degrees of freedom to account for the effects of aspect conditions).

As in computer graphics, all shape description is done in object-centered coordinates in 3-D, if possible, to take full advantage of a decoupled motion description relative to other objects.

2.2.2 Stationary Objects, Buildings

In road traffic, the road network and vegetation as well as buildings near the road are the stationary objects of most interest; roads will be discussed in Chapters 7 to 10. Highly visible large structures may be used as landmarks for orientation. The methods for shape representation are the same as for mobile objects; they do not need a motion description, however. These techniques are well known from computer graphics and are not treated here.

2.2.3 Mobile Objects in General

In this introductory chapter, only the basic ideas for object representation in the 4-D approach will be discussed. Detailed treatment of several object classes is done in connection with the application domain (Chapter 14). As far as possible, the methods used are taken from computer graphics to tap the large experience accumulated in that area. The major difference is that in computer vision, the actual model both with respect to shape and to motion is not given but has to be inferred from the visual appearance in the image sequence. This makes the use of complex shape models with a large number of tesselated surface elements (*e.g.*, triangles) obsolete; instead, simple encasing shapes like rectangular boxes, cylinders, polyhedra, or convex hulls are preferred. Deviations from these idealized shapes such as rounded edges or corners are summarized in fuzzy symbolic statements (like "rounded") and are taken into account by avoiding measurement of features in these regions.

2.2.4 Shape and Feature Description

With respect to shape, objects and subjects are treated in the same fashion. Only rigid objects and objects consisting of several rigid parts linked by joints are treated here; for elastic and plastic modeling see, *e.g.*, [Metaxas, Terzepoulos 1993]. Since objects may be seen at different distances, the appearance in the image may vary considerably in size. At large distances, the 3-D shape of the object, usually, is of no importance to the observer, and the cross section seen contains most of the information for tracking. However, this cross section may depend on the angular aspect conditions; therefore, both coarse-to-fine and aspect-dependent modeling of shape is necessary for efficient dynamic vision. This will be discussed for simple rods and for the task of perceiving road vehicles as they appear in normal road traffic.

2.2.4.1 Rods

An idealized rod (like a geometric line) is an object with an extension in just one direction; the cross section is small compared to its length, ideally zero. To exist in the real 3-D world, there has to be matter in the second and third dimensions. The simplest shapes for the cross section in these dimensions are circles (yielding a thin cylinder for a constant radius along the main axis) and rectangles, with the square as a special case. Arbitrary cross sections and arbitrary changes along the main axis yield generalized cylinders, discussed in [Nevatia, Binford 1977] as a flexible generic 3-D-shape (sections of branches or twigs from trees may be modeled this way). In many parts of the world, these "sticks" are used for marking the road in winter when snow may eliminate the ordinary painted markings. With constant cross–sections as circles and triangles, they are often encountered in road traffic also: Poles carrying traffic signs (at about 2 m elevation above the ground) very often have circular cross sections. Special poles with cross sections as rounded triangles (often with reflecting glass inserts of different shapes and colors near the top at about 1 m) are in use for alleviating driving at night and under foggy conditions. Figure 2.12 shows some shapes of rods as used in road traffic. No matter what the shape, the rod will appear in an image as a line with intensity edges, in general. Depending on the shape of the cross section, different shading patterns may occur. Moving around a pole with cross section (b) or (c) at constant distance R, the width of the line will change; in case (c), the diagonals will yield maximum line width when looked at orthogonally.

Under certain lighting conditions, due to different reflection angles, the two sides potentially visible may appear at different intensity values; this allows recognizing the inner edge. However, this is not a stable feature for object recognition in the general case.

The length of the rod can be recognized only in the image directly when the angle between the optical axis and the main axis of the rod is known. In the special case where both axes are aligned, only the cross section as shown in (a) to (c) can be seen and rod length is not at all observable. When a rod is thrown by a human, usually, it has both translational and rotational velocity components. The rotation occurs around the center of gravity (marked in Figure 2.12), and rod length in the image will oscillate depending on the plane of rotation. In the special case where the plane of rotation contains the optical axis, just a growing and shrinking line appears. In all other cases, the tips of the rod describe an ellipse in the image plane (with different excentricities depending on the aspect conditions on the plane of rotation).

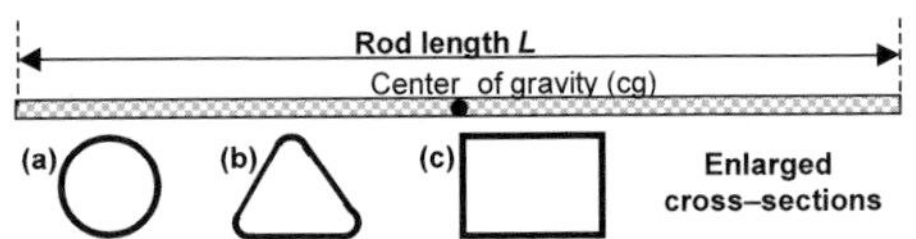

Figure 2.12. Rods with special applications in road traffic

2.2.4.2 Coarse-to-fine 2-D Shape Models

Seen from behind or from the front at a large distance, any road vehicle may be adequately described by its encasing rectangle. This is convenient since this shape has just two parameters, width *B* and height *H*. Precise absolute values of these parameters are of no importance at large distances; the proper scale may be inferred from other objects seen such as the road or lane width at that distance. Trucks (or buses) and cars can easily be distinguished. Experience in real-world traffic scenes tells us that even the upper boundary and thus the height of the object may be omitted without loss of functionality. Reflections in this spatially curved region of the car body together with varying environmental conditions may make reliable tracking of the upper boundary of the body very difficult. Thus, a simple U-shape of unit height (corresponding to about 1 m turned out to be practically viable) seems to be sufficient until 1 to 2 dozen pixels on a line cover the object in the image. Depending on the focal length used, this corresponds to different absolute distances.

Figure 2.13a shows this very simple shape model from straight ahead or exactly from the rear (no internal details). If the object in the image is large enough so that details may be distinguished reliably by feature extraction, a polygonal shape approximation of the contour as shown in Figure 2.13b or even with internal details (Figure 2.13c) may be chosen. In the latter case, area-based features such as the license plate, the dark

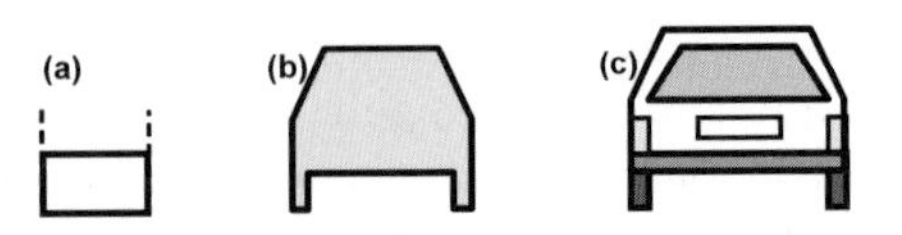

Figure 2.13. Coarse-to-fine shape model of a car in rear view: (a) encasing rectangle of width *B* (U-shape); (b) polygonal silhouette; (c) silhouette with internal structure

tires, or the groups of signal lights (usually in orange or reddish color) may allow more robust recognition and tracking.

2.2.4.3 Coarse-to-fine 3-D Shape Models

If multifocal vision allows tracking the silhouette of the entire object (*e.g.*, a vehicle) and of certain parts, a detailed measurement of tangent directions and curves may allow determining the curved contour. Modeling with Ferguson curves [Shirai 1987], "snakes" [Blake 1992], or linear curvature models easily derived from tangent directions at two points relative to the chord direction between those points [Dickmanns 1985] allows efficient piecewise representation. For vehicle guidance tasks, however, this will not add new functionality.

If the view onto the other car is from an oblique direction, the depth dimension (length of the vehicle) comes into play. Even with viewing conditions slightly off the axis of symmetry of the vehicle observed, the width of the car in the image will start increasing rapidly because of the larger length L of the body and due to the sine-effect in mapping.

Usually, it is very hard to determine the lateral aspect angle, body width B and length L simultaneously from visual measurements. Therefore, switching to the body diagonal D as a shape representation parameter has proven to be much more robust and reliable in real-world scenes [Schmid 1993]. Figure 2.14 shows the generic description for all types of rectangular boxes. For real objects with rounded shapes such as road vehicles, the encasing rectangle often is a sufficiently precise description for many purposes. More detailed shape descriptions with sub–objects (such as wheels, bumper, light groups, and license plate) and their appearance in the image due to specific aspect conditions will be discussed in connection with applications.

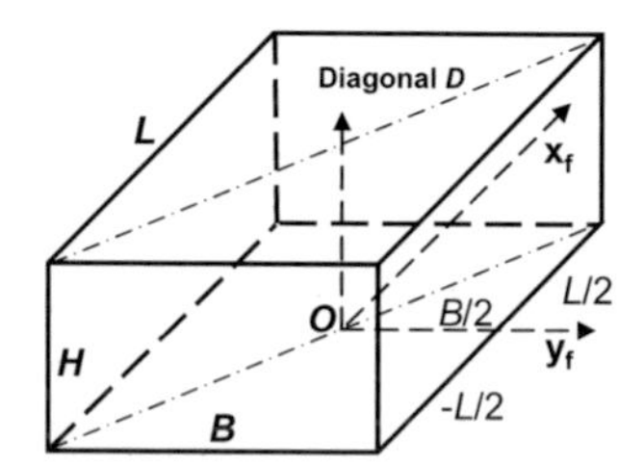

Figure 2.14. Object-centered representation of a generic box with dimension L, B, H; origin in center of ground plane

3-D models with different degrees of detail: Just for tracking and relative state estimation of cars, taking one of the vertical edges of the lower body and the lower bound of the object into account has proven sufficient in many cases [Thomanek 1992, 1994, 1996]. This, of course, is domain specific knowledge, which has to be introduced when specifying the features for measurement in the shape model. In general, modeling of highly measurable features for object recognition has to depend on aspect conditions.

Similar to the 2-D rear silhouette, different models may also be used for 3-D shape. Figure 2.13a corresponds directly to Figure 2.14 when seen from behind. The encasing box is a coarse generic model for objects with mainly perpendicular surfaces. If these surfaces can be easily distinguished in the image and their separation line may be measured precisely, good estimates of the overall body dimensions can be obtained for oblique aspect conditions even from relatively small image sizes. The top part of a truck and trailer frequently satisfies these conditions.

Polyhedral 3-D shape models with 12 independent shape parameters (see Figure 2.15 for four orthonormal projections as frequently used in engineering) have been investigated for road vehicle recognition [Schick 1992]. By specializing these parameters within certain ranges, different types of road vehicles such as cars, trucks, buses, vans, pickups, coupes, and sedans may be approximated sufficiently well for recognition [Schick, Dickmanns 1991; Schick 1992; Schmid 1993]. With these models, edge measurements should be confined to vehicle regions with small curvatures, avoiding the idealized sharp 3-D edges and corners of the generic model.

Aspect graphs for simplifying models and visibility of features: In Figure 2.15, the top-down the side view and the frontal and rear views of the polygonal model are given. It is seen that the same 3-D object may look completely different in these special cases of aspect conditions. Depending on them, some features may be visible or not. In the more general case with oblique viewing directions, combined features from the views shown may be visible. All aspect conditions that allow seeing the same set of features (reliably) are collected into one class. For a rectangular box on a plane and the camera at a fixed elevation above the ground, there are eight such aspect classes (see Figures 2.15 and 2.16): Straight from the front, from each side, from the rear, and an additional four from oblique views. Each can contain features from two neighboring groups.

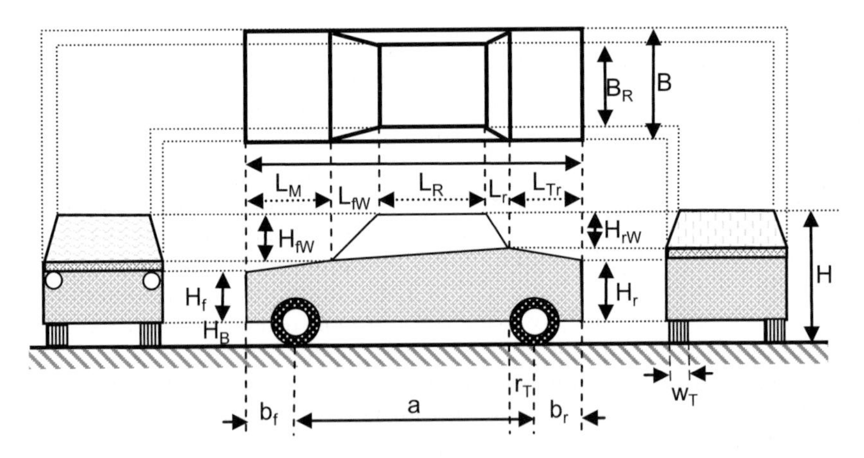

Figure 2.15. More detailed (idealized) generic shape model for road vehicles of type "car" [Schick 1992]

Due to this fact, a single 3-D model for unique (forward perspective) shape representation has to be accompanied by a set of classes of aspect conditions, each class containing the same set of highly visible features. These allow us to infer the presence of an object corresponding to this model from a collection of features in the image (inverse 3-D shape recognition including rough aspect conditions, or – in short – "hypothesis generation in 3-D").

This difficult task has to be solved in the initialization phase. Within each class of aspect conditions hypothesized, in addition, good initial estimates of the relevant state variables and parameters for recursive iteration have to be inferred from the relative distribution of features. Figure 2.16 shows the features for a typical car; for each vehicle class shown at the top, the lower part has special content.

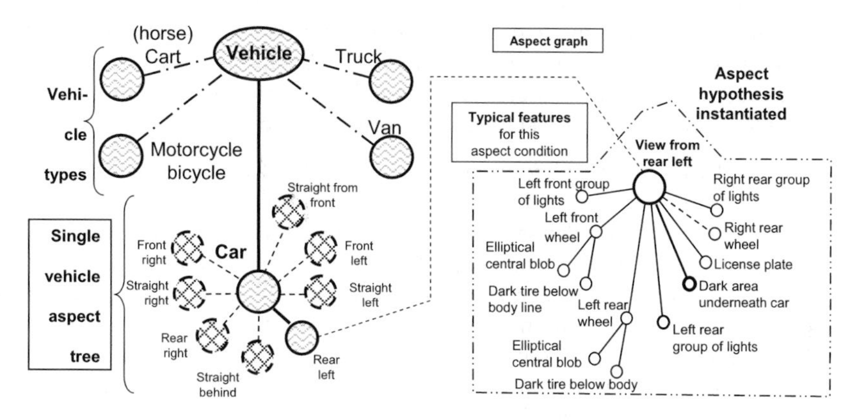

Figure 2.16. Vehicle types, aspect conditions, and feature distributions for recognition and classification of vehicles in road scenes

In Figure 2.17, a sequence of appearances of a car is shown driving in simulation on an oval course. The car is tracked from some distance by a stationary camera with gaze control that keeps the car always in the center of the image; this is called fixation-type vision and is assumed to function ideally in this simulation, *i.e.*, without any error).

The figure shows but a few snapshots of a steadily moving vehicle with sharp edges in simulation. The actual aspect conditions are computed according to a motion model and graphically displayed on a screen, in front of which a camera observes the motion process. To be able to associate the actual image interpretation with the results of previous measurements, a motion model is necessary in the analysis process also, constraining the actual motion in 3-D; in simulation, of course, the generic dynamical model is the same as in simulation. However, the actual control input is unknown and has to be reconstructed from the trajectory driven and observed (see Section 14.6.1).

2.2.5 Representation of Motion

The laws and characteristic parameters describing motion behavior of an object or a subject along the fourth dimension, time, are the equivalent to object shape representations in 3-D space. At first glance, it might seem that pixel position in the image plane does not depend on the actual speed components in space but only on the actual position. For one time this is true; however, since one wants to understand 3-D motion in a temporally deeper fashion, there are at least two points requiring modeling of temporal aspects:

1. Recursive estimation as used in this approach starts from the values of the state variables predicted for the next time of measurement taking.
2. Deeper understanding of temporal processes results from having representational terms available describing these processes or typical parts thereof in symbolic form, together with expectations of motion behavior over certain time-scales.

A typical example is the maneuver of lane changing. Being able to recognize these types of maneuvers provides more certainty about the correctness of the perception process. Since everything in vision has to be hypothesized from scratch, recognition of processes on different scales simultaneously helps building trust in the hypotheses pursued. Figure 2.17 may have been the first result from hardware-in-the-loop simulation where a technical vision system has determined the input

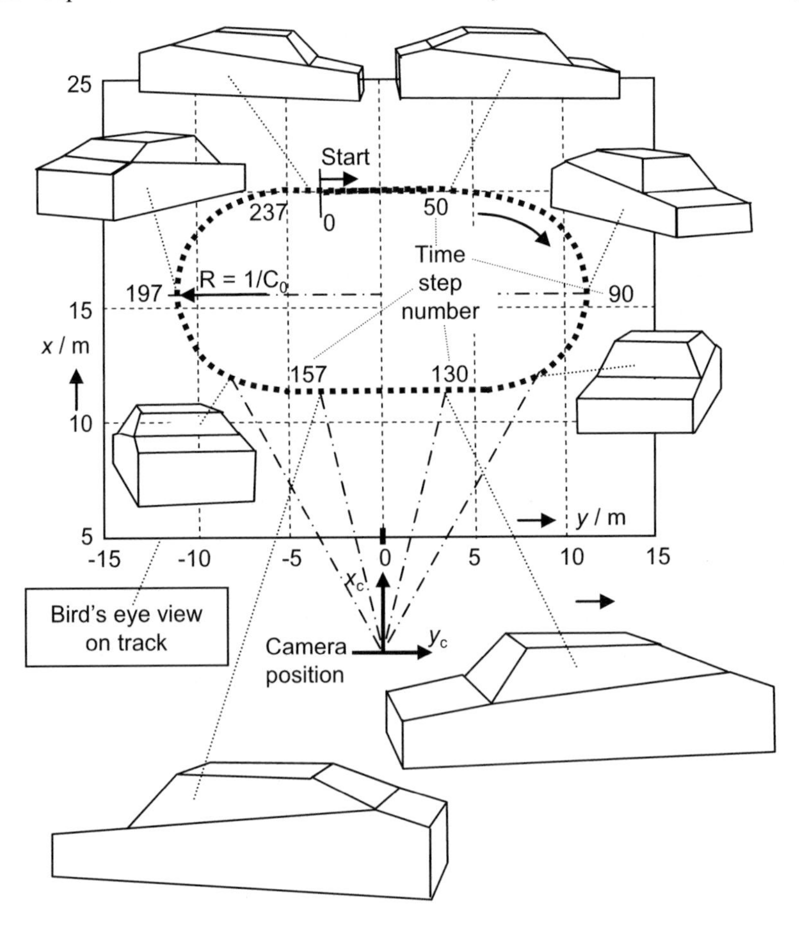

Figure 2.17. Changing aspect conditions and edge feature distributions while a simulated vehicle drives on an oval track with gaze fixation (smooth visual pursuit) by a stationary camera. Due to continuity conditions in 3-D space and time, "catastrophic events" like feature appearance/disappearance can be handled easily.

control time history for a moving car from just the trajectory observed, but, of course, with a motion model "in mind" (see Section 14.6.1).

The translations of the center of gravity (cg) and the rotations around this cg describe the motion of objects. For articulated objects also, the relative motion of the components has to be represented. Usually, the modeling step for object motion results in a (nonlinear) system of n differential equations of first order with n state components $\underline{X}$, q (constant) parameters $\underline{p}$ and r control components $\underline{U}$ (for subjects see Chapter 3).

2.2.5.1 Definition of State and Control Variables

A set of

- *State variables* is a collection of variables for describing temporal processes, which allows decoupling future developments from the past. State variables cannot be changed at one time. (This is quite different from "states" in computer science or automaton theory. Therefore, to accentuate this difference, sometimes use will be made of the terms *s-state* for systems dynamics states and *a-state* for automaton-state to clarify the exact meaning.) The same process may be described by different state variables, like Cartesian or polar coordinates for positions and their time derivatives for speeds. Mixed descriptions are possible and sometimes advantageous. The minimum number of variables required to completely decouple future developments from the past is called the order n of the system. Note that because of the second-order relationship between forces or moments and the corresponding temporal changes according to Newton's law, velocity components are state variables.
- *Control variables* are those variables in a dynamic system, that may be changed at each time "at will". There may be any kind of discontinuity; however, very frequently control time histories are smooth with a few points of discontinuity when certain events occur.

Differential equations describe constraints on temporal changes in the system. Standard forms are n equations of first order ("state equations") or an n-th order system, usually given as a transfer function of nth order for linear systems. There are an infinite variety of (usually nonlinear) differential equations for describing the same temporal process. System parameters $\underline{p}$ allow us to adapt the representation to a class of problems

$$dX / dt = f(X, p, t) . \tag{2.26}$$

Since real-time performance, usually, requires short cycle times for control, linearization of the equations of motion around a nominal set point (index N) is sufficiently representative of the process if the set point is adjusted along the trajectory. With the substitution

$$X = X_N + x , \tag{2.27}$$

one obtains

$$dX / dt = dX_N / dt + dx / dt . \tag{2.28}$$

The resulting sets of differential equations then are for the nominal trajectory:

$$dX_N / dt = f(X_N, p, t) ; \tag{2.29}$$

for the linearized perturbation system follows:

$$dx/dt = F \cdot x + v'(t)\,, \tag{2.30}$$

with $$F = df/dX_N \tag{2.31}$$

as an (n × n)-matrix and $\underline{v}'(t)$ an additive noise term.

2.2.5.2 Transition Matrices for Single Step Predictions

Equation 2.30 with matrix F may be transformed into a difference equation with cycle time T for grid point spacing by one of the standard methods in systems dynamics or control engineering. (Precise numerical integration from 0 to T for $v = 0$ may be the most convenient one for complex right–hand sides.) The resulting general form then is

$$x[(k+1)T] = A \cdot x[kT] + v[kT]$$

$$\text{or in short-hand} \qquad x_{k+1} = A \cdot x_k + v_k\,, \tag{2.32}$$

with matrix A of the same dimension as F. In the general case of local linearization, all entries of this matrix may depend on the nominal state variables. Procedures for computing the elements of matrix A from F have to be part of the 4-D knowledge base for the application at hand.

For objects, the trajectory is fixed by the initial conditions and the perturbations encountered. For subjects having additional control terms in these equations, determination of the actual control output may be a rather involved procedure. The wide variety of subjects is discussed in Chapter 3.

2.2.5.3 Basic Dynamic Model: Decoupled Newtonian Motion

The most simple and yet realistic dynamic model for the motion of a rigid body under external forces F_e is the Newtonian law

$$d^2x/dt^2 = \Sigma F_e(t)/m\,. \tag{2.33}$$

With unknown forces, colored noise $\underline{v(t)}$ is assumed, and the right–hand side is approximated by first–order linear dynamics (with time constant $T_C = 1/\alpha$ for acceleration a). This general third-order model for each degree of freedom may be written in standard state space form [BarShalom, Fortmann 1988]

$$d/dt \begin{pmatrix} x \\ V \\ a \end{pmatrix} = \underbrace{\begin{pmatrix} 0 & 1 & 0 \\ 0 & 0 & 1 \\ 0 & 0 & -\alpha \end{pmatrix}}_{F} \cdot \begin{pmatrix} x \\ V \\ a \end{pmatrix} + \underbrace{\begin{pmatrix} 0 \\ 0 \\ 1 \end{pmatrix}}_{g} \cdot v(t). \tag{2.34}$$

For the corresponding discrete formulation with sampling period T and $e^{-\alpha T} = \gamma$, the transition matrix $\underline{A}$ becomes

$$A = \begin{pmatrix} 1 & T & [T/\alpha - (1-\gamma)/\alpha] \\ 0 & 1 & (1-\gamma)/\alpha \\ 0 & 0 & \gamma \end{pmatrix};\ for\ \alpha = 0: A = \begin{pmatrix} 1 & T & T/2 \\ 0 & 1 & T \\ 0 & 0 & 1 \end{pmatrix}. \tag{2.35}$$

The perturbation input vector is modeled by

$$b_k \cdot v_k \quad \text{with} \quad b_k^T = [T^2/2,\ T,\ 1], \tag{2.36}$$

which yields the discrete model

$$x_{k+1} = A \cdot x_k + b_k \cdot v_k \,. \tag{2.37}$$

The value of the expectation is $\mathrm{E}[v_k] = 0$, and the variance is $\mathrm{E}[v_k^2] = \sigma_q^2$ (essential for filter tuning).The covariance matrix Q for process noise is given by

$$Q = b_k \cdot \sigma_q^2 \cdot b_k^T = \begin{pmatrix} T^4/4 & T^3/2 & T^2/2 \\ T^3/2 & T^2 & T \\ T^2/2 & T & 1 \end{pmatrix} \cdot \sigma_q^2 . \tag{2.38}$$

This model may be used independently in all six degrees of freedom as a default model if no more specific knowledge is given.

2.3 Points of Discontinuity in Time

The aspects discussed above for smooth parts of a mission with nice continuity conditions alleviate perception; however, sudden changes in behavior are possible, and sticking to the previous mode of interpretation would lead to disaster.

Efficient dynamic vision systems have to take advantage of continuity conditions as long as they prevail; however, they always have to watch out for discontinuities in object motion observed to adjust readily. For example, a ball flying on an approximately parabolic trajectory through the air can be tracked efficiently using a simple motion model. However, when the ball hits a wall or the ground, elastic reflection yields an instantaneous discontinuity of some trajectory parameters, which can nonetheless be predicted by a different model for the motion event of reflection. So the vision process for tracking the ball has two distinctive phases which should be discovered in parallel to the primary vision task.

2.3.1 Smooth Evolution of a Trajectory

Flight phases (or in the more general case, smooth phases of a dynamic process) in a homogeneous medium without special events can be tracked by continuity models and low-pass filtering components (like Section 2.2.5.3). Measurement values with oscillations of high frequency are considered to be due to noise; they have to be eliminated in the interpretation process. The natural sciences and engineering have compiled a wealth of models for different domains. The least-squares error model fit has proven very efficient both for batch processing and for recursive estimation. Gauss [1809] opened up a new era in understanding and fitting motion processes when he introduced this approach in astronomy. He first did this with the solution curves (ellipses) for the differential equations describing planetary motion.

Kalman [1960] derived a *recursive formulation* using *differential models* for the motion process when the statistical properties of error distributions are known. These algorithms have proven very efficient in space flight and many other applications. Meissner, Dickmanns [1983]; Wuensche [1987] and Dickmanns [1987] extended this approach to perspective projection of motion processes described in physical space; this brought about a quantum leap in the performance capabilities of real-time computer vision. These methods will be discussed for road vehicle applications in later sections.

2.3.2 Sudden Changes and Discontinuities

The optimal settings of parameters for smooth pursuit lead to unsatisfactory tracking performance in case of sudden changes. The onset of a harsh braking maneuver of a car or a sudden turn may lead to loss of tracking or at least to a strong transient motion estimated. If the onsets of these discontinuities can be predicted, a switch in model or tracking parameters at the right moment will yield much better results. For a bouncing ball, the moment of discontinuity can easily be predicted by the time of impact on the ground or wall. By just switching the sign of the angle of incidence relative to the normal of the reflecting surface and probably decreasing speed by some percentage, a new section of a smooth trajectory can be started with very likely initial conditions. Iteration will settle much sooner on the new, smooth trajectory arc than by continuing with the old model disregarding the discontinuity (if this recovers at all).

In road traffic, the compulsory introduction of the braking (stop) lights serves the same purpose of indicating that there is a sudden change in the underlying behavioral mode (deceleration), which can otherwise be noticed only from integrated variables such as speed and distance. The pitching motion of a car when the brakes are applied also gives a good indication of a discontinuity in longitudinal motion; it is, however, much harder to observe than braking lights in a strong red color.

Conclusion:

> As a general scheme in vision, it can be concluded that partially smooth sections and local discontinuities have to be recognized and treated with proper methods both in the 2-D image plane (object boundaries) and on the time line (events).

2.4 Spatiotemporal Embedding and First-order Approximations

After the rather lengthy excursion to object modeling and how to embed temporal aspects of visual perception into the recursive estimation approach, the overall vision task will be reconsidered in this section. Figure 2.7 gave a schematic survey of the way features at the surface of objects in the real 3-D world are transformed into features in an image by a properly defined sequence of "homogeneous coordinate transformations" (HCTs). This is easily understood for a static scene.

To understand a dynamically changing scene from an image sequence taken by a camera on a moving platform, the temporal changes in the arrangements of objects also have to be grasped by a description of the motion processes involved.

Therefore, the general task of real-time vision is to achieve a compact internal representation of motion processes of several objects observed in parallel by evaluating feature flows in the image sequence. Since egomotion also enters the content of images, the state of the vehicle carrying the cameras has to be observed simultaneously. However, vision gives information on *relative* motion only between objects, unfortunately, in addition, with appreciable time delay (several tenths of a second) and no immediate correlation to inertial space. Therefore, conventional sensors on the body yielding relative motion to the stationary environment (like odometers) or inertial accelerations and rotational rates (from inertial sensors like accelerometers and angular rate sensors) are very valuable for perceiving egomotion and for telling this apart from the visual effects of motion of other objects. Inertial sensors have the additional advantage of picking up perturbation effects from the environment before they show up as unexpected deviations in the integrals (speed components and pose changes). All these measurements with differing delay times and trust values have to be interpreted in conjunction to arrive at a consistent interpretation of the *situation* for making decisions on appropriate behavior.

Before this can be achieved, perceptual and behavioral capabilities have to be defined and represented (Chapters 3 to 6). Road recognition as indicated in Figures 2.7 and 2.9 while driving on the road will be the application area in Chapters 7 to 10. The approach is similar to the human one: Driven by the optical input from the image sequence, an internal *animation process in 3-D space and time* is started with members of generically known object and subject classes that are to duplicate the visual appearance of "the world" by prediction-error feedback. For the next time for measurement taking (corrected for time delay effects), the expected values in each measurement modality are predicted. The prediction errors are then used to improve the internal state representation, taking the Jacobian matrices and the confidence in the models for the motion processes as well as for the measurement processes involved into account (error covariance matrices).

For vision, the concatenation process with HCTs for each object-sensor pair (Figure 2.7) as part of the physical world provides the means for achieving our goal of understanding dynamic processes in an integrated approach. Since the analysis of the next image of a sequence should take advantage of all information collected up to this time, temporal prediction is performed based on the actual best estimates available for all objects involved and based on the dynamic models as discussed. Note that no storage of image data is required in this approach, but only the parameters and state variables of those objects instantiated need be stored to represent the scene observed; usually, this reduces storage requirements by several orders of magnitude.

Figure 2.9 showed a road scene with one vehicle on a curved road (upper right) in the viewing range of the egovehicle (left); the connecting object is the curved road with several lanes, in general. The mounting conditions for the camera in the vehicle (lower left) on a platform are shown in an exploded view on top for clarity. The coordinate systems define the different locations and aspect conditions for object mapping. The trouble in vision (as opposed to computer graphics) is that the entries in most of the HCT-matrices are the unknowns of the vision problem (relative distances and angles). In a tree representation of this arrangement of objects (Figure 2.7), each edge between circles represents an HCT and each node (circle) represents an object or sub–object as a movable or functionally separate part. Objects may be inserted or deleted from one frame to the next (dynamic scene tree).

This scene tree represents the mapping process of features on the surface of objects in the real world up to hundreds of meters away into the image of one or more camera(s). They finally have an extension of several pixels on the camera chip (a few dozen micrometers with today's technology). Their motion on the chip is to be interpreted as body motion in the real world of the object carrying these features, taking body motion affecting the mapping process properly into account. Since body motions are smooth, in general, spatiotemporal embedding and first-order approximations help making visual interpretation more efficient, especially at high image rates as in video sequences.

2.4.1 Gain by Multiple Images in Space and/or Time for Model Fitting

High–frequency temporal embedding alleviates the correspondence problem between features from one frame to the next, since they will have moved only by a small amount. This reduces the search range in a top-down feature extraction mode like the one used for tracking. Especially, if there are stronger, unpredictable perturbations, their effect on feature position is minimized by frequent measurements. Doubling the sampling rate, for example, allows detecting a perturbation onset much earlier (on average). Since tracking in the image has to be done in two dimensions, the search area may be reduced by a square effect relative to the one-dimensional (linear) reduction in time available for evaluation. As mentioned previously for reference, humans cannot tell the correct sequence of two events if they are less than 30 ms apart, even though they can perceive that there are two separate events [Pöppel, Schill 1995]. Experimental experience with technical vision systems has shown that using every frame of a 25 Hz image sequence (40 ms cycle time) allows object tracking of high quality if proper feature extraction algorithms to subpixel accuracy and well-tuned recursive estimation processes are applied. This tuning has to be adapted by knowledge components taking the situation of driving a vehicle and the lighting conditions into account.

This does not include, however, that all processing on the higher levels has to stick to this high rate. Maneuver recognition of other subjects, situation assessment, and behavior decision for locomotion can be performed on a (much) lower scale without sacrificing quality of performance, in general. This may partly be due to the biological nature of humans. It is almost impossible for humans to react in less than several hundred milliseconds response time. As mentioned before, the unit "second" may have been chosen as the basic timescale for this reason.

However, high image rates provide the opportunity both for early detection of events and for data smoothing on the timescale with regard to motion processes of interest. Human extremities like arms or legs can hardly be activated at more than 2 Hz corner frequency. Therefore, efficient vision systems should concentrate computing resources to where information can be gained best (at expected feature locations of known objects/subjects of interest) and to regions where new objects may occur. Foveal–peripheral differentiation of spatial resolution in connection with fast gaze control may be considered an optimal vision system design found in nature, if a corresponding management system for gaze control, knowledge application and interpretation of multiple, piecewise smooth image sequences is available.

2.4.2 Role of Jacobian Matrix in the 4-D Approach to Dynamic Vision

It is in connection with 4-D spatiotemporal motion models that the sensitivity matrix of perspective feature mapping gains especial importance. The dynamic models for motion in 3-D space link feature positions from one time to the next. Contrary to perspective mapping in a single image (in which depth information is completely lost), the partial first-order derivatives of each feature with respect to all variables affecting its appearance in the image do contain spatial information. Therefore, linking the temporal motion process in 4-D with this physically meaningful Jacobian matrix has brought about a quantum leap in visual dynamic scene understanding [Dickmanns, Meissner 1983, Wünsche 1987, Dickmanns 1987, Dickmanns, Graefe 1988, Dickmanns, Wuensche 1999]. This approach is fundamentally different from applying some (arbitrary) motion model to features or objects in the image plane as has been tried many times before and after 1987. It was surprising to learn from a literature review in the late 1990s that about 80 % of so-called Kalman–filter applications in vision did not take advantage of the powerful information available in the Jacobian matrices when these are determined, *including egomotion and the perspective mapping process*.

The nonchalance of applying Kalman filtering in the image plane has led to the rumor of brittleness of this approach. It tends to break down when some of the (unspoken) assumptions are not valid. Disappearance of features by self-occlusion has been termed a *catastrophic event*. On the contrary, Wünsche [1986] was able to show that not only temporal predictions in 3-D space were able to handle this situation easily, but also that it is possible to determine a limited set of features allowing optimal estimation results. This can be achieved with relatively little additional effort exploiting information in the Jacobian matrix. It is surprising to notice that this early achievement has been ignored in the vision literature since. His system for visually perceiving its state relative to a polyhedral object (satellite model in the laboratory) selected four visible corners fully autonomously out of a much larger total number by maximizing a goal function formed by entries of the Jacobian matrix (see Section 8.4.1.2).

Since the entries into a row of the Jacobian matrix contain the partial derivatives of feature position with respect to all state variables of an object, the fact that all the entries are close to zero also carries information. It can be interpreted as an indication that this feature does not depend (locally) on the state of the object; therefore, this feature should be discarded for a state update.

If all elements of a column of the Jacobian matrix are close to zero, this is an indication that all features modeled do not depend on the state variable corresponding to this column. Therefore, it does not make sense to try to improve the estimated value of this state component, and one should not wonder that the mathematical routine denies delivering good data. Estimation of this variable is not possible under these conditions (for whatever reason), and this component should be removed from the list of variables to be updated. It has to be taken as a standard case, in general in vision, that only a selection of parameters and variables describing another object are observable at one time with the given aspect conditions. There has to be a management process in the object recognition and tracking procedures, which takes care of these particular properties of visual mapping (see later section on system integration).

If this information in properly set up Jacobian matrices is observed during tracking, much of the deplored brittleness of Kalman filtering should be gone.

3 Subjects and Subject Classes

Extending representational schemes found in the literature up to now, this chapter introduces a concept for visual dynamic scene understanding centered on the phenomenon of *control variables* in dynamic systems. According to the international standard adopted in mathematics, natural sciences, and engineering, control variables are those variables of a dynamic system, which can be changed at any moment. On the contrary, *state variables* are those, which cannot be changed instantaneously, but have to evolve over time. State variables de-couple the future evolution of a system from the past; the minimal number required to achieve this is called the order of the system.

It is the existence of control variables in a system that separates subjects from objects (proper). This fact contains the kernel for the emergence of a "free will" and consciousness, to be discussed in the outlook at the end of the book. Before this can be made understandable, however, this new starting point will be demonstrated to allow systematic access to many terms in natural language. In combination with well-known methods from control engineering, it provides the means for solving the symbol grounding problem often deplored in conventional AI [Winograd, Flores 1990]. The decisions made by subjects for control application in a given task and under given environmental conditions are the driving factors for the evolution of goal-oriented behavior. This has to be seen in connection with performance evaluation of populations of subjects. Once this loop of causes becomes sufficiently well understood and explicitly represented in the decision-making process, emergence of "intelligence" in the abstract sense can be stated.

Since there are many factors involved in understanding the actual situation given, those that influence the process to be controlled have to be separated from those that are irrelevant. Thus, perceiving the situation correctly is of utmost importance for proper decision-making. It is *not* intended here to give a general discussion of this methodical approach *for all kinds of subjects*; rather, this will be confined to vehicles with the sense of vision just becoming realizable for transporting humans and their goods. It is our conviction, however, that all kinds of subjects in the biological and technical realm can be analyzed and classified this way.

Therefore, without restrictions, ***subjects*** *are defined as bodily objects with the capability of measurement intake and control output depending on the measured data as well as on stored background knowledge.*

This is a very general definition subsuming all animals and technical devices with these properties.

3.1 General Introduction: Perception – Action Cycles

Remember the definition of control variables given in the previous chapter: They encompass all variables describing the dynamic process, which can be changed at any moment. Usually, it is assumed as an idealization that a mental or computational decision for a control variable can be implemented without time delay and distortion of the time history intended. This may require high gains in the implementation chain. In addition, fast control actuation relative to slow body motion capabilities may be considered instantaneous without making too large an error. If these real-world effects cannot be neglected, these processes have to be modeled by additional components in the dynamic system and taken into account by increasing the order of the model.

The same is true for the sensory devices transducing real-world state variables into representations on the information processing level. Situation assessment and control decision-making then are computational activities on the information processing level in which measured data are combined with stored background knowledge to arrive at an optimal (or sufficiently good) control output. The quality of realization of this desired control and the performance level achieved in the mission context may be monitored and stored to allow us to detect discrepancies between the mental models used and the real-world processes observed. The motion-state of the vehicle's body is an essential part of the situation given, since both the quality of measurement data intake and control output may depend on this state.

Therefore, the closed loop of perception, situation assessment/decision–making and control activation of a moving vehicle always has to be considered in conjunction. Potential behavioral capabilities of subjects can thus be classified by first looking at the capabilities in each of these categories separately and then by stating which of these capabilities may be combined to allow more complex maneuvering and mission performance. All of this is not considered a sequence of quasi-static states of the subject that can be changed in no time (as has often been done in conventional AI). Rather, it has to be understood as a dynamic process with alternating smooth phases of control output and sudden changes in behavioral mode due to some (external or internal) event. Spatiotemporal aspects predominate in all phases of this process.

3.2 A Framework for Capabilities

To link image sequences to understanding motion processes in the real world, a few basic properties of control application are mentioned here. Even though control variables, by definition, can be changed arbitrarily from one time to the next, for energy and comfort reasons one can expect sequences of smooth behaviors. For the same reason, it can even be expected that there are optimal sequences of control application (however "optimal" is defined) which occur more often than others. These *stereotypical* time histories for achieving some state transition efficiently constitute valuable knowledge not only for controlling movements of the vehicle's body, but also for understanding motion behavior of other subjects. Human language has special expressions for these capabilities of motion control, which often are performed sub-consciously: They are called *maneuvers* and have a temporal extension in the seconds-to-minutes range.

Other control activities are done to maintain an almost constant state relative to some desired one, despite unforeseeable disturbances encountered. These are called *regulatory* control activities, and there are terms in human language describing them. For example, *"lane keeping"* when driving a road vehicle is one such activity where steering wheel input is somehow linked to road curvature, lateral offset, and yaw angle relative to the road. The speed V driven may depend on road curvature since lateral acceleration depends on V^2/R, with R the radius of the curve. When driving on a straight road, it is therefore also essential to recognize the onset of a beginning curvature sufficiently early so that speed can be reduced either by decreasing fuel injection or by activating the brakes. The deceleration process takes time, and it depends on road conditions too [dry surface with good friction coefficient or wet (even icy) with poor friction]. Vision has to provide this input by concentrating attention on the road sections both nearby and further away. Only knowledgeable agents will be able to react in a proper way: They know where to look and for what (which types of features yield reliable and good hints). This example shows that there are situations where a more extended task context has to be taken into account to perform the vision task satisfactorily.

Another example is given in Figure 3.1. If both vehicles have just a radar (or laser) sensor on board, which is not able to recognize the road and lane boundaries, the situation perceived seems quite dangerous. Two cars are moving toward each other at high speed (shown by the arrows in front) on a common straight line.

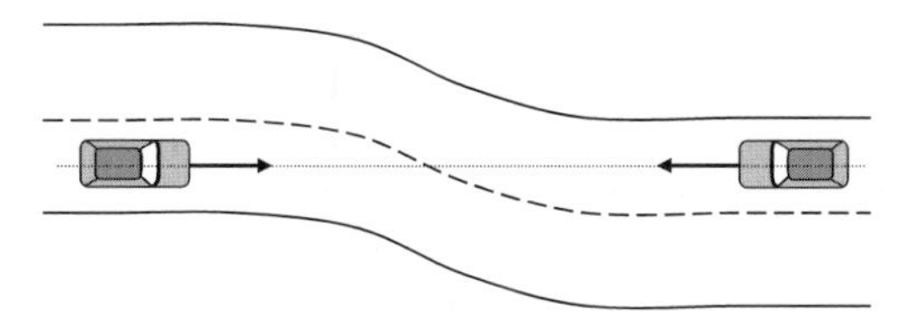

Figure 3.1. Judgment of a situation depends on the environmental context and on knowledge about behavioral capabilities and goals

Humans and advanced technical vision systems seeing the S-shaped road curvature conclude that the other vehicle is going to perform lane keeping as an actual control mode. The subject vehicle doing the same will result in no stress and a harmless passing maneuver. The assumption of a suicidal driver in the other car is extremely unlikely. This shows, however, that the decision process from basic vision "here and now" to judgment of a situation and coming up with a reasonable or optimal solution for one's own behavior may be quite involved. Intelligent reactions and defensive driving require knowledge about classes of subjects encountered in a certain domain and about likely perceptual and behavioral capabilities of the participants.

Driving in dawn or dusk near woods on minor roads may lead to an encounter with animals. If a vehicle has killed an animal previously and the cadaver lies on the road, there may be other animals including birds feeding on it. The behavior to be expected of these animals is quite different depending on their type.

Therefore, for subjects and proper reactions when encountering them, a knowledge base should be available on

1. How to recognize members of classes of subjects.
2. Which type of reaction may be expected in the situation given. Biological subjects, in general, have articulated bodies with some kind of elasticity or plasticity. This may complicate visual recognition in a snapshot image. In real life or in a video stream, typical motion sequences (even of only parts of the body) may alleviate recognition considerably. Periodic motion of limbs or other body parts is such an example. This will not be detailed here; we concentrate on typical motion behaviors of vehicles as road traffic participants, controlled by humans or by devices for automation.

Before this is analyzed in the next section, Table 3.1 ends this general introduction to the concept of subjects by showing a collection of different categories of capabilities (not complete).

Table 3.1. Capabilities characterizing subjects (type: road vehicles)

Categories of capabilities	Devices/algorithms	Capabilities
Sensing	odometry, inertial sensor set, radar, laser range finder, body-fixed imaging sensors, active vertebrate-type vision.	measure distance traveled, speed; 3 linear accelerations, 3 rotational rates; range to objects, bearing; body-fixed fields of view, gaze controlled vision
Perception (data association with knowledge stored)	data processing algorithms, data fusion, data interpretation, knowledge representation	motion understanding, scene interpretation, situation assessment
Decision-making	rule bases, integration methods, value systems	prediction of trajectories, evaluation of goal oriented behaviors;
Motion control	controllers, feed-forward and feedback algorithms, actuators	locomotion, viewing direction control, articulated motion
Data logging and retrieval, statistical evaluation	storage media, algorithms	remembrance, judge data quality, form systematic databases
Learning	value system, quality criteria, application rules	improvement and extension of own behavior
Team work, cooperation	communication channels, visual interpretation	joint (coordinated) solution of tasks and missions, increase efficiency
Reasoning	AI software	group planning

The concept of explicitly represented capabilities allows systematic structuring of subject classes according to the performance expected from its members. Beside shape in 3-D space, subjects can be recognized (and sometimes even identified as individual) by their stereotypical behavior over time. To allow a technical vision system to achieve this level of performance, the corresponding visually observable motion and gaze control behaviors should be modeled into the knowledge base. It has to be allocated at a higher perceptual level for deeper understanding of dynamic scenes.

The *basic* capabilities of a subject are

1. Sensing (measuring) some states of environmental conditions, of other objects/subjects in the environment, and of components of the subject state.
2. Storing results of previous sensing activities and linking them to overall situational aspects, to behavior decisions, and to resulting changes in states observed.
3. Behavior generation depending on 1 and 2.

Step 2 may already require a higher developmental level not necessarily needed in the beginning of an evolutionary biological process. For the technical systems of interest here, this step is included right from the beginning by goal-oriented engineering, since later capabilities of learning and social interaction have to rely on it. Up to now, these steps are mostly provided by the humans developing the system. They perform adaptations to changing environmental conditions and expand the rule base for coping with varying environments. In these cases, only data logging is performed by the system itself; the higher functions are provided by the developer on this basis. Truly autonomous systems, however, should be able to perform more and more of these activities by themselves; this will be discussed in the outlook at the end of the book. The suggestion is that all rational mental processes can be derived on this basis.

The decisive factors for these learning activities are (a) availability of time scales and the scales for relations of interest, like spatial distances; (b) knowledge about classes of objects and of subjects considered; (c) knowledge about performance indices; and (d) about value systems for behavior decisions; all these enter the decision-making process.

3.3 Perceptual Capabilities

For biological systems, five senses have become proverbial: Seeing, hearing, smelling, tasting, and touching. It is well known from modern natural sciences that there are a lot more sensory capabilities realized in the wide variety of animals. The proprioceptive systems telling the actual state of an articulated body and the vestibular systems yielding information on a subject's motion-state relative to inertial space are but two essential ones widely spread. Ultrasound and magnetic and infrared sensors are known to exist for certain species.

The sensory systems providing access to information about the world to animals of a class (or to each individual in the class by its specific realization) are characteristic of their potential behavioral capabilities. Beside body shape and the specific locomotion system, the sensory capabilities and data processing as well as knowledge association capabilities of a subject determine its behavior.

Perceptual capabilities will be treated separately for conventional sensors and the newly affordable imaging sensors, which will receive most attention later on.

3.3.1 Sensors for Ground Vehicle Guidance

In ground vehicles, speed sensors (tachometers) and odometers (distance traveled) are the most common sensors for vehicle guidance. Formerly, these signals were derived from sensing at just one wheel. After the advent of antilock braking systems (ABS), the rotational speed of each wheel is sensed separately. Because of the availability of a good velocity signal, this state variable does not need to be determined from vision but can be used for motion prediction over one video cycle.

Measuring oil or water temperature and oil pressure, rotational engine speeds (revolutions per minute) and fuel remaining mainly serves engine monitoring. In connection with one or more inertial rotational rate sensors and the steering angle measured, an "electronic stability program" (ESP or similar acronym) can help avoid dangerous situations in curve steering. A few top-range models may be ordered with range measurement devices to objects in front for distance keeping (either by radar or laser range finders). Ultrasound sensors for (near-range) parking assistance are available, too. Video sensors for lane departure warning just entered the car market after being available for trucks since 2000.

Since the U.S. Global Positioning System (GPS) is up and open to the general public, the absolute position on the globe can be determined to a few meters accuracy (depending on parameters set by the military provider). The future European Galileo system will make global navigation more reliable and precise for the general public.

The angular orientations of the vehicle body are not measured conventionally, in general, so that these state variables have to be determined from visual motion analysis. This is also true for the slip (drift) angle in the horizontal plane stating the difference in azimuth as angle between the vehicle body and the trajectory tangent at the location of the center of gravity (cg).

Though ground vehicles did not have any inertial sensors till lately, modern cars have some linear accelerometers and angular rate sensors for their active safety systems like airbags and electronic stability programs (ESP); this includes measurement of the steering angle. Since full sets of inertial sensors have become rather inexpensive with the advent of microelectronic devices, it will be assumed that in the future at least coarse acceleration and rotational rate sensors will be available in any car having a vision system. This allows the equivalent of vestibular – ocular data communication in vertebrates. As discussed in previous chapters, this considerably alleviates the vision task under stronger perturbations, since a subject's body orientation can be derived with sufficient accuracy before visual perception starts analyzing data on the object level. Slow inertial drifts may be compensated for by visual feedback. External thermometers yield data on the outside temperature which may have an important effect on visual appearance of the environment around the freezing point. This may sometimes help in disambiguating image data not easily interpretable.

For a human driver guiding a ground vehicle, the sense of vision is the most important source of information, especially in environments with good look-ahead ranges and sudden surprising events. Over the last two decades, the research community worldwide has started developing the sense of vision for road vehicles, too. [Bertozzi *et al.* 2000] and [Dickmanns 2002 a, b] give a review on the development.

3.3.2 Vision for Ground Vehicles

Similar to the differences between insect and vertebrate vision systems in the biological realm, two classes of technical vision systems can also be found for ground vehicles. The more primitive and simple ones have the sensory elements directly mounted on the body. Vertebrate vision quickly moves the eyes (with very little inertia by themselves) relative to the body, allowing much faster gaze pointing control independent of body motion.
The performance levels achievable with vision systems depend very much on the field of view (f.o.v.) available, the angular resolution within the f.o.v., and the capability of pointing the f.o.v. in certain directions. Figure 3.2 gives a summary of the most important performance parameters of a vision system. Data and knowledge processing capabilities available for real-time analysis are the additional important factors determining the performance level in visual perception.

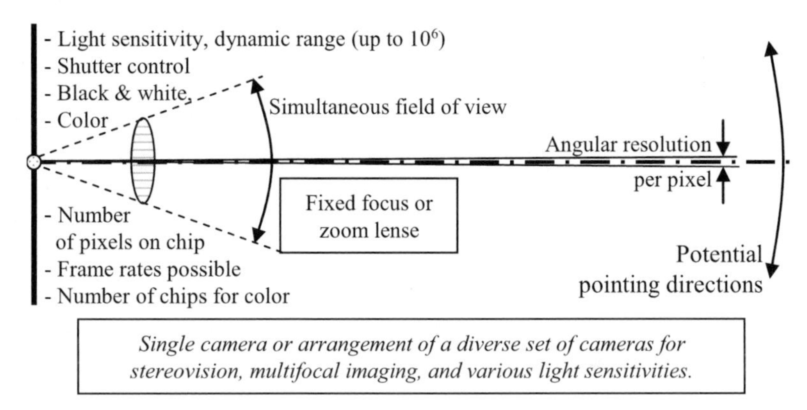

Figure 3.2. Performance parameters for vision systems

Cameras mounted directly on a vehicle body are subjected to any motion of the entire vehicle; they can be turned towards an object of interest only by turning the vehicle body. Note that with typical "Ackermann"-type steering of ground vehicles (front wheels on the tips of the front axle can be turned around an almost vertical axis), the vehicle cannot change viewing direction when stopped, and only in a very restricted manner otherwise. In AI-literature, this is called a nonholonomic constraint.

Resolution within the field of view is homogeneous for most vision sensors. This is not a good match to the problem at hand, where looking almost parallel to a planar surface from an observation point at small elevation above the surface means that distance on the ground in the real world changes with the image row from the bottom to the horizon. Non-homogeneous image sensors have been researched [*e.g.*, Debusschere *et al.* 1990] but have not found wider application yet. Using two cameras with different focal lengths and almost parallel optical axes has also been studied [Dickmanns, Mysliwetz 1992]; the results have led to the *MarVEye*–concept to be discussed in Chapter 12.

Since most of the developments of vision systems for road vehicles are using the simple approach of mounting cameras directly on the vehicle body, some of the implications will be discussed first so that the limitations of this type of visual sensor are fully understood. Then, the more general and much more powerful vertebrate-type active vision capabilities will be discussed.

3.3.2.1 Eyes Mounted Directly on the Body

Since spatial resolution capabilities improve with elevation above the ground, most visual sensors are mounted at the top of the front windshield. Figure 3.3 shows an example of stereovision. A single camera or two cameras with different focal lengths, not requiring a large stereo base, can be hidden nicely behind the rear-view mirror inside the vehicle. The type of vision system may thus be discovered by visual inspection when the underlying principles are known.

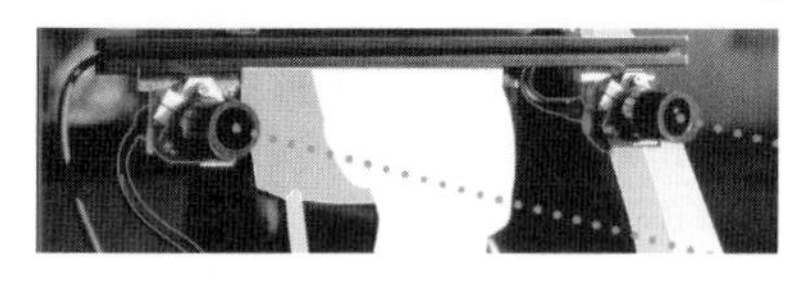

Figure 3.3. Two cameras mounted fix on vehicle body

Pitch effects: When driving on smooth surfaces, pitch perturbations on the body are small (less than 1°), usually. Strong braking actions and accelerations may lead to pitch angle changes of up to 3 or 4°. Pitch angles have an influence on the vertical range of rows to be evaluated when searching for objects on a planar ground at a given distance in the 3-D world. For a camera with a strong telelens (f.o.v. of ~ same size as the perturbation, 3 to 4°), this means that an object of interest previously tracked may no longer be visible at all in a future image! In a camera with a normal lens of ~ 35° vertical f.o.v., this corresponds to a shift of only ~ 10 % of the total number of rows (~ 50 rows in absolute terms). This clearly indicates that body-fixed vision sensors are limited to cameras with small focal lengths. They may be manageable for seeing large objects in the near range; however, they are unacceptable for tracking objects of the same size further away.

When the front wheels of a car drive over an obstacle on the ground of about 10 cm height with the rear wheels on the flat ground (at a typical axle distance ~ 2.5 m), an oscillatory perturbation in pitch with amplitude of about 2° will result. At 10 meters distance, this perturbation will shift the point, where an optical ray through a fixed pixel hits an object with a vertical surface, by almost half a meter up and down. However, at 200 meters distance, the vertical shift will correspond to plus or minus 10 meters! Assuming that the highly visible car body height is ~ 1 m, this perturbation in pitch (min. to max.) will lead to a shift in the vertical direction of 1 unit (object size) at 10 m distance, while at 200 m distance, this will be 20 units. This shows that object-oriented feature extraction *under perturbations* requires a much larger search range further away for this type of vision system. Looking almost parallel to a flat ground, the shift in look-ahead distance L for a given image line z is much greater. To be precise, for a camera elevation of 1.5 m above the ground, a perturbation of 50 mrad (~ 3°) upward shifts the look-ahead distance from 30 m to infinity (to the horizon).

If the pitch rate could be measured inertially, the gaze-controllable eye would allow commanding the vertical gaze control by the negative value of the pitch rate measured. Experiments with inexpensive rate sensors have shown that perturbations in the pitch angle amplitude of optical rays can be reduced by at least one order of magnitude this way (inertial angular rate feedback, see Figure 12.2).

Driving cross-country on rough terrain may lead to pitch amplitudes of ± 20° at frequencies up to more than 1 Hz. Pitch rates up to ~ 100°/s may result. In addition to pitch, bank and yaw angles may also have large perturbations. Visual orientation with cameras mounted directly on the vehicle body will be difficult (if not impossible) under these conditions. This is especially true since vision, usually, has a rather large delay time (in the tenths of a second range) until the situation has been understood purely based on visual perception.

If a subject's body motion can be perceived by a full set of inertial sensors (three linear accelerometers and three rate sensors), integration of these sensor signals as in "strap-down navigation" will yield good approximations of the true angular position with little time delay (see Figure 12.1). Note however, that for cameras mounted directly on the body, the images always contain the effects of motion blur due to integration time of the vision sensors! On the other hand, the drift errors accumulating from inertial integration have to be handled by visual feedback of low-pass filtered signals from known stationary objects far away (like the horizon).

In a representation with a scene tree as discussed in Chapter 2, the reduction in complexity by mounting the cameras directly on the car body is only minor. Once the computing power has been there for handling this concept, there is almost no advantage in data processing compared to active vision with gaze control. Hardware costs and space for mounting the gaze control system are the issues keeping most developers away from taking advantage of a vertebrate type eye. As soon as high speeds with large look-ahead distances or dynamic maneuvering are required, the visual perception capabilities of cameras mounted directly on body will no longer be sufficient.

Yaw effects: For roads with small radii of curvature R, another limit shows up. For example, for R = 100 m, the azimuth change along the road is curvature $C = 1/R$ (0.01 m^{-1}) times arc–length l. The lateral offset y at a given look-ahead range is given by the second integral of curvature C (assumed constant here, see Figure 3.4) and can be approximated for small angles by the term to the right in Equation 3.1.

$$\chi = \chi_0 + C \cdot l; \quad y = y_0 + \int \sin\chi \, \mathrm{d}l \approx \chi_0 \cdot l + C \cdot l^2 / 2. \tag{3.1}$$

For a horizontal f.o.v. of 45° (± 22.5°), the look-ahead range up to which other vehicles on the road are still in the f.o.v. is ~ 73 m ($\chi_0 = 0$). (Note that the distance traveled on the arc is 45° · π/ 180° · 100 m = 78.5 m.) At this point, the heading angle of the road is 0.785 radian (~ 45°), and the lateral offset from the tangent vector to the subject's motion is ~ 30 m; the bearing angle is 22.5°, so that the aspect angle of the other vehicle is 45° – 22.5° = 22.5° from the rear right-hand side. Increasing the f.o.v. to 60° (+ 33%) increases the look-ahead range to 87 m (+ 19%) with a lateral range of 50 m (+67%). The aspect angle of the other vehicle then is 30°. This numerical example clearly shows the limitations of fixed camera arrangements. For roads with even smaller radii of curvature, look-ahead ranges decrease rapidly (see circles 50 and 10 m radius on lower right in Figure 3.4).

Especially tight maneuvering with radii of curvature R down to ~ 6 m (standard for road vehicles) requires active gaze control if special sensors for these rather rare opportunities are to be avoided. By increasing the range of yaw control in gaze azimuth to about 70° relative to the vehicle body, all cases mentioned can be handled easily.

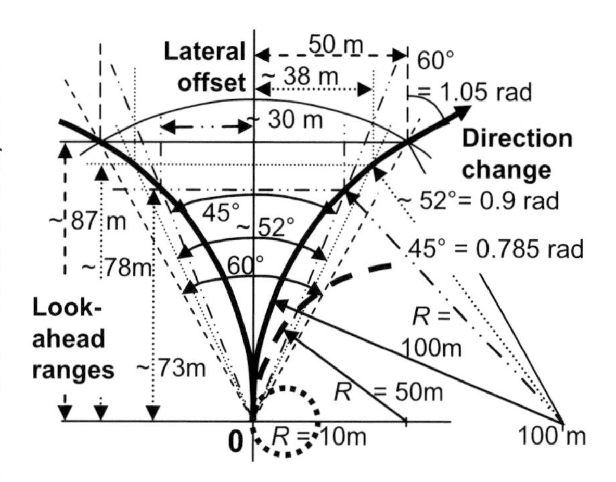

Figure 3.4. Horizontal viewing ranges

In addition, without active gaze control, all angular perturbations from rough ground are directly inflicted upon the camera viewing conditions leading to motion blur. Centering of other objects in the image may be impossible if this is in conflict with the driving task.

3.3.2.2 Active Gaze Control

The simplest and most effective degree of freedom for active gaze control of road vehicles on smooth surfaces with small look-ahead ranges is the pan (yaw) angle (see Figure 1.3). Figure 3.5 shows a solution with the pan as the outer and the tilt degree of freedom as the inner axis for the test vehicle VaMoRs, designed for driving on uneven ground. This allows a large horizontal viewing range and improves the problem due to pitching motion by inertial stabilization; inertial rate sensors for a single axis are mounted directly on the platform so that pitch stabilization is independent from gaze direction in yaw. Beside the possibility of view stabilization, active gaze control brings new degrees of freedom for visual perception. The potential gaze directions enlarge the total field of view. The pointing ranges in yaw and pitch characterize the design. Typical values for automotive applications are ± 70° in yaw (pan) and 25° in pitch (tilt). They yield a very much enlarged potential field of view for a given body orientation. Depending on the missions to be performed, the size of and the magnification factor between the simultaneous fields of view (given one viewing direction) as well as the potential angular viewing ranges have to be selected properly. Of course, only features appear-

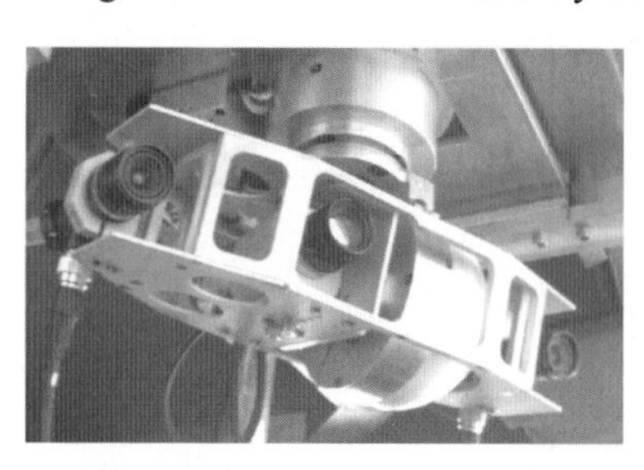

Figure 3.5. Two-axes gaze control platform with large stereo base of ~ 30 cm for VaMoRs. Angular ranges: Pan (yaw) ≈ ± 70°, tilt (pitch) ≈ ± 25°. It is mounted behind the upper center of the front windshield, about 2 m above the ground

ing in the actual simultaneous field of view can be detected and can attract attention (if there is no other sensory modality like hearing in animals, *calling* for attention in a certain direction). If the entire potential field of view has to be covered for detecting other objects, this can be achieved only by time-slicing attention with the wide field of view through sequences of viewing direction changes (scans). Usually, in most applications there are mission elements and maneuvers for which the viewing area of interest can be determined from the mission plan for the task to be solved next. For example, turning off onto a crossroad to the right or left automatically requires shifting the field of view in this direction (Chapter 10 and Section 14.6.5).

The request for economy in vision data leads to foveal-peripheral differentiation, as mentioned above. The size and the increase in resolution of the foveal f.o.v. are interesting design parameters to be discussed in Chapter 12. They should be selected such that several seconds of reaction time for avoiding accidents can be guaranteed. The human fovea has a f.o.v. from 1 to 2°. For road vehicle applications, a ratio of focal lengths from 3 to 10 as compared to wide-angle cameras has proven sufficient for the same size of imaging chips in all cameras.

Once gaze control is given, the modes of operation available are characteristic of the system. Being able to perform very fast gaze direction changes reduces the time delays in saccadic vision. In order to achieve this, usually, nonlinear control modes taking advantage of the maximal power available are required. Maximum angular speeds of several hundred degrees per second are achievable in both biological and technical systems. This allows reducing the duration of saccades to a small fraction of a second even for large amplitudes.

For visual tracking of certain objects, keeping their image centered in the field of view by visual feedback reduces motion blur (at least for this object of special interest). With only small perturbations remaining, the relative direction to this object can be read directly from the angle encoders for the pointing platform (solving part of the so-called "where"-problem by conventional measurements). The overall vision process will consist of sequences of saccades and smooth pursuit phases.

Search behavior for surveillance of a certain area in the outside world (3-D space) is another mode of operation for task performance. For optimal results, the parameters for search should depend on the distance to be covered.

When the images of the camera system (vehicle eye) are analyzed by several detection and recognition processes, there may be contradictory requirements for gaze control from these specialists for certain object classes. Therefore, there has to be an expert for the optimization of viewing behavior taking the information gain for mission performance of the overall system into account. If the requirements of the specialist processes cannot be satisfied by a single viewing direction, sequential phases of attention with intermediate saccadic gaze shifts have to be chosen [Pellkofer 2003]; more details will be discussed in Chapter 14. It is well known that the human vision system can perform up to five saccades per second. In road traffic environments, about one to two saccades per second may be sufficient; coarse tracking of the object not viewed by the telecamera may be done by one of the wide-angle cameras meanwhile.

If the active vision system is not able to satisfy the needs of the specialists for visual interpretation, it has to notify the central decision process to adjust mission performance to this situation (see Figure 14.1). Usually in ground vehicle guidance, slowing down or stopping under safe conditions is the way out for buying more time for perception.

3.3.2.3 Capability Network for Active Vision

The perceptual capabilities discussed above can be grouped according to signal flows required during execution and according to the complexity needed for solving typical classes of tasks. No general survey on active vision is intended here. A number of publications dealing with this problem are [Aloimonos *et al.* 1987; Ballard 1991; Blake and Yuille 1992; more recent ones]. Here, we will follow the approach developed by [Pellkofer 2003] (see also [Pellkofer *et al.* 2001, 2002]).

Figure 3.6 shows a graphical representation of the capabilities available for gaze control in the EMS-vision system (to be discussed in more detail in Chapters 12 and 14). The lowest row in the figure contains the hardware for actuation in two degrees of freedom and the basic software for gaze control (box, at right).

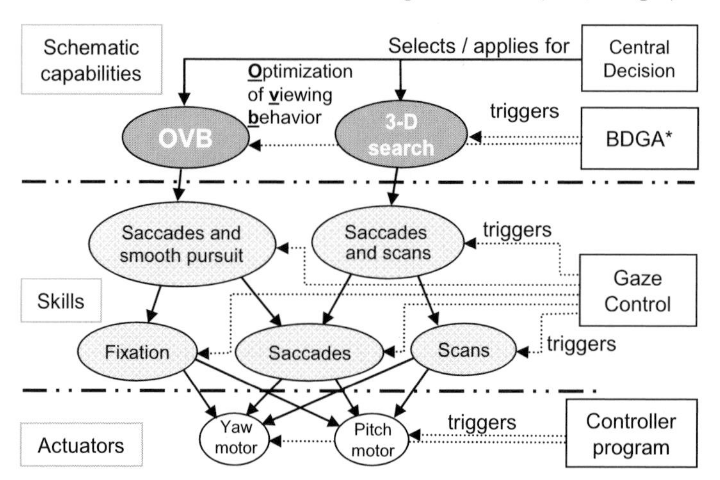

Figure 3.6. Capability network for active gaze control (after [Pellkofer 2003])
*BDGA = behavior decision for gaze and attention

On the second level from the bottom, the basic skills are represented with the expert for gaze control (GC) in the box to the right. This process runs on the processor closest to the hardware to minimize delay times. It receives its commands from the process for behavior decision for gaze and attention (BDGA). By combining two of its basic skills in a sequence with proper transitions, more complex skills on the third level originate. Scans differ from fixation (visual feedback) in that they are performed with constant angular speed (a parameter set by BDGA). GC is the process executing the commands from BDGA; these may be given partly by symbols and partly by just specifying the parameters needed.

The algorithms for planning sequences of saccades and phases of smooth pursuit are represented in the capability network on the upper level by the capability node "optimization of viewing behavior" (OVB, upper left). By representing these capabilities explicitly on the abstract level in the capability network, the process "central decision" (CD, upper right) can parameterize and use it in a flexible way. The capability OVB depends on the availability of the complex skill *saccade and smooth pursuit.* It possesses as parameters the maximal number of saccades, the planning horizon, eventually, a constant angular position for one of the platform axes, and the potential for initiating a new plan for gaze direction control. The demand of attention and the combination of regions of attention for certain objects to be observed are communicated by the specialists for recognizing and tracking these objects (see Chapters 13 and 14). Complex patterns in visual perception may emerge this way depending on the priorities set in the system. The second schematic capability beside OVB on this level is *3-D search*. This allows scanning a certain area in the environment of the vehicle in the real world by sequences of *saccades and scans*. The scans are performed with constant angular speeds so that image evaluation is possible; saccades may be interspersed so that the scanning direction is always the same. Scan rate may depend on the distance viewed.

Central decision (CD), the process in charge of achieving the goals of the mission, has contact only with BDGA, not to the lower levels directly. This modularization alleviates system development and naturally leads to multiple scales (coarse-to-fine differentiation). It should have become clear that this scheme allows characterizing vision systems to a relatively fine degree. Compact representation schemes for a wide variety of vision systems are possible and left open for future developments. The concept has been designed to be flexible and easily expandable.

3.3.2.4 Feature Extraction Capabilities

Beside the capabilities of gaze control, the capabilities of visual feature extraction characterize the performance level achievable by a subject. Thresholds in perception of edge and corner features are as important as recognizing shades of gray values or colors in a stable way. Recognizing shapes originating from boundaries of homogeneous image areas or from smooth or connected boundary sections allows inferences for hypothesis generation of objects or subjects seen, especially when continuity conditions over time can be discovered and also tracked. This will be one of the major topics of this book.

Biological vision systems have developed a high standard in recognizing textures, even two different ones simultaneously, as when one surface moves behind a partially obscuring other object (for example, an animal behind a tree or bush). The state of development of processing power of computers does not yet allow this in technical systems. In biological evolution, in certain situations like a predator approaching prey, maybe only those prey animals had a chance to survive which were able to solve this problem sufficiently well. For many applications of technical systems, this high level of visual capabilities is probably not necessary.

3.3.3 Knowledge Base for Perception Including Vision

At least as important for high-performance vision systems as the bottom-up capabilities for sensor data acquisition and processing is the knowledge that can be made readily available to the interpretation process for integration of information. Deeper understanding of complex situations and the case decisions necessary for proper reaction can be achieved only if relevant knowledge is available to a sufficient degree. This will be discussed in later chapters to some extent because of its importance, after the notions of subjects and situations have been fully introduced in this and the next chapter. This broad topic is considered a major area of development for intelligent systems with the sense of vision.

3.4 Behavioral Capabilities for Locomotion

The behavior most easily detectable by a vision system is motion of other objects or subjects. Therefore, this will be treated here ahead of decision making, even though decisions have to precede egomotion internally after signals from sensors have been received.

Motion capabilities depend very much on the basic shape of the body and on the means for locomotion of the subject. Legged motion widely spread in biological systems is hardly found in technical systems. On the other hand, the “axle-and-wheels” solution for locomotion abundant in technical systems cannot be found in biological systems because nature has not been able to solve the maintenance problems of these devices with soft tissue and blood vessels. Also, special preparation of the natural environment needed for using wheels efficiently could not be provided; humans solved this problem by road building, one of the outstanding achievements of human civilization. Tracked vehicles for going cross-country also use wheels, but have a special device for smoothing the surface these wheels roll on (tracks with articulated chain members).

Birds are able to walk on two legs and to hop, and, in addition, most species have the capability to fly by flapping their wings which they can fold to the body and unfold for flying. On the contrary, human technology again uses the “axle-and-wheels” solution with blades mounted to the wheel for generating propulsion (in propellers and partly in jet engines) or lift (in helicopters). Both principles are directly reflected in the visual appearance of these subjects. In the realm of insects, many more locomotion solutions can be found. Snakes solved their locomotion problem by typical wave-like sliding motion. For locomotion in vertical structures like trees, many-legged solutions may be of advantage.

The most highly developed creatures in biology with four limbs have developed special skills with their backward “legs” for running on almost flat ground. As soon as vertical structures have to be dealt with, the forward “arms” may support locomotion by grasping and swinging. This multiple use of extremities in connection with the wide variety of image processing needed for this purpose (including evaluation of data from inertial sensor of their body) may have led to the development of the most powerful brain found on our planet. Some species even use their tail to improve climbing and swinging performance in trees.

Without locomotion of the body, subjects with articulated bodies in both the biological and the technical realm are able to move their limbs for some kind of behavior. Grasping for objects nearby may be found in both areas (arm motion of animals or of industrial robots). Humans may use their arms for conveying information to a partner or to opponents. Cranes move their arms for loading and unloading vehicles or for lifting goods.

This may suffice to show the generality of the approach for understanding dynamic scenes by body shapes, their articulations, and their degrees of freedom for motion, controlled by some actuators with constrained motion capabilities, which get their commands from some data and knowledge processing device. Species may be recognized by their stereotypical motion behaviors, beside their appearance with 3-D shape and surface properties.

3.4.1 The General Model: Control Degrees of Freedom

To enable the link between image sequence interpretation and understanding motion processes with subjects in the real world, a few basic properties of control application are discussed here. Again, it is not intended to treat all possible classes of subjects but to concentrate on just one class of technical subjects from which a lot of experience has been gained by our group over the last two decades: Road vehicles (and air vehicles not treated here).

3.4.1.1 Differential Equations with Control Variables

As mentioned in the introduction, *control variables* are the variables in a dynamic system that distinguishes subjects from objects (proper). Control variables may be changed at each time "at will". Any kind of discontinuities are allowed; however, very frequently control time histories are smooth with a few points of discontinuity when certain events occur.

Differential equations describe constraints on temporal changes in the system, including the effects of control input. Again, standard forms are n equations of first order ("state equations") for an nth order system. In the transformed frequency domain, they are usually given as a set of transfer functions of nth order for linear systems. There is an infinite variety of (usually nonlinear) differential equations for describing the same temporal process. System parameters $\underline{p}$ allow us to adapt the representation to a class of problems

$$d\underline{X}/dt = \underline{f}(\underline{X}, \underline{U}, \underline{p}, t). \qquad (3.2)$$

Since real-time performance usually requires short cycle times for control, linearization of the equations of motion around a nominal set point (index N) is sufficiently representative of the process, if the set point is adjusted along the trajectory. With the substitutions

$$\underline{X} = \underline{X}_N + \underline{x}, \qquad \underline{U} = \underline{U}_N + \underline{u}, \qquad (3.3)$$

one obtains

$$d\underline{X}/dt = d\underline{X}_N/dt + d\underline{x}/dt\,. \tag{3.4}$$

The resulting sets of differential equations for the nominal trajectory then are

$$d\underline{X}_N\ /dt = \underline{f}\,(\,\underline{X}_N\,,\,\underline{U}_N\,,\,\underline{p}\,,\,t\,)\,, \tag{3.5}$$

and for the linearized perturbation system,

$$d\underline{x}/dt = F\cdot\ \underline{x} + G\cdot\underline{u} + \underline{v}'(t), \tag{3.6}$$

$$\text{with}\quad F = d\underline{f}\,/\,d\underline{X}|_N\ ;\qquad G = d\underline{f}\,/\,d\underline{U}|_N$$

as $(n \times n)$- respectively $(n \times r)$-matrices and $\underline{v}'(t)$ an additive noise-term. In systems with feedback components, the local feedback component simultaneously ensures (or at least improves) the validity of the linearized model, if the loop is stable.

Figure 3.7 shows this approximation of a nonlinear process with perturbations by a nominal nonlinear part (without perturbations), superimposed by a linear

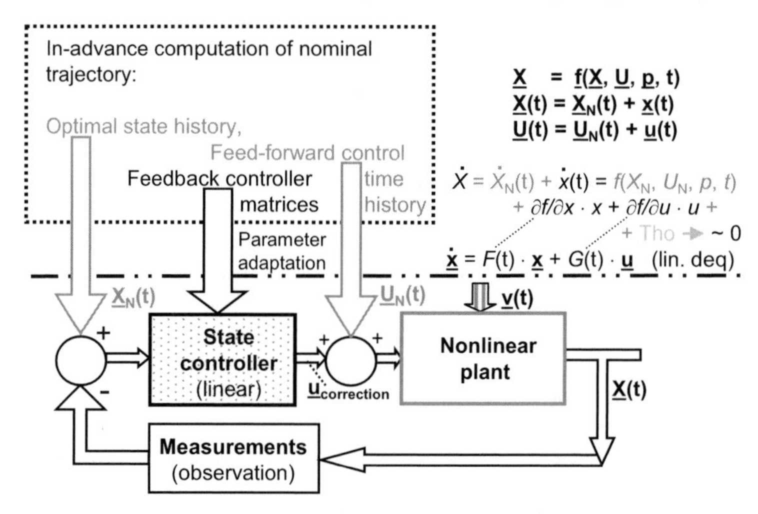

Figure 3.7. Approximation of a nonlinear process by superposition of a nominal nonlinear part and a superimposed linear part with a vector of perturbations $\underline{v}(t)$

(usually time-varying) feedback part taking care of unpredictable perturbations. The nominal nonlinear part is numerically optimized off-line in advance of the nominal conditions exploiting powerful numerical optimization methods derived from the calculus of variation. Along the optimal trajectory, the time histories of the partial derivative matrices F and G are stored; this is the basis for time-varying feedback with the linear perturbation system.

This approach is very common in engineering (for example in aero/space trajectory control) since the advent of digital computers in the second half of the last century. From here on, underlining of vectors will be dropped; the context will make the actual meaning clear.

3.4.1.2 Transition Matrices for Single Step Predictions

Equation 3.6 with matrices F and G may be transformed into a difference equation with the cycle time T for grid point spacing by one of the standard methods. (Precise numerical integration from 0 to T for $v = 0$ may be the most convenient one for complex right-hand sides.) The resulting general form then is

$$x[(k+1)T] = A \cdot x[kT] + B \cdot u[kT] + v[kT]$$

$$\text{or in short-hand notation,} \quad x_{k+1} = A \cdot x_k + B \cdot u_k + v_k, \tag{3.7}$$

where the matrices A, B have the same dimensions as F, G. In the general case of local linearization, all entries of these matrices may depend on the nominal state and control variables (X_N, U_N). The procedures for computing the elements of A and B have to be part of the "4-D knowledge base" for the application at hand. Software packages for these transformations are standard in control engineering.

For deeper understanding of motion processes of subjects observed, a knowledge base has to be available linking the actual state and its time history to goal-oriented behaviors and to stereotypical control outputs on the time line. This will be discussed in Section 3.4.3.

Once the initial conditions of the state are fixed or given, the evolving trajectory will depend both on this state (through matrix A, the so-called homogeneous part) and on the controls applied (the non-homogeneous part). Of course, this part also has to take the initial conditions into account to achieve the goals set in a close-to-optimal way. The collection of conditions influencing the decision for control output is called the "situation" (to be discussed in Chapters 4 and 13).

3.4.2 Control Variables for Ground Vehicles

A wheeled ground vehicle has three control variables, usually, two for longitudinal control and one for lateral control, the steering system. Longitudinal control is achieved by actuating either fuel injection (for acceleration or mild decelerations) or brakes (for decelerations up to ≈ -1 g (Earth gravity acceleration ≈ 9.81 m s^{-2})). Ground vehicles are controlled through proper time histories of these three control variables. In synchronization with the video signal this is done 25 (PAL-imagery) or 30 times a second (NTSC). Characteristic maneuvers require corresponding stereotypical temporal sequences of control output. The result will be corresponding time histories of changing state variables. Some of these can be measured directly by conventional sensors, while others can be observed from analyzing image sequences.

After starting a maneuver, these expected time histories of state variables form essential knowledge for efficient guidance of the vehicle. The differences between expectations and actual measurements give hints on the situation with respect to perturbations and can be used to apply corrective feedback control with little time delay; the lower implementation level does not have to wait for the higher system levels to respond with a change in the behavioral mode running. To a first degree of approximation, longitudinal and lateral control can be considered decoupled (not affecting each other). There are very sophisticated dynamic models available in automotive engineering in the car industry and in research for simulating and analyzing dynamical motion in response to control input and perturbations; only a very brief survey is given here. Mitschke (1988, 1990) is the standard reference in this field in German. (The announced reference [Giampiero 2007] may become a counterpart in English.)

3.4.2.1 Longitudinal Control Variables

For longitudinal acceleration, the following relation holds:

$$d^2x/dt^2 \approx \{-F_a - F_r - F_g - F_b - F_c + F_p\}/m. \tag{3.8}$$

F_a = aerodynamic forces proportional to velocity squared (V^2),
F_r = roll-resistance forces from the wheels,
F_g = weight component in hilly terrain ($-m \cdot g \cdot \sin(\gamma)$; γ = slope angle);
F_b = braking force, depends on friction coefficient μ (tire – ground), normal force on tire, and on brake pressure applied (control u_{lon1});
F_c = longitudinal force due to curvature of trajectory,
F_p = propulsive forces from engine torque through wheels (control u_{lon2}),
m = vehicle mass.

Figure 3.8 shows the basic effects of propulsive forces F_p at the rear wheels. Adding and subtracting the same force at the cg yields torque-free acceleration of the center of gravity and a torque around the cg of magnitude $H_{cg} \cdot F_p$ which is balanced by the torque of additional vertical forces ΔV at the front and rear axles. Due to spring stiffness of the body suspension, the car body will pitch up by $\Delta\theta_p$, which is easily noticed in image analysis.

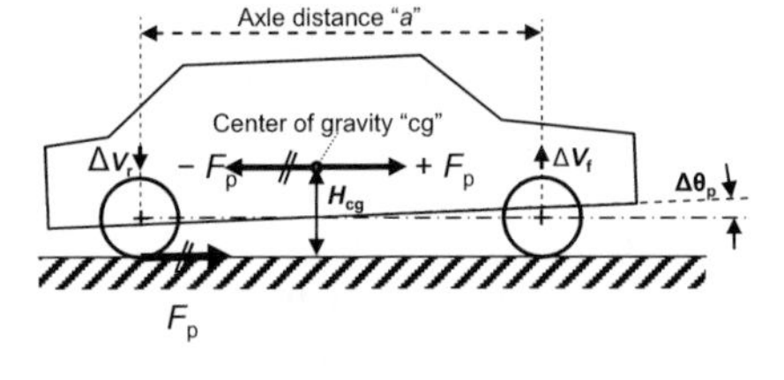

Figure 3.8. Propulsive acceleration control: Forces, torques and orientation changes in pitch

Similarly, the braking forces at the wheels will result in additional vertical force components of opposite sign, leading to a downward pitching motion $\Delta\Theta_b$, which is also easily noticed in vision. Figure 3.9 shows the forces, torque, and change in pitch angle. Since the braking force is proportional to the normal (vertical) force on the tire, it can be seen that the front wheels will take more of the braking load than the rear wheels. Since vehicle acceleration and deceleration can be easily measured by linear accelerometers mounted to the car body, the effects of control application can be directly "felt" by conventional sensors. This allows predicting expected values for several sensors. Tracking the difference between predicted and measured values helps gain confidence in motion models and their assumed parameters, on the one hand, and monitoring environmental conditions, on the other hand. The change in visual appearance

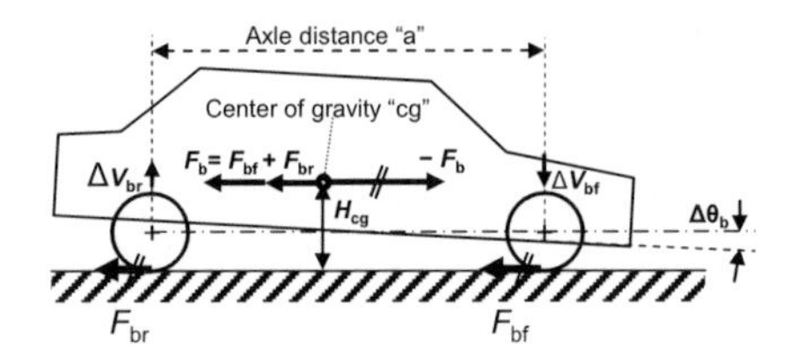

Figure 3.9. Longitudinal deceleration control: Braking

of the environment due to pitching effects must correspond to accelerations sensed. A downward pitch angle leads to a shift of all features upward in the images. [In humans, perturbations destroying this correspondence may lead to "motion sickness". This may also originate from different delay times in the sensor signal paths (*e.g.*, "simulator sickness") or from additional rotational motion around other axes disturbing the vestibular apparatus in humans which delivers the inertial data.]

For a human driver, the direct feedback of inertial data after applying one of the longitudinal controls is essential information on the situation encountered. For example, when the deceleration felt after brake application is much lower than expected the friction coefficient to the ground may be smaller than expected (slippery or icy surface). With a highly powered car, failing to meet the expected acceleration after a positive change in throttle setting may be due to wheel spinning. If a rotation around the vertical axis occurs during braking, the wheels on the left- and right-hand sides may have encountered different frictional properties of the local ground. To counteract this immediately, the system should activate lateral control with steering, generating the corresponding countertorque.

3.4.2.2 Lateral Control of Ground Vehicles

A generic steering model for lateral control is given in Figure 3.10; it shows the so-called Ackermann–steering, in which (in an idealized quasi-steady state) the axes of rotation of all wheels always point to a single center of rotation on the extended rear axle. The simplified "bicycle model" (shown) has an average steering angle λ at the center of the front axle and a turn radius $R \approx R_f \approx R_r$. The curvature C of the trajectory driven is given by $C = 1/R$; its relation to the steering angle λ is shown in the figure.

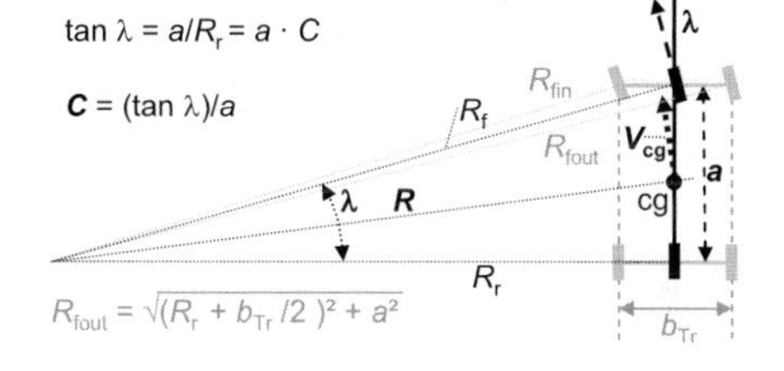

Figure 3.10. Ackermann steering for ground vehicles: Steer angle λ, turn radius R, curvature $C = 1/R$, axle distance a

Setting the cosine of the steering angle equal to 1 and the sine equal to the argument for magnitudes λ smaller than 15° leads to the simple relation $\lambda = a / R = a \cdot C$, or

$$C = \lambda / a. \tag{3.9}$$

Since curvature C is defined as "heading change over arc length" ($d\chi/dl$), this simple (idealized) model neglecting tire softness and drift angles yields a direct indication of heading changes due to steering control:

$$d\chi / dt = d\chi / dl \cdot dl / dt = C \cdot V \simeq V \cdot \lambda / a. \tag{3.10}$$

Note that the trajectory heading angle χ is rarely equal to the vehicle heading angle ψ; the difference is called the slip angle β. The simple relation Equation 3.10 yields an expected turn rate depending linearly on speed V multiplied by the steering angle. The vehicle heading angle ψ can be easily measured by angular rate sensors (gyros or tiny modern electronic devices). Turn rates also show up in image sequences as lateral shifts of all features in the images.

Simple steering maneuvers: Applying a constant steering rate A (considered the standard lateral control input and representing a good approximation to the behavior of real vehicles) over a period T_{SR} yields the final steering angle and path curvature

$$\lambda = \lambda_0 + A \cdot t, \qquad C = (\lambda_0 + A \cdot t)/a = C_0 + A \cdot t/a;$$
$$\lambda_f = \lambda_0 + A \cdot T_{SR}, \qquad C_f = C_0 + A \cdot T_{SR}/a. \tag{3.11}$$

Integrating Equation 3.10 with the top relation 3.11 for C yields the (idealistic!) change in heading angle for constant speed V

$$\Delta\chi = \int (C \cdot V)dt = V \cdot \int [C_0 + A \cdot t/a]dt$$
$$= V \cdot [C_0 \cdot T_{SR} + A \cdot T_{SR}^2 /(2\, a)]. \tag{3.12}$$

The first term on the right-hand side is the heading change due to a constant steering angle (corresponding to C_0); a constant steering angle for the duration τ thus leads to a circular arc of radius $1/C_0$ with a heading change of magnitude

$$\Delta\chi_C = V \cdot C_0 \cdot \tau. \tag{3.13a}$$

The second term (after the plus sign) in Equation 3.12 describes the contribution of the ramp-part of the steering angle. For initial curvature $C_0 = 0$, there follows

$$\Delta\chi_{ramp} = V \cdot \int [A\, t/a]dt = 0.5 \cdot V \cdot A\, t^2/a. \tag{3.13b}$$

Turn behavior of road vehicles can be characterized by their minimal turn radius ($R_{min} = 1/C_{max}$). For cars with axle distance "a" from 2 to 3.5 m, R may be as low as 6 m, which according to Figure 3.10 and Equation 3.9 yields λ_{max} around 30°. This means that the linear approximation for the equation in Figure 3.10 is no longer valid. Also the bicycle model is only a poor approximation for this case. The largest radius of all individual wheel tracks stems from the outer front wheel R_{fout}. For this radius, the relation to the radius of the center of the rear axle R_r, the width of the vehicle track b_{Tr} and the axle distance are given at the lower left of Figure 3.10. The smallest radius for the rear inner wheel is R_r - $b_{Tr}/2$. For a track width of a typical car $b_{Tr} = 1.6$ m, $a = 2.6$ m, and $R_{fout} = 6$ m, the rear axle radius for the bicycle model would be 4.6 m (and thus the wheel tracks would be 3.8 m for the inner and 5.4 m for the outer rear wheel) while the radius for the inner front wheel is also 4.6 m (by chance here equal to the center of the rear axle). This gives a feeling for what to expect from standard cars in sharp turns. Note that there are four distinct tracks for the wheels when making tight turns, e.g., for avoiding negative obstacles (ditches). For maneuvering with large steering angles, the linear approximation of Equation 3.9 for the bicycle model is definitely not sufficient!

Another property of curve steering is also very important and easily measurable by linear accelerometers mounted on the vehicle body with the sensitive axis in the direction of the rear axle (y-axis in vehicle coordinates). It measures centrifugal accelerations a_y which from mechanics are known to obey the law of physics:

$$a_y = V^2/R = V^2 \cdot C. \tag{3.14}$$

For a constant steering rate A over time t this yields with Equation 3.11 a constantly changing curvature C, assuming no other effects due to dynamics, time delays, bank angle or soft tires:

$$a_y = V^2 \cdot (\lambda_0 + A \cdot t)/a. \tag{3.15}$$

At the end of a control input phase starting from $\lambda_0 = 0$ with constant steering rate over a period T_{SR}, the maximal lateral acceleration is

$$a_{y,\ \max} = V^2 \cdot A \cdot T_{SR} / a\,. \tag{3.16}$$

For passenger comfort in public transportation, horizontal accelerations are usually kept below 0.1 $g \approx 1$ m/s². In passenger cars, levels of 0.2 to 0.4 g are commonly encountered. With a typical steering rate of $|A| \approx 1.15$ °/s = 0.02 rad/s, the lateral acceleration level of $\approx 0.2\ g$ (2 m/s²) is achieved in a maneuver-time dubbed T_2. For the test vehicle "VaMP", a Mercedes sedan 500-SEL with an axle distance $a = 3.14$ m, this maneuver time T_2 (divided by a factor of 2 for scaling in the figure) is shown in Figure 3.11 as a curved solid line. Table 3.2 contains some numerical values for low speeds and precise values for higher speeds.

It can be seen that for low speeds this maneuver time is relatively large (row 3 of the table); a large steering angle (line with triangles and row four) has to be built up until the small radius of curvature (line with stars, third row from bottom) yields the lateral acceleration set as limit. For very low speeds, of course, this limit cannot be reached because of the limited steering angle. At a speed of 15 m/s (54 km/h, a typical maximal speed for city traffic) the acceleration level of 0.2 g is reached after $\approx$ 1.4 seconds. The (idealized) radius of curvature then is $\approx$ 113 m; this shows that the speed is too high for tight curving. Also when the heading angle reaches the lateral acceleration limit (falling dashed curved line in Figure 3.11), the (idealized) lateral speed at that point (dashed curved line) and the lateral positions (dotted line) become small rapidly with higher speeds V driven.

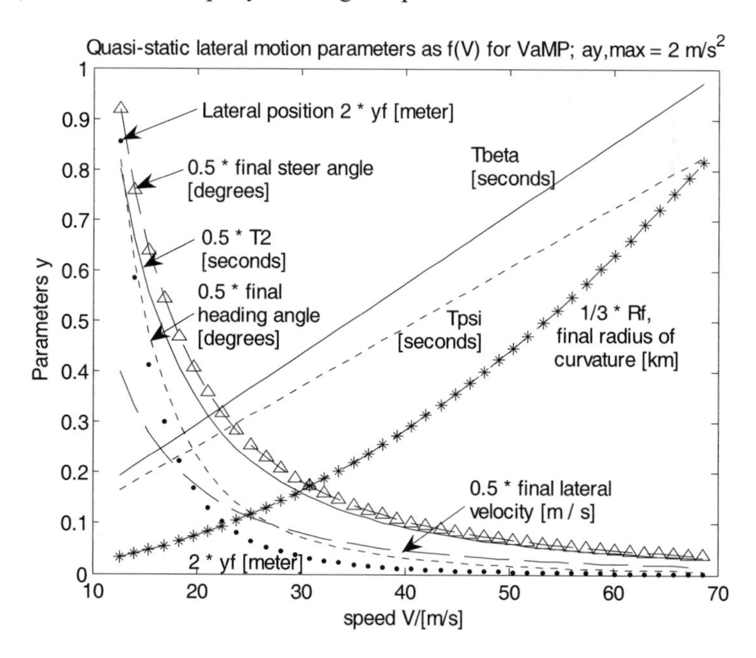

Figure 3.11. Idealized motion parameters as function of speed V for a steering rate step input of $A = 0.02$ rad/s until the lateral acceleration level of 2 m/s² is reached (quasi-static results for a first insight into lateral dynamics)

These numbers may serve as a first reference for grasping the real-world effects when the corresponding control output is used with a real vehicle in testing. In Section 3.4.5, some of the most essential effects stemming from systems dynamics neglected here will be discussed.

Table 3.2. Typical final state variables as function of speed V for a steering maneuver with constant control output (steering-rate A = 0.02 rad/s) starting from $\lambda = 0$ until a centrifugal acceleration of 0.2 g is reached (idealized with infinite cornering stiffness)

0	1	2	3	4	5	6	7	8
V (m/s)	5.278	7.5	10	15	20	30	40	70
T_2 (s)	11.27	5.58	3.14	1.396	0.785	0.349	0.196	0.064
$\Delta\lambda_f$ (°)	12.9	6.40	3.60	1.60	0.89	0.40	0.225	0.073
$\Delta\chi_f$ (°)	122.	42.6	18.0	5.33	2.25	0.666	0.281	0.0525
R_f (m)	13.9	28.1	50	113	200	450	800	2.450
v_f (m/s)	(-)	(5.58)	(3.14)	1.396	0.785	0.349	0.196	0.064
y_f (m)	-	(10.4)	(3.29)	0.65	0.205	0.041	0.013	0.0014

Column 1 (for about 19 km/h) marks the maximal steering angle for which the linearization for the relation $C(\lambda)$ (Equation 3.10) is approximately correct; the following columns show the rapid decrease in maneuver time until 0.2 g is reached. Columns 2, 3, and 4 correspond to speeds for driving in urban areas (27, 36, and 54 km/h), while 30 m/s ≈ 67.5 mph ≈ 108 km/h (column 6) is typical for U.S. highways; average car speed on a free German Autobahn is around 40 m/s (≈ 145 km/h), and the last column corresponds to the speed limit electronically set in many premium cars (≈ 250 km/h). Of course, the turn rate A at high speeds has to be reduced for increased accuracy in lateral control. Notice that for high speeds, the lateral acceleration level of 2 m/s² is reached in a small fraction of a second (row 3) and that the heading angles χ_f (row 5) are very small.

Real-world effects of tire stiffness (acting like springs in the lateral direction in combination with the vector of the moment of momentum) will change the results dramatically for this type of control input as a function of speed. This will be discussed in Section 3.4.5. To judge the changes in behavior due to speed driven by these types of vehicles, these results are important components of the knowledge base needed for safe driving. High-speed driving requires control inputs quite different from those for low-speed driving; many drivers missing corresponding experience do not know this. Section 3.4.5.2 is devoted to high-speed driving with impulse-like steering control inputs.

For small steering and heading (χ) angles, lateral speed v_f and lateral position y_f relative to a straight reference line can be determined as integrals over time. For $\lambda_0 = 0$, the resulting final lateral speed and position of this simple model according to Equation 3.14 would be

$$v_f \simeq V \cdot \Delta\chi_{ramp} = 0.5 \cdot V^2 \cdot A \cdot T_{SR}^2 / a.$$

$$y_f = \int (V \cdot \Delta\chi_{ramp}) dt \simeq 0.5 \cdot V^2 \cdot A \cdot \int t^2 dt / a = \frac{V^2 \cdot A \cdot T_{SR}^3}{6 \cdot a}. \tag{3.17}$$

Row 7 (second from the bottom) in Table 3.2 shows lateral speed v_f and row 8 lateral distance y_f traveled during the maneuver. Note that for speeds $V < 10$ m/s (columns 1 to 3), the heading angle (row 5) is so large that computation with the linear model (Equation 3.17) is no longer valid (see terms in brackets in the dotted area at bottom left of the table). On the other hand, for higher speeds (> ≈ 30 m/s), both lateral speed and position remain quite small when the acceleration limit is reached; at top speed (last column), they remain close to zero. This indicates again quite different behavior of road vehicles in the lower and upper speed ranges. The full nonlinear relation replacing Equation 3.17 for large heading angles is, with Equation 3.13b,

$$v(t) = V \cdot \sin(\Delta\chi_{\text{ramp}}) = V \cdot \sin(0.5 \cdot V \cdot A \cdot t^2 / a). \quad (3.18)$$

Since the cosine of the heading angle can no longer be approximated by 1, there is a second equation for speed and distances in the original x-direction:

$$dx / dt = V \cdot \cos(\Delta\chi_{ramp}) = V \cdot \cos(0.5 \cdot V \cdot A \cdot t^2 / a). \quad (3.19)$$

The time integrals of these equations yield the lateral and longitudinal positions for larger heading angles as needed in curve steering; this will not be followed here. Instead, to understand the consequences of one of the simplest maneuvers in lateral control, let us adjoin a negative ramp of equal magnitude directly after the positive ramp. This so-called "doublet" is shown in Figure 3.12.

The integral of this doublet is a triangular "pulse" in steering angle time history (dashed line). Scaling time by T_{SR} leads to the general description given in the figure. Since the maneuver is locally symmetrical at around point "1" and since the steering angle is zero at the end, this maneuver leads to a change in heading direction.

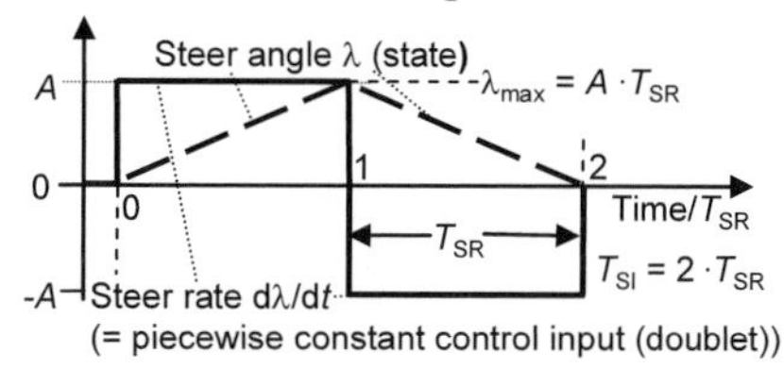

Figure 3.12. Doublet in constant steering rate $U_{ff}(t) = d\lambda/dt$ as control time history over two periods T_{SR} with opposite sign ± A yields a "pulse" in steer angle for heading change

Pulses in steering angle: Mirroring the steering angle time history at $T_{SR} = T_2$ (when a lateral acceleration of 0.2 g is reached), that is, applying a constant negative steering rate $-A$ from T_2 to $2T_2$ yields a heading change maneuver (idealized) with maximum lateral acceleration of ≈ 2 m/s².

The steering angle is zero at the end, and the heading angle is twice the value given in row 5 of Table 3.2 for infinite tire stiffness. From column 2, row 5 it can be seen that for a speed slightly lower than 7.5 m/s ≈ 25 km/h a 90°-turn should result with a minimal turn radius of about 28 m (row 6). For exact computation of the trajectory driven, the sine– and cosine–effects of the heading angle χ (according to Equations 3.18/3.19) have to be taken into account.

For speeds higher than 50 km/h (≈ 14 m/s), all angles reached with a "pulse"–maneuver in steering and moderate maximum lateral acceleration will be so small that Equation 3.17 is valid. The last two rows in Table 3.2 indicate for this speed range that a driving phase with constant λ_f (and thus constant lateral acceleration) over a period of duration τ should be inserted at the center of the pulse to decrease the time for lane changing (lane width is typically 2.5 to 3.8 m) achievable by a proper sequence of two opposite pulses. This maneuver, in contrast, will be called an "extended pulse" (Figure 3.13). It leads to an increased heading angle and thus to higher lateral speed at the end of the extended pulse. However, tire stiffness not taken into account here will change the picture drastically for higher speeds, as will be discussed below; for low speeds, the magnitude of the steering rate A and the absolute duration of the pulse or the extended pulse allow a wide range of maneuvering, taking other limits in lateral acceleration into account.

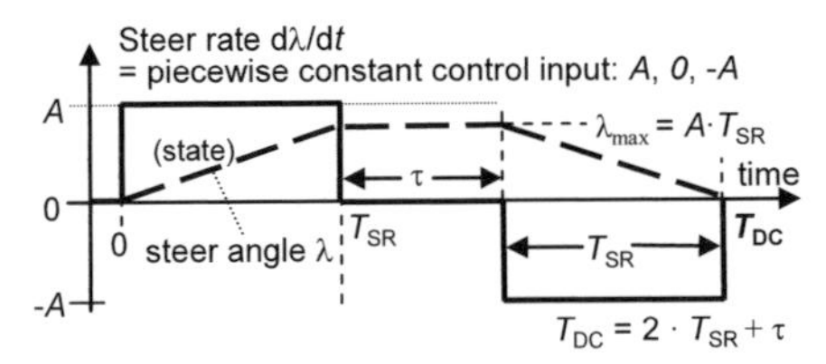

Figure 3.13. "Extended pulse" steering with central constant lateral acceleration level as maneuver control time history $u_{ff}(t)$ = $d\lambda/dt$ for controlled heading changes at higher speeds

Steering by extended pulses at moderate speeds: In the speed range beyond about 20 m/s ($\approx$ 70 km/h), lateral speed v_f and offset y_f (last two rows in Table 3.2) show very small numbers when reaching the lateral acceleration limit of $a_{y,max}$ = 0.2 g with a ramp. A period of constant lateral acceleration with steering angle λ_f (infinite tire stiffness assumed again!) and duration τ is added (see Figure 3.13) to achieve higher lateral speeds. To make a smooth lane change (of lane width $w_L \approx$ 3.6 m lateral distance) in a reasonable time, therefore, a phase with constant λ_f over a duration τ (*e.g.*, $\tau = 0.5$ seconds) at the constant (quasi-steady) lateral acceleration level of $a_{y,max}$ (2 m/s²) increases lateral speed by $\Delta v_C = a_{y,max} \cdot \tau$ (= 1 m/s for $\tau = 0.5$ s). The lateral distance traveled in this period due to the constant steering angle is $\Delta y_{C0} \approx a_{y,max} \cdot \tau^2/2$ ($= 2 \cdot 0.5^2/2 = 0.25$ m in the example chosen). Due to the small angles involved (sine $\approx$ argument), the total "extended pulse" builds up a lateral velocity v_{EP} (v_f from Equation 3.17, row 7 in Table 3.2) and a lateral offset y_{EP} at the end of the extended pulse (y_f from row 8 of the table) of

$$v_{EP} = (\Delta v_C + 2 \cdot v_f); \qquad y_{EP} \approx \Delta y_{C0} + 2 \cdot y_f. \tag{3.20}$$

Lane change maneuver: A generic lane change maneuver can be derived from two extended pulses in opposite directions. In the final part of this maneuver, an extended pulse similar to the initial one is used (steering rate parameter $-A$); it will need the same space and time to turn the trajectory back to its original direction. Subtracting the lateral offset gained in these phases ($2\ y_{EP}$) from lane width w_L yields the lateral distance to be passed in the intermediate straight line section between the two extended pulses; dividing this distance by the lateral speed v_{EP} at the end of the first pulse yields the time τ_{LC} spent driving straight ahead in the center section.

$$\tau_{LC} = (w_L - 2 \cdot y_{EP}) / v_{EP}. \tag{3.21}$$

Turning the vehicle back to the original driving direction in the new lane requires triggering the opposite extended pulse at the lateral position $-y_{EP}$ from the center of the new lane (irrespective of perturbations encountered or not precisely known lane width). This (quasi-static) maneuver will be compared later on to real ones taking dynamic effects into account.

Learning parameters of generic steering maneuvers: Performing this "lane change maneuver" several times at different speeds and memorizing the parameters as well as the real outcome constitutes a learning process for car driving. This will be left open for future developments. The essential point here is that knowledge about these types of maneuvers can trigger a host of useful (even optimal) behavioral components and adaptations to real-world effects depending on the situation encountered. Therefore, the term "maneuver" is very important for subjects: Its implementation in accordance with the laws and limits of physics provides the behavioral skills of the subject. Its compact representation with a few numbers and a symbolic name is important for planning, where only the (approximate) left and right boundary values of the state variables, the transition time, and some extreme values in between (quasi-static parameters) are sufficient for decision-making. This will be discussed in Section 3.4.4.1.

Effects of maneuvers on visual perception: The final effects to be discussed here are the centrifugal forces in curves and their influence on measurement data, including vision. The centrifugal forces proportional to curvature of the trajectory $C \cdot V^2$ may be thought to attack at the center of gravity. The counteracting forces keeping the vehicle on the road occur at the points where the vehicle touches the ground.

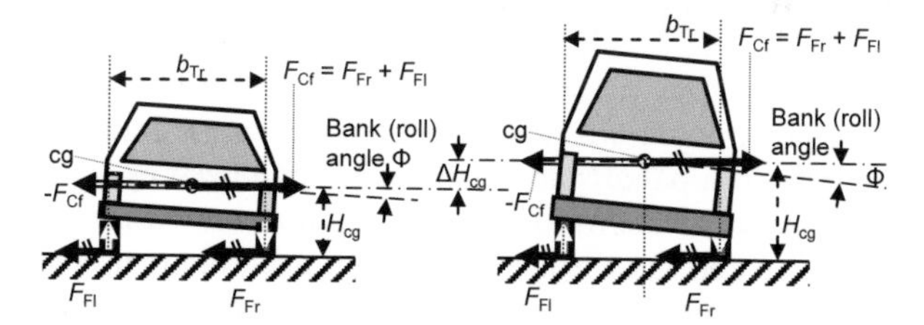

Figure 3.14. Vehicle banking in a curve due to centrifugal forces ~ $C \cdot V^2$; influence of elevation of cg

Figure 3.14 shows the balance of forces and torques leading to a bank angle Φ of the vehicle body in the outward direction of the curve driven. Therefore, the elevation H_{cg} of the cg above the ground is an important factor determining the inclination to banking of a vehicle in curves. Sports utility vehicles (SUV) or vans (Figure 3.14 right) tend to have a higher cg than normal cars (left) or even racing cars. Their bank angle Φ is usually larger for the same centrifugal forces; as a consequence, speed in curves has to be lower for these types of vehicles. However, suspension system design allows reducing this banking effect by some amount.

Critical situations may occur in dynamic maneuvering when both centrifugal and braking forces are applied. In the real world, the local friction coefficients at the wheels may be different. In addition, the normal forces at each wheel also differ due to the torque balance from braking and curve steering. Figure 3.15 shows a qualitative representation in a bird's-eye view. Unfortunately, quite a few accidents occur because human drivers are not able to perceive the environmental conditions and the inertial forces to be expected correctly. Vehicles with autonomous perception capabilities could help reduce the accident rate. A first successful step in this direction has been made with the device called ESP (electronic stability program or similar acronym, depending on the make). Up to now, this unit looks just at the yaw rate (maybe linear accelerations in addition) and the individual wheel speeds. If these values do not satisfy the conditions for a smooth curve, individual braking forces are applied at proper wheels. This device has been introduced as a mass product (especially in Europe) after the infamous "moose tests" of a Swedish journalist with a brand new type of vehicle.

He was able to topple over this vehicle toward the end of a maneuver intended to avoid collision with a moose on the road; the first sharp turn did not do any serious harm. Only the combination of three sharp turns in opposite directions at a certain frequency in resonance with the eigenfrequencies of the car suspension produced this effect. Again, this indicates how important knowledge of dynamic behavior of the car and "maneuvers" as stereotypical control sequences can be.

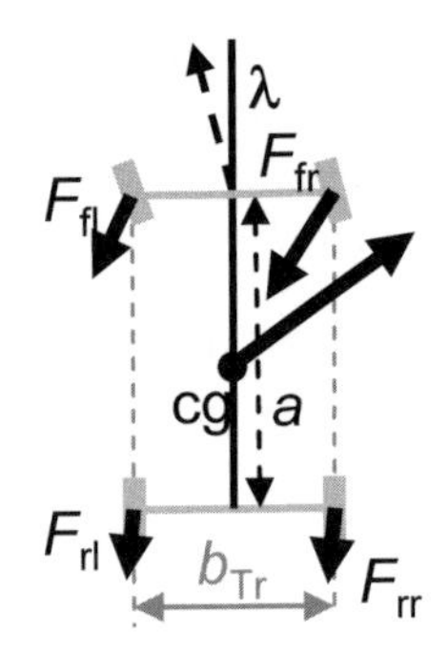

Figure 3.15. Frictional and inertial forces yield torques around all axes; in curves,

3.4.3 Basic Modes of Control Defining Skills

In general, there are two components of control activation involved in intelligent systems. If a payoff function is to be optimized by the maneuver, previous experience will have shown that certain control time histories perform better than others. It is essential knowledge for good or even optimal control of dynamic systems to know, in which situations what type of maneuver should be performed with which set of parameters; usually, the maneuver is defined by certain time histories of (coordinated) control input. The unperturbed trajectory corresponding to this nominal feed-forward control time history is also known, either stored or computed in parallel by numerical integration of the dynamic model exploiting the given initial conditions and the nominal control input. If perturbations occur, another important knowledge component is how to link additional control inputs to the deviations from the nominal (optimal) trajectory to counteract the perturbations effectively (see Figure 3.7). This has led to the classes of feed-forward and feedback control in systems dynamics and control engineering:

1. Feed-forward components $\underline{U}_{ff}$ derived from a deeper understanding of the process controlled and the maneuver to be performed.
2. Feedback components $\underline{u}_{fb}$ to force the trajectory toward the desired one despite perturbations or poor models underlying step 1.

3.4.3.1 Feed-forward Control: Maneuvers

There are classes of situations for which the same (or similar) kinds of control laws are useful; some parameters in these control laws may be adaptable depending on the actual states encountered.

Heading change maneuvers: For example, to perform a change in driving direction, the control time history input displayed in Figure 3.13 is one of a generic class of realizations. It has three phases with constant steering rate, two of the same magnitude A, but with opposite signs and one with zero output in between. The two characteristic time durations are T_{SR} for $\pm A$ and τ for the central zero-output.

$A \cdot T_{SR}$ yields the maximum steering angle λ_f (fixing the turn radius), with which a circular arc of duration τ is driven (see Table 3.2); the total maneuver time T_{DC} for a change in heading direction then is $2 \cdot T_{SR} + \tau$. The total angular change in heading is the integral of curvature over the arc length and depends on the axle distance of the car (see Figure 3.10 for the idealized case of infinitely stiff tires). Proper application of Equation 3.12 yields the (idealized) numerical values.

A special case is the 90° heading change for turning off onto a crossroad. If the vehicle chosen drives at 27 km/h ($V \approx 7.5$ m/s, column 2 in Table 3.2) then $T_{SR} = T_2$ is ≈ 5.6 seconds, and the limit of 2 m/s² for lateral acceleration is reached with $\Delta\lambda_f = 6.4°$ and $\Delta\chi_f \approx 42.6°$. The radius of curvature R is 28.1 m (C = 0.0356 m^{-1}, Equation 3.9); this yields a turn rate $C \cdot V$ (Equation 3.10) of 15.3°/s. Steering back to straight-ahead driving on the crossroad with the mirrored maneuver for the steering angle leaves almost no room for a circular arc with radius R_f [$\tau = (90 - 2 \cdot 42.6)/15.3 \approx 0.3$ s]; the total turn–off–duration then is ≈ 11.2 s and the total distance traveled is about 84 m.

For tight turns on narrow roads, either the allowed lateral acceleration has to be increased, or lower speed has to be selected. A minimal turn radius of 6 m driven at $V = 7$ m/s yields an ideal turn rate V/R of about 67°/s and a (nominal) lateral acceleration V^2/R of about 0.82 g (~ 8 m/s²); this is realizable only on dry ground with good homogeneous friction coefficients at all wheels. Slight variations will lead to slipping motion and uncontrollable behavior. For the selected convenient limit of maximum lateral acceleration of 2 m/s² with the minimal turn radius possible (6 m), a speed of $V \approx 3.5$ m/s ($\approx$ 12.5 km/h or 7.9 mph)should be chosen. These effects have to be kept in mind when planning turns.

The type of control according to Figure 3.13 is often used at higher speeds with smaller values for A and T_{SR} (τ close to 0) for heading corrections after some perturbation. Switching the sequence of the sign of A results in a heading change in the opposite direction.

Lane change maneuvers: Combining two extended pulses of opposite sign with proper control of magnitude and duration results in a "lane change maneuver" discussed above and displayed in Figure 3.16.

The numerical values and the temporal extensions of these segments for a lateral translation of one lane width depend on the speed driven and the maximum lateral acceleration level acceptable. The behavioral capability of lane changing may thus be represented symbolically by a name and the parameters specifying this control output (just a few numbers, as given in the legend of the figure). Together with the initial and final boundary values of the state variables and maybe some extreme values in between, this is sufficient for the (abstract) planning and decision level. Only the processor directly controlling the actuator needs to know the details of how the maneuver is realized. For very high speeds, maneuver times for the pulses become very small [see T2–curve (solid) in Figure 3.11]. In these cases, tire stiffness effects play an important role; there will be additional dynamic responses which interact with vehicle dynamics. This will be discussed in Section 3.4.5.2.

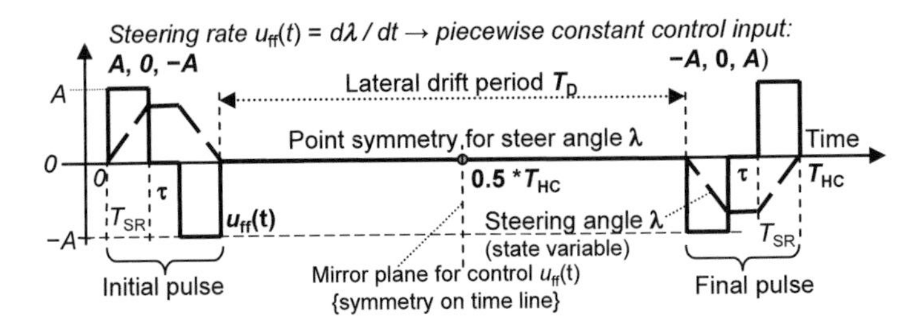

Figure 3.16. High-speed lane change maneuver with two steering "pulses", including a central constant lateral acceleration phase of duration τ at the beginning and end, as well as a straight drift period T_D in between; the duration T_D is adapted such that at the end of the second (opposite) pulse, the vehicle is at the center of the neighboring lane driving tangentially to the road. The *maneuver control time history* $u_{ff}(t) = d\lambda/dt$ for lane change at higher speeds is [legend: magnitude(duration)]: $A(T_{SR})$, $0(\tau)$, $-A(T_{SR})$, $0(T_D)$, $-A(T_{SR})$, $0(\tau)$, $-A(T_{SR})$

Table 3.3 shows in column 2 a list of standard maneuvers for ground vehicles (rows 1 – 6 for longitudinal, 7 – 11 for lateral, and 12 –18 for combined longitudinal and lateral control). Detailed realizations have been developed by [Zapp 1988, Bruedigam 1994; Mueller 1996; Maurer 2000; and Siedersberger 2003]. Especially the latter two references elaborate the approach presented here.

The development of behavioral capabilities is an ongoing challenge for autonomous vehicles and will need attention for each new type of vehicle created. It should be a long–term goal that each new autonomous vehicle is able to adapt to its own design parameters at least some basic generic behavioral capabilities from a software pool by learning via trial and error. Well-defined payoff functions (quality and safety measures) should guide the learning process for these maneuvers.

3.4.3.2 Feedback Control

Suitable feedback control laws are selected for keeping the state of the vehicle close to the ideal reference state or trajectory; different control laws may be necessary for various types and levels of perturbations. The general control law for state feedback with gain matrix K and $\Delta\underline{x} = \underline{x}_C - \underline{x}$ (the difference between commanded and actual state values) is

$$\underline{u}_{fb}(kT) = K^T \cdot \Delta\underline{x}(kT). \tag{3.22}$$

For application to the subject vehicle, either the numerical values of the elements of the matrix K directly or procedures for determining them from values of the actual situation and/or state have to be stored in the knowledge base. To achieve better long-term precision in some state variable, the time integral of the error $\Delta x_i = x_{Ci} - x_i$ may be chosen as an additional state with a commanded value of zero.

For observing and understanding behaviors of other subjects, realistic expected perturbations of trajectory parameters are sufficient knowledge for decision–

making with respect to safe behavior; the exact feedback laws used by other subjects need not be known.

Table 3.3. Typical behavioral capabilities (skills) needed for road vehicles

Longi-tudinal	**Feed-forward control (maneuver)**	**Feedback control**
1	Acceleration from standstill to speed set	Drive at constant speed
	Transition to convoy driving from higher speed	Distance keeping to vehicle ahead (average values, fluctuations)
2	Observe right of way at intersections	
3	Braking to a preset speed	Safe convoy driving with distance = f(speed)
4	Braking to stop at reasonable distance (moderate, early onset)	Halt at preset location
5	Stop and Go driving	
6	Emergency stops	
Lateral		
7	Lane changing [ranges and maneuver times as *f* (speed)]	Lane keeping (accuracy), Road-running, Line following
8	Follow vehicle ahead (in maneuvers recognized)	Follow vehicle ahead in same track
9	Obstacle avoidance	Keep safety margin to moving obstacle
10	Handling of road forks	Distance keeping to border line
11	Proper setting of turn lights before start of maneuver	
Longit. +lateral		
12	Turning off onto crossroad	Moving into lane with flowing traffic
13	Entering and leaving a traffic circle	Entering and driving in a traffic circle
14	Overtaking behavior [safety margins as *f* (speed)]	Observe safety margins
15	Negotiating “hairpin” curves (switchbacks)	Proper reaction to animals detected on or near the driveway
16	U-turns on bidirectional roads	
17	Observing traffic regulations (max. speed, passing interdiction)	Proper reaction to static obstacles detected in own lane
18	Parking in a parking bay	Parking alongside the road

More detailed treatment of modeling will be given in the application domains in later chapters. To aid practical understanding, a simple example of modeling ground vehicle dynamics will be given in Section 3.4.5. Depending on the situation and maneuver intended, different models may be selected. In lateral control, a third-order model is sufficient for smooth and slow control of lateral position of a vehicle when tire dynamics does not play an essential role. A fifth-order model taking tire stiffness and rotational dynamics into account will be shown as contrast for demonstrating the effects of short maneuver times on dynamic behavior.

Depending on the situation and maneuver intended, different models may be selected. In lateral control, a third-order model is sufficient for smooth and slow control of lateral position of a vehicle when tire dynamics does not play an essential role. A fifth-order model taking tire stiffness and rotational dynamics into account will be shown as contrast for demonstrating the effects of short maneuver times on dynamic behavior.

Instead of full state feedback, often simple output feedback with a PD- or PID-controller is sufficient. Taking visual features in 2-D as output variable even works sometimes (in relatively simple cases like lane following on planar high-speed roads). Typical tasks solved by feedback control for ground vehicles are given in the right-hand column of Table 3.3. Controller design for automotive applications is a well–established field of engineering and will not be detailed here.

3.4.4 Dual Representation Scheme

To gain flexibility for the realization of complex systems and to accommodate the established methods from both systems engineering (SE) and artificial intelligence (AI), behaviors are represented in duplicate form: (1) in the way they are implemented on real-time processors for controlling actuators in the real vehicle, and (2) as abstracted entities for supporting the process of decision making on the mental representation level, as indicated above (see Figure 3.17).

In the case of simple maneuvers, even approximate analytical solutions of the dynamic maneuver are available; they will be discussed in more detail in Section 3.4.5 and can be used twofold:

1. For computing reference time histories of some state variables or measurement values to be expected, like heading or lateral position or accelerometer and gyro readings at each time, and
2. for taking the final boundary values of the predicted maneuver as base for maneuver planning on the higher levels. Just transition time and the state variables achieved at that time, altogether only a few (quasi-static) numbers, are sufficient (symbolic) representations of the *process* treated, lasting several seconds in general.

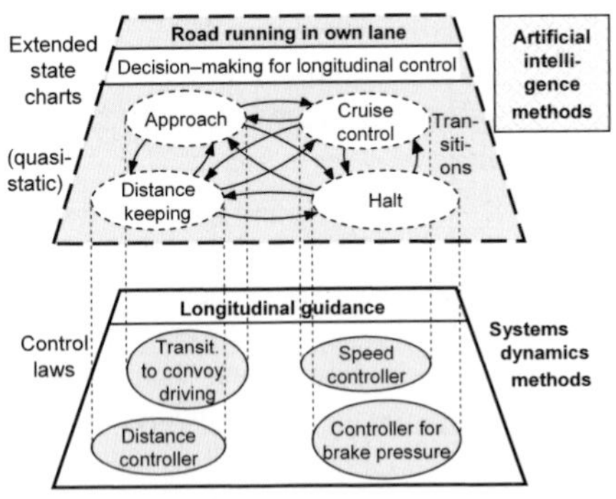

Figure 3.17. Dual representation of behavioral modes: 1. Decision level (dashed), quasi-static AI-methods, extended state charts [Harel 1987] with conditions for transitions between modes. 2. Realization on (embedded, distributed) processors close to the actuators through feed-forward and feedback control laws [Maurer 2000; Siedersberger 2004]

3.4.4.1 Representation for Supporting the Process of Decision-Making

Point 2 constitutes a sound grounding of linguistic situation aspects. For example, the symbolic statement: *The subject is performing a lane change* (lateral offset of one lane width) is sufficiently precise for decision-making if the percentage of the maneuver already performed and vehicle speed are known. With respect to the end of this maneuver, two more linguistic aspects can be predicted: *The subject will have the same heading direction as at the start of the maneuver* and *the tangential velocity vector will be at the center of the neighboring lane being changed to.*

In more complicated situations without analytical solutions available, today's computing power allows numerical integration of the corresponding equations over the entire maneuver time within a fraction of a video cycle and the use of the numerical results in a way similar to analytical solutions.

Thus, a general procedure for combining control engineering and AI methods may be incorporated. Only the generic nominal control time histories $\underline{u}_{ff}(\cdot)$ and feedback control laws guaranteeing stability and sufficient performance for this specific maneuver have to be stored in a knowledge base for generating these "behavioral competencies". Beside dynamical models, given by Equation 3.6 and 3.8 for each generic maneuver element, the following items have to be stored:

1. The situations when it is applied (started and ended), and
2. the feed-forward control time histories $\underline{u}_{ff}(\cdot)$; together with the dynamic models. This includes the capability of generating reference trajectories (commanded state time histories) when feedback control is applied in addition to deal with unpredictable perturbations.

All these maneuvers can be performed in different fashions characterized by some parameters such as total maneuver time, maximum acceleration or deceleration allowed, rate of control actuation, *etc.* For example, lane change may either be done in 2, 6, or in 10 seconds at a given speed. The characteristics of a lane change maneuver will differ profoundly for the speed range of modern vehicles when all real-world dynamic effects are taken into account. Therefore, the concept of *maneuvers* may be quite involved from the point of view of systems dynamics. Maneuver time need not be identical with the time of control input; it is rather defined as the time until all state variables settle down to their (quasi-) steady values. These real-world effects will be discussed in Section 3.4.5; they have to be part of the knowledge base and have to be taken into account during decision-making. Otherwise, the discrepancies between internal models and real-world processes may lead to serious problems.

It also has to be ensured that the models for prediction and decision-making on the abstract (AI-) level are equivalent – with respect to their outcome – to those underlying the implementation algorithms on the systems engineering level. Figure 3.17 shows a visualization of the two levels for behavior decision and implementation [Maurer 2000, Siedersberger 2004].

3.4.4.2 Implementation for Control of Actuator Hardware

In modern vehicles with specific digital microprocessors for controlling the actuators (qualified for automotive environments), there will be no direct access to actuators for processors on higher system levels. On the contrary, it is more likely that after abstract decision-making, there will be several processors in the down-link chain to the actuators. To achieve efficient system architectures, the question then is which level should be assigned which task. Here, it is assumed that (as in the EMS–implementation for VaMoRs and VaMP, see Figure 14.7), a PC-type processor forms the interface between the perception- and evaluation level (PEL), on one hand, and specific microprocessors for actuator control, on the other hand. This processor has direct access to conventional measurement data and can close loops from measurements to actuator output with minimal time delay.

The control process has to know what to do with the symbolic commands coming from the PEL for implementing basic strategic decisions, taking the actual state of the vehicle into account. It has more up-to-date information available on local aspects and should, therefore, not be forced to work as a slave, but should have the freedom to choose how to optimally achieve the goals set by the strategic decision received from the PEL. For example, quick reactions to unforeseen perturbations should be performed under the subject's responsibility. Of course, these cases have to be communicated back to the higher levels for more thorough and in-depth evaluation.

It is on this level that all control time histories for standard maneuvers and all feedback laws for regulation of desired states have to be decided in detail. This is the usual task of controller design and of proper triggering in systems dynamics. In Figure 3.17, this is represented by the lower level shown for longitudinal control.

3.4.5 Dynamic Effects in Road Vehicle Guidance

Due to the relatively long delay times associated with visual scene interpretation it is important for instant correct appreciation of newly developing situations that two facts mentioned above already are taken into account: First, inertial sensing allows immediate perception of effects of perturbations onto the own body. It also immediately reflects actual control implementation in most degrees of freedom. Second, exploiting the dynamical models in connection with measured control outputs, expectations for state variable time histories can be computed. Comparing these to actually measured or observed ones allows checking the correctness of conditions for which the behavioral decisions have been made. If discrepancies exceed threshold values, careful and attentive checking of the developing states may help avoiding dangerous situations.

A typical example is a braking action on a winter road. In response to a commanded brake pressure with steering angle zero, a certain deceleration level with no rotations around the longitudinal and the vertical axes are expected. There will be a small pitching motion due to the distance between the points where forces act (see Figure 3.9 above). With body suspension by springs and dampers, a second-order (oscillatory or critically damped) rotational motion can be expected. Very often in winter, road conditions are not homogeneous for all wheels. Assume that the wheels on one side move on snow or ice while on the other side the wheels run on asphalt (MacAdam, concrete). This yields different friction coefficients and thus different braking forces on both sides of the vehicle. Since total friction has decreased, the measured longitudinal deceleration ($a_x < 0$) will be lower than expected. However, due to the torque developed by the different braking forces on both sides of the vehicle, there also will be a rotational onset around the vertical axis and maybe a slight banking (rolling) motion around the longitudinal axis. This situation is rather common, and therefore, one standard automotive test procedure is the so-called "μ-split braking" behavior of vehicles (testing exactly this).

Because of the importance of these effects for safe driving, they have to be taken into account in visual scene interpretation. The 4-D approach to vision has the advantage of allowing us to integrate this knowledge into visual perception right from the beginning. Typical motion behaviors are represented by generic models that are available to the recursive estimation processes for prediction–error feedback when interpreting image sequences (see Chapter 6). This points to the fact that humans developing dynamic vision systems for ground vehicles should have a good intuition with respect to understanding how vehicles behave after specific control inputs; maybe they should have experience, at least to some degree, in test driving.

3.4.5.1 Longitudinal Road Vehicle Guidance

The basic differential equation for locomotion in longitudinal degrees of freedom (dof) has been given in a coarse form in Equation 3.8. However, longitudinal dof encompass one more translation (vertical motion or "heave"), dominated by Earth gravity, and an additional rotation (pitch) around the y-axis (parallel to the rear axle and going through the cg).

Vertical curvature effects: Normally, Earth gravity ($g \approx 9.81$ m/s²) keeps the wheels in touch with the ground and the suspension system compressed to an average level. On a flat horizontal surface, there will be almost no vertical wheel and body motion (except for acceleration and deceleration). However, due to local surface slopes and curvatures, the vertical forces on a wheel will vary individually. Depending on the combination of local slopes and bumps, the vehicle will experience all kinds of motion in all degrees of freedom. Roads are designed as networks of surface "bands" having horizontal curvatures (in vertical projection) in a limited range of values. However, for the vertical components of the surface, minimal curvatures in both lateral and longitudinal directions are attempted by road building. In hilly terrain and in mountainous areas, vertical curvatures C_V may still have relatively large values because of the costs of road building. This will limit top speed allowed on hilly roads since at the lift-off speed V_L, the centrifugal acceleration will compensate for gravity. From $V_L^2 \cdot C_V = g$ there follows

$$V_L = \sqrt{g / C_V}\,. \qquad (3.23)$$

Driving at higher speed, the vehicle will lift off the ground (lose instant controllability). Only a small fraction of weight is allowed to be lost due to vertical centrifugal forces $V^2 \cdot C_V$ for safe driving. At $V = 30$ m/s (108 km/h), the vertical radius of curvature for liftoff will be $R_V = 1/C_V \approx 92$ m; to lose at most 20% of normal weight as contact force, the maximal vertical radius of curvature would be 450 m. Going cross-country at 5 m/s (18 km/h), local vertical radii of curvature of about

2.5 m would have the local (say, the front) wheels leave the ground. Since, in general, there will be forces on the rear wheels, pitch acceleration downward will also result. These are conditions well-known from rallye-driving. Vertical curvatures can be recognized by vision in the look-ahead range so that these dynamic effects on vehicle motion can be foreseen and will not come by surprise.

Autonomous vehicles going cross-country have to be aware of these conditions to select proper speed as well as the shape and location of the track for steering. This is a complex optimization task: On the tracks, for the wheels on both sides of the vehicle the vertical surface profiles have to be recognized at least approximately correctly. From this information, the vertical and rotational perturbations (heave, pitch, and roll) to be expected can be estimated. Since lateral control leaves a degree of freedom in the curvature of the horizontal track through steering, a compromise allowing a safe trajectory at a good speed with acceptable perturbations from uneven terrain has to be found. This will remain a challenging task for some time to come.

Slope effects on longitudinal motion: Figure 3.18 is a generalization of the horizontal case with acceleration, shown in Figure 3.8, to the case of driving on terrain that slopes in the driving direction. Now, the all dominating gravity vector has a component of magnitude ($m \cdot g \cdot \sin\sigma$) in the driving direction. Going uphill, it will oppose vehicle acceleration, and downhill it will push in the driving direction. It will be measured by an accelerometer sensitive in this direction even when standing still. In this case, it may be used for determining the orientation in pitch of the vehicle body. Also when driving, this part will not correspond to real vehicle acceleration (dV/dt) relative to the environment. The gravity component has to be subtracted from the reading of the accelerometer (sensor signal) for interpretation. The effect of slopes on speed control is tremendous for normal road vehicles. Uphill, speeds achievable are much reduced; for example, a vehicle of 2000 kg mass driving at 20 m/s on a slope of 10% (5.7°) needs ≈ 40 kW power just for weight lifting. Going downhill, at a slope angle of 11.5° the braking action has to correspond to $0.2 \cdot g$ in order not to gain speed.

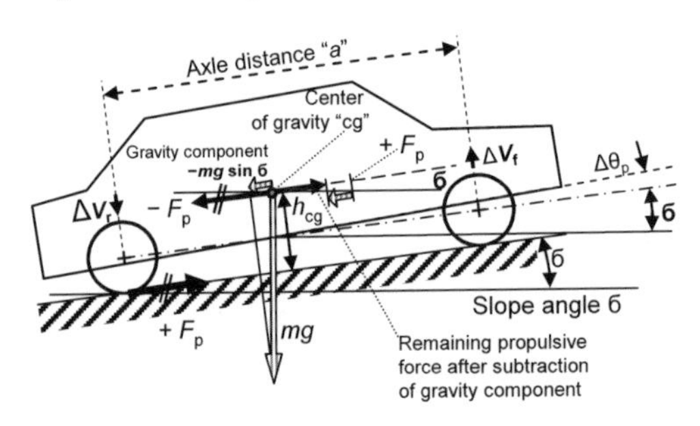

Figure 3.18. Longitudinal acceleration components going uphill: Forces, torques, and orientation changes in pitch

Driving not in the direction of maximum slope (angle $\Delta\psi$ relative to gradient direction) will complicate the situation, since there will be a lateral force component acting in addition to the longitudinal one. Note that the longitudinal component a_{Lon} will remain almost constant for small deviations $\Delta\psi$ from the gradient direction (cosine-effect $\cos\Delta\psi \approx 1$ up to 15°, that is, $a_{Lon} \approx a_{grad} = g \cdot \sin\sigma$), while the lateral component will increase linearly according to the sine of the relative heading angle $\Delta\psi$ ($a_{lat} \approx g \cdot \sin\sigma \cdot \sin\Delta\psi$, yielding $\approx 0.2\, g \cdot \sin\sigma$ for an angle $\Delta\psi$ of 11.5°). At $\Delta\psi = 45°$ (midway between the vertical gradient and horizontal direction), both longitudinal and lateral components of gravity acceleration will be 0.7 · a_{grad}, yielding a sum of the components of ≈ 141% of a_{grad}. These facts lead to the rule that going uphill or downhill should preferably be done in gradient direction, especially since the width of the vehicle track, usually, is smaller than the axle distance so that the danger of toppling over after hitting a perturbation is minimized.

Horizontal longitudinal acceleration capabilities: An essential component for judging vehicle performance is the acceleration capability from one speed level to another, measured in seconds. Standard tests for cars are acceleration from rest to 100 km/h and from 80 to 120 km/h (*e.g.*, for passing). Assuming a constant acceleration level of "1 g" (9.81 m/s²) would yield 2.83 seconds from 0 to 100 km/h. Since the friction coefficient is, usually, less than 1, this value may be considered a lower limit for the acceleration time from 0 to 100 km/h of very-high-performance vehicles. Racing cars with downward aerodynamic lift can exploit higher normal forces on the tires and thus higher acceleration levels at higher speeds if engine power permits. Today's premium cars typically achieve values from 4 to 8 seconds, while standard cars and vans show values between 10 and 20 seconds from 0 to 100 km/h.

Figure 3.19 shows test results of our test vehicle VaMoRs, a 5-ton van with top speed of around 90 km/h (≈ 25 m/s). It needed about 40 s to accelerate from 1.5 to 10 m/s (left plot) and ≈ 55 s from 13 to 20 m/s (right-hand plot). The approach of top speed is very slow (as usual in low-powered vehicles). Taking the throttle position back decelerates the vehicle at a rate of about 0.3 m/s² at 10 m/s (left) and about 0.45 m/s² at 20 m/s (right). To achieve higher deceleration levels, the brakes have to be used.

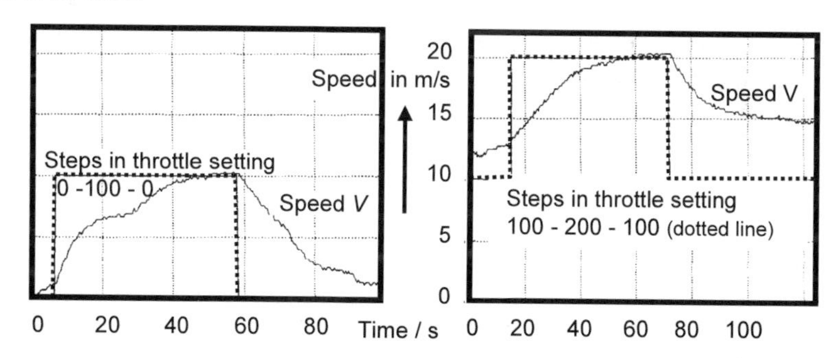

Figure 3.19. Speed over time as a step response to throttle setting: Experimental results for test vehicle **VaMoRs**, 5-ton van [Brüdigam 1994]

Braking capability: A big advantage of ground vehicles as compared to aquatic or air vehicles is the fact that large deceleration forces can easily be generated by braking. Modern cars on dry surfaces achieve braking decelerations close to "– 1*g*" (gravity acceleration). This corresponds to a braking distance of about 38.6 m (in 2.83 s) from V = 100 km/h to halt. Here the friction coefficient is close to 1, and the measured total acceleration magnitude including gravity is $\sqrt{2} \cdot g$ (45° downward to the rear). It is immediately clear that all objects lying loosely in the vehicle body will experience a large acceleration relative to the body; therefore, they have

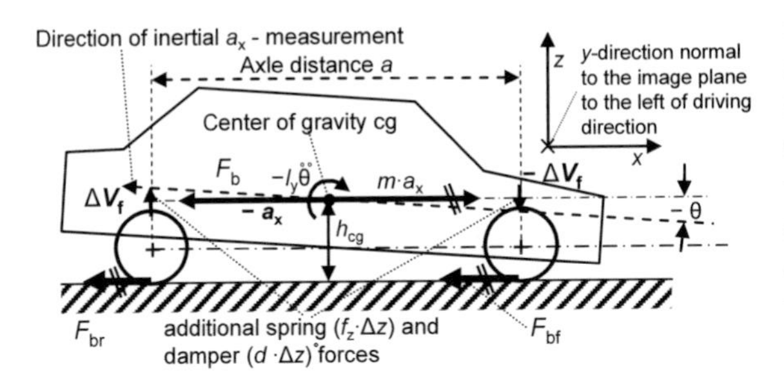

Figure 3.20. Deceleration by braking: Forces, torques, and orientation change in pitch

to be fastened to the vehicle body (*e.g.*, by seat belts or nets). On most realistic surfaces, deceleration will certainly be smaller. In normal traffic conditions, a realized friction coefficient of $\mu = 0.5$ is considered harsh braking (deceleration $a_x \approx -5$ m/s², that is, from 100 km/h to a stop in 5.56 s). Figure 3.20 shows the components for judging the dynamic effects of braking.

Since the center of gravity is at elevation h_{cg} above the point where the braking forces of the wheels attack (F_{bf} at front and F_{br} at the rear wheels in the contact region with the ground), there will be an additional torque in the vertical plane, counteracted initially by the moment of inertia in pitch ($-I_y \cdot d^2\theta/dt^2$). This leads to a downward pitch acceleration (with $I_y = m \cdot i_y^2$) via $h_{cg} \cdot m \cdot a_x = -I_y \cdot d^2\theta/dt^2$ of

$$d^2\theta/dt^2 = h_{cg} \cdot a_x / i_y^2. \tag{3.24}$$

Now, due to the suspension system of the body relative to the wheels with springs and damping elements, vertical forces ΔV_f in wheel suspension will build up, counteracting the torque from the braking forces. Spring force components ΔV_f are proportional to vertical displacements ($f_z \cdot \Delta z \sim \theta$), and damping force components are proportional to displacement speed ($d(\Delta z)/dt \sim d\theta/dt$). Usually, the resulting motion will be a damped rotational oscillation (second–order system). Since this immediately affects vision when the cameras are mounted directly on the vehicle body, the resulting visual effects of (self-initiated) braking actions should be taken into account at all interpretation levels. This is the reason that expectations of motion behavior are so beneficial for vision with its appreciable, unavoidable delay times of several video cycles.

In a steady deceleration phase ($-a_x$ = constant), the corresponding change in pitch angle θ_b can be determined from the equilibrium condition of the additional horizontal and vertical forces acting at axle distance a, taking into account that the vertical motion at the axles is $\theta_b \cdot a/2 = \Delta z$ (cg at $a/2$) and $h_{cg} \cdot m \cdot a_x = -a \cdot \Delta V_f =$ $-a \cdot f_z \cdot \theta_b \cdot a/2$ which yields

$$\theta_b = -[2h_{cg} \cdot m/(f_z \cdot a^2)] \cdot a_x = -p_b \cdot a_x \tag{3.25}$$

The term in brackets is a proportionality factor p_b between constant linear deceleration ($-a_x$) and resulting stationary additional pitch angle θ_b (downward positive here). The time history of θ after braking control initiation will be constrained by a second-order differential equation taking into account the effects discussed in connection with Equation 3.24. In visual state estimation to be discussed in Chapter 9, this knowledge will be taken into account; it is directly exploited in the recursive estimation process. Figure 3.21 shows, in the top left graph, the pitch rate response to a step input in acceleration. The softness of the suspension system in combination with inertia of the body lead to the oscillation extending to almost 2 seconds after the change in control input. The general second-order dynamic model for an arbitrary excitation $f[a_x(t)]$ for braking is given by

$$d^2\theta / dt^2 + D \cdot d\theta / dt + f_{Sp} \cdot \theta = f[a_x(t)]. \qquad (3.26)$$

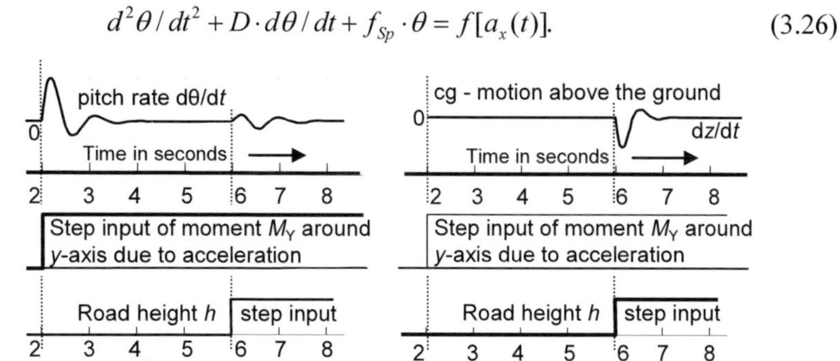

Figure 3.21. Simulation of vehicle suspension model: Pitch rate (top left) and heave response (top right) of ground vehicle suspension after step input in acceleration (center) as well as the height profile of the ground (bottom)

Since the eigenfrequency of the vehicle does not change over time and since it is characteristic of the vehicle in a given loading state, this oscillation over as many as 50 video cycles can be expected for a certain control input. This alleviates image sequence interpretation when properly represented in the perception system.

Pitching motion due to partial loss of wheel support: This topic would also fit under "vertical curvature effects" (above). However, the eigenmotion in pitch after a step input in wheel support may be understood more easily after the step input in deceleration has been discussed. Figure 3.22 shows a vehicle that just lost ground under the front wheels due to a negative step input of the supporting surface while driving at a certain speed.

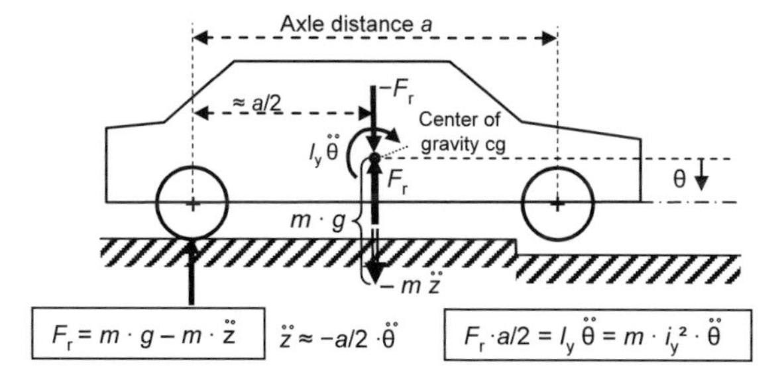

Figure 3.22. Pitch and downward acceleration after losing ground contact with the front wheels (cg assumed at center of axle distance)

The weight ($m \cdot g$) of the vehicle together with the forces at the rear axle will produce both a downward acceleration of the cg and a rotational acceleration around the cg. The relations given at the bottom of the figure (including D'Alembert inertial forces and moments) yield the differential equation

$$d^2\theta / dt^2 \cdot (a^2 / 4 + i_y^2) = g \cdot a / 2 . \qquad (3.27)$$

Normalizing the inertial radius i_y by half the axle distance $a/2$ to the non-dimensional inertial radius i_{yN} finally yields the initial rotational acceleration

$$d^2\theta / dt^2 = 2 \cdot g / [a \cdot (1 + i_{yN}^2)] . \tag{3.28}$$

With i_{yN} in the range of 0.8 to 0.9, usually, and axle distances between 2 and 3.5 m for cars, angular accelerations to be expected are in the range from about 6 to 12 rad/s², that is 350 to 700°/s squared resulting in a build–up of angular speed of about 14 to 28°/s per video cycle time of 40 ms. Inertial sensors will immediately measure this crisply way, while image interpretation will be confused initially; this is a strong argument in favor of combined inertial/visual dynamic scene interpretation. Nature, of course, has discovered these complementarities early and continues to use them in vertebrate type vision. Figure 3.21 has shown pitch rate and heave motion after a step input in surface elevation in the opposite direction in the top two graphs (right-hand part); the response extends over many video cycles (~ 1.5 seconds, *i.e.,* about 35 cycles). Due to tire softness, the effects of a positive or negative step input will not be exactly the same, but rather similar, especially with respect to duration.

Pitching and rolling motion due to wheel – ground interaction: A very general approach to combined visual/inertial perception in ground vehicle guidance would be to mount linear accelerometers in the vertical direction at each suspension point of the (four) wheels and additional angular rate sensors around all body axes. The sum of the linear accelerations measured, integrated over time, would yield heave motion. Integrals of pairwise sums of accelerometer signals (front vs. rear and left vs. right-hand side) would indicate pitch and roll accelerations which could then be fused with the rate sensor data for improved reliability. Their integral would be available with almost no time delay compared to visual interpretation and could be of great help when driving in rough terrain, since at least the high-frequency part of the body orientation would be known for visual interpretation. The (low-frequency) drift errors of inertial integrals can be removed by results from visual perception.

Remember the big difference between inertial and visual data interpretation: Inertial data are "lead" signals (*measured* time derivatives) containing the influence of all kinds of perturbations, while visual interpretation relies very much on models containing (time-integrated) state variables. In vision, perturbations have to be discovered "in hindsight" when assumptions made do not show up to be valid (after considerable delay time).

3.4.5.2 Lateral Road Vehicle Guidance

To demonstrate some dynamic effects of details in modeling of the behavior of road vehicles, the lane change maneuvers with the so-called "bicycle model" (see Figure 3.10) of a different order are discussed here. First, let us consider an *idealized* maneuver (completely decoupled translational motion and no rotations). Applying a constant lateral acceleration a_y (of, say, 2 m/s²) in a symmetrical positive and negative fashion, we look for the time T_{LC} in which one lane width W_L of 3.6 m can be traversed with lateral speed v_y back to zero again at the end. One obtains

$$T_{LC} = 2 \cdot \sqrt{W_L / a_y} . \tag{3.29}$$

For the data mentioned, the lane change time is T_{LC} = 2.68 seconds, and the maximum speed at the center of the idealized maneuver is $v_{ymaxLCi}$ $\{t = T_{LC}/2\}$ = 2.68 m/s. Since Ackermann steering is nonholonomic, real cars cannot perform this type of lane change maneuver; however, it is nice as a reference for realizable maneuvers to be discussed in the following. For a_y = 4 m/s², lane change time would be 1.9 seconds and maximum lateral speed $v_y\,(T_{LC}/2)$ = 3.8 m/s.

Fifth-order dynamic model for lateral road vehicle guidance: The very busy Figure 3.23 shows the basic properties of a simple but full order (linear) bicycle

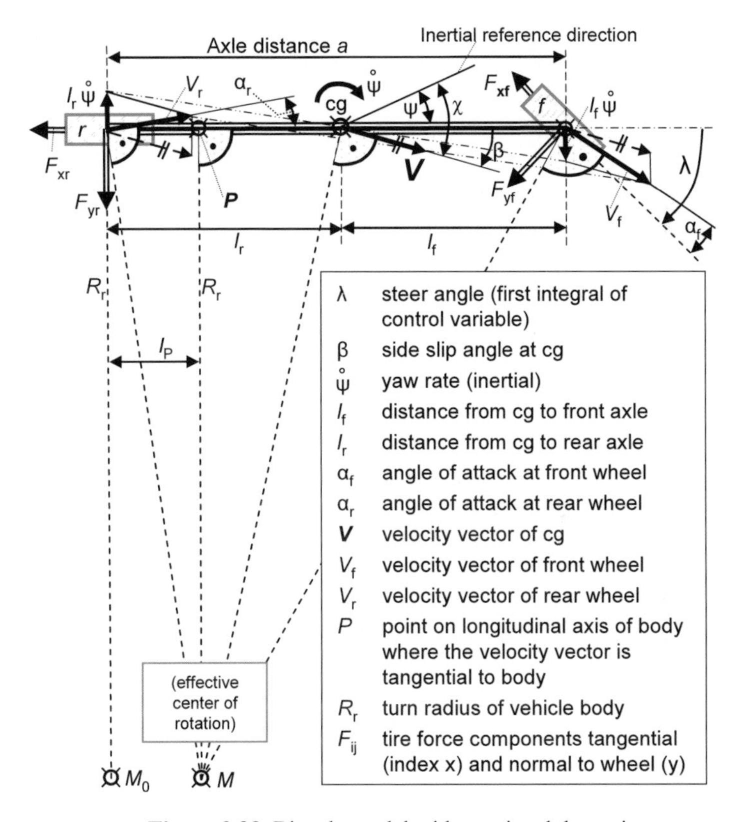

Figure 3.23. Bicycle model with rotational dynamics

model, taking combined tire forces from both the left- and right-hand side as well as translational and rotational dynamics into account. [The full model with all nonlinearities and separately modeled dynamics for the wheel groups and the body is too complex to allow analytical solutions; these are used in numerical simulations.] Here, interest lies in some major effects of lateral maneuvering for turns and lane changes. More involved models may be found in [Mitschke 1990; Giampiero 2007].

Side forces on the wheels (index y) are generated by introducing an angle of attack at the front wheel(s) through a steering angle λ. Tires may be considered to act as springs in the lateral direction with an approximately linear characteristic for small angles of attack ($|\alpha| < \approx 3°$); only this regime is considered here. For the test vehicle **VaMoRs**, this allows lateral accelerations up to about 0.4 g = 4 m/s² in the linear range.

With k_T as the lateral tire force coefficient linking vertical tire force $F_N = m_{WL} \cdot g$ (wheel load due to gravity) via angle of attack to lateral tire force F_y, there follows

$$F_{yf} = k_T \cdot \alpha_f \cdot F_{Nf}; \qquad F_{yr} = k_T \cdot \alpha_r \cdot F_{Nr}. \tag{3.30}$$

If the vehicle weight is distributed almost equally onto all wheels of a four-wheel vehicle, m_{WL} is close to one quarter of total vehicle mass; in the bicycle model, it is close to one half the total mass both on the front and rear axle. Defining the mass related lateral force coefficient k_{ltf}

$$k_{ltf} = F_y /(m_{WL} \cdot \alpha_f) = k_T \cdot g \text{ (in m/s}^2\text{/rad)}, \tag{3.31}$$

and multiplying this coefficient with both the actual wheel load (in terms of mass) and the angle of attack yields the lateral tire force F_y. The sum of all torques (including the inertial D'Alembert-term with $I_z = m \cdot i_z^2$ as the moment of inertia around the vertical axis) yields (see Figure 3.23)

$$I_z \cdot \ddot{\psi} - F_{yr} \cdot l_r + (F_{xf} \cdot \sin\lambda + F_{yf} \cdot \cos\lambda) \cdot lf = 0. \tag{3.32}$$

The force balance normal to the vehicle body yields with $d\chi/dt = d\chi/ds \cdot ds/dt$ = (curvature C of the trajectory driven times speed V), and thus with the centrifugal force at the cg: $C \cdot V^2 = m \cdot V \cdot d\chi/dt$

$$\begin{aligned} &m \cdot V \cdot d\chi / dt \cdot \cos\beta + m \cdot d / dt \cdot \sin\beta + \\ &\quad + F_{yr} + F_{xf} \cdot \sin\lambda + F_{yf} \cdot \cos\lambda = 0. \end{aligned} \tag{3.33}$$

From the center of Figure 3.23, it can be seen that trajectory heading χ is the sum of vehicle body heading ψ and side slip angle β ($\chi = \psi + \beta$) and thus

$$d\chi / dt = d\psi / dt + d\beta / dt. \tag{3.34}$$

For small angles of attack at the wheels, the following relations hold after [Mitschke 1990]:

$$\alpha_f = \beta - \lambda + d\psi / dt \cdot l_f / V; \qquad \alpha_r = \beta - d\psi / dt \cdot l_r / V. \tag{3.35}$$

For further simplification of the relations, the cg is assumed to lie at the center between the front and rear axles ($l_f = l_r = a/2$), so that half of the vehicle mass rests on each axle (wheel of bicycle model: $F_{Nr} = F_{Nf} = mg/2$). Then, the following linear fifth-order dynamic model for lateral control of a vehicle with Ackermann-steering at constant speed and with the state vector $\underline{x}_{La}$ (steering angle λ, inertial yaw rate $d\psi/dt$, slip angle β, body heading angle ψ, and lateral position y) results:

$$\underline{x}_{La}^T = [\lambda, \ \dot{\psi}, \ \beta, \ \psi, \ y]. \tag{3.36}$$

With the following abbreviations:

$$\begin{aligned} &i_{zB}^2 = [i_z /(a/2)]^2; \\ &T_\psi = V \cdot i_{zB}^2 / k_{ltf}; \quad \text{and} \\ &T_\beta = V / k_{ltf}, \end{aligned} \tag{3.37}$$

the set of first-order differential equations is written

$$\frac{d}{dt}\begin{pmatrix}\lambda\\ \dot{\psi}\\ \beta\\ \psi\\ y\end{pmatrix}=\begin{pmatrix}0 & 0 & 0 & 0 & 0\\ V/(aT_\psi) & -1/T_\psi & 0 & 0 & 0\\ 1/(2T_\beta) & -1 & -1/T_\beta & 0 & 0\\ 0 & 1 & 0 & 0 & 0\\ 0 & 0 & V & V & 0\end{pmatrix}\cdot\begin{pmatrix}\lambda\\ \dot{\psi}\\ \beta\\ \psi\\ y\end{pmatrix}+\begin{pmatrix}1\\ 0\\ 0\\ 0\\ 0\end{pmatrix}\cdot u \tag{3.38}$$

$$\mathrm{d}\underline{\mathrm{x}}_{La}/\mathrm{d}t = \qquad \Phi \qquad \cdot\underline{\mathrm{x}}_{La} \quad + \mathrm{b}\cdot\mathrm{d}\lambda/\mathrm{d}t.$$

For the test vehicle VaMP, a 240 kW (325 HP) powered sedan Mercedes 500 SEL, the parameters involved are (average representative values) $m = 2650$kg (that is, $m_{WL} = m/2 = 1325$ kg for the bicycle model), $k_T = 96$ kN/rad, $I_z = 5550$ kg m²; and $a = 3.14$ m. This leads to $i_{zB}^2 = 0.85$ and $k_{ltf} \approx 72$ (m/s² per rad) = 1.25 (m/s² per degree wheel angle of attack), and finally to the following *speed-dependent time constants* for lateral motion (V in m/s):

$$\begin{aligned} T_\psi &= V/84.7 = 0.01389\cdot V\ (s)\,,\\ T_\beta &= V/72 \quad = 0.0118\cdot V\ (s). \end{aligned} \tag{3.39}$$

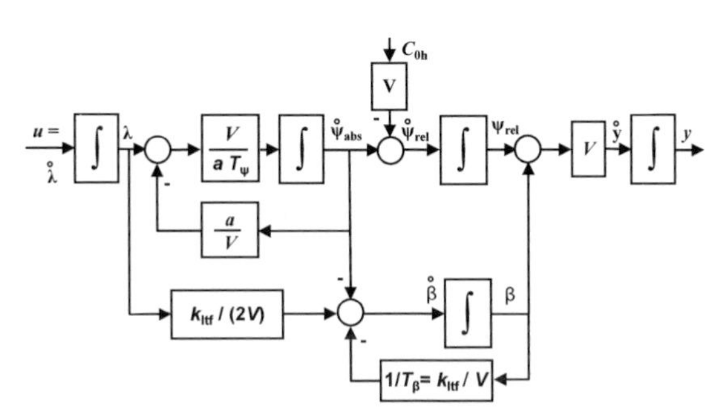

Figure 3.24. Block diagram of fifth-order (bicycle) model for lateral control of road vehicles taking rotational dynamics around the vertical axis into account

These values as a function of speed V already have been shown in Figure 3.11 for the test vehicle VaMP (top right). They increase up to values of 0.9 seconds at maximum speed. The block diagram corresponding to Equation 3.38 is shown in Figure 3.24.

From the fact that the systems dynamics matrix Φ has only zeros above the diagonal and the negative inverse values of the two time constants on the diagonal, the specialist in systems dynamics immediately recognizes that the system has three eigenvalues at the origin of the Laplace-transform "s"-plane (integrators) and two first-order subsystems with eigenvalues as inverse time constants on the negative real axis; the corresponding time histories are exponentials of the natural number e = 2.71828 of the form $c_i \cdot \exp(-t/T_i)$. Since the maximum speed of VaMP is 70 m/s, the eigenvalue $-1/T_\beta$ will range from $-\infty$ to $-1\ s^{-1}$. Since amplitudes of first-order systems have diminished to below 5% in the time range of three time constants, it can be seen that for speeds above about 1 m/s (≈ 3.6 km/h), the dynamic effects should be noticeable in image sequence analysis at video rate (40 ms cycle time). On the other hand, four video cycles (160 ms) are typical delay times for recognition of complex scenes by vision including proper reaction so that up to speeds of 4 m/s (≈ 14 km/h), neglecting the dynamic effects may be within the noise level ($1/T_{\beta 3} \approx 18\ s^{-1}$).

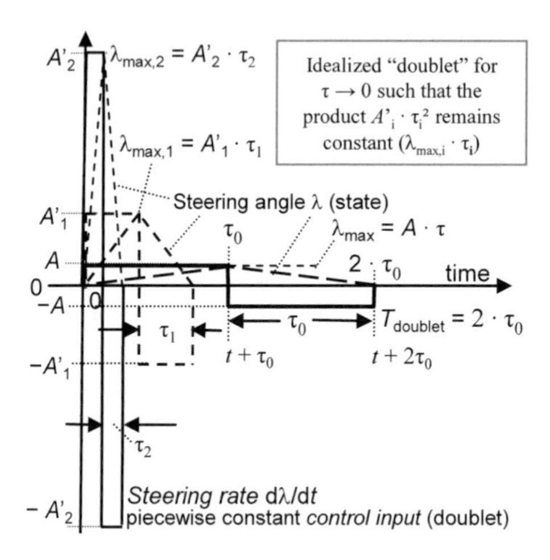

Figure 3.25. Doublet in constant steer rate $u_{ff}(\cdot) = d\lambda/dt$ as *control time history* over two periods τ with opposite sign of amplitude ± A' yields an "ideal impulse" in steer angle for heading change and $\tau \rightarrow 0$

The eigenfrequency of human arms and legs is in the 2 Hz range ($\omega = 12.6\ s^{-1}$) so that the first-order delay effects at lower speeds will hardly be noticeable by humans too. However, when speed increases, there will be strong dynamical effects. This will be shown with an idealized maneuver: The doublet as shown in Figure 3.12 is redrawn in Figure 3.25 on an absolute timescale with control output beginning at zero. From Table 3.2, it can be seen that time T_2, in which a preset acceleration limit can be reached with constant control output A, decreases rapidly with speed V. Let us, therefore, look at the limiting case for the doublet when its duration 2τ goes to zero.

The doublets in Figure 3.25 can be generated as a sum of three step functions. From 0 to τ the only step function $u_1(t) = A' \cdot 1(t)$ is active; from τ to 2τ a superposition of two step functions, the second one delayed by τ, yields $u_2(t) = A' \cdot [1(t) - 2 \cdot 1(t - \tau)]$. For the third phase from 2τ forward, the previous function plus a step delayed by 2τ is valid: $u(t) = u_2(t) + A' \cdot 1(t - 2\tau)$.

This yields the control input time history of superimposed delayed step functions shown in the figure, which can be summarized as control function with the two parameters A' and τ:

$$u(t) = A' \cdot [1(t) - 2(t - \tau) + 1(t - 2\tau)] . \qquad (3.40)$$

For making the transition to distribution theory [Papoulis 1962] when the period of the doublet τ goes to zero, we rewrite the amplitude A' in Equation 3.40 under the side constraint that the product ($A_i' \cdot \tau_i^2$) is kept constant when duration τ_i is decreased to zero

$$A'(t, \tau_i) = A_i / \tau_i^2 . \qquad (3.41)$$

This (purely theoretical) time function has a simple description in the frequency domain; Equation 3.40 can now be written with A = constant

$$u(t, \tau) = A \cdot [1(t) - 2 \cdot 1(t - \tau) + 1(t - 2 \cdot \tau)] / \tau^2 . \qquad (3.42)$$

As a two-step difference approximation based on step functions, there follows

$$u(t, \tau) = A \cdot \left[\frac{1(t) - 1(t - \tau)}{\tau} - \frac{1(t - \tau) - 1(t - 2 \cdot \tau)}{\tau} \right] \Big/ \tau . \qquad (3.43)$$

Recognizing that each expression in the square bracket describes a Dirac impulse for τ toward 0, nice theoretical results for the (ideal) doublet and doublet responses are obtained easily.

In the (idealized) limit, when τ decreases to 0 with the product $A' \cdot \tau^2 = A$ kept constant (that is, A' increases strongly, see Figure 3.25), the doublet input function becomes the derivative of the Dirac impulse:

$$u_{\text{idd}} = \lim_{\tau \to 0} u(t,\tau) = \frac{\delta(t) - \delta(t-\tau)}{\tau} \Big|_{\tau \to 0} = \dot{\delta}(t)\,. \tag{3.44}$$

This shows that the "idealized" doublet is the second derivative of the step function and the first derivative of the Dirac impulse; since in the Lapace-domain forming the derivative means multiplication by s, there follows $u_{\text{idd}}(s) = A \cdot s$.

Applying the Laplace transform to Equation 3.38 and grouping terms yields, with the initial values $\underline{x}_{\text{La}}(0)$,

$$(sI - \Phi) \cdot \underline{x}_{\text{La}}(s) = b \cdot u(s) + \underline{x}_{\text{La}}(0). \tag{3.45}$$

As derived in Appendix B.2, the time responses to the idealized doublet in steering rate $u_{\text{idd}}(s) = A \cdot s$ as input are simple products of the transfer function with this input function. Figure 3.26 shows results not scaled on the time axis with data for

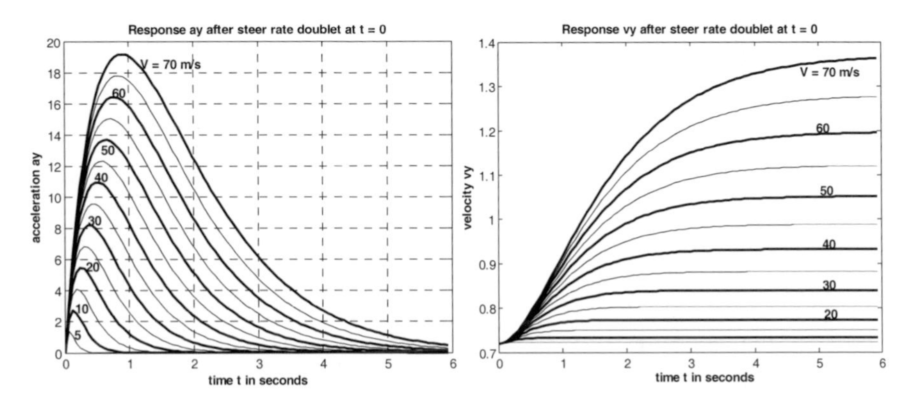

Figure 3.26. Lateral acceleration and speed after (an idealized) doublet in steer rate (Dirac impulse in steer angle) at $t = 0$. This leads to a change in driving direction (heading). Due to tire dynamics, maximum lateral acceleration occurs at later times (between T_β and T_ψ) up to about 1 second after the control input (for VaMP) with increasing forward velocity V (parameter on curves). Lateral acceleration effects after a steer angle impulse extend over more than 100 video cycles with no control activity. At higher driving speeds V, a large percentage of the lateral velocity v_y develops over time due to dynamic effects from tires and rotational motion.

the test vehicle VaMP. It can be seen that for small velocities V, the effect of the doublet on steering rate is concentrated around $t = 0$. For higher speeds, the energy stored initially in the tires due to their stiffness (spring property) has an effect on lateral acceleration over several seconds. Maximal lateral acceleration occurs between T_β and T_ψ (see Figure B.1); the absolute value for the same input amplitude A at high speeds is about an order of magnitude larger than at low speeds. This requires adaptation in behavior. The control input A has to be adjusted with speed so that the underlying conditions for linearity and for comfort are not violated.

At top speed of $V = 70$ m/s, about half of the final lateral speed results from this effect, while for $V = 5$ m/s (lowest thin line in bottom figure), this part is negligible. These differences in dynamic behavior have to be taken into account when computing expectations after a control input in steering angle. They play an important role in decision-making, for example, when the amplitude of a control input as a function of speed driven has to be determined as one of the situational aspects.

Lane changes with realistic control input: The idealized relations discussed above yield a good survey of the basic behavioral properties of road vehicles as a function of speed driven. Real maneuvers have to take the saturation effects in steering rate into account. Analytical solutions become rather complex for these conditions. Simulation results with the nonlinear set of equations are easily available today. Table 3.4 and Figure 3.27 show some simulation results for a typical vehicle. Three different characteristic control input times T_C have been selected: 6, 4, and 2 seconds (column 1, Table 3.4) for two speeds. Column 2 shows the steering rates leading to a final lateral offset of one lane width (3.75 m) after the transients have settled, with the other parameters pre–selected (duration of central 0-input time beside T_C).

Table 3.4. Expectations for lane change maneuvers ($\Delta y = 3.75$ m) with preset control input times according to the 5th-order bicycle model with piecewise constant steering rates

Column Speed V/km/h	1 T_C control time /s	2 $d\lambda/dt$ / °/s	3 λ_{max} / °	4 ψ_{max} / °	5 β_{max} / °	6 a_{ymax} / m/s^2	7 v_{ymax} / m/s	8 state transit. time /s	Re- mark: Figure 3.27
40	6	0.9	1.4	6.4	0.16	0.7	1.2	6.5	
40	4	3.1	3.1	9.6	0.36	1.5	1.8	4.5	
40	2	24	12	18.5	1.7	5.2	3.2	2.7	
100	6	0.16	0.22	2.5	0.5	0.6	1.2	7.9	(a)
100	4	0.5	0.5	3.7	0.95	1.1	1.6	6.0	
100	2	3.9	2	6.5	2.6	2.7	2.35	4.4	(b)
100	6	0.18	0.23	2.2	0.5	0.6	1	7.9	
100	4	0.64	0.51	3.2	0.9	1 1	1.45	6	
100	2	5.1	2	6	2.2	2.3	2.26	4.4	(c)

It is seen that state transition time (column 8) increases relative to T_C with increased speed; the ratio of these times increases with decreasing T_C. Figure 3.27a (corresponding to row 4 in Table 3.4, maneuver time 6 seconds) shows that the state transitions are almost finished when the control input is finished. Figure 3.27b (corresponding to row 6 in Table 3.4, maneuver time 2 seconds) shows in the lower right part, that maximum (negative) lateral acceleration occurs after control input has been finished (is back to 0). The maximum positive acceleration has increased 4.5-fold compared to a), while the maximal lateral speed has almost doubled. Steering rate (column 2) for the 2-second maneuver is 24 times as high as for the 6-second maneuver while the maximum steering angle increased by a factor of 9 (column 3). Maximum slip angle (column 5) increases fivefold while the maximum heading angles (column 4) differ by a factor of only 2.6. In the short maneuver with $T_C = 2$ [1/3 of case a)], state transition needs about twice the control input time

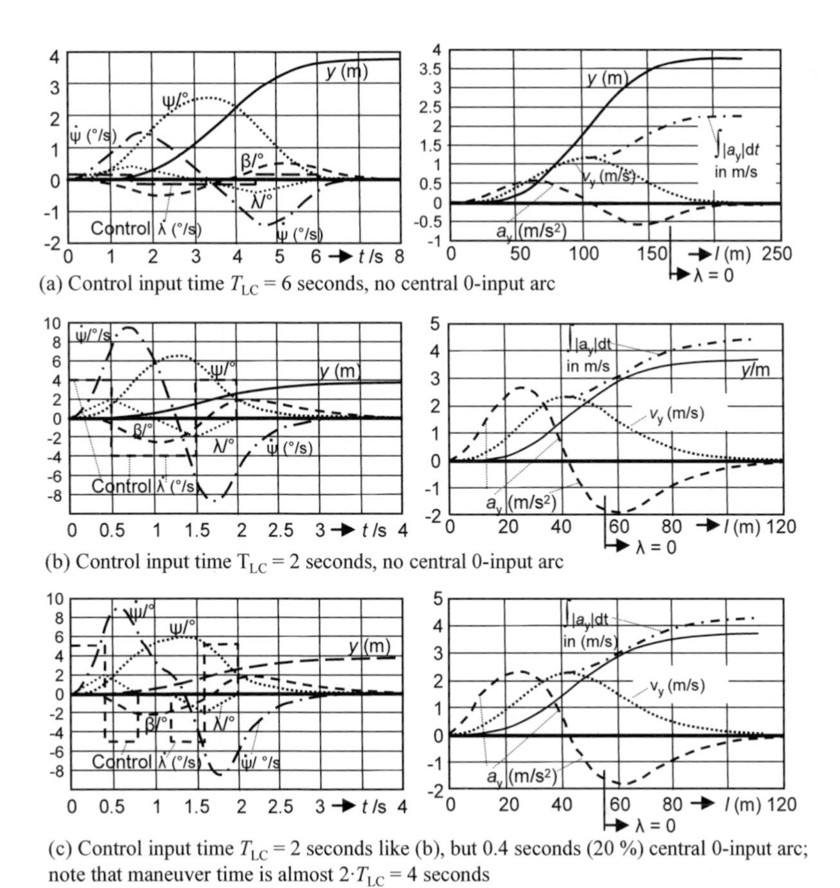

(a) Control input time T_{LC} = 6 seconds, no central 0-input arc

(b) Control input time T_{LC} = 2 seconds, no central 0-input arc

(c) Control input time T_{LC} = 2 seconds like (b), but 0.4 seconds (20 %) central 0-input arc; note that maneuver time is almost $2 \cdot T_{LC}$ = 4 seconds

Figure 3.27. Comparison of lane change maneuvers with fifth-order dynamic model at a speed of V = 100 km/h (27.8 m/s) and realistic steer rate inputs. The amplitude A is adjusted such that the lateral offset y_f is exactly one lane width of 3.75 m after the transients have settled. *Left column shows the variables (over time):* $d\lambda/dt$ = steer rate (— —) in (°/s); angle λ (·········); slip angle β (- - -) in (°); $d\psi/dt$ = yaw rate (— · —) in (°/s), angle ψ (·········). *Right column the variables (over distance):* lateral acceleration a_y (- - -) in (m/s²) and the integral of its magnitude over time (— · —); lateral velocity v_y (·········), offset y (——); integral of magnitude of a_y (dash-dotted line).

(column 8): compared to (a), maneuver time for the complete state transition is reduced from 7.9 to 4.4 s (by ~ 45%).

Inserting a central section of 20% duration (0.4 seconds) with zero control input (last row in the table corresponding to Figure 3.27c requires an increase in steering rate by about 30% for the same lateral offset y_f = 3.75 m at the end of the maneuver. This leads to the same maximal steering angle (column 2) as the case without the central 0-input section. However, the maximal values of lateral acceleration and of the other state variables (ψ, β, v_y) are reduced slightly.

Since perturbations are abundant in road traffic, toward the end of the lane change maneuver after taking the new lane as reference for perception, a feedback component is superimposed leading to automatic centering in the new lane. This also takes care of curvature onsets during the lane change maneuver.

Driving on curved roads: The assumption is made that longitudinal speed V is controlled in dependence of road curvature in order not to exceed lateral acceleration limits. Speed acts as a parameter in selecting lateral behaviors as discussed above.

The heading angle of the vehicle body with respect to inertial space is designated by ψ_{abs} and with respect to the local road tangent by ψ_{rel}. Between ψ_{abs} and ψ_{rel} is the heading angle of the road χ. The temporal change in road heading at speed V is (see Equation 3.10)

$$d\chi / dt = C_{0h} \cdot V \, . \tag{3.46}$$

The visually recognizable curvature C_{0h} of the road at the actual location of vehicle cg can be introduced as an additional term in the dynamic model (see Chapter 6). In the block diagram Figure 3.24 (center top), this has been used to decouple local roadrunning from the absolute geodetic direction. Local heading ψ_{rel} times speed V yields the lateral speed v on the road.

With representations like these, the linguistic symbol "lane keeping" is activated by organizing the feedback control output computed by Equation 3.22 with a proper matrix K to be used for the steering rate λ-dot. Note that the visually determined quantities "road curvature C_{0h}", lateral position in the lane y, relative heading angle ψ_{rel} as well as the conventionally measured value of vehicle speed V are used in the closed-loop action-perception cycle taking a dynamic model for the motion process into account. It has been shown in linear control theory that complete state feedback yields optimal control laws with respect to a chosen payoff function.

This feedback control constitutes the behavioral capability *"roadrunning"* made up of the perceptual capability *road (lane) recognition with relative egostate* (including reconstruction of the slip angle β not directly measurable) and the locomotion capability *lane keeping* by state feedback. Since visual evaluation of the situation and control computation as well as implementation take their time (a few tenths of a second), this *time delay between measurement taking and control output has to be taken into account* when determining the control output. The spatio-temporal models of the process allow doing this with well-known methods from control engineering. Tuning all the parameters such that the abstract symbolic capabilities for roadrunning coincide with real-world behavior of subjects is the equivalent of *"symbol grounding"*, often deplored in AI as missing.

3.4.6 Phases of Smooth Evolution and Sudden Changes

Similar to what has been discussed for "lane keeping" (by feedback control) and for "lane change" (by feed-forward control), corresponding control laws and their abstract representation in the system have to be developed for all behavioral capabilities like turningoff, *etc.* This is not only true for locomotion but also for gaze control in an active vision system. By extending these types of explicit representations to all processes for perception, decision-making, and mission planning as well as mission performance and monitoring, a very flexible overall system will result. These aspects have been discussed here to motivate the need for both smooth parts of mission performance with nice continuity conditions alleviating perception, and sudden changes in behavior where sticking to the previous mode would lead to failure (or probably disaster).

Efficient dynamic vision systems have to take advantage of continuity conditions as long as they prevail; however, they always have to watch out for discontinuities in motion, both of the subject's body and of other object observed, to be able to adjust readily. For example, a vehicle following the rightmost lane on a road can be tracked efficiently using a simple motion model. However, when an obstacle occurs suddenly in this lane, for example, a ball or an animal running onto the road, there may be a harsh reaction to one side. At this moment, a new motion phase begins, and it cannot be expected that the filter tuning for optimal tracking remains the same. So the vision process for tracking (similar to the bouncing ball example in Section 2.3.2) has two distinctive phases which should be handled in parallel.

3.4.6.1 Smooth Evolution of a Trajectory

Continuity models and low-pass filtering components can help to easily track phases of a dynamic process in an environment without special events. Measurement values with high-frequency oscillations are considered due to noise, which has to be eliminated in the interpretation process. The natural sciences and engineering have compiled a wealth of models for different domains. The methods described in this book have proven to be well suited for handling these cases on networks of roads.

However, in road traffic environments, continuity is interrupted every now and then due to initiation of new behavioral components by subjects and maybe by weather.

3.4.6.2 Sudden Changes and Discontinuities

The optimal settings of parameters for smooth pursuit lead to unsatisfactory tracking performance in cases of sudden changes. The onset of a harsh braking maneuver of a car or a sudden turn may lead to loss of tracking or at least to a strong transient motion estimated, especially so, if delay times in the visual perception process are large. If the onsets of these discontinuities could be well predicted, a switch in model or tracking parameters at the right time would yield much better results. The example of a bouncing ball has already been mentioned.

In road traffic, the compulsory introduction of the braking (stop) lights serves the same purpose of indicating that there is a sudden change in the underlying behavioral mode (deceleration). Braking lights have to be detected by vision for defensive driving; this event has to trigger a new motion model for the car at which it is observed. The level of braking is not yet indicated by the intensity of the braking lights. There are some studies under way for the new LED-braking lights to couple the number of LEDs lighting up to the level of braking applied; this could help finding the right deceleration magnitude for the hypothesis of the observed braking vehicle and thus reduce transients.

Sudden onsets of lateral maneuvers are supposed to be preceded by warning lights blinking at the proper side. However, the reliability of behaving according to this convention is rather low in many parts of the world.

As a general scheme in vision, it can be concluded that partially smooth sections and local discontinuities have to be recognized and treated with proper methods both in the 2-D image plane (object boundaries) and on the time line (events).

3.4.6.3 A Capability Network for Locomotion

The capability network shows how more complex behaviors depend on more basic ones and finally on the actuators available. The timing (temporal sequencing) of their activation has to be learned by testing and corresponding feedback of errors occurring in the real world. Figure 3.28 shows the capability network for locomotion of a wheeled ground vehicle. Note that some of the parameters determining the trigger point for activation depend on visual perception and on other measurement values. The challenges of system integration will be discussed in later chapters after the aspects of knowledge representation have been discussed.

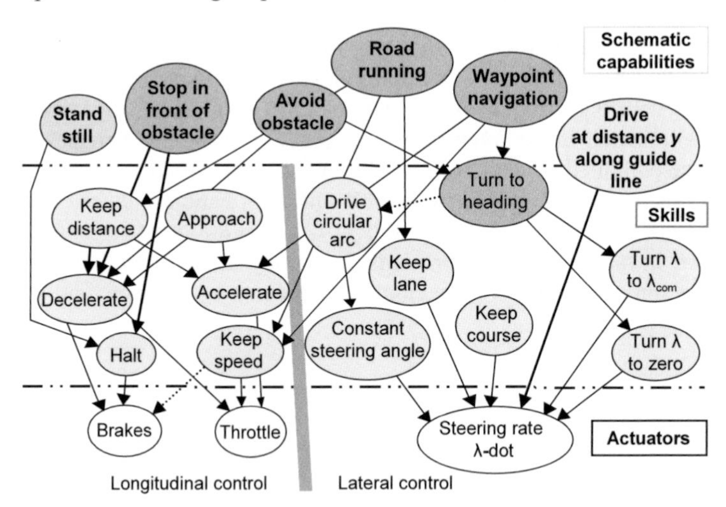

Figure 3.28. Network of behavioral capabilities of a road vehicle: Longitudinal and lateral control is fully separated only on the hardware level with three actuators; many basic skills are realized by diverse parameterized feed-forward and feedback control schemes. On the upper level, abstract schematic capabilities as triggered from "central decision" are shown [Maurer 2000, Siedersberger 2004]

3.5 Situation Assessment and Decision-Making

Subjects differ from objects (proper) in that they have perceptual impressions from the environment and the capability of decision-making with respect to their control options. For subjects, a control term appears in the differential equation constraints on their motion activities, which allows them to influence their motion; this makes subjects basically different from objects.

If decisions on control selection are not implicitly given in the code implementing subject behavior, but may be made according to some explicit goal criteria, something like *free will* occurs in the behavior decision process of the subject. Because of the fundamentally new properties of subjects, these require separate methods for knowledge representation and for combining this knowledge with actual perception to achieve their goals in an optimal fashion (however defined). The collection of all facts of relevance for decision-making is called the *situation*. It is especially difficult if other subjects, who also may behave *at will* to achieve their goals, form part of this process; these behaviors are unknown, usually, but may be guessed sometimes from reasoning as for own decision-making.

Some expectations for future behavior of other subjects can be derived from trying to understand the situation as it might look for oneself in the situation supposed to be given for the other subject. At the moment, this is beyond the actual state of the art of autonomous systems. But the methods under development for the subject's decision-making will open up this avenue. In the long run, capabilities of situation assessment of other subjects may be a decisive factor in the development of really intelligent systems. Subjects may group together, striving for common goals; this interesting field of group behavior taking real-world constraints into account is even further out in the future than individual behavior. But there is no doubt that the methods will become available in the long run.

3.6 Growth Potential of the Concept, Outlook

The concept of subjects characterized by their capabilities in sensory perception, in data processing (taking large knowledge bases for object/subject recognition and situation assessment into account), in decision-making and planning as well as in behavior generation is very general. Through an explicit representation of these capabilities, avenues for developing autonomous agents with new mental capabilities of learning and cooperation in teams may open up. In preparation for this long-term goal, representing humans with all their diverse capabilities in this framework should be a good exercise. This is especially valuable for mixed teams of humans and autonomous vehicles as well as for generating intelligent behavior of these vehicles in environments abounding with activities of humans, which will be the standard case in traffic situations.

In road traffic, other subjects frequently encountered (at least in rural environments) beside humans are four-legged animals of different sizes: horses, cattle, sheep, goats, deer, dogs, cats, *etc.*; birds and poultry are two-legged animals, many of which are able to fly.

Because of the eminent importance of humans and four-legged animals in any kind of road traffic, autonomous vehicles should be able to understand the motion capabilities of these living beings in the long run. This is out into the future right now; the final section of this chapter shows an approach and first results developed in the early 1990s for recognition of humans. This field has seen many activities since the early work of Hogg (1984) in the meantime and has grown to a special area in technical vision; two recent papers with application to road traffic are [Bertozzi *et al.* 2004; Franke *et al.* 2005]

3.6.1 Simple Model of Human Body as Traffic Participant

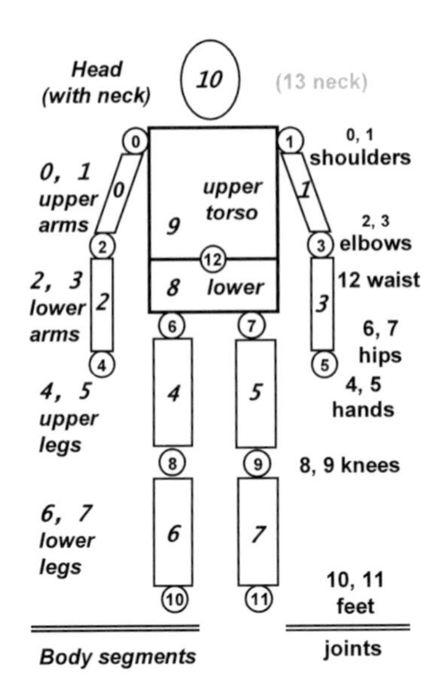

Figure 3.29. Simple generic model for human shape with 22 degrees of freedom, after [Kinzel 1994]

Elaborate models for the motion capabilities of human bodies are available in different disciplines of physiology, sports, and computer animation [Alexander 1984; Bruderlin, Calvert 1989; Kroemer 1988]. Humans as traffic participants with the behavioral modes of walking, running, riding bicycles or motor bikes as well as modes for transmitting information by waving their arms, possibly with additional instruments, show a much reduced set of stereotypical movements. Kinzel (1994a, b), therefore, selected the articulated body model shown in Figure 3.29 to represent humans in traffic activities in connection with the 4-D approach to dynamic vision. Visual recognition of moving humans becomes especially difficult due to the vast variety of clothing encountered and of objects carried. For normal Western style clothing the cyclic activities of extremities are characteristic of humans moving. Motion of limbs should be separated from body motion since they behave in different modes and at different eigenfrequencies, usually.

Limbs tend to be used in typical cyclic motion, while the body moves more steadily. The rotational movements of limbs may be in the same or in opposite direction depending on the style and the phase of grasping or running.

Figure 3.30 shows early results achieved with the lower part of the body model from Figure 3.29; cyclic motion of the upper leg (hip angle, amplitude $\approx 60°$, upper graph) and the lower leg (knee angle, amplitude $\approx 100°$, bottom graph) has been recognized roughly in a computer simulation with real-time image sequence

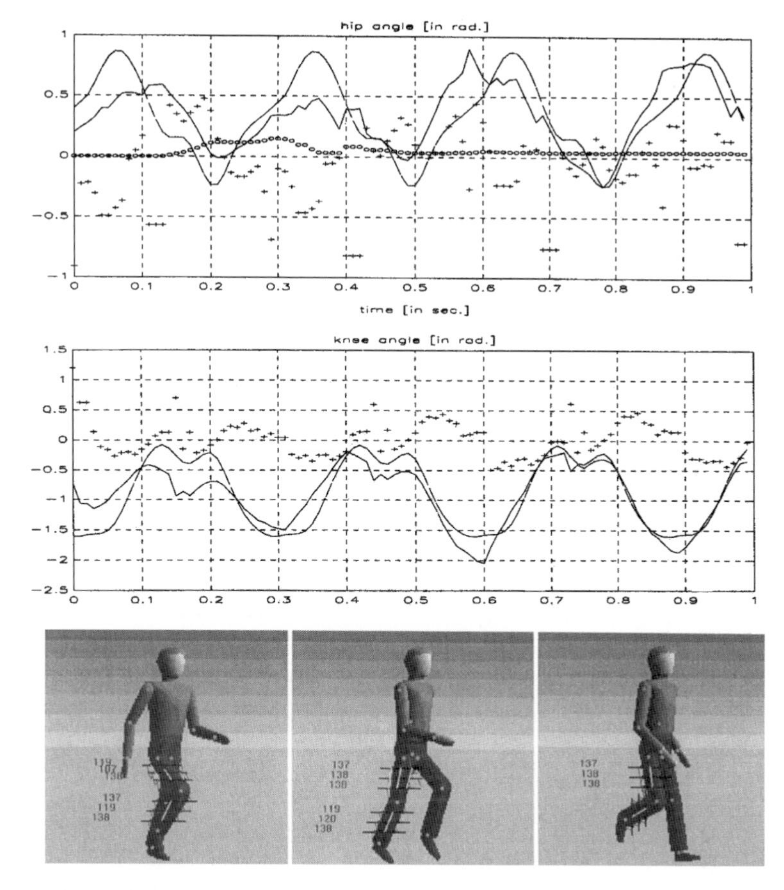

Figure 3.30. Quantitative recognition of motion parameters of a human leg while running: simulation with real image sequence processing, after [Kinzel 1994].

evaluation and tracking. At that time, microprocessor resources were not sufficient to do this onboard a car in real time (at least a factor of 5 was missing). In the meantime, computing power has increased by more than two orders of magnitude per processor, and human gesture recognition has attracted quite a bit of attention. Also the wide-spread activities in computer animation with humanoid robots, and especially the demanding challenge of the humanoid robo-cup league have advanced this field considerably, lately.

From the field last-mentioned and from analysis of sports as well as dancing activities there will be a pressure towards automatically recognizing human (-oid) motion. This field can be considered developing on its own; application within semi-autonomous road or autonomous ground vehicles will be more or less a side product. The knowledge base for these application areas of ground vehicles has to be developed as a specific effort, however. In case of construction sites or accident areas with human traffic regulation, future (semi-) autonomous vehicles should also have the capability of proper understanding of regulatory arm gestures and of proper behavior in these unusual situations. Recognizing grown-up people and children wearing various clothing and riding bicycle or carrying bulky loads will remain a challenging task.

3.6.2 Ground Animals and Birds

Beside humans, two superclasses of other animals play a role in rural traffic: Four-legged animals of various sizes and with various styles of running, and birds (from crows, hen, geese, turkeys, to ostrich), most of which can fly and run or hop on the ground. This wide field of subjects has hardly been touched for technical vision systems. In principle, there is no basic challenge for successful application of the 4-D approach. In practice, however, a huge volume of work lies ahead until technical vision systems will perceive animals reliably.

4 Application Domains, Missions, and Situations

In the previous chapters, the basic tools have been treated for representing objects and subjects with homogeneous coordinates in a framework of the real 3-D world and with spatiotemporal models for their motion. Their application in combination with procedural computing methods will be the subject of Chapters 5 and 6. The result will be an estimated state of single objects/subjects for the point "here and now" during the visual observation process. These methods can be applied multiple times in parallel to n objects in different image regions representing different spatial angles of the world around the set of cameras.

Vision is not supposed to be a separate exercise of its own but to serve some purpose in a task or mission context of an acting individual (subject). For deeper understanding of what is being seen and perceived, the goals of egomotion and of other moving subjects as well as the future trajectories of objects tracked should be known, at least vaguely. Since there is no information exchange between oneself and other subjects, usually, their future behavior can only be hypothesized based on the situation given and the behavioral capabilities of the subjects observed. However, out of the set of all objects and subjects perceived in parallel, generally only a few are of direct relevance to their own plans of locomotion.

To be efficient in perceiving the environment, special attention and thus perceptual resources and computing power for understanding should be concentrated on the most important objects/subjects. The knowledge needed for this decision is quite different from that one needed for visual object and state recognition. The decision has to take into account the mission plan and the likely behavior of other subjects nearby as well as the general environmental conditions (like quality of visual perception, weather conditions and likely friction coefficient for maneuvering, as well as surface structure). In addition, the sets of rules for traffic regulation valid in the part of the world, where the vehicle is in operation, have to be taken into account.

4.1 Structuring of Application Domains

To survey where the small regime, onto which the rest of the book will be concentrating, fits in the overall picture, first (contributions to) a loosely defined ontology for ground vehicles will be given. Appendix A shows a structured proposal which, of course, is only one of many possible approaches. Here, only some aspects of certain missions and application domains are discussed to motivate the items selected for presentation in this book. An all-encompassing and complete ontology for ground vehicles would be desirable but has not yet been assembled in the past.

From the *general environmental conditions* grouped under A.1, up to now only a few have been perceived explicitly by sensing, relying on the human operator to take care for the rest. More autonomous systems have to have perceptual capabilities and knowledge bases available to be able to recognize more of them by themselves. Contrary to humans, intelligent vehicles will have much more extended access to satellite navigation (such as *GPS* now or *Galileo* in the future). In combination with digital maps and geodetic information systems, this will allow them improved mission planning and global orientation.

Obstacle detection both on roads and in cross-country driving has to be performed by local perception since temporal changes are too fast, in general, to be reliably represented in databases; this will presumably also be the fact in the future. In cross-country driving, beside the vertical surface profiles in the planned tracks for the wheels, the support qualities of the ground for wheels and tracks also have to be estimated from visual appearance. This is a very difficult task, and decisions should always be on the safe side (avoid entering uncertain regions).

Representing *national traffic rules and regulations* (Appendix A.1.1) is a straightforward task; their ranges of validity (national boundaries) have to be stored in the corresponding databases. One of the most important facts is the general rule of right- or left-hand traffic. Only a few traffic signs like *stop* and *one-way* are globally valid. With speed signs (usually a number on a white field in a red circle) the corresponding dimension has to be inferred from the country one is in (km/h in continental Europe or mph in the United Kingdom or the United States, *etc.*).

Lighting conditions (Appendix A.1.2) affect visual perception directly. The dynamic range of light intensity in bright sunshine with snow and harsh shadows on dark ground can be extremely large (more than six orders of magnitude may be encountered). Special high-dynamic-range cameras (HDRC) have been developed to cope with the situation. The development is still going on, and one has to find the right compromise in the price-performance trade-off. To perceive the actual situation correctly, representing the recent time history of lighting conditions and of potential disturbances from the environment may help. Weather conditions (*e.g.*, blue skies) and time of day in connection with the set of buildings in the vicinity of the trajectory planned (tunnel, underpass, tall houses, *etc.*) may allow us to estimate expected changes which can be counteracted by adjusting camera parameters or viewing directions. The most pleasant weather condition for vision is an overcast sky without precipitation.

In normal visibility, contrasts in the scene are usually good. Under foggy conditions, contrasts tend to disappear with increasing distance. The same is true at dusk or dawn when the light intensity level is low. Features linked to intensity gradients tend to become unreliable under these conditions. To better understand results in state estimation of other objects from image sequences (Chapters 5 and 6), it is therefore advantageous to monitor average image intensities as well as maximal and minimal intensity gradients. This may be done over entire images, but computing these characteristic values for certain image regions in parallel (such as sky or larger shaded regions) gives more precise results.

It is recommended to have a steady representation available of intensity statistics and their trends in the image sequence: Averages and variances of maximum and minimum image intensities and of maximum and minimum intensity gradients in representative regions. When surfaces are wet and the sun comes out, light reflections may lead to highlights. Water surfaces (like puddles) rippled by wind may exhibit relatively large glaring regions which have to be excluded from image interpretation for meaningful results. Driving toward a low standing sun under these conditions can make vision impossible. When there are multiple light sources like at night in an urban area, regions with stable visual features have to be found allowing tracking and orientation by avoiding highlighted regions.

Headlights of other vehicles may also become hard to deal with in rainy conditions. Backlights and stoplights when braking are relatively easy to handle but require color cameras for proper recognition. In RGB-color representation, stop lights are most efficiently found in the R-image, while flashing blue lights on vehicles for ambulance or police cars are most easily detected in the B-channel. Yellow or orange lights for signaling intentions (turn direction indicators) require evaluation of several RGB channels or just the intensity signal. Stationary flashing lights at construction sites (light sequencing, looking like a hopping light) for indication of an unusual traffic direction require good temporal resolution and correlation with subject vehicle perturbations to be perceived correctly.

Recognition of weather conditions (Appendix A.1.3) is especially important when they affect the interaction of the vehicle with the ground (acceleration, deceleration through friction between tires and surface material). Recognizing and adjusting behavior to rain, hail, and snow conditions may prevent accidents by cautious driving. Slush and loose or wet dirt or gravel on the road may have similar effects and should thus be recognized. Heavy winds and gusts can have a direct effect on driving stability; however, they are not directly visible but only by secondary effects like dust or leaves whirling up or by moving grass surfaces and plants or branches of trees. Advanced vision systems should be able to perceive these weather conditions (maybe supported by inertial sensors directly feeling the accelerations on the body). Recognizing fine shades of texture may be a capability for achieving this; at present, this is beyond the performance level of microprocessors available at low cost, but the next decade may open up this avenue.

Roadway recognition (Appendix A.2) has been developed to a reasonable state since recursive estimation techniques and differential geometry descriptions have been introduced two decades ago. For freeways and other well-kept, high-speed roads (Appendices A.2.1 and A.2.2), lane and road recognition can be considered state of the art. Additional developments are still required for surface state recognition, for understanding the semantics of lane markings, arrows, and other lines painted on the road as well as detailed perception of the infrastructure along the road. This concerns repeating poles with different reflecting lights on both sides of the roadway, the meaning of which may differ from one country to the next, and guiderails on road shoulders and many different kinds of traffic and navigation signs which have to be distinguished from advertisements. On these types of roads there is only unidirectional traffic (one-way), usually, and navigation has to be done by proper lane selection.

On *ordinary state roads with two-way traffic* (Appendix A.2.3) the perceptual capabilities required are much more demanding. Checking free lanes for passing has to take oncoming traffic with high speed differences between vehicles and the type of central lane markings into account. With speeds allowed of up to 100 km/h in each direction, relative speed can be close to 60 m/s (or 2.4 m per video cycle of 40 ms). A 4-second passing maneuver thus requires about 250 m look-ahead range, way beyond what is found in most of today's vision systems. With the resolution required for object recognition and the perturbation level in pitch due to nonflat ground, inertial stabilization of gaze direction seems mandatory.

These types of roads may be much less well kept. Lane markings may be reduced to a central line indicating by its type whether passing is allowed (dashed line) or not (solid line). To the sides of the road, there may be potholes to be avoided; sometimes these may be found even on the road itself.

On all of these types of road, for short periods after (re-) construction there may be no lane markings at all. In these cases, vehicles and drivers have to orient themselves according to road width and to the distance from "their" side of the sealed surface. "Migrating construction sites" like for lane marking may be present and have to be dealt with properly. The same is true for maintenance work or for grass cutting in the summer.

Unmarked country roads (Appendix A.2.4) are usually narrow, and oncoming traffic may require slowing down and touching the road shoulders with their outer wheels. The road surface may not be well kept, with patches of dirt and high-spatial frequency surface perturbations. The most demanding item, however, may be the many different kinds of *subjects* on the road: People and children walking, running and bicycling, carrying different types of loads or guarding animals. Wild animals range from hares to deer (even moose in northern countries) and birds feeding on cadavers.

On *unsealed roads* (Appendix A.2.5) where speed driven is much slower, usually, in addition to the items mentioned above, the vertical surface structure becomes of increasing interest due to its unstable nature. Tracks impressed into the surface by heavily loaded vehicles can easily develop, and the likelihood of potholes (even large ones into which wheels of usual size will fit) requires stereovision for recognition, probably with sequential view fixation on especially interesting areas.

Driving cross-country, *tracks* (Appendix A.2.6) can alleviate the task in that they show where the ground is sufficiently solid to support a vehicle. However, due to non-homogeneous ground properties, vertical curvature profiles of high spatial frequency may have developed and have to be recognized to adjust speed so that the vehicle is not bounced around losing ground contact. After a period of rain when the surface tends to be softer than usual, it has to be checked whether the tracks are not so deep that the vehicle touches the ground with its body when the wheels sink into the track. Especially, tracks filled with water pose a difficult challenge for decision-making.

In Appendix A.2.7, all *infrastructure* items for all types of roads are collected to show the gamut of figures and objects which a powerful vision system for traffic application should be able to recognize. Some of these are, of course, specific to certain regions of the world (or countries). There have to be corresponding data bases and algorithms for recognizing these items; they have to be swapped when entering a zone with new regulations.

In section Appendix A.3 the different *types of vehicles* are listed. They have to be recognized and treated according to their *form (shape), appearance and function* of the vehicle (Appendix A.4). This type of structuring may not seem systematic at first glance. There is, of course, one column like A.4 for each type of vehicle under A.3. Since this book concentrates on the most common wheeled vehicles (cars and trucks), only these types are discussed in more detail here. Geometric size and 3-D shape (Appendix A.4.1) have been treated to some extent in Section 2.2.3 and will be revisited for recognition in Chapters 7 to 10.

Subpart hierarchies (Appendix A.4.2) are only partially needed for vehicles driving, but when standing, open doors and hoods may yield quite different appearances of the same vehicle. The property of glass with respect to mirroring of light rays has a fundamental effect on features detected in these regions. Driving through an environment with tall buildings and trees at the side or with branches partially over the road may lead to strongly varying features on the glass surfaces of the vehicle, which have nothing to do with the vehicle itself. These regions should, therefore, be discarded for vehicle recognition, in general. On the other hand, with low light levels in the environment, the glass surfaces of the lighting elements on the front and rear of the vehicle (or even highlights on windscreens) may be the only parts discernible well and moving in conjunction; under these environmental conditions, these groups are sufficient indication for assuming a vehicle at the location observed.

Variability of image shape over time depending on the 3-D aspect conditions of the 3-D object "vehicle" (Appendix A.3) is important knowledge for recognizing and tracking vehicles. When machine vision was started in the second half of the last century, some researchers called the appearance or disappearance of features due to self-occlusion a "catastrophic event" because the structure of their (insufficient) algorithm with fixed feature arrangements changed. In the 4-D approach where objects and aspect conditions are represented as in reality and where temporal changes also are systematically represented by motion models, there is nothing exciting with the appearance of new or disappearance of previously stable features. It has been found rather early that whenever the aspect conditions bring two features close to each other so that they may be confused (wrong feature correspondence), it is better to discard these features altogether and to try to find unambiguous ones [Wünsche 1987]. The recursive estimation process to be discussed in Chapter 6 will be perturbed by wrong feature correspondence to a larger extent than by using slightly less well-suited, but unambiguous features. Grouping regimes of aspect conditions with the same highly recognizable set of features into classes is important knowledge for hypothesis generation and tracking of objects. When detecting new feature sets in a task domain, it may be necessary to start more than one object hypothesis for fast recognition of the object observed. Such 4-D object hypotheses allow predicting other features which should be easily visible; in case they cannot be found in the next few images, the hypothesis can be discarded immediately. An early jump to several 4-D hypotheses thus has advantages over too many feature combinations before daring an object hypothesis (known as a combinatorial explosion in the vision literature).

Photometric appearance (Appendix A.4.4) can help in connection with the aspect conditions to find out the proper hypothesis. Intensity and color shading as well as high resolution in texture discrimination contribute positively to eliminating false object hypotheses. Computing power and algorithms are becoming available now for using these region-based features efficiently. The last four sections discussed are concerned with single object (vehicle) recognition based on image sequence analysis. In our approach, this is done by specialist processes for certain object classes (roads and lanes, other vehicles, landmarks, *etc.*).

When it comes to understanding the semantics of processes observed, the *functionality* aspects (Appendix A.4.5) prevail. For proper recognition, observations have to be based on spatially and temporally more extended representation. Trying to do this with data-intensive images is not yet possible today, and maybe even not desirable in the long run for data efficiency and corresponding delay times involved. For this reason, the results of perceiving single objects (subjects) "here and now" directly from image sequence analysis with spatiotemporal models are collected in a "dynamic object database" (DOB) in symbolic form. Objects and subjects are represented as members of special classes with an identification number, their time of appearance, and their relative state defined by homogeneous coordinates, as discussed in Section 2.1.1. Together with the algorithms for homogeneous coordinate transformations and shape computation, this represents a very compact but precise state and shape description. Data volumes required are decreased by two to three orders of magnitude (KB instead of MB). Time histories of state variables are thus manageable for several (the most important) objects/subjects observed.

For subjects, this allows recognizing and understanding *maneuvers and behaviors* of which one knows members of this type of subject class are capable (Appendix A.4.6). Explicit representations of perceptual and behavioral capabilities of subjects are a precondition for this performance level. Tables 3.1 and 3.3 list the most essential capabilities and behavioral modes needed for road traffic participants. Based on data in the ring-buffer of the DOB for each subject observed, this background knowledge now allows guessing the intentions of the other subject. This qualitatively new information may additionally be stored in special slots of the subject's representation. Extended observations and comparisons to standards for decisions–making and behavior realization now allows attributing additional characteristic properties to the subject observed. Together with the methods available for predicting movements into the future (fast-in-advance simulation), this allows predicting the likely movements of the other subject; both results can be compared and assessed for dangerous situations encountered. Thus, real-time vision as propagated here is an animation process with several individuals based on previous (actual) observations and inferences from a knowledge base of their intentions (expected behavior).

This demanding process cannot be performed for all subjects in sight but is confined to the most relevant ones nearby. Selecting and perceiving these most relevant subjects correctly and focusing attention on them is one of the decisive tasks to be performed steadily. The judgment, which subject is most relevant, also depends on the task to be performed. When just cruising with ample time available, the situation is different from the same cruising state in the leftmost of three lanes, but an exit at the right is to be taken in the near future. On a state road, cruising in the rightmost lane but having to take a turnoff to the left from the leftmost lane yields a similar situation. So the situation is not just given by the geometric arrangement of objects and subjects but also depends on the task domain and on the intentions to be realized.

Making predictions for the behavior of other subjects is a difficult task, especially when their *perceptual capabilities* (Appendix A.4.7) and those for *planning and decision-making* (Appendix A.4.8) are not known. This may be the case with respect to animals in unknown environments. These topics (Appendix A.6) and the well-known but very complex appearance and behavior of humans (Appendix A.5) are not treated here.

Appendix A.7 is intended to clarify some notions in vehicle and traffic control for which different professional communities have developed different terminologies. (Unfortunately, it cannot be assumed that, for example, the terms "dynamic system" or "state" will be understood with the same meaning by one person from the computer science and a second one from the control engineering communities.)

4.2 Goals and Their Relations to Capabilities

To perform a mission efficiently under perturbations, both the goal of the mission together with some quality criteria for judging mission performance and the capabilities needed to achieve them have to be known.

The main goal of road vehicle traffic is to transport humans or goods from point A to point B safely and reliably, observing some side constraints and maybe some optimization criteria. A smooth ride with low values of the time integrals of (longitudinal and lateral) acceleration magnitudes (absolute values) is the normal way of driving (avoiding hectic control inputs). For special missions, *e.g.*, on ambulance or touring sightseers, these integrals should be minimized.

An extreme type of mission is racing, exploiting vehicle capabilities to the utmost and probably reducing safety by taking more risks. Minimal fuel consumption is the other extreme where travel time is of almost no concern.

Safety and collision avoidance even under adverse conditions and in totally unexpected situations is the most predominant aspect of vehicle guidance. Driving at lower speed very often increases safety; however, on high-speed roads during heavy traffic, it can sometimes worsen safety. Going downhill, the additional thrust from gravity has to be taken into account which may increase braking distance considerably. When entering a crossroad or when starting a passing maneuver on a road with two-way traffic, estimation of the speed of other vehicles has to be done with special care, and an additional safety margin for estimation errors should be allowed. Here, it is important that the acceleration capabilities of the subject vehicle under the given conditions (actual mass, friction coefficient, power reserves) are well known and sufficient.

When passing on high-speed roads with multiple lanes, other vehicles in the convoy being passed sometimes start changing into your lane at short distances, without using indication signs (blinker); even these critical situations not conforming to standard behavior have to be coped with successfully.

4.3 Situations as Precise Decision Scenarios

The definition for "situation" used here is the following: *A situation encompasses all aspects of relevance for decision-making in a given scenario and mission context.* This includes environmental conditions affecting perception and limit values for control application (such as wheel to ground friction coefficients) as well as the set of traffic regulations actually valid that have been announced by traffic signs (maximum speed allowed, passing prohibited, *etc.*). With respect to other objects/subjects, a situation is not characterized by a single relation to *one* other unit but to the *total number of objects of relevance.* Which of those detected and tracked are relevant is a difficult decision. Even the selected regions of special attention are of importance. The objects/subjects of relevance are not necessarily the nearest ones; for example, driving at higher speed, some event happening at a farther look-ahead distance than the two preceding vehicles may be of importance: A patch of dense fog or a front of heavy rain or snow can be detected reliably at relatively long distance. One should start reacting to these signs at a safe distance according to independent judgment and not only when the preceding vehicles start their reactions.

Some situational aspects can be taken into account during mission planning. For example, driving on roads heading *into the low-standing sun* at morning or evening should be avoided by proper selection of travel time. Traffic congestion during rush hour also may be avoided by proper timing. Otherwise, the driver/autonomous vehicle has to perceive the indicators for situational aspects, and from a knowledge base, the proper behavior has to be selected. The three components required to perform this reliably are discussed in the sections below: Environmental background, objects/subjects of relevance, and the rule systems for decision-making. Beside the rules for handling planned missions, another set of perceptual *events* has to be monitored which may require another set of rules to be handled for selecting proper reactions to these events.

4.3.1 Environmental Background

This has not received sufficient attention in the recent past since, at first, the basic capabilities of perceiving roads and lanes as well as other vehicles had to be demonstrated. Computing power for including at least some basic aspects of environmental conditions at reasonable costs is now coming along. In Section 4.1 and Appendix A.1.2 (lighting conditions)/A.1.3 (weather conditions), some aspects have already been mentioned. Since these environmental conditions change rather slowly, they may be perceived at a low rate (in the range of seconds to minutes). An economical way to achieve this may be to allot remaining processing time per video cycle of otherwise dedicated image processing computers to this "environmental processing" algorithm. These low-frequency results should be made available to all other processes by providing special slots in the DOB and depositing the values with proper time stamps. The situation assessment algorithm has to check these values for decision-making regularly.

The specialist processes for visual perception should also have a look at them to adjust parameters in their algorithms for improving results. In the long run, a direct feedback component for learning may be derived. Perceiving weather conditions through textures may be very computer-intensive; once the other basic perception tasks for road and other vehicles run sufficiently reliable, additional computing power becoming available may be devoted to this task, which again can run at a very low rate. Building up a knowledge base for the inference from distributed textures in the images toward environmental conditions will require a large effort. This includes transitions in behavior required for safe mission performance.

4.3.2 Objects/Subjects of Relevance

A first essential step is to direct attention (by gaze control and corresponding image evaluation) to the proper environmental regions, depending on the mission element being performed. This is, of course, different for simple roadrunning, for preparing lane changes, or for performing a turnoff maneuver. Turning off to the left on roads with oncoming (right-hand) traffic is especially demanding since their lane has to be crossed.

Driving in urban environments with right-of-way for vehicles on crossroads coming from the right also requires special attention (looking into the road). Entering traffic circles requires checking traffic in the circle, because these vehicles have the right-of-way. Especially difficult are *4-way-stops* in use in some countries; here the right-of-way depends on the time of reaching the stop–lines on all four incoming roads.

Humans may be walking on roads through populated areas and in *stop-and-go* traffic. On state, urban and minor roads, humans may ride bicycles, may be roller skating, jogging, walking, or leisurely strolling. Children may be playing on the road. Recognizing these situations with their semantic context is actually out of range for machine vision. However, detecting and recognizing moving volumes (partially) filled with massive bodies is in the making and will become available soon for real-time application. Avoiding these areas with a relatively large safety margin may be sufficient for driver assistance and even for autonomous driving. Some nice results for assistance in recognizing humans crossing in front of the vehicle (walking or biking) have been achieved in the framework of the project "Invent" [Franke *et al.* 2005].

With respect to animals on the road, there are no additional principal difficulties for perception except the perhaps erratic motion behavior some of these animals may show. Birds can both move on the ground and lift off for flying; in the transition period there are considerable changes in their appearance. Both their shapes and the motion characteristics of their limbs and wings will change to a large extent.

4.3.3 Rule Systems for Decision-Making

Perception systems for driver assistance or for autonomous vehicle guidance will need very similar sets of rules for the perception part (maybe specialized to some task of special interest). Once sufficient computing power for visual scene analysis and understanding is affordable, the information anyway in the image streams can be fully exploited, since both kinds of application will gain from deeper understanding of motion processes observed. This tends to favor three separate rule bases in a modular system: The first one for perception (control of gaze direction and attention) has to be available for both types of systems. In addition, there have to be two different sets, one for assistance systems and one for autonomous driving (locomotion, see Chapters 13 and 14).

Since knowledge components for these task domains may differ widely, they will probably be developed by different communities. For driver assistance systems, the human-machine-interface with many psychological aspects poses a host of challenges and interface parameters. Especially, if the driver is in charge of all safety aspects for liability reasons, the choice of interface (audio, visual, or tactile) and the ways of implementing the warnings are crucial. Quite a bit of effort is going into these questions in industry at present (see the proceedings of the yearly *International Symposium on Intelligent Vehicles* [Masaki 1992–1999]). Tactile inputs may even include motion control of the whole vehicle. Horseback riders develop a fine feeling for slight reactions of the animal to its own perceptions. The question is whether similar types of special motion are useful for the vehicle to direct attention of the driver to some event the vehicle has noticed. Introducing vibrations at the proper side of the driver seat when the vehicle approaches one of the two lane markers left or right too closely is a first step done in this direction [Citroen 2004]. First correcting reactions in the safe direction or slight resistance to maneuvers intended may be further steps; because of the varying reactions from the population of drivers, finding the proper parameters is a delicate challenge.

For autonomous driving, the relatively simple task is to find the solution when to use which maneuvers or/and feedback algorithms with which set of optimal parameters. Monitoring the process initiated is mandatory for checking actual performance achieved in contrast to the nominal one expected. Statistics should be kept on the behavior observed, for learning reasons.

In case some unexpected "event" occurs (like a vehicle changing into your lane immediately in front of you without giving signs), this situation has to be handled by a transition in behavior; reducing throttle setting or hitting the brakes has to be the solution in the example given. These types of transitions in behavior are coded in extended state charts [Harel 1987; Maurer 2000]; actual implementation and results will be discussed in later chapters. The development of these algorithms and their tuning, taking delay times of the hardware involved into account is a challenging engineering task requiring quite a bit of effort.

Note that in the solution chosen here, the rule base for decision–making does not contain the control output for the maneuvers but only the conditions, when to switch from one maneuver or driving state to another one. Control implementation is done at a lower level with processors closer to the actuators (see Section 3.4.4).

4.4 List of Mission Elements

Planning entire missions is usually done before the start of the mission. During this process, the mission is broken down into mission elements which can be performed with the same set of behavioral modes. The list of mission elements is the task description for the process governing situation assessment and behavior decision. It also calls for implementation of behavioral capabilities actually available. In case some nominal behavioral capability is actually not available because of some hardware failure, this fact is detected by this process (by polling corresponding bits in the hardware monitoring system), and mission replanning has to take this new situation into account.

The duration of mission elements may be given by local timescales or by some outside event; for example, lane following should be done until a certain geographical position has been reached at which a turnoff at an intersection has to be taken. This is independent of the time it took to get there.

During these mission elements defined by nominal "strategic" aspects of mission performance, tactical deviations from the nominal plan are allowed such as lane changing for passing slower traffic or convoy driving at speeds lower than planned for the mission element. To compensate for the corresponding time losses, the vehicle may increase travel speed for some period after passing the convoy (parameter adjustment) [Gregor 2002; Hock 1994].

In the region of transition between two mission elements, the perception system may be alerted to detect and localize the relative position so that a transient maneuver can be started in time, taking time delays for implementation into account. A typical example is when to start the steer rate maneuver for a turnoff onto a crossroad. Sections 14.6.5 and 14.6.6 will discuss this maneuver as one important element of mission performance as implemented on the test vehicle VaMoRs. Figure 14.15 shows the graphical visualization of the overall mission. The corresponding list of mission elements (coarse resolution) is as follows:

1. Perform roadrunning from start till *GPS* signals the approach of a crossroad onto which a turnoff to the left shall be made.
2. While approaching the crossroad, determine by active vision the precise width and orientation of the cross road as well as the distance to the intersection.
3. Perform the turnoff to the left.
4. At a given GPS-waypoint on the road, leave the road at a right angle to the right for cross-country driving.
5. Drive toward a sequence of landmarks (*GPS*-based); while driving, detect and perceive negative obstacles (ditches) visually and avoid them through bypassing on the most convenient side. [There is no one, actually in this part of the mission.]
6. Visually detect and recognize a road being approached during cross-country driving (point 6 in Figure 14.5). Estimate intersection angle, road width, and distance to the road (continually while driving).
7. Turn onto road to the left from cross-country driving; adjust speed to surface inclination encountered.
8. Perform roadrunning, recognizing crossroad as landmarks (points 6 to 8).

9. Cross both intersections (T-junction left and right).
10. Leave road to the left at GPS-waypoint (9) for cross-country driving.
11. While driving toward a sequence of landmarks (*GPS*-based), detect and perceive negative obstacles (ditches) visually and avoid them through bypassing on the most convenient side. [In the case of Figure 14.15 there is one ditch on trajectory-arc 10; it should be avoided by evading to the right. If a ditch is encountered, there is a new list of tasks to be performed in a similar manner for this complex evasive maneuver consisting of several perceptual tasks and sequences of motion control.]
12. Finish mission at GPS-landmark X (in front of a road).

In driver assistance systems, similar mission elements exist, such as roadrunning (with "lane departure warning"), convoy driving (with "adaptive cruise control"), or "stop-and-go" traffic. The assistance functions can be switched on or off separately by the human operator. A survey on this technology including the human-machine-interface (HMI) may be found in [Maurer, Stiller 2005].

5 Extraction of Visual Features

In Chapters 2 and 3, several essential relations among features appearing in images and objects in the real world have been discussed. In addition, basic properties of members of the classes "objects" and "subjects" have been touched upon to enable efficient recognition from image sequences. Not only spatial shape but also motion capabilities have been described as background for understanding image sequences of high frequency (video rate). This complex task can be broken down into three consecutive stages (levels), each requiring specialized knowledge with some overlap. Since the data streams required for analysis are quite different in these stages, namely (1) whole images, (2) image regions, and (3) symbolic descriptions, they should be organized in specific data bases.

The first stage is to discover the following items in the entire fields of view (images): (a) what are characteristic image parameters of influence for interpreting the image stream, and (b) where are regions of special interest in the images?

The answer to question (a) has to be determined to tap background knowledge which allows deeper understanding of the answers found under (b). Typical questions to be answered by the results to complex (a) are (1) What are the lowest and the highest image intensities found in each image? It is not so much the value of a single pixel of interest here [which might be an outlier (data error)] but of small local groups of pixels, which can be trusted more. (2) What are the lowest and highest intensity gradients (again evaluated by *receptive fields* containing several pixels)? (3) Are these values drastically different in different parts of the images? Here, an indication of special image regions such as 'above and below the horizon', or 'near a light source or further away from it' may be of importance. (4) Are there large regions with approximately homogeneous color or texture distribution (representing areas in the world with specific vegetation or snow cover, *etc.*)? At what distance are they perceived?

Usually, the answer to (b) will show up in collections of certain features. Which features are good indicators for objects of interest is, of course, domain specific. Therefore, the knowledge base for this stage 1 concentrates on types and classes of image features for certain task domains and environmental conditions; this will be treated in Section 5.1.

At this level, only feature data are to be computed as background material for the higher levels, which try to associate environmental aspects with these data sets by also referring to the mission performed and to knowledge about the environment, taking time of day and year into account.

In the second stage, the question is asked 'What type of object is it, generating the feature set detected', and 'what is its relative state at the present time'? Of course, this can be answered for only one object/subject at one time by a single interpretation process. So, the amount of image data to be touched is reduced drastically, while background knowledge on the object/subject class allows asking specific questions in image interpretation with correspondingly tuned feature extractors. Continuity conditions over time play an important role in state estimation from image data. In a complex scene, many of these questions have to be answered in parallel during each video cycle. This can be achieved by time slicing attention of a single processor/software combination or by operating with several or many processors (maybe with special software packages) in parallel. Increasing computing power available per processor will shift the solution to the former layout. Over the last decade, the number of processors in the vision systems of UniBwM has been reduced by almost an order of magnitude (from 46 to 6) while at the same time the performance level increased considerably. The knowledge bases for recognizing single objects/subjects and their motion over time will be treated in Chapters 6 and 12.

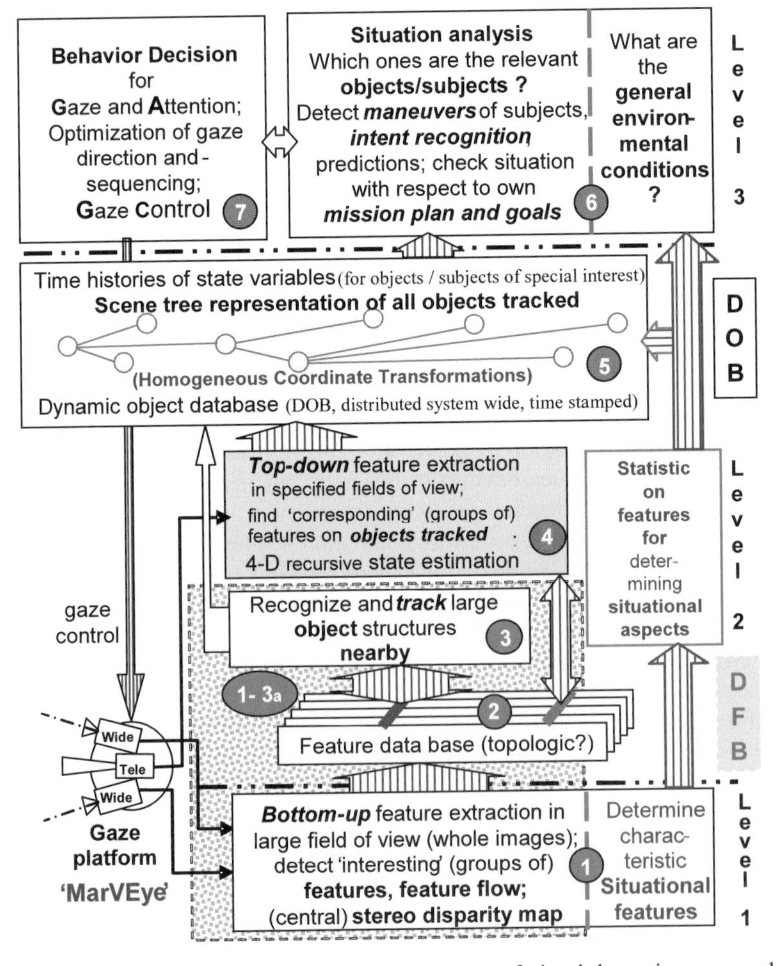

Figure 5.1. Structured knowledge base for three stages of visual dynamic scene understanding in expectation-based, multi-focal, saccadic (EMS) vision"

The results of all of these single-object–recognition processes have to be presented to the situation assessment level in unified form so that relative motion between objects and movements of subjects can be appreciated on a larger spatial and temporal scale. The dynamic object database (DOB) solves this task. On the situation level, working on huge volumes of image data is no longer possible. Therefore, the DOB also serves the purpose of presenting the scene recognized in an object-oriented, symbolic way. Figure 5.1 shows the three levels for image sequence processing and understanding. The results of the right-hand branch from level 1 are fed into this scheme to provide background information on the lighting and other environmental conditions.

The situation to be assessed on the decision level has to include all of this and the trajectory planned for the subject body in the near future. Both safety aspects and mission goals have to be taken into account here; a selection has to be made between more and less relevant objects/subjects by judging hazard potentials from their trajectories/behaviors. This challenge will be discussed in Chapter 13. Figure 5.1 visualizing the stages mentioned for visual dynamic scene interpretation will be discussed in more detail after the foundations for feature extraction and object/subject recognition have been laid down.

5.1 Visual Features

The discussion of the topic of feature extraction will be done here in an exemplary fashion only for road scenes. Other domains may require different feature sets; however, edge and corner features are very robust types of features under a wide range of varying lighting and aspect conditions in many domains. Additional feature sets are gray value or color blobs, certain intensity or color patterns, and textures. The latter cover a wide range; they are very computer-intensive, in general.

In biological vertebrate vision, edge features of different size and under different orientations are one of the first stages of visual processing (in V1 [Hubel and Wiesel 1962]). There are many algorithms available for extracting these features (see [Duda, Hart 1973; Ballard, Brown 1982; Canny 1983; http://iris.usc.edu/Vision-Notes/bibliography/contents.html]. A very efficient algorithm especially suited for road-scene analysis has been developed by Kuhnert (1988) and Mysliwetz (1990). Search directions or patterns are also important for efficient feature extraction. A version of this well-proven algorithm, the workhorse of the 4-D approach over two decades, will be discussed in detail in Section 5.2. Computing power in the 1980s did not allow more computer-intensive features for real-time applications at that time. Now that four orders of magnitude in computing power per microprocessor have been gained and are readily available, a more general feature extraction method dubbed "UBM", the basic layout of which has been developed by Hofmann (2004) and the author, will be discussed in Section 5.3. It unifies the extraction of the following features in a single pass: Nonplanar regions of the image intensity function, linearly shaded blobs, edges of any orientation, and corners.

5.1.1 Introduction to Feature Extraction

The amount of data collected by an imaging sensor is the same when looking at a uniformly gray region or at a visually complex colored scene. However, the amount of information perceived by an intelligent observer is considerably different. A human would characterize the former case exhaustively by using just three words: "uniformly gray", and possibly a term specifying the gray tone (intensity). The statement "uniformly" may be the result of rather involved low-level parallel computations; but this high-level representational symbol in combination with the intensity value contains all the information in the image. In contrast, if several homogeneously colored or textured subregions are being viewed, the borderlines between these regions and the specification of the color value per region contain all the information about the scene (see Figure 5.2).

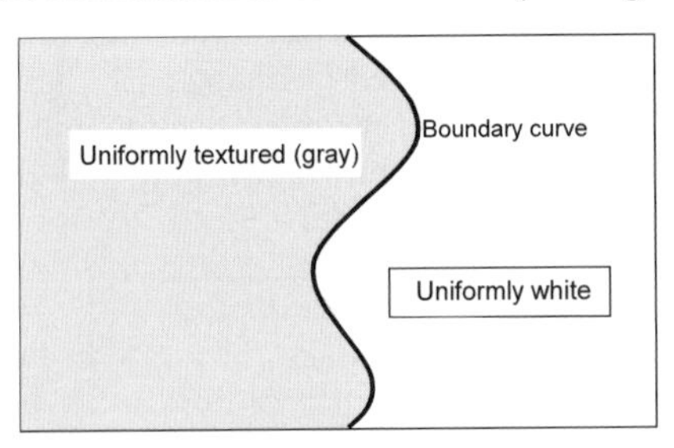

Figure 5.2. Two homogeneous regions; most information is in the boundary curve

Instead of having to deal with all the color values of all pixels, this number of data may be considerably reduced by just listing the coordinates of the boundary elements; depending on the size of the regions, this may be orders of magnitude less data for the same amount of information. This is the reason that sketches of boundary lines are so useful and widely spread.

Very often in images of the real world, line elements change direction smoothly over arc-length, except at discrete points called "corners". The direction change per unit arc-length is termed curvature and is the basis for differential geometry [Spivak 1970]. The differential formulation of shapes is coordinate-free and does not depend on the position and angular orientation of the object described. The same 2-D shape on different scales can be described in curvature terms by the same function over arc length and one scaling factor. Measurement of the tangent direction to a region, therefore, is a basic operation for efficient image processing. For measuring tangent directions precisely at a given scale, a sufficiently large environment of the tangent point has to be taken into account to be precise as a function of scale level and to avoid "spurious details" [Florack *et al.* 1992]. Direction coding over arc length is a common means for shape description [Freeman 1974; Marshall 1989].

Curvature coding over arc length is less widely spread. In [Dickmanns 1985], an approximate, general, efficient, coordinate-free 2-D shape description scheme in differential-geometry terms has been given, based on local tangent direction measurements relative to the chord line linking two consecutive boundary points with limited changes in tangent direction (< 0.2 radians). It is equivalent to piecewise third-order Hermite polynomial approximations based on boundary points and their tangent directions.

However, sticking to the image plane for shape description of 3-D bodies in the real world may not be the best procedure; rigid 3-D bodies and curves yield an infinite number of 2-D views by perspective mapping (at least theoretically), depending on the aspect conditions. The shape invariance in this case can be captured only by using 3-D shape models and models for mapping by central projection. This approach is much better suited for visually recognizing the environment during ego–motion and for tracking other (massive) objects over time, than for single snapshot interpretation. This is true since massive bodies move smoothly over time, and invariance properties with respect to time, such as eigen–frequencies, damping, and stereotypic motion characteristics (like style of walking), may be exploited as knowledge about specific objects/subjects in the real world. Therefore, embedding the image analysis task in a temporal continuum and exploiting known motion characteristics in an object-oriented way will alleviate the image sequence interpretation task (extended idea of *gestalt*). It requires, however, that the internal representation be in four dimensions right from the beginning: in 3-D space and time for single objects. This is the essence of the 4-D approach to dynamic machine vision developed in the early 1980s [Meissner, Dickmanns 1983; Dickmanns 1987; Wünsche 1987].

By embedding (a) simple feature extraction with linear edge elements, (b) regions with linear shading models, and (c) horizontal and vertical image stripes into this framework of spatio–temporal object orientation, these methods gain considerably in power and useful range of application. By exploiting a knowledge base on dynamic motion derived from previous experience in observing motion processes in 3-D space of specific 3-D objects carrying highly visible features on their surface, scene understanding is considerably alleviated. Specific groups of linearly extended edge feature sets and adjacent homogeneous areas of gray, color (or in the future texture) values are interpreted as originating from these spatial objects under specific aspect conditions. This background may also be one of the reasons, beside their robustness to changing lighting conditions, that in highly developed biological vision systems (like the mammalian ones) edge element operators abound [Hubel, Wiesel 1962; Koenderink, van Doorn 1990]. Without 3-D invariance and without knowledge about motion processes and about perspective projection (implicit or explicit), the situation would be quite different with respect to the usefulness of these operators.

The edge-based approach has an advantage over region-based approaches if invariance under varying lighting conditions is considered. Even though the intensities and color values may change differently with time in adjacent image regions, the position of the boundaries between them does not, and the edge will remain visible as the locus of highest intensity or color gradient. In natural environments, changing lighting conditions are more the rule than an exception. Therefore, Section 5.2 will be devoted to edge-based methods.

However, for robust interpretation of complex images, region-based image evaluation is advantageous. Since today's processors do not allow full scale area-based processing of images in real time, a compromise has to be sought. Some aspects of region-based image evaluation may be exploited by confining the regional operations to the vicinity of edges. This is done in conjunction with the edge-based approach, and it can be of help in establishing feature correspondence for object recognition using a knowledge base and in detecting occlusions by other objects.

A second step toward including area-based information in the 4-D scheme under the constraint of limited computing power is to confine the evaluation areas to stripes, whose orientation and width have to be chosen intelligently ad hoc, exploiting known continuity conditions in space and time. These stripes will be condensed to one-dimensional (averaged over the width of the stripes) representation vectors by choosing proper schemes of symbolic descriptions for groups of pixel values in the stripe direction.

Coming back to the difference between data and information, in knowledge-based systems, there may be a large amount of information in relatively few data, if these data allow a unique retrieval access to a knowledge base containing information on the object recognized. This, in turn, may allow much more efficient recognition and visual tracking of objects by attention focusing over time and in image regions of special interest (window concept).

According to these considerations, the rest of Chapter 5 will be organized as follows: In Section 5.1.2, proper scaling of fields of view in multi-focal vision and in selecting scales for templates is discussed. Section 5.2 deals with an efficient basic edge feature extraction operator optimized for real-time image sequence understanding. In Sections 5.3, (2-D) region-based image evaluation is approached as a sequence of one-dimensional image stripe evaluations with transition to symbolic representations for alleviating data fusion between neighboring stripes and for image interpretation. An efficient method with some characteristics of both previous approaches is explained in Section 5.4; it represents a trade-off between accuracy achievable in perception and computational expense.

Contrasting the feature extraction methods oriented towards single-object recognition, Section 5.5 gives an outlook on methods and characteristic descriptors for recognizing general outdoor environments and situations. Computing power available in the past has not allowed applying this in real-time onboard vehicles; the next decade should allow tackling this task for better and more robust scene understanding.

5.1.2 Fields of View, Multi-focal Vision, and Scales

In dealing with real-world tasks of surveillance and motion control very often coverage of the environment with a large field of view is needed only nearby. For a vehicle driving at finite speed, only objects within a small distal range will be of interest for collision avoidance. When one is traveling at high speed, other low-speed objects become of interest for collision avoidance only in a rather small angular region around the subject's velocity vector. [For several high-speed vehicles interacting in the same space, special rules have to be established to handle the situations, such as in air traffic control, where different altitudes are assigned to airplanes depending on their heading angle (in discrete form by quadrants).]

Central projection is the basic physical process of imaging; depending on the distal range of the object mapped into the image, this results in one pixel representing areas of different size on objects in the real world. Requiring a certain resolution normal to the optical axis for objects in the real world, therefore, requires range-dependent focal lengths for the imaging system.

Biological systems have mastered this problem by providing different pixel and receptive field sizes in the sensor hardware *eye*. In the foveal area designed for long-range, high-resolution viewing, the density of sensor elements is very high; in the peripheral areas added for large angular viewing range, element density is small. By this combination, a large viewing range can be combined with high resolution at least in some area with relatively moderate overall data rates. The area of high resolution can be shifted by active viewing direction (gaze) control by both the eye and the head.

In technical systems, since inexpensive sensors are available only with homogeneous pixel distributions, an equivalent mapping system is achieved by mounting two or mores cameras with lenses of different focal lengths fixed relative to each other on a platform. The advantage of a suite of sensors covering the same part of the scenery is that this part is immediately available to the system in a multi-scale data set. If the ratio of the focal lengths is four, the image produced by the shorter focal length represents (coarse) information on the second pyramid level of the image taken with the higher resolution (larger focal length). This dual scale factor may sometimes be advantageous in real-time vision where time delays are critical. On one hand, efficient handling of objects requires that a sufficiently large number of pixels be available on each object for recognition and identification; on the other hand, if there are too many pixels on a single object, image processing becomes too involved and slow.

As mentioned in the introduction, in complex scenes with many objects or with some objects with a complex pattern of sub-objects, relying solely on edge features may lead to difficulties and ambiguities. Combining the interpretation of edge features with area-based features (average intensity, color, or texture) often allows easy disambiguation. Figure 5.3 shows a case of efficient real-time image sequence processing. Large homogeneous areas can be tracked by both edge features and region-based features. In the near range, the boundaries between the regions are not sharp but fuzzy (strongly perturbed, unsealed country road with grass spreading onto the road). For initialization from a normal driving situation, searching edges with large receptive fields in most likely areas is very efficient.

The area-based method covering the entire image width would improve robustness to road parameters other than expected, but would also be costly because of

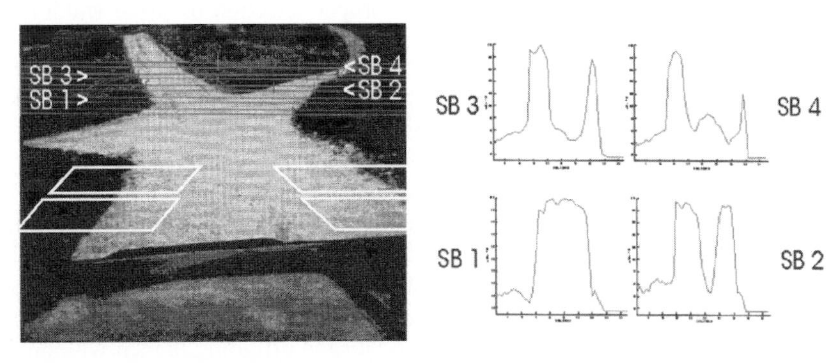

Figure 5.3. Combining edge and area-based features for robust object detection and recognition. Near range: Only edge detection in regions and with parameters selected according to a normal driving situation. Far range: Four stripes covering the entire width of the image; determine steep edges and intensity plateaus (lower part) to discover road forks.

the many pixels to be touched. Working with large receptive fields has proven to be reasonably reliable and efficient here. However, if computing power will allow color and texture processing with new area-based features, a new quality of recognition and higher robustness will result. Therefore, a compromise has been found that allows using some of the advantages of area-based features efficiently in connection with the 4-D approach.

In road vehicle guidance where the viewing direction is essentially parallel to the ground, this method offers some advantages. Due to the scaling effect of range (distance x) in perspective mapping, features further away will be reduced in size; this may cause trouble in the interpretation process for a stereotypical application of pyramid-methods over larger image regions. In the upper rows, each pixel covers a much larger distance in range than in the lower ones.

Figure 5.4 with the inserted table shows the effect of distance in a vertical stripe of an image, scaled by the camera elevation *H* above the ground; the same stripe width in the real world on the ground shows up in a decreasing number of rows with distance.

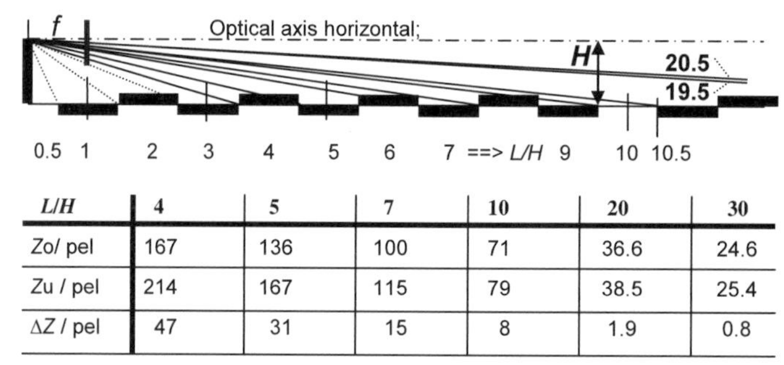

L/H	4	5	7	10	20	30
Zo/ pel	167	136	100	71	36.6	24.6
Zu / pel	214	167	115	79	38.5	25.4
ΔZ / pel	47	31	15	8	1.9	0.8

Figure 5.4. Mapping of a horizontal slice at distance *L/H* (from Zu = (*L/H* – 0.5) to *Zo* = (*L/H* + 0.5) into the image plane (focal length f = 750 pixel)

Confining regional representations to image slices or stripes at almost constant distance, these problems may be reduced by proper selection of stripe width (see Figure 5.3, upper part). Due to unknown road curvature, the road may appear anywhere in the image, and it may have a forking point somewhere. Therefore, the horizontal stripes SB1 to SB4 in Figure 5.3 are selected as a bunch of regions extending over the entire image width. The resulting image intensity distributions are shown in the lower part. In SB1, the road fork does not yet show up. SB2 has a small dark section between two brighter ones (with almost the same total width in the image between the outer edges, even though further away), indicating that the road may have branched. This is confirmed in SB3 with a widened dark area in between. The value of stripe SB4 is doubtful in this case since the branched-off road fills only a few pixels; with the hypothesis of a road fork from SB2, 3, it would be more meaningful to search in a separate stripe for the off-going branch with properly adapted parameters in the next image, if possible with higher image resolution.

In other cases, the stripes need not cover the entire image width right from the beginning but may be confined to some meaningful fraction depending on object size expected (domain knowledge) or known from previous images, thereby reducing the computational load.

It can be seen from Figure 5.4 that the same width at scaled distance 5 (depth in viewing direction, 31 pixels wide) is reduced to about 2 pixels at a scaled distance of 20. Clearly, features in the real world will change in appearance accordingly! In the stripe-based method to be discussed in Section 5.3, these facts may be taken into account by proper parameter specification depending on the actual situation encountered. At longer distances, often the center of homogeneous regions in a stripe can be more reliably detected than the exact position of edges belonging to the same region. For example, crossroad detection as shown in Figure 5.5 is much more robust based on area-based features than based on edge features since these may abound in the region along the road. Checking aggregations of homogeneous areas in road direction, as needed in the next step for hypothesis generation, is much more efficient than checking edge aggregations with their combinatorial explosion.

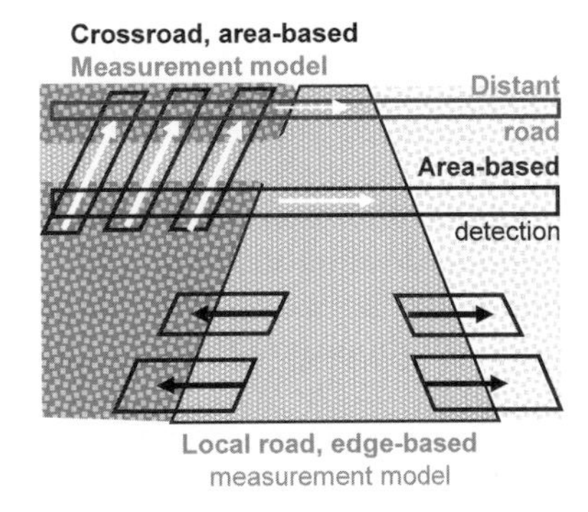

Figure 5.5. Crossroad detection in look-ahead region further away.

5.2 Efficient Extraction of Oriented Edge Features

In general, multiple scales are advantageous in image processing to optimally exploit information in noise corrupted images [Florack *et al.* 1992]. To avoid spurious details, the finest scale should be much larger than pixel size; a mask size of three or four pixels is considered a practical lower bound for the extraction of linearly extended features. The upper scale limit is given by image size, since during the search process, no image boundary should be hit; a maximum mask size of one-half or one-third of the image size seems to be a meaningful upper limit. Spacing mask sizes by a factor of 2 then results in seven or eight mask sizes for 1K×1K pixel images.

An alternative approach to working with large mask sizes is to compute pyramid images [Burt *et al.* 1981] by averaging four neighboring pixels on the ith level to one pixel at the $(i+1)$th pyramid level and then applying a smaller mask size to the higher level image. Note however, that these two operations are not exactly the same, since in the latter case resolution in the image plane has been lost. Therefore, to preserve high resolution at the small scale, mask sizes up to 17 pixels in width have been generated and implemented. Masks of larger size have not been necessary in the problem areas treated up to now. If they are needed, either four or five pyramid levels have to be generated or, for the sake of saving computing time, simple sub-sampling may be done, substituting one intensity value of level i for the mean of four neighboring ones on level $i-1$ in the exact pyramid. Thus, about half the scale range is covered by different mask sizes while the other half is handled on larger scales by image size reduction.

Literature on edge feature extraction abounds (see, *e.g.*, World Wide Web: *http://iris.usc.edu/Vision-Notes/bibliography/contents.html*; detection and analysis of edges and lines (Chapter 5 there); 2-D feature analysis, extraction, and representations (Chapter 7); Chapter 3 there gives a survey on books (3.2), collections, overviews, and surveys).

5.2.1 Generic Types of Edge Extraction Templates

A substantial reason for the efficiency of the methods developed at UniBwM for edge feature extraction stems from the fact that both low-pass filtering in one direction and accurate search in an almost normal direction were combined. By proper partitioning of the overall task, simple but efficient pixel processing in just one dimension has been achieved consecutively. It is assumed that the direction of the edge to be found is known approximately. While this is true for tracking, it turned out that the algorithms are also useful for initialization if proper parameter settings are chosen. The software packages developed for edge extraction are generalizations of the Prewitt operator [Ballard, Brown 1982]; they have matured over three generations of coding in different computer languages. The original ideas of Kuhnert (1988) have been refined and coded first in the second half of the 1980s in the language FORTRAN by Mysliwetz (1990); the current version was developed in the early 1990s under the name KRONOS D.Dickmanns (1997) in the language *Occam* for transputers. For the next generation of processors, it has been converted to *C* under the name CRONOS and polished by S. Fuerst.

5.2.1.1 Low-pass Filtering of an Oriented Pixel Field into a Vector

To obtain good correlation values for a local edge extraction operator, its orientation should be almost tangential to the edge in the image. Figure 5.6 shows a trapezoidal dark area in front of a pixel grid, the edges of which are to be detected. The mask for edge detection shown to the left in the pixel grid has approximately the same inclination as the left edge of the area to be detected; the mask is $n_w = 17$ pixels wide and $m_d = 5$ pixels deep. To be independent of absolute light intensity, the number of plus signs and minus signs in the mask has to be equal; in the case shown, there are two pixel formations along the edge (called "mask elements" henceforth, one pixel wide) with minus and two with plus signs. Separating these two blocks is a mask element with zeros which reduces sensitivity to slightly deviating edge direction

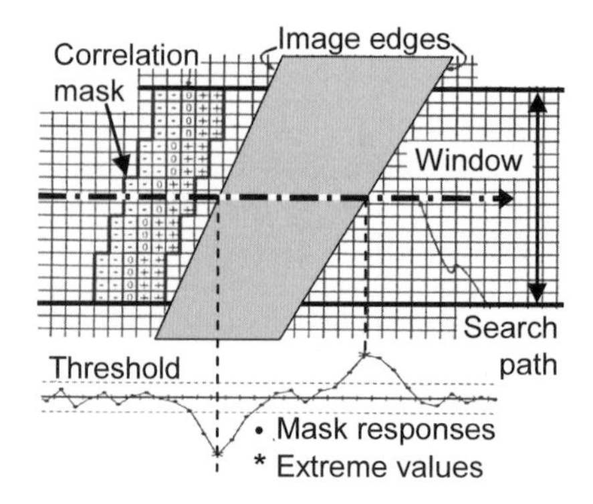

Figure 5.6. Edge localization by shifting a *ternary* correlation mask

from the mask direction or to slightly curving edges; image pixels under the zero-mask elements need not be touched, of course, since their weight in the mask vanishes.

For efficient computation of correlations, search direction should be either horizontal (in the y-direction, rows, *dash-dotted arrow* in the figure) or vertical (in the z-direction, columns); diagonal searches have also been used, initially, but because of the square-root-of-2 effect in spacing, when proceeding in diagonal search direction, they have not gained acceptance. The wider the masks chosen, the more angular orientations may be specified. Figure 5.7 shows the basic pixel alignments for mask elements of different mask widths n_w (3, 5, 9, 17) and the orientation angles achievable.

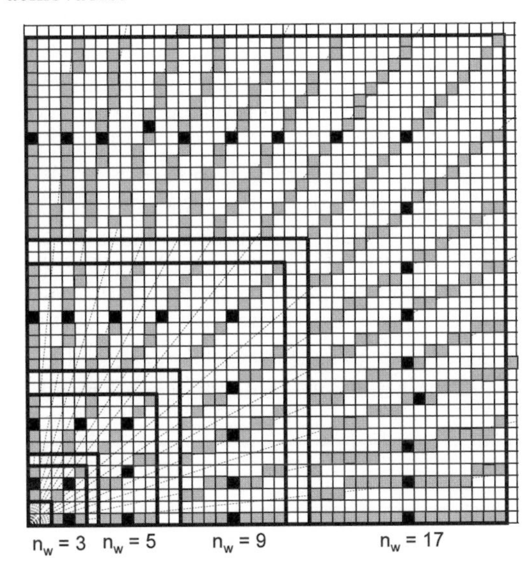

Figure 5.7. Basic edge directions as used in the edge extractor "CRONOS": four mask widths of $n_w = 3, 5, 9, 17$ pixels ($n_w = 2^i + 1, i = 1, \ldots 4$)

These numbers also give the angular resolution achievable per quadrant (45° for $n_w = 3$ down to 5.6° for $n_w = 17$). Mirroring at the horizontal and/or vertical boundaries yields all directions. In a first step of the algorithm, all pixels for one mask element (directions shown gray in Figure 5.7) in the entire search range are summed up and stored at the center position (black pixel in the figure). Mask correlation then only has to deal with a vector instead of a 2-D pixel array, independent of mask width n_w. This corresponds to low-pass filtering in the edge direction. Due to the discrete pixel size, the spacing between the edge element orientations is not exactly the same; this is indicated in Figure 5.7 for an extreme case with $n_w = 17$ by shifting the black pixel by one unit. Since the results are used in recursive estimation with a high sampling rate (25 Hz) this has not been detrimental; it is one minor component in measurement noise.

The zero-direction is defined as horizontal to the right; clockwise counting is applied. The set of edge directions shown in Figure 5.8 is generated from the basic set for horizontal search with edge directions between 270 and 315° (above the diagonal); the set for the angles from 225 to 270° is obtained by mirroring at the vertical axis. The sets needed for vertical search (315 to 360° and 180 to 225°) are obtained by mirroring those given at the 315° respectively, at the 225°-line. Mirroring all of these at the horizontal line completes the full set from 0 to 360°. (However, in the way these elements will be used, only one half set is needed since the other is just an inversion of the sign, see below).

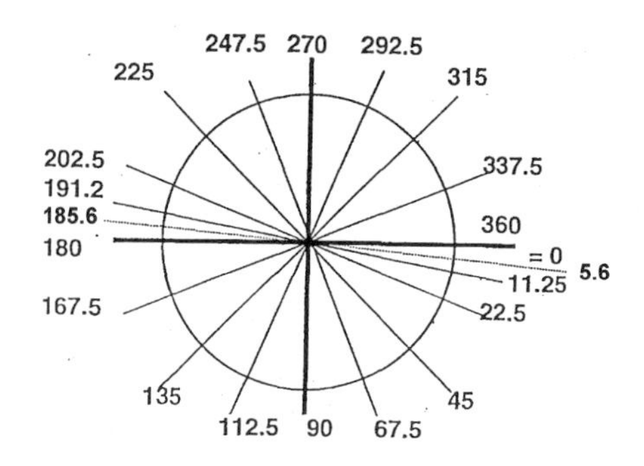

Figure 5.8. Definition of edge orientation as used in CRONOS: Starting from the horizontal direction to the right, angular increments are counted clockwise

After specification of search range, mask width, and orientation to be used, the first computational step is to sum up all the pixel values over the mask width n_w and, thus, collapse the width to a single vector component. This vector spans over the search range (named PathLen in CRONOS, see Figure 5.9).

It represents the average intensity values in the direction of mask orientation; this corresponds to low-pass filtering in this mask direction. With more than 16-bit processors and 8-bit intensity values for each pixel, there is no need to divide by the number of pixels summed, thus saving computing time. (If intensity values close to the original ones are preferred, shift operations may be used.) In the software packages in use at UniBwM, the options for mask widths are $n_w = 2^i + 1$, with preference for $i = 2$ to 4; this means that the smallest mask width with only three pixels is not used.

This odd value of n_w has been chosen initially to have a symmetrical distribution of the image stripe represented around the center of the nominal pixel position, which is convenient if no subpixel resolution is used. Using subpixel resolution, defining $n_w = 2^i$ is the cleaner solution for working on different scales.

It is seen from Figure 5.9 that part of the search path length is lost at the boundaries for oblique mask orientations; this has to be taken into account when specifying the search range.

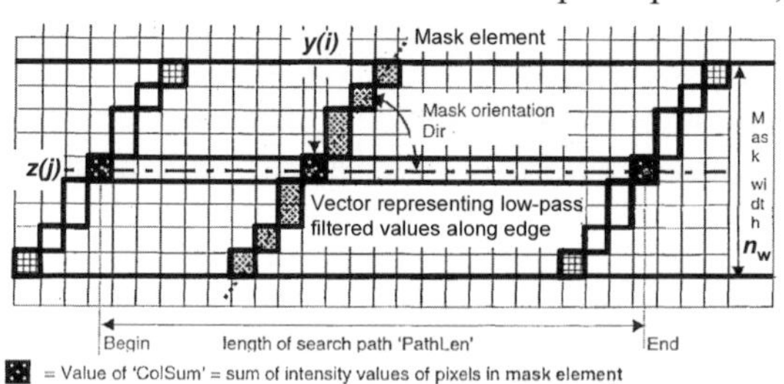

Figure 5.9. Low-pass (high spatial frequency) filtering orthogonal to the expected edge direction reduces the search stripe to a vector, independent of mask width n_w for efficient computation of correlation values

5.2.1.2 Computation of Ternary Correlation Values

The vector obtained in the previous section is the basis for edge localization by ternary correlation. By subtracting two consecutive vector components from each other, gradient information in the search direction for the given angular orientation of the mask is obtained (see Figure 5.10a, upper left). At the point where this difference is largest in amplitude, the gradient over two consecutive mask elements is maximal.

However, due to local perturbations, this need not correspond to an actual extreme gradient on the scale of interest. Experience with images from natural environments has shown that two additional parameters may considerably improve the results obtained:

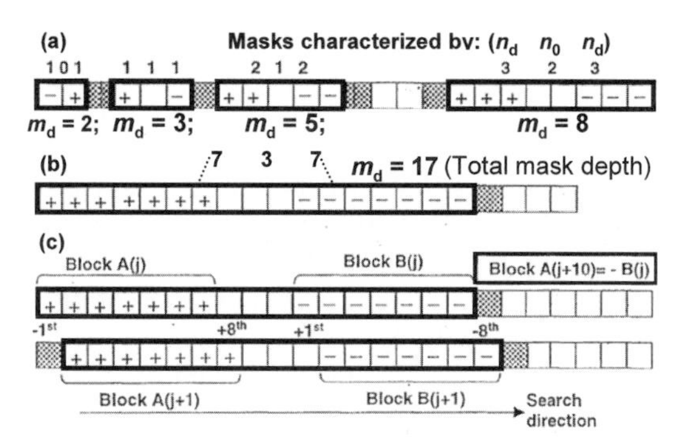

Figure 5.10. Efficient mask evaluation with the "Colsum"-vector; the n_d-values given are typical for sizes of "receptive fields" formed

1. By allowing a yet to be specified number n_0 of entries in the mask center to be dropped, the results achieved may be more robust. This can be immediately appreciated when taking into account that either the actual edge direction may deviate from the mask orientation used or the edge is not straight but curved; by setting central elements of the mask to zero, the extreme intensity gradient becomes more pronounced. The rest of Figure 5.10 shows typical mask parameters with $n_0 = 1$ for masks three and five pixels in depth ($m_d = 3$ or 5), and with $n_0 = 2$ for $m_d = 8$ as well as $n_0 = 3$ for $m_d = 17$ (rows b, c).
2. Local perturbations are suppressed by assigning to the mask a significant depth n_d, which designates the number of pixels along the search path in each row or column in each positive and negative field. The total mask depth m_d then is $m_d = 2\, n_d + n_0$. Figure 5.10 shows the corresponding mask schemes. In line (b) a rather large mask for finding the transition between relatively large homogeneous areas with ragged boundaries is given ($m_d = 17$ pixels wide and each field with seven elements, so that the correlation value is formed from large averages; for a mask width n_w of 17 pixels, the correlation value is formed from $7 \cdot 17 = 119$ pixels). With the number of zero-values in between chosen as $n_0 = 3$, the total *receptive field* (= mask) size is $17 \cdot 17 = 289$ pixels. The sum formed from n_d mask elements (vector values "ColSum") divided by ($n_w \cdot n_d$) represents the average intensity value in the oblique image region adjacent to the edge. At the maximum correlation value found, this is the average gray value on one side of the edge. This information may be used for recognizing a specific edge feature in consecutive images or for grouping edges in a scene context.

For larger mask depths, it is more efficient when shifting the mask along the search direction, to subtract the last mask element (ColSum-value) from the summed field intensities and add the next one at the front in the search direction, see line (c) in Figure 5.10); the number of operations needed is much lower than for summing all ColSum elements anew in each field.

The optimal value of these additional mask parameters n_d and n_0 as well as the mask width n_w depend on the scene at hand and are considered knowledge gained by experience in visually similar environments. From these considerations, generic edge extraction mask sets for specific problems have resulted. In Figure 5.11, some representative receptive fields for different tasks are given. The mask parameters can be changed from one video frame to the next, allowing easy adaptation to changing scenes observed continuously, like driving on a curved road.

The large mask in the center top of Figure 5.11 may be used on dirt roads in the near region with ragged transitions from road to shoulder. For sharp, pronounced edges like well-kept lane markings, a receptive field like that in the upper right corner (probably with $n_d = 2$, that is, $m_d = 5$) will be most efficient. The further one looks ahead, the more the mask width n_w should be reduced (9 or 5 pixels); part (c) in the lower center shows a typical mask for edges on the right-hand side of a straight road further away (smaller and oblique to the right).

The 5 × 5 (2, 1, 2) mask at the left hand side of Figure 5.11 has been the standard mask for initial detection of other vehicles and obstacles on the road through horizontal edges; collections of horizontal edge elements are good indicators for objects torn by gravity to the road surface. Additional masks are then applied for checking object hypotheses formed.

If narrow lines like lane markings have to be detected, there is an optimal mask width depending on the width of the line in the image: If the mask depth n_d chosen is too large, the line will be low-pass-filtered and extreme gradients lose in magnitude; if mask depth is too small, sensitivity to noise increases.

As an optional step, while adding up pixel values for mask elements "ColSum" or while forming the receptive fields, the extreme intensity values of pixels in ColSum and of each ColSum vector component (max. and min.) may be determined. The former gives an indication of the validity of averaging (when the extreme values are not too far apart), while the latter may be used for automatically adjusting threshold parameters. In natural environments, in addition, this gives an indication

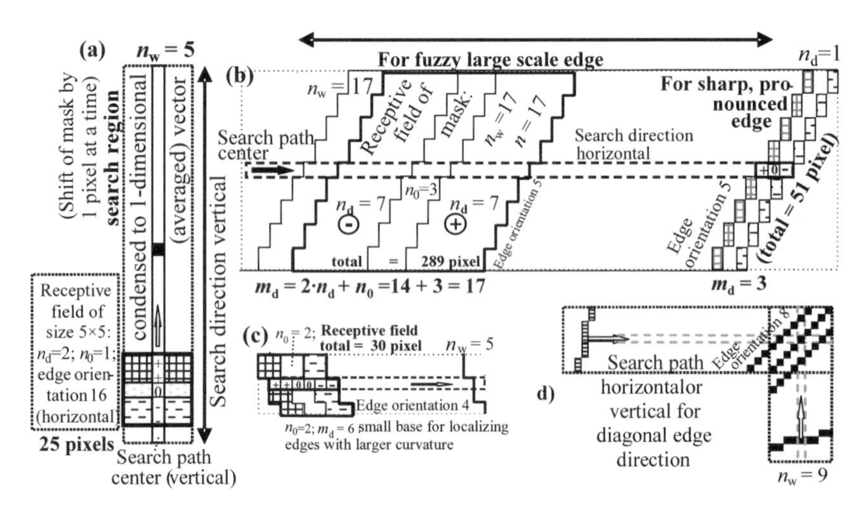

Figure 5.11. Examples of receptive fields and search paths for efficient edge feature extraction; mask parameters can be changed from one video-frame to the next, allowing easy adaptation to changing scenes observed continuously

of the contrasts in the scene. These are some of the general environmental parameters to be collected in parallel (right-hand part of Figure 5.1).

5.2.2 Search Paths and Subpixel Accuracy

The masks defined in the previous section are applied to rectangular search ranges to find all possible candidates for an edge in these ranges. The smaller these search ranges can be kept, the more efficient the overall algorithm is going to be. If the high-level interpretation via recursive estimation is stable and good information on the variances is available, the search region for specific features may be confined to the 3 σ region around the predicted value, which is not very large, usually (σ = standard variation). It does not make sense first to perform the image processing part in a large search region fixed in advance and afterwards sort out the features according to the variance criterion. In order not to destabilize the tracking process, prediction errors > 3 σ are considered outliers and are usually removed when they appear for the first time in a sequence.]

Figure 5.6 shows an example of edge localization with a ternary mask of size $n_w = 17$, $n_d = 2$, and $n_0 = 1$ (*i.e.*, mask depth $m_d = 5$). The mask response is close to zero when the region to which it is applied is close to homogeneously gray (irrespective of the gray value); this is an important design factor for abating sensitivity to light levels. It means that the plus– and minus regions have to be the same size.

The lower part of the figure shows the resulting correlation values (mask responses) which form the basis for determining edge location. If the image areas within each field of the mask are homogeneous, the response is maximal at the location of the edge. With different light levels, only the magnitude of the extreme value changes but not its location. Highly discernible extreme values are obtained also for neighboring mask orientations. The larger the parameter n_0, the less pronounced is the extreme value in the search direction, and the more tolerant it is to deviations in angle. These robustness aspects make the method well suited for natural outdoor scenes.

Search directions (horizontal or vertical) are automatically chosen depending on the feature orientation specified. The horizontal search direction is used for mask orientations between 45 to 135° as well as between 225 and 315°; vertical search is applied for mask directions between 135 to 225° and 315 to 45°. To avoid too frequent switching between search directions, a hysteresis (dead zone of about one direction–increment for the larger mask widths) is often used that means switching is actually performed (automatically) 6 to 11° beyond the diagonal lines, depending on the direction from which these are approached.

5.2.2.1 Subpixel Accuracy by Second-Order Interpolation

Experience with several interpolation schemes, taking up to two correlation values on each side of the extreme value into account, has shown that the simple second-order parabola interpolation is the most cost-effective and robust solution (Figure 5.12). Just the neighboring correlation values around a peak serve as a basis.

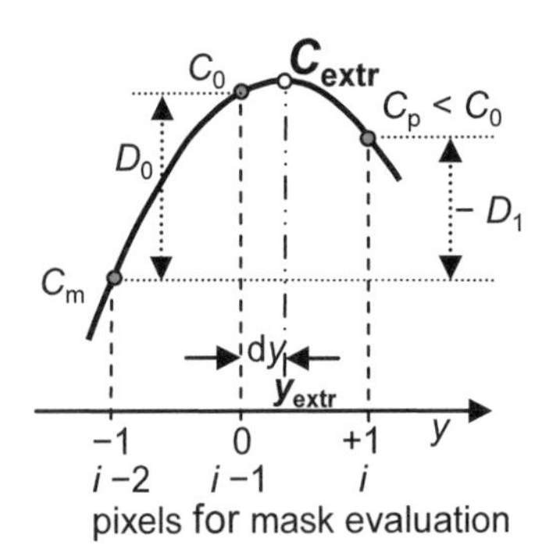

Figure 5.12. Subpixel edge localization by parabolic interpolation after passing a maximum in mask response

If an extreme value of the magnitude of the mask response above the threshold level (see Figure 5.6) has been found by stating that the new value is smaller than the old one, the last three values are used to find the interpolating parabola of second order. Its extreme value yields the position y_{extr} of the edge to subpixel accuracy and the corresponding magnitude C_{extr}; this position is obtained at the location where the derivative of the parabolic unction is zero. Designating the largest correlation value found as C_0 at pixel position 0, the previous one C_m at −1, and the last correlation value C_p at position +1 (which indicated that there is an extreme value by its magnitude $C_p < C_0$), the following differences

$$D_0 = C_0 - C_m; \qquad D_1 = C_m - C_p \tag{5.1}$$

yield the location of the extreme value at distance

$$dy = -0.5/(2 \cdot D_0 / D_1 + 1)$$

from pixel position 0, such that: $$y_{extr} = y_0 + dy \tag{5.2}$$

with the value $$C_{extr} = C_0 - 0.25 \cdot D_1 \cdot dy.$$

From the last expressions of Equation 5.1 and 5.2 it is seen that the interpolated value lies on the side of C_0 on which the neighboring correlation value measured is larger. Experience with real-world scenes has shown that subpixel accuracy in the range of 0.3 to 0.1 may be achieved.

5.2.2.2 Position and Direction of an Optimal Edge

Determining precise edge direction by applying, additionally, the two neighboring mask orientations in the same search path and performing a bi–variant interpolation has been investigated, but the results were rather disappointing. Precise edge direction can be determined more reliably by exploiting results from three neighboring search paths with the same mask direction (see Figure 5.13).

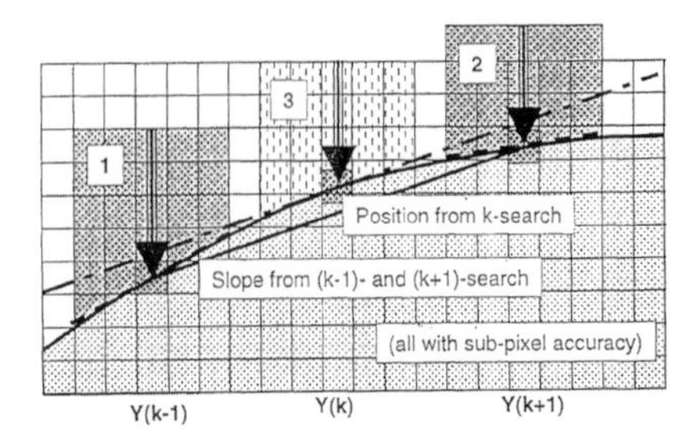

Figure 5.13. Determination of the tangent direction of a slightly curved edge by sub-pixel localization of edge points in three neighboring search paths and parabolic interpolation

The central edge position to subpixel accuracy yields the position of the tangent point, while the tangent direction is determined from the straight line connecting the positions of the (equidistant) neighboring edge points; this is the result of a parabolic interpolation for the three points.

Once it is known that the edge is curved – because the edge point at the center does not lie on the straight line connecting the neighboring edge points – the question arises whether the amount of curvature can also be determined with little effort (at least approximately). This is the case.

5.2.2.3 Approximate Determination of Edge Curvature

When applying a series of equidistant search stripes to an image region, the method of the previous section yields to each point on the edge also the corresponding edge direction that is its tangent. Two points and two slopes determine the coefficients of a third-order polynomial, dubbed *Hermite*-interpolation after a French mathematician. As a third-order curve, it can have at most one inflection point. Taking the connecting line (*dash-dotted* in Figure 5.14) between the two tangent points P_{-d} and P_{+d} as reference (chord line or *secant*), a simple linear relationship for a smooth curve with small angles ψ relative to the chord line can be derived. Tangent directions are used in differential-geometry terms, yielding a linear curvature model; the reference is the slope of the straight line connecting the tangent points (secant). Let m_{-d} and m_{+d} be the slopes of the tangents at points P_{-d} and P_{+d} respectively; s be the running variable in the direction of the arc (edge line); and ψ the angle between the local tangent and the chord direction ($|\psi| < 0.2$ radian so that $\cos(\psi) \approx 1$).

The linear curvature model in differential-geometry terms with s as running variable along the arc s from $x \approx -d$ to $x \approx +d$ is:

$$C = C_0 + C_1 \cdot s \,; \qquad d\psi = C \cdot ds \,. \tag{5.3}$$

Since curvature is a second-order concept with respect to Cartesian coordinates, lateral position y results from a second integral of the curvature model. With the origin at the center of the chord, x in the direction of the chord, y normal to it, and $\psi_{-d} = \arctan(m_{-d}) \approx m_{-d}$ as the angle between the tangent and chord directions at point P_{-d}, the equation describing the curved arc then is given by Equation 5.4 below [with ψ in the range $\pm$ 0.2 radian ($\sim$ 11°), the cosine can be approximated by 1 and the sine by the argument ψ]:

$$\begin{aligned} &x = -d + s; \quad \psi(s) = \psi_0 + \int_0^s C(\sigma)\, d\sigma = \psi_0 + C_0 \cdot s + C_1 \cdot s^2/2 \,; \\ &y(s) = y_0 + \int_0^s \sin[\psi(\sigma)]\, d\sigma = y_0 + \psi_0 \cdot s + C_0 \cdot s^2/2 + C_1 \cdot s^3/6. \end{aligned} \tag{5.4}$$

At the tangent points at the ends of the chord ($\pm\, d$), there is

$$\begin{aligned} &m_{-d} \approx \psi_{-d} = \psi_0 - C_0 \cdot d + C_1 \cdot d^2/2; && \text{(a)} \\ &m_{+d} \approx \psi_{+d} = \psi_0 + C_0 \cdot d + C_1 \cdot d^2/2. && \text{(b)} \end{aligned} \tag{5.5}$$

At the points of intersection of chord and curve, there is, by definition $y(\pm\, d) = 0$,

$$\begin{aligned} &y(-d) = y_0 - \psi_0 \cdot d + C_0 \cdot d^2/2 - C_1 \cdot d^3/6 = 0; && \text{(a)} \\ &y(+d) = y_0 + \psi_0 \cdot d + C_0 \cdot d^2/2 + C_1 \cdot d^3/6 = 0. && \text{(b)} \end{aligned} \tag{5.6}$$

Equations 5.5 and 5.6 can be solved for the curvature parameters C_0 and C_1 as well as for the state values y_0 and m_0 (ψ_0) at the origin $x = 0$ to yield

$$\begin{aligned} C_0 &\simeq (m_{+d} - m_{-d})/(2 \cdot d), \\ C_1 &\simeq 1.5 \cdot (m_{+d} + m_{-d})/d^2, \\ \psi_0 &\simeq -0.25 \cdot (m_{+d} + m_{-d}), \\ y_0 &\simeq 0.25 \cdot (m_{+d} - m_{-d}) \cdot d. \end{aligned} \tag{5.7}$$

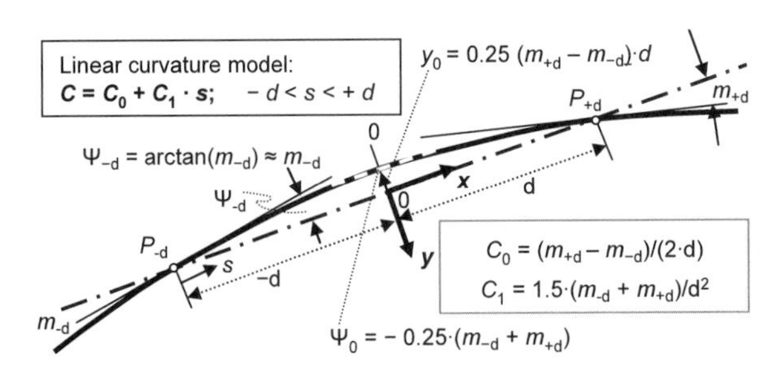

Figure 5.14. Approximate determination of curvature of a slightly curved edge by sub-pixel localization of edge points and tangent directions: *Hermite*-interpolation of a third order parabola from two tangent points

The linear curvature model can be computed easily from the tangent directions relative to the chord line and the distance ($2 \cdot d$) between the tangent points. Of course, this distance has to be chosen such that the angle constraint ($|\psi| < 0.2$ radian) is not violated. On smooth curves, this is always possible; however, for large curvatures, the distance d allowed becomes small and the scale for measuring edge locations and tangent directions probably has to be adapted. Very sharp curves have to be isolated and jumped over as "corners" having large directional changes over small arc lengths. In an idealized but simple scheme, they can be approximated by a *Dirac* impulse in curvature with a finite change in direction over zero arc length.

Due to the differencing process unavoidable for curvature determination, the results tend to be noisy. When basic properties of objects recognized are known, a post–processing step for noise reduction exploiting this knowledge should be included.

Remark: The special advantage of subscale resolution *for dynamic vision* lies in the fact that the onset of changes in motion behavior may be detected earlier, yielding better tracking performance, crucial for some applications. The aperture problem inherent in edge tracking will be revisited in Section 9.5 after the basic tracking problem has been discussed.

5.2.3 Edge Candidate Selection

Usually, due to image noise there are many insignificant extreme values in the resulting correlation vector, as can be seen in Figure 5.6. Positioning the threshold properly (and selecting the mask parameters in general) depends very much on the scene at hand, as may be seen in Figure 5.15, due to shadow boundaries and scene noise, the largest gradient values may not be those looked for in the task context (road boundary). Colinearity conditions (or even edge elements on a smoothly

Figure 5.15. The challenge of edge feature selection in road scenes: Good decisions can be made only by resorting to higher level knowledge. Road scenes with shadows (and texture); extreme correlation values marking road boundaries may not be the absolutely largest ones.

curved line) may be needed for proper feature selection; therefore, threshold selection in the feature extraction step should not eliminate these candidates. Depending on the situation, these parameters have to be specified by the user (now) or by a knowledge-based component on the higher system levels of a more mature version. Average intensity levels and intensity ranges resulting from region-based methods (see Section 5.3) will yield information for the latter case.

As a service to the user, in the code CRONOS, the extreme values found in one function call may be listed according to their correlation values; the user can specify how many candidates he wants presented at most in the function call. As an extreme value of the search either the pixel position with the largest mask response may be chosen (simplest case with large measurement noise), or several neighboring correspondence values may be taken into account allowing interpolation.

5.2.4 Template Scaling as a Function of the Overall "Gestalt"

An additional degree of freedom available to the designer of a vision system is the focal length of the camera for scaling the image size of an object to its distance in the scene. To analyze as many details as possible of an object of interest, one tends to assume that a focal length, which lets the object (in its largest dimension) just fill the image would be optimal. This may be the case for a static scene being observed from a stationary camera. If either the object observed or the vehicle carrying the camera or both can move, there should be some room left for searching and tracking over time. Generously granting an additional space of the actual size of the object to each side results in the requirement that perspective mapping (focal length) should be adjusted so that the major object dimension in the image is about one third of the image. This leaves some regions in the image for recognizing the environment of the object, which again may be useful in a task context.

To discover essential shape details of an object, the smallest edge element template should not be larger than about one-tenth of the largest object dimension. This yields the requirement that the size of an object in the image to be analyzed in some detail should be about 20 to 30 pixels. However, due to the poor angular resolution of masks with a size of three pixels, a factor of 2 (60 pixels) seems more comfortable. This leads to the requirement that objects in an image must be larger than about 150 pixels. Keep in mind that objects imaged with a size (region) of only about a half dozen pixels still can be noticed (discovered and roughly tracked), however, due to spurious details from discrete mapping (rectangular pixel size) into the sensor array, no meaningful shape analysis can be performed.

This has been a heuristic discussion of the effects of object size on shape recognition. A more operational consideration based on straight edge template matching and coordinate-free differential geometry shape representation by piecewise functions with linear curvature models is to follow.

A lower limit to the support region required for achieving accuracy of about one-tenth of a pixel in a tangent position and about 1° in the tangent direction (order of magnitude) by subpixel resolution is about eight to ten pixels. The efficient scheme given in [Dickmanns 1985] for accurately determining the curvature parameters is limited to a smooth change in the tangent direction of about 20 to 25°; for recovering a circle (360°). This means that about $n_{elef} \approx$ 15 to 18 elemental edge features have to be measured. Since the ratio of circumference to diameter is π for a circle, the smallest circle satisfying these conditions for non–overlapping support regions is n_{elef} times (mask size = 8 to 10 pixels) divided by π. This yields a required size of about 40 to 60 pixels in linear extension of an object in an image.

Since corners (points of finite direction change) can be included as curvature impulses measurable by adjacent tangent directions, the smallest (horizontally aligned) measurable square is ten pixels wide while the diagonal is about 14 pixels; more irregularly shaped objects with concavities require a larger number of tangent measurements. The convex hull and its dimensions give the smallest size measurable in units of the support region. Fine internal structures may be lost.

From these considerations, for accurate shape analysis down to the percent range, the image of the object should be between 20 and 100 pixel in linear extension, in general. This fits well in the template size range from 3 (or 5) to 17 (or 33) pixels. Usual image sizes of several hundred lines allow the presence of several well-recognizable objects in each image; other scales of resolution may require different focal lengths for imaging (from microscopy to far ranging telescopes).

Template scaling for line detection: Finally, choosing the right scale for detecting (thin) lines will be discussed using a real example [Hofmann 2004]. Figure 5.16 shows results for an obliquely imaged lane marking which appears 16 pixels wide in the search direction (top: image section searched, width $n_w = 9$ pixel). Summing up the mask elements in the edge direction corresponds to rectifying the image

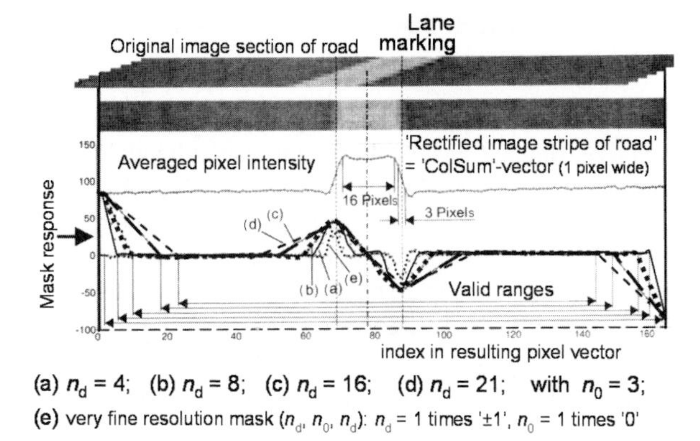

Figure 5.16. Optimal mask size for line recognition: For general scaling, mask size should be scaled by line width (= 16 pixels here)

stripe, as shown below in the figure; however, only one intensity value remains, so that for the rest of the pixel-operations with different mask sizes in the search direction, about one order of magnitude in efficiency is gained. All five masks investigated (a) to (e) rely on the same "ColSum"-vector; depending on the depth of the masks, the valid search ranges are reduced (see double-arrows at bottom).

The averaged intensity profile of the mask elements is given in the vertical center (around 90 for the road, and ~130 for the lane marker); the lane marking clearly sticks out. Curve (e) shows the mask response for the mask of highest possible resolution (1, 0, 1); see legend. It can be seen that the edge is correctly detected with respect to location, but due to the smaller extreme value, sensitivity to noise is higher than that for the other masks. All other masks have been chosen with $n_0 = 3$ for reducing sensitivity to slightly different edge directions including curved edges. In practical terms, this means that the three central values under the mask shifted over the ColSum–vector need not be touched; only n_d values to the left and to the right need be summed.

Depth values for the two fields of the mask of $n_d = 4$, 8, and 16 (curves a, b, c) yield the same gradient values and edge location; the mask response widens with increasing field width. By scaling the field depth n_d of the mask by the width of the line l_w to be detected, the curves can be generalized to scaled masks of depths n_d/l_w = ¼, ½, and 1. Case (d) shows with $n_d/l_w = 21/16 = 1.3$ that for field depths larger than line width, the maximal gradient decreases and the edge is localized at a wrong position. So, the field width selected should always be smaller than the line to be detected. The number of zeros at the center should be less than the field depth, probably less than half that value for larger masks; values between 1 and 3 have shown good results for n_d up to 7. For the detection of dirt roads with jagged edges and homogeneous intensity values on and off the road, large n_0 are favorable.

5.3 The Unified Blob-edge-corner Method (UBM)

The approach discussed above for detecting edge features of single (sub-) objects based on receptive fields (masks) has been generalized to a feature extraction method for characterizing image regions and general image properties by oriented edges, homogeneously shaded areas, and nonhomogeneous areas with corners and texture. For characterizing textures by their statistical properties of image intensities in real time (certain types of textures), more computing power is needed; this has to be added in the future. In an even more general approach, stripe directions could be defined in any orientation, and color could be added as a new feature space. For efficiency reasons, here, only horizontal and vertical stripes in intensity images are considered, for which only one matrix index and the gray values vary at a time). To achieve reusability of intermediate results, stripe widths are confined to even numbers and are decomposed into two half-stripes.

5.3.1 Segmentation of Stripes through Corners, Edges, and Blobs

In this image evaluation method, the goal is to start from as few assumptions on intensity distributions as possible. Since pixel noise is an important factor in outdoor environments, some kind of smoothing has to be taken into account, however. This is done by fitting models with planar intensity distribution to local pixel values if they exhibit some smoothness conditions; otherwise, the region will be characterized as nonhomogeneous. Surprisingly, it has turned out that the planarity check for local intensity distribution itself constitutes a nice feature for region segmentation.

5.3.1.1 Stripe Selection and Decomposition into Elementary Blocks

The field size for the least-squares fit of a planar pixel-intensity model is $(2{\cdot}m) \times (2{\cdot}n)$, and is called the "model support region" or mask region. For reusability of intermediate results in computation, this support region is subdivided into basic (elementary) image regions (called *m*ask *el*ement*s* or briefly "mels") that can be defined by two numbers: The number of pixels in the row direction m, and the number of pixels in the column direction n. In Figure 5.17, m has been selected as 4 and n as 2; the total stripe width for row search thus is 4 pixels. For $m = n = 1$, the highest possible image resolution will be obtained; however, strong influence of noise on the pixel level may show up in the results in this case.

When working with video fields (sub–images with only odd or even row–indices, as is often done in practical applications), it makes sense for horizontal stripes to choose $m = 2n$; this yields averaging of pixels at least in row direction for $n = 1$. Rendering these mels as squares, finally yields the original rectangular image shape with half the original full-frame resolution. By shifting stripe evaluation by only half the stripe width, all intermediate pixel results in one half-stripe can be reused directly in the next stripe by just changing sign (see below). The price to be paid for this convenience is that the results obtained have to be represented at the center point of the support region which is exactly at pixel boundaries. However, since subpixel accuracy is looked for anyway, this is of no concern.

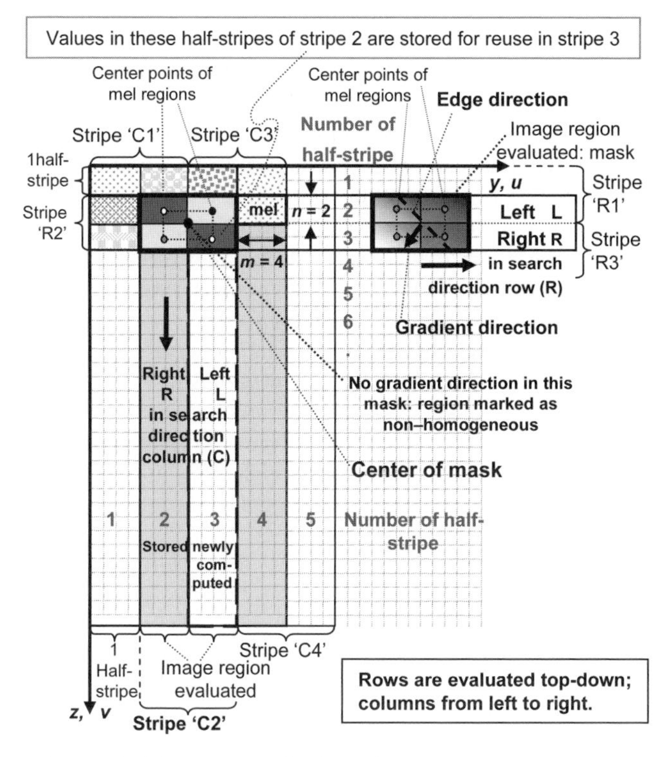

Figure 5.17. Stripe definition (row = horizontal, column = vertical) for the (multiple) feature extractor 'UBM' in a pixel-grid; mask elements (*mels*) are defined as basic rectangular units for fitting a planar intensity model

Still open is the question of how to proceed within a stripe. Figure 5.17 suggests taking steps equal to the width of mels; this covers all pixels in the stripe direction once and is very efficient. However, shifting mels by just 1 pixel in the stripe direction yields smoother (low-pass filtered) results [Hofmann 2004]. For larger mel-lengths, intermediate computational results can be used as shown in Figure 5.18.

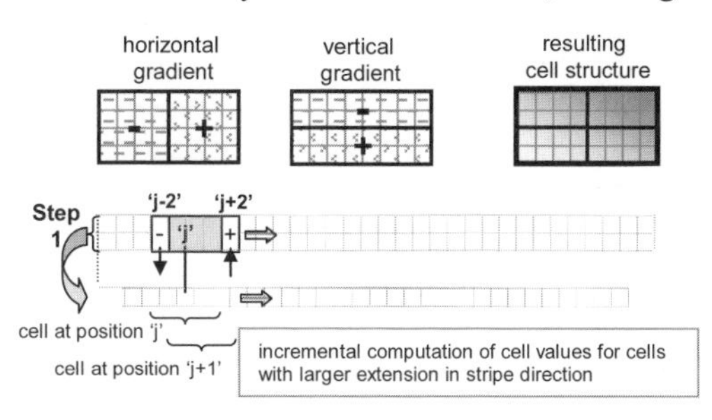

Figure 5.18. Mask elements (mels) for efficient computation of gradients and average intensities

This corresponds to the use of Colsum in the method CRONOS (see Figures 5.9 and 5.10). The new summed value for the next mel can be obtained by subtracting the value of the last column and adding the one of the next column [(j−2) and (j+2) in the example shown, bottom row in Figure 5.18].

For the vertical search direction, image evaluation progresses top-down within the stripe and from left to right in the sequence of stripes. Shifting of stripes is always done by mel-size m or n (width of half-stripe), while shifting of masks in the search direction can be specified from 1 to m or n (see Figure 5.19b below); the latter number m or n means pure block evaluation, however, only coarse resolution. This yields the lowest possible computational load with all pixels used just once in one mel. For objects in the near range, this may still be sufficient for tracking.

The goal was to obtain an algorithm allowing easy adaptation to limited computing power; since high resolution is required in only a relatively small part of images, in general in outdoor scenes, only this region needs to be treated with more finely tuned parameters (see Figure 5.37 below). Specifying a rectangular region of special interest by its upper left and lower right corners, this sub-area can be precisely evaluated in a separate step. If no view stabilization is available, the decision for the corner points may even be based on actual evaluation results with coarse resolution. The initial analysis with coarse resolution guarantees that only the most promising subregions of the image are selected despite angular perturbations stemming from motion of the subject body, which shifts around the inertially fixed scene in the image. This *attention-focusing* avoids unnecessary details in regions of less concern.

Figure 5.19 shows the definitions necessary for performing efficient multiple-scale feature evaluation. The left part (a) shows the specification of masks of different sizes (with mel-sizes from 1×1 to 4×2 and 4×4, *i.e.*, two pyramid stages). Note that the center of a pixel or of mels does not coincide with the origin O of the masks, which is for all masks at (0, 0). The mask origin is always defined as the point where all four quadrants (mels) meet. The computation of the average inten-

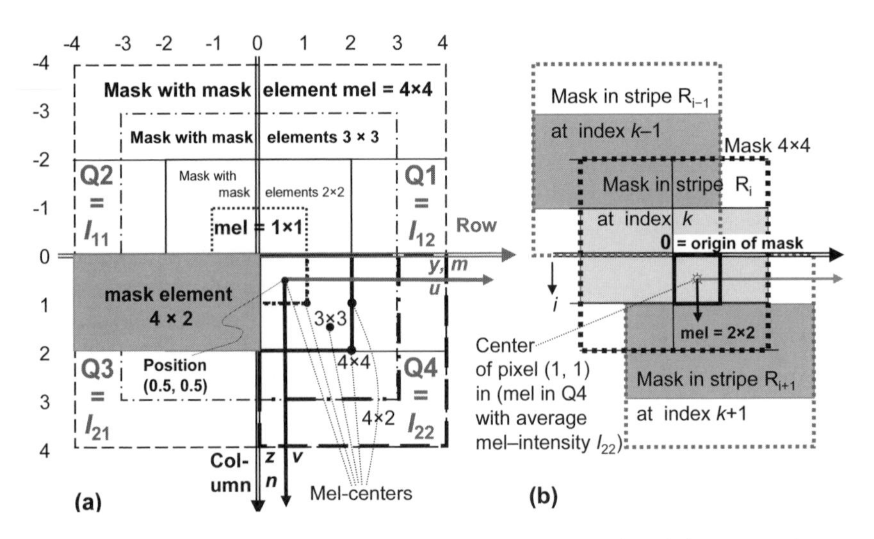

Figure 5.19. With the reference points chosen here for the mask and the average image intensities in quadrants Qi, fusing results from different scales becomes simple; (a) basic definitions of mask elements, (b) progressive image analysis within stripes and with sequences of stripes (shown here for rows)

sities in each mel (I_{12}, I_{11}, I_{21}, I_{22} in quadrants Q1 to Q4) is performed with the reference point at (0.5, 0.5), the center of the first pixel nearest to the mask origin in the most recent mel; this yields a constant offset for all mask sizes when rendering pixel intensities from symbolic representations. For computing gradients, of course, the real mel centers shown in quadrant Q4 have to be used.

The reconstruction of image intensities from results of one stripe is done for the central part of the mask (± half the size of the width normal to the search direction of the mask element). This is shown in the right part (b) of the figure by different shading. It shows (low-frequency) shifting of the stripe position by $n = 2$ (index i) and (high-frequency) shifting of the mask position in search direction by 1 (index k). Following this strategy in both row and column direction will yield nice low-pass-filtered results for the corresponding edges.

5.3.1.2 Reduction of the Pixel Stripe to a Vector with Attributes

The first step is to sum up all n pixel or cell values in the direction of the width of the half-stripe (lower part in Figure 5.18). This reduces the half-stripe for search to a vector, irrespective of stripe width specified. It is represented in Figure 5.18 by the bottom row (note the reduction in size at the boundaries). Each and every further computation is based on these values that represent the average pixel or cell intensity at the location in the stripe if divided by the number of pixels summed. However, these individual divisions are superfluous computations and can be spared; only the final results have to be scaled properly for image intensity.

In our example with $m = 4$ in Figure 5.18, the first mel value has to be computed by summing up the first four values in the vector. When the mels are shifted by one pixel or cell length for smooth evaluation of image intensities in the stripe (center row), the four new mel values are obtained by subtracting the trailing pixel or cell value at position $j-2$ and by adding the leading one at $j+2$ (see lower left in Figure 5.18). The operations to be performed for gradient computation in horizontal and vertical directions are shown in the upper left and center parts of the figure. Summing two mel values (vertically in the left and horizontally in the center subfigure) and subtracting the corresponding other two sums yields the difference in (average) intensities in the horizontal and vertical directions of the support region. Dividing these numbers by the distances between the centers of the mels yields a measure of the (averaged) horizontal and vertical image intensity gradient at that location. Combining both results allows computing the absolute gradient direction and magnitude. This corresponds to determining a local plane tangent to the image intensity distribution for each support region (mask) selected.

However, it may not be meaningful to enforce a planar approximation if the intensities vary irregularly by a large amount. For example, the intensity distribution in the mask top left of Figure 5.17 shows a situation where averaging does not make sense. Figure 5.20a shows the situation with intensities as vectors above the center of each mel. For simplicity, the vectors have been chosen of equal magnitude on the diagonals. The interpolating plane is indicated by the dotted lines; its origin is located at the top of the central vector representing the average intensity I_C. From the dots at the center of each mel in this plane, it can be recognized that two diagonally adjacent vectors of average mel intensity are well above, respec-

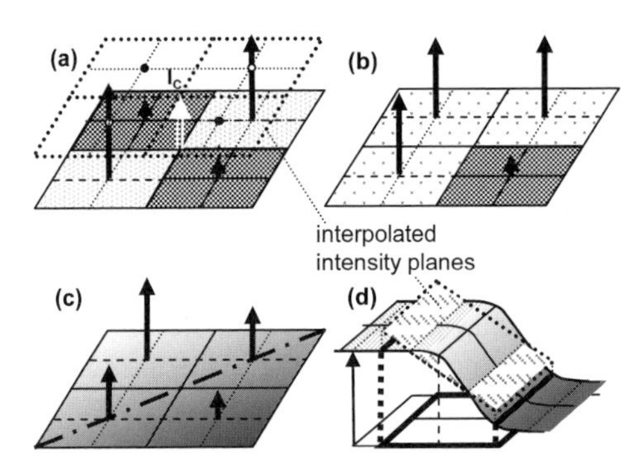

Figure 5.20. Feature types detectable by UBM in stripe analysis

tively, below the interpolating plane. This is typical for two corners or a textured area (*e.g.*, four checkerboard fields or a saddle point).

Figure 5.20b represents a perfect (gray value) corner. Of course, the quadrant with the differing gray value may be located anywhere in the mask. In general, all gray values will differ from each other. The challenge is to find algorithms allowing reasonable separation of these feature types versus regions fit for interpolation with planar shading models (lower part of Figure 5.20) at low computational cost. Well known for corner detection among many others are the "Harris"- [Harris, Stephens 1988], the KLT- [Tomasi, Kanade 1991] and the "Haralick"- [Haralick, Shapiro 1993] algorithms, all based on combinations of intensity gradients in several regions and directions. The basic ideas have been adapted and integrated into the algorithm UBM. The goal is to segment the image stripe into regions with smooth shading, corner points, and extended nonhomogeneous regions (textured areas). It will turn out that nonplanarity is a new, easily computable feature on its own (see Section 5.3.2.1).

Corner points are of special value in tracking since they often allow determining optical feature flow in image sequences (if robustly recognizable); this is one important hint for detecting moving objects before they have been identified on higher system levels. These types of features have shown good performance for detecting pedestrians or bicyclists in the near range of a car in urban traffic [Franke *et al.* 2005].

Stripe regions fit for approximation by sequences of shading models are characterized by their average intensities and their intensity gradients over certain regions in the stripe; Figure 5.20c shows such a case. However, it has to be kept in mind that a planar fit to intensity profiles with nonlinear intensity changes in only one direction can yield residues of magnitude zero with the four symmetric support points in the method chosen (see Figure 5.20d); this is due to the fact that three points define a plane in space, and the fourth point (just one above the minimal number required for fixing a plane) is not sufficient for checking the real spatial structure of the surface to be approximated. This has to be achieved by combining results from a sequence of mask evaluations.

By interpolation of results from neighboring masks, extreme values of gradients including their orientation are determined to subpixel accuracy. Note that, contrary to the method CRONOS, no direction has to be specified in advance; the direction of the maximal gradient is a result of the interpolation process. For this reason the method UBM is called "direction-sensitive" (instead of "direction selective" in the case of CRONOS). It is, therefore, well suited for initial (strictly "bottom-up") image analysis [Hofmann 2004], while CRONOS is very efficient once predominant edge directions in the image are known and their changes can be estimated by the 4-D approach (see Chapter 6).

During these computations within stripes, some statistical properties of the images can be determined. In step 1, all pixel values are compared to the lowest and the highest values encountered up to then. If one of them exceeds the actual extreme value, the actual extreme is updated. At the end of the stripe, this yields the maximal ($I_{\text{max-st}}$) and the minimal ($I_{\text{min-st}}$) image intensity values in the stripe. The same statistic can be run for the summed intensities normal to the stripe direction ($I_{\text{wmax-st}}$ and $I_{\text{wmin-st}}$) and for each mel ($I_{\text{cmax-st}}$ and $I_{\text{cmin-st}}$); dividing the maximal and minimal value within each mel by the average for the mel, these scaled values will allow monitoring the appropriateness of averaging. A reasonable balance between computing statistical data and fast performance has to be found for each set of problems.

Table 5.1 summarizes the parameters for feature evaluation in the algorithm UBM; they are needed for categorizing the symbolic descriptions within a stripe, for selecting candidates, and for merging across stripe boundaries. Detailed meanings will be discussed in the following sections.

Table 5.1. Parameters for feature evaluation in image stripes

Parameter	Description
ErrMax	Maximally allowed percent error of interpolated intensity plane through centers of four mels (typically 3 to 10%); note that the errors at all mel centers have same magnitude! (see Section 5.3.2.2)
CircMin (qmin)	Minimal "circularity" required, threshold value on scaled second eigenvalue for corner selection [0.75 corresponds to an ideal corner (Figure 5.20b), the maximal value 1 to an ideal double–corner (checkerboard, Figure 5.20a)]; (see section 5.3.3)
traceNmin	(Alternate) threshold value for selection of corner candidates; useful for adjusting the number of corner candidates.
IntensGradMin	Threshold value for intensity gradients to be accepted as edge candidates; (see Section 5.3.2.3)
AngleFactHor	Factor for limiting edge directions to be found in horizontal search direction (rows); (see Section 5.3.2.3)
AngleFactVer	Factor for limiting edge directions to be found in vertical search direction (columns); (see Section 5.3.2.3)
VarLim	Upper bound on variance allowed for a fit on both ends of a linearly shaded blob segment
Lsegmin	Minimum length required of a linearly shaded blob segment to be accepted (suppression of small regions)
DelIthreshMerg	Tolerance in intensity for merging adjacent regions to 2-D blobs
DelSlopeThrsh	Tolerance for intensity gradients for merging adjacent regions to 2-D blobs

The five feature types treated with the method UBM are (1) textured regions (see Section 5.3.2.1), (2) edges from extreme values of gradients in the search direction (see Section 5.3.2.3), (3) homogeneous segments with planar shading mod-

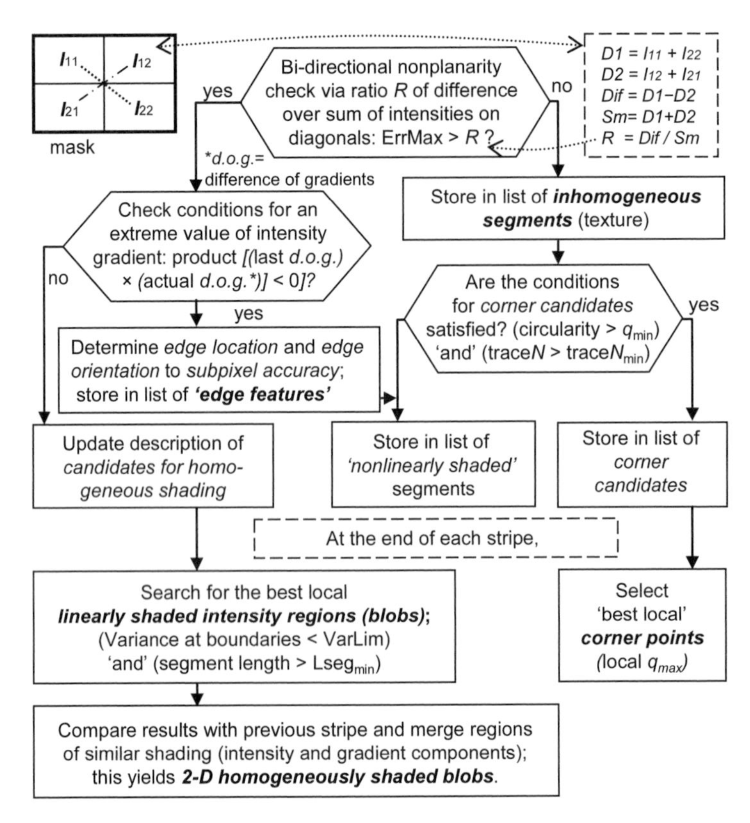

Figure 5.21. Decision tree for feature detection in the unified blob-edge-corner method (UBM) by local and global gradients in the mask region

els (see Section 5.3.2.4), (4) corners (see Section 5.3.3), and (5) regions nonlinearly shaded in one direction, which, however, will not be investigated further here. They have to lie between edges and homogeneously shaded areas and may be merged with class 1 above.

The sequence of decisions in the unified approach to all these features, exploiting the same set of image data evaluated in a stripe-wise fashion, is visualized in Figure 5.21. Both horizontal and vertical stripes may be searched depending on the orientation of edges in the image. Localization of edges is best if they are oriented close to orthogonal to the search direction; therefore, for detecting horizontal edges and horizontal blob boundaries, a vertical search should be preferred. In the general case, both search directions are needed, but edge detection can then be limited to (orthogonal to the search direction) ± 50°. The advantage of the method lies in the fact that (1) the same feature parameters derived from image regions are used throughout, and (2) the regions for certain features are mutually exclusive. Compared to investigating the image separately for each feature type, this reduces the computer workload considerably.

5.3.2 Fitting an Intensity Plane in a Mask Region

For more efficient extraction of features with respect to computing time, as a first step the sum of pixel intensities I_{cs} is formed within rectangular regions, so-called *cells* of size $m_c \times n_c$; this is a transition to a coarser scale. For $m_c = n_c$, *pyramid levels* are computed, especially for $m_c = n_c = 2$ the often used step-2 pyramids. With I_{ij} as pixel intensity at location $u = i$ and $v = j$ there follows for the cell

$$I_{cs} = \sum_{i=1}^{m_c} \sum_{j=1}^{n_c} I_{ij}. \tag{5.8}$$

The average intensity $\overline{I}_c$ of all pixels in the cell region then is

$$\overline{I}_c = I_{cs} /(m_c \cdot n_c). \tag{5.9}$$

The normalized pixel intensity I_{pN} for each pixel in the cell region is

$$I_{pN} = I_p / \overline{I}_c \quad \text{around the average value } \overline{I}_c . \tag{5.10}$$

Cells of different sizes may be used to generate multiple scale images of reduced size and resolution for efficient search of features on a larger scale. When working with video fields, cells of size 2 in the row and 1 in the column direction will bring some smoothing in the row direction and lead to much shorter image evaluation times. When coarse-scale results are sufficient, as for example with high-resolution images for regions nearby, cells of size 4 × 2 efficiently yield scene characteristics for these regions, while for regions further away, full resolution can be applied in much reduced image areas; this foveal–peripheral differentiation contributes to efficiency in image sequence evaluation. The region of high-resolution image evaluation may be directed by an attention focusing process on a higher system level based on results from a first coarse analysis (in a present or previous image).

The second step is building mask elements ("mels") from cells; they contain

$$\begin{aligned} m \cdot m_c &= m_p \quad \text{pixels in the search direction, and} \\ n \cdot n_c &= n_p \quad \text{pixels normal to the search direction.} \end{aligned} \tag{5.11}$$

Define

$$I_{MEs} = \sum_{i=1}^{m} \sum_{j=1}^{n} \overline{I}_{cs,ij} \quad \text{(sum of cell intensities) } I_{CS} , \tag{5.12}$$

then the average intensity of cells and thus also of pixels in the mask element is

$$\overline{I}_{ME} = I_{MEs} /(m \cdot n) . \tag{5.13}$$

In the algorithm UBM, mels are the basic units on which efficiency rests. Four of those are always used to form masks (see Figures 5.17-5.19) as support regions for image intensity description by symbolic terms:

> *Masks are support regions for the description and approximation of local image intensity distributions by parameterized symbols (image features): (1) 'Textured areas' (nonplanar elements), (2), 'oriented edges' (3) 'linearly shaded regions', and (4) 'corners'. Masks consist of four mask elements (mels) with average image intensities* $I_{11s}, I_{12s}, I_{21s}, I_{22s}$.

The average intensity of all mels in the mask region is

$$I_{Mean,s} = (I_{11s} + I_{12s} + I_{21s} + I_{22s})/4. \tag{5.14}$$

To obtain intensity elements of the order of magnitude 1, the normalized intensity in mels is formed by division by the mean value of the mask:

$$I_{ijN} = I_{ijs} / I_{Mean,s}. \tag{5.15}$$

This means that

$$\left[\sum I_{ijN}\right]/4 = 1. \tag{5.16}$$

The (normalized) gradient components in a mask then are given as the difference of intensities divided by the distance between mel-centers:

$$\left.\begin{aligned}
f_{r_{1N}} &= (I_{12N} - I_{11N})/m && \text{(upper row direction)} && \text{(a)}\\
f_{r_{2N}} &= (I_{22N} - I_{21N})/m && \text{(lower row direction)} && \text{(b)}\\
f_{c_{1N}} &= (I_{21N} - I_{11N})/n && \text{(left column direction)} && \text{(c)}\\
f_{c_{2N}} &= (I_{22N} - I_{12N})/n && \text{(right column direction).} && \text{(d)}
\end{aligned}\right\} \tag{5.17}$$

The first two are local gradients in the row-direction (index r) and the last two in the column direction (index c). The global gradient components of the mask are

$$\begin{aligned}
f_{r_N} &= \left(f_{r_{1N}} + f_{r_{2N}}\right)/2 && \text{(global row direction)}\\
f_{c_N} &= \left(f_{c_{1N}} + f_{c_{2N}}\right)/2 && \text{(global column direction).}
\end{aligned} \tag{5.18}$$

The normalized global gradient g_N and its angular orientation ψ then are

$$g_N = \sqrt{f_{r_N}^2 + f_{c_N}^2} \tag{5.19}$$

$$\psi = \arctan\left(f_{c_N} / f_{r_N}\right). \tag{5.20}$$

ψ is the gradient direction in the (u, v)-plane. The direction of the vector normal to the tangent plane of the intensity function (measured from the vertical) is given by

$$\gamma = \arctan\left(g_N\right). \tag{5.21}$$

5.3.2.1 Adaptation of a Planar Shading Model to the Mask Area

The origin of the local coordinate system used is chosen at the center of the mask area where all four mels meet. The model of the planar intensity approximation with least sum of the squared errors in the four mel–centers has the yet unknown parameters I_0, g_y, and g_z (intensity at the origin, and gradients in the y- and z-directions). According to this linear model, the intensities at the mel–centers are

$$\begin{aligned}
I_{11Np} &= I_0 - g_y \cdot m/2 - g_z \cdot n/2,\\
I_{12Np} &= I_0 + g_y \cdot m/2 - g_z \cdot n/2,\\
I_{21Np} &= I_0 - g_y \cdot m/2 + g_z \cdot n/2,\\
I_{22Np} &= I_0 + g_y \cdot m/2 + g_z \cdot n/2.
\end{aligned} \tag{5.22}$$

Let the measured values from the image be $I_{11N\mu}, I_{12N\mu}, I_{21N\mu}$ and $I_{22N\mu}$. Then the errors e_{ij} can be written:

$$\begin{bmatrix} e_{11} \\ e_{12} \\ e_{21} \\ e_{22} \end{bmatrix} \begin{matrix} = I_{11Np} - I_{11N\mu} = \\ = I_{12Np} - I_{12N\mu} = \\ = I_{21Np} - I_{21N\mu} = \\ = I_{22Np} - I_{22N\mu} = \end{matrix} \begin{bmatrix} 1 & -m/2 & -n/2 \\ 1 & +m/2 & -n/2 \\ 1 & -m/2 & +n/2 \\ 1 & +m/2 & +n/2 \end{bmatrix} \begin{bmatrix} I_0 \\ g_y \\ g_z \end{bmatrix} - \begin{bmatrix} I_{11N\mu} \\ I_{12N\mu} \\ I_{21N\mu} \\ I_{22N\mu} \end{bmatrix}. \quad (5.23)$$

To minimize the sum of the squared errors, this is written in matrix form:

$$\underline{e} = \underline{\underline{A}} \cdot \underline{g} - \underline{I}_{N\mu} \quad \text{(a)}$$

where
$$A^T = \begin{bmatrix} 1 & 1 & 1 & 1 \\ -m/2 & m/2 & -m/2 & m/2 \\ -n/2 & -n/2 & n/2 & n/2 \end{bmatrix} \quad \text{(b)} \quad (5.24)$$

and
$$I_{N\mu}^T = \begin{bmatrix} I_{11N\mu} & I_{12N\mu} & I_{21N\mu} & I_{22N\mu} \end{bmatrix}. \quad \text{(c)}$$

The sum of the squared errors is $e^T e$ and shall be minimized by proper selection of $\begin{bmatrix} I_0 & g_y & g_z \end{bmatrix} = p^T$. The necessary condition for an extreme value is that the partial derivative $d(e^T e)/dp = 0$; this leads to

$$A^T I_{N\mu} = A^T A \cdot p, \quad (5.25)$$

with solution (pseudo–inverse) $\quad p = (A^T A)^{-1} \cdot A^T \cdot I_{N\mu}$. $\quad$ (5.26)

From Equation 5.24b follows

$$A^T A = \begin{bmatrix} 4 & 0 & 0 \\ 0 & m^2 & 0 \\ 0 & 0 & n^2 \end{bmatrix}$$

and $\quad$ (5.27)

$$(A^T A)^{-1} = \begin{bmatrix} 1/4 & 0 & 0 \\ 0 & 1/m^2 & 0 \\ 0 & 0 & 1/n^2 \end{bmatrix},$$

and with Equations 5.24c and 5.14

$$A^T I_{N\mu} = \begin{bmatrix} 4 \\ \left(f_{r_{1N}} + f_{r_{2N}}\right) m^2/2 \\ \left(f_{c_{1N}} + f_{c_{2N}}\right) n^2/2 \end{bmatrix}. \quad (5.28)$$

Inserting this into Equation 5.26 yields, with Equation 5.17, the solution

$$p^T = [I_0 \;\; g_y \;\; g_z] = [1 \;\; f_{rN} \;\; f_{cN}]. \quad (5.29)$$

5.3.2.2 Recognizing Textured Regions (limit for planar approximations)

By substituting Equation 5.29 into 5.23, forming $(e_{12} - e_{11})$ and $(e_{22} - e_{21})$ as well as $(e_{21} - e_{11})$ and $(e_{22} - e_{12})$, and by summing and differencing the results, one finally obtains

$$e_{21} = e_{12} \quad (5.30)$$

and $$e_{11} = e_{22} ;$$
this means that the errors on each diagonal are equal. Summing up all errors e_{ij} yields, with Equation 5.16, $\sum e_{ij} = 0$.

This means that the errors on the diagonals have opposite signs, but their magnitudes are equal! These results allow an efficient combination of feature extraction algorithms by forming the four local gradients after Equation 5.17 and the two components of the gradient within the mask after Equations 5.18 and 5.29. All four errors of a planar shading model can thus be determined by just one of the four Equations 5.23. Even better, inserting proper expressions for the terms in Equation 5.23, the planar interpolation error with Equation 5.12 turns out to be

$$\text{ErrInterp} = [(I_{11} + I_{22}) - (I_{12} + I_{21})] / I_{\text{MEs}} . \tag{5.31}$$

Efficiently programmed, its evaluation requires just one additional difference and one ratio computation. The planar shading model is used only when the magnitude of the residues is sufficiently small

$$|\,\text{ErrInterp}\,| < \varepsilon_{\text{pl,max}} \text{ (dubbed } \textit{ErrMax}\text{).} \tag{5.32}$$

Figure 5.22. *Nonplanarity* features in central rectangle of an original video–field (cell size $m_c = 1$, $n_c = 1$, mel size $1{\times}1$, finest possible resolution, MaxErr = 4%)

Figure 5.22 shows a video field of size 288 × 768 pixels with nonplanar regions for $m_c = n_c = m = n = 1$ (highest possible resolution) and ErrMax = 4% marked by white dots; regions at the left and right image boundaries as well as on the motor hood (lower part) and in the sky have been excluded from evaluation.

Only the odd or even rows of a full video frame form a field; fields are transmitted every 40 ms in 25 Hz video. Confining evaluation to the most interesting parts of fields leaves the time of the complementary field (20 ms) as additional computing time; the image resolution lost in the vertical direction is hardly felt with subpixel feature extraction. [Interleaved video frames (from two fields) have the additional disadvantage that for higher angular turn rates in the row direction while taking the video, lateral shifts result between the fields.]

The figure shows that not only corner regions but also – due to digitization effects – some but not all edges with certain orientations (lane markings on the left and parts of silhouettes of cars) are detected as nonplanar. The number of features detected strongly depends on the threshold value ErrMax.

Figure 5.23 shows a summary of results for the absolute number of masks with nonplanar intensity distribution as a function of a variety of cell and mask parameters as well as of the threshold value ErrMax in percent. If this threshold is selected too small (say, 1%), very many nonplanar regions are found. The largest number of over 35 000 is obtained if a mask element is selected as a single pixel; this corresponds to ~ 17% of all masks evaluated of the image.

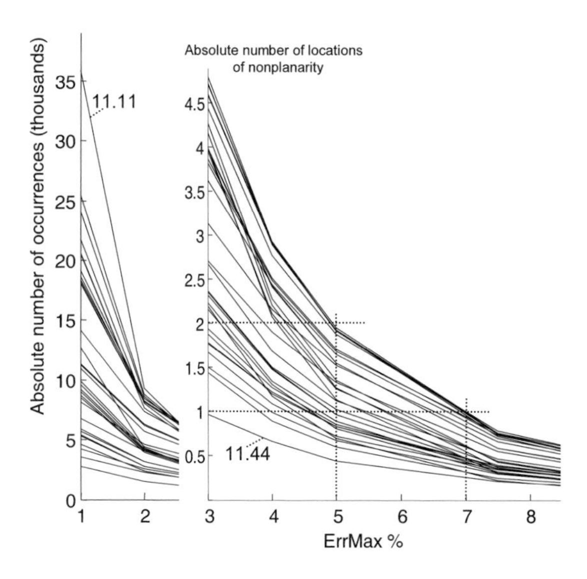

Figure 5.23. Absolute number of mask locations with residues exceeding ErrMax for a wide variety of cell and mask parameters m, n, m_c, n_c. For ErrMax ≥ 3% a new scale is used for better resolution. For ErrMax ≥ 5% at most 2000 nonplanar intensity regions are found out of at most ~ 200 000 mask locations for highest resolution with mel = pixel in a video–field.

The number of nonplanar regions comes down rapidly for higher values of ErrMax. For ErrMax = 2%, this number drops to less than 1/3; for higher values of ErrMax, the scale has been changed in the figure for better resolution. For ErrMax = 5%, the maximum number of nonplanar masks is less than 2000, that is less than 1% of the number of original pixels; on the other hand, for all cell and mask parameters investigated in the range $[1 \leq (m_c, n_c) \leq 2$ and $1 \leq (m, n) \leq 4]$, the number of nonplanar intensity regions does not drop below ~ 600 (for ErrMax = 5%). This is an indication that there is significant nonplanarity in intensity distribution over the image which can be picked up by any set of cell and mask parameters and also by the computer-efficient ones with higher parameter values that show up in the lower curves of Figure 5.23. Note that these curves include working on the first pyramid level $m_c = n_c = 2$ with mask elements $m \leq 4$ and $n \leq 4$; only the lowest curve 11.44 ($m = n = 1$, $m_c = n_c = 4$), for which the second pyramid level has been formed by preprocessing during cell computation, shows ~ 450 nonplanar regions for ErrMax = 5%. The results point in the direction that a combination of features from different pyramid scales will form a stable set of features for corner candidates.

For the former set of parameters (first pyramid level), decreasing the threshold value ErrMax to 3% leaves at least ~ 1500 nonplanar features; for curve 11.44, to reach that number of nonplanar features, ErrMax has to be lowered to 2%; however, this corresponds to ~ 50 % of all cell locations in this case. Averaging over cells or mask elements tends to level-off local nonplanar intensity distributions; it may therefore be advisable to lower threshold ErrMax in these cases in order not to

Figure 5.24. Visualization of %–threshold values in image intensity for separating planar from nonplanar local intensity models: In the rectangles, all pixel values have been increased by a factor corresponding to the percentage indicated as inset

lose sharp corners of moderate intensity differences. On the contrary, one might guess that for high resolution images, analyzed with small parameter values for cells and masks, it would be advantageous to increase ErrMax to get rid of edges but to retain corners. To visualize the 5%-threshold, Figure 5.24 shows the video field with intensities increased within seven rectangular regions by 2, 3, 4, 5, 7.5, 10, and 15% respectively; the manipulation is hardly visible in darker regions for values less than 5%, indicating that this seems to be a reasonable value for the threshold ErrMax from a human observer's point of view. However, in brighter image regions (*e.g.*, sky), even 3% is very noticeable.

The effect of lifting the threshold value ErrMax to 7.5% for planar intensity approximation for highest resolution (all parameters = 1, in shorthand notation (11.11) for the sequel) is shown in Figure 5.25.

In comparison to Figure 5.22, it can be seen that beside many edge positions many corner candidates have also been lost, for example, on tires and on the dark

Figure 5.25. Nonplanar features superimposed on original videofield for the threshold values MaxErr = 4% (left) and 7.5% (right); cell size $m_c = 1$, $n_c = 1$, mel size 1×1 (rows compressed after processing). More than 60% of features are lost in right image.

truck in front. This indicates that to keep candidates for real corners in the scene, ErrMax should not be chosen too large. The threshold has to be adapted to the scene conditions treated. There is not yet sufficient experience available to automate this threshold adaptation which should certainly be done based on results from several sets of parameters (m_c , n_c , m, n) and with a payoff function yet to be defined. Values in the range 2 ≤ ErrMax ≤ 5% are recommended as default for reducing computer load, on the one hand, and for keeping good candidates for corners, on the other.

Larger values for mel and cell parameters should be coupled with smaller values of ErrMax. Working on the first (2×2) pyramid level of pixels (cell size $m_c = 2$, $n_c = 2$) reduces the number of mask evaluations needed by a factor of 8 compared to working on the pixel level. The third power of 2 is due to the fact that now the half-stripe in the number of pixels is twice as wide as in the number of cells; in total, this is roughly a reduction of one power of 10 in computing effort (if the pyramid image is computed simultaneously with frame grabbing by a special device).

Figure 5.26 shows a juxtaposition of results on the pixel level (left) and on the first pyramid level (right). For ErrMax = 2% on the pixel level (left part), the 4% of occurrences of the number of nonplanar regions corresponds to about 8000 locations, while on the first pyramid level, 16% corresponds to about 4000 locations (see reference numbers on the vertical scale). Thus, the absolute number of nonplanar elements decreases by about a factor of 2 on the first pyramid level while the relative frequency in the image increases by about a factor of 4. Keeping in mind that the number of image elements on the first pyramid level has decreased by the same factor of 4, this tells us that on this level, most nonplanar features are preserved. On the pixel level, many spurious details cause the frequency of this feature to increase; this is especially true if the threshold ErrMax is reduced (see leftmost ordinate in Figure 5.26 for ErrMax = 1%).

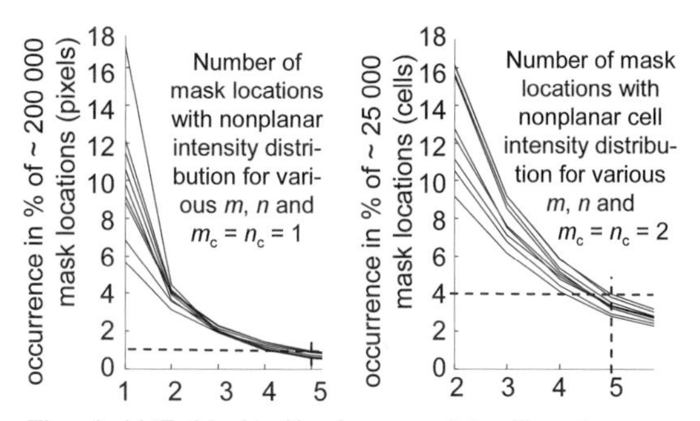

Figure 5.26. Relative number of *nonplanar* mask locations as a function of parameters for mask element formation m, n, m_c , n_c. Note that in the right part ErrMax starts at 2%; in the left part, which represents working on the pixel level, the relative frequency of nonplanar regions is down to ~ 4% for this value.

For the largest part of standard images from roads, therefore, working on the first pyramid level with reduced resolution is sufficient; only for the farther look-ahead regions on the road, which appear in a relatively small rectangle around the center, is full resolution recommended. Quite naturally, this yields a foveal–peripheral differentiation of image sequence analysis with much reduced computing resources needed. Figure 5.27 demonstrates that when working with video fields, a further reduction by a factor of 2 is possible without sacrificing detection of significant nonplanar features.

The right-hand picture is based on cell size (4×2); 4 pixels each in two rows are summed to yield cell intensity, while on the left, the cell is a pixel on the (2×2) first pyramid level. The subfigures are superpositions of all pixels found to belong to

Figure 5.27. Nonplanar features for parameter set ErrMax = 2.5%, mask elements $m = n = 2$ and cell size $n_c = m_c = 2$ (*i.e.*, first pyramid level, left; after processing compressed 2:1 in rows for better comparison). Changing m_c to 4 (right) yields about the same number of features : ~ 2500.

nonplanar regions, both in row search (horizontal white bars) and in column search (vertical white bars); it can be seen that beside corner candidates many edge candidates are also found in both images. For similar appearance to the viewer, the left picture has been horizontally compressed after finishing all image processing. From the larger number of local vertical white bars on the left, it can be seen that nonplanarity still has a relatively large spread on the first pyramid level; the larger cell size of the right-hand image cuts the number of masks to be analyzed in half (compare results in corner detection in Figure 5.39, 5.40 below). Note that even the reflections on the motor hood are detected. The locations of the features on different scales remain almost the same. These are the regions where *stable corner features for tracking* can be found, avoiding the aperture problem (sliding along edges). All significant corners for tracking are among the nonplanar features. They can now be searched for with more involved methods, which however, have to be applied to candidate image regions at least an order of magnitude smaller (see Figure 5.26). After finding these regions of interest on a larger scale first, for precise localization full resolution may be applied in those regions.

5.3.2.3 Edges from Extreme Values of Gradients in Search Direction

Gradient values of the intensity function have to be determined for least-squares fit of a tangent plane in a rectangular support region (Equation 5.29). Edges are defined by extreme values of the gradient function in the search direction (see Figure 5.28). These can easily be detected by multiplying two consecutive values of their differences in the search direction $[(g_0 - g_m)\cdot(g_p - g_0)]$; if the sign of the product is negative, an extreme value has been

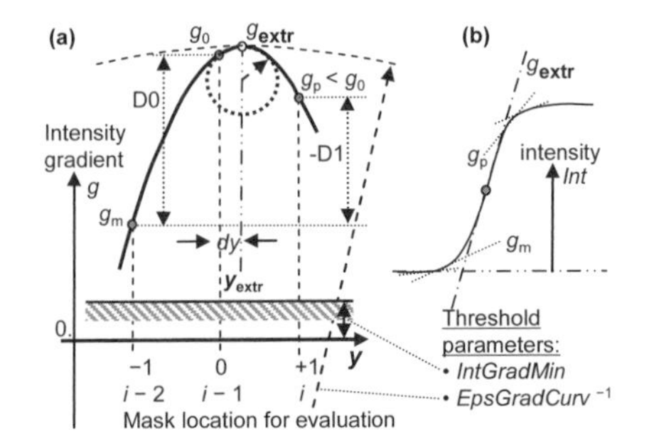

Figure 5.28. Localization of an edge to subpixel accuracy by parabolic interpolation after passing a maximum value g_0 of the intensity gradient

passed. Exploiting the same procedure shown in Figure 5.12, the location of the extreme value can be determined to sub-cell accuracy. This indicates that accuracy is not necessarily lost when cell sizes are larger than single pixels; if the signals are smooth (and they become smoother by averaging over cells), the locations of the extreme values may be determined to better than one-tenth the cell size. Mel sizes of several pixels in length and width (especially in the search direction), therefore, are good candidates for efficient and fast determination of edge locations with this gradient method.

Figure 5.28a shows three gradient values g_m, g_0, and g_p determining the parabola to be interpolated. The second-order coefficient of this parabolic curve, dubbed "mintcurv" is given by

$$\text{mintcurv} = 0.5 \cdot (g_m + g_p) - g_0 . \tag{5.33}$$

To eliminate noise effects from data, two threshold values are introduced before an edge is computed:

1. The magnitude of mintcurv has to be larger than a threshold value $\varepsilon_{intcurv}$; this eliminates very shallow extremes (large radii of the osculating circles, dashed lines in 5.28a). Leaving this threshold out may often be acceptable.
2. The absolute value of the maximum gradient encountered has to be larger than a threshold value "IntGradMin"; this admits only significant gradients as candidates for edges. The larger the mel size, the smaller this threshold should be chosen.

Proper threshold values for classes of problems have to be found by experimentation; in the long run, the system should be capable of doing this by itself, given corresponding payoff functions. Since edges oriented directly in the search direction are prone to larger errors, they can be excluded by limiting the ratio of the gradient components for acceptance of candidates. Figure 5.29 shows the principal idea. When both gradient components $|g_y|$ and $|g_z|$ are equal, the edge direction is 45°. Excluding all cases where Equation 5.34 is valid, a selection of k_α around 1 will allow finding all edges by a combined row and column search.

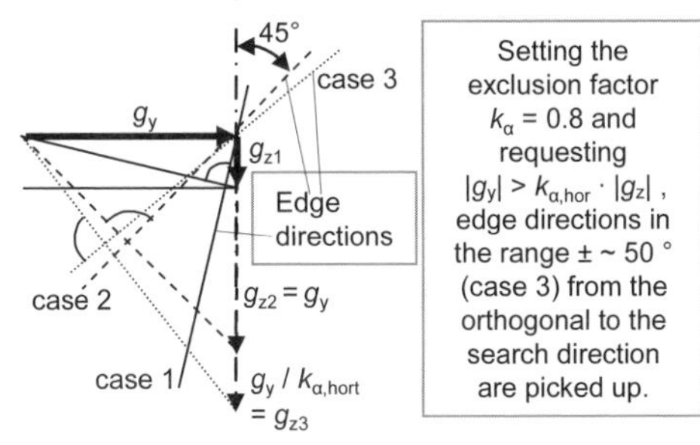

Figure 5.29. Limiting edge directions to be found in row–search around the vertical can be achieved by introducing limits for the ratio of gradient components ($|g_y| / |g_z| > k_{\alpha,hor}$); (analogous for column search: $k_{\alpha,hor}$ need not be equal to $k_{\alpha,vert}$)

$$\begin{aligned} |g_y| &> k_{\alpha,hor} \cdot |g_z| \text{ in row search and} \\ |g_z| &> k_{\alpha,vert} \cdot |g_y| \text{ in column search ,} \end{aligned} \tag{5.34}$$

(Edges with orientation close to diagonal should be detected in both search directions, leading to redundancy for cross checking.) Sub-mel localization of edges is performed only when all conditions mentioned are satisfied. The extreme value is found where the derivative of the gradient is zero. Defining

$$\text{slopegr} = 0.5 \cdot (g_p - g_m) \tag{5.35}$$

for a row search leads to the results: (a) sub-mel location dy from the central reference 0 and (b) the magnitude of the extreme gradient component g_{ymax}:

$$\begin{aligned} dy &= -\ 0.5 \cdot \text{slopegr/mintcurv}, \\ g_{y\max} &= g_0 + (\text{slopegr} + 0.5 \cdot \text{mintcurv} \cdot dy) \cdot dy. \end{aligned} \tag{5.36}$$

To determine the edge direction precisely, it is necessary to compute the value of the orthonormal gradient component (here $g_{z,edge}$) with a second-order interpolation from the corresponding g_z values and with the same dy also. The expressions for column search (dz, g_{zmax}, and $g_{y,edge}$) are obtained in analog fashion. Figure 5.30 shows results of edge finding on the cell level ($m_c = 2$, $n_c = 1$, *i.e.*, pure horizontal field compression) with mels of size $m = n = 3$.

Even edges in reflections on the motor–hood (bottom of image) and shadow boundaries are detected. Road areas are almost free of edge elements originating from noise, but lane markings are fully detected in a column search; the lane marking to the right is partially detected in a row search also, because of the steepness of the edge around $\pi/4$ in the image. It was surprising to see the left front wheel of the truck to the right detected so well with the new computation of edge directions from the interpolated gradient components. Such results can hardly be expected from radar or even from laser range finders; for this reason, once the computing power is available at low cost, machine vision is considered the favorite sense for perceiving the road environment in detail.

Figure 5.30. Edges from extreme values in gradient components in row (yellow) and column search (red, parameters m, n . m_c, n_c = 33.21), IntGradMin = 0.016; white regions are *bi-directionally nonplanar* features.

Distance to other vehicles can easily be estimated sufficiently accurate by the "flat ground"-assumption and by finding the lower bounds of collections of features moving in conjunction [Thomanek 1996]; especially, the dark areas underneath the vehicles and their boundary lines are highly visible in the figure. Taking the regions of the road blobs into account, seven vehicle hypotheses can be derived immediately (two in the left lane, two upfront in the own lane, and three in the lane to the right). Tracking these hypotheses over time will clarify the situation, in general.

Due to the small size of Figure 5.30, it is hard to appreciate the details; the original visualization on a large screen with thinner lines for the results from row and column search yields vivid impressions to the human observer. The white line elements in the figure show the locations of *bi-directionally nonplanar* features (nonhomogeneous areas determined in the first step of the UBM). Edges can be regarded as *uni-directionally nonplanar* features. The cross-section shown in Figure 5.28b is more clearly visualized in perspective projection in Figure 5.31; there is a transition between two different intensity levels with the *edge* at the position of steepest gradient (not emphasized here; it would lie approximately at the position of the right mel centers I_{12} and I_{22}, where the plane fitted cuts the nonlinear intensity surface in the figure).

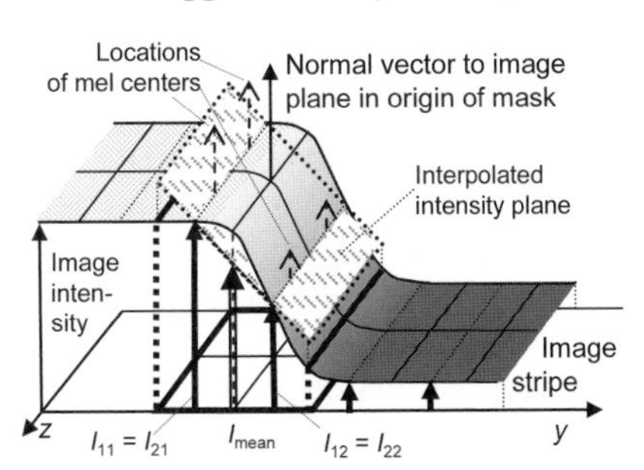

Figure 5.31. Visualization of the fact that curvature in only one direction of the intensity function (orthonormal curvature = 0) yields residues ≡ 0, using four support points; the mask passes the (weak) nonplanarity test of Section 5.3

Usually, edges are boundaries of homogeneously shaded, textured, or colored regions. Textured regions exhibit nonplanar features all over (*e.g.*, see tree in upper left center of Figure 5.30, left of the dark truck); usually, they do not have simple characteristic parameters; trees, text on traffic or navigation signs, and strongly curved edges of vehicles reflecting the environment are typical textured areas in road scenes (see Figure 5.30). These regions are grouped as inhomogeneous and are discarded from further analysis, at present, except testing for corner features (see Section 5.33 below).

The rest of the image between nonplanar regions and edge features now has to be checked for *linearly shaded blobs*. These candidate regions are analyzed first within a single stripe only; to be efficient, linearly shaded segments satisfying certain variance conditions have to have a certain minimal length L_{segmin} to qualify as *1-D blob features*; this suppresses spurious details right from the beginning. After finding a blob-centered description for the *1-D blob*, merging into larger regions with the results of previously analyzed stripes is performed, if certain threshold conditions for fusion are met; this leads to so-called *2-D blobs*. Stable blobs and their centers are very nice features for tracking. Contrary to edges, their shading characteristics (average gray value and gradient components) allow easier recognition of objects from one image to the next, both within a temporal sequence for tracking and between images from a pair of cameras for stereo interpretation.

5.3.2.4 Extended Shading Models in Image Stripes

Figure 5.32 shows a result of intensity analysis in a vertical stripe [Hofmann 2004]. Regions within stripes with similar shading parameters (average intensity, gradient magnitude and gradient direction) are grouped together as homogeneously shaded segments with a set of common parameters; they are called *(1-D) blob-features* here. The vertical image stripe marked by a white rectangle (at the right-hand side) yields the feature set shown graphically to the left, with some dotted lines indicating correspondences.

It is proposed here, to group regions satisfying the threshold conditions on the variance, but of small segment length < L_{segmin}, into a special class; depending on the

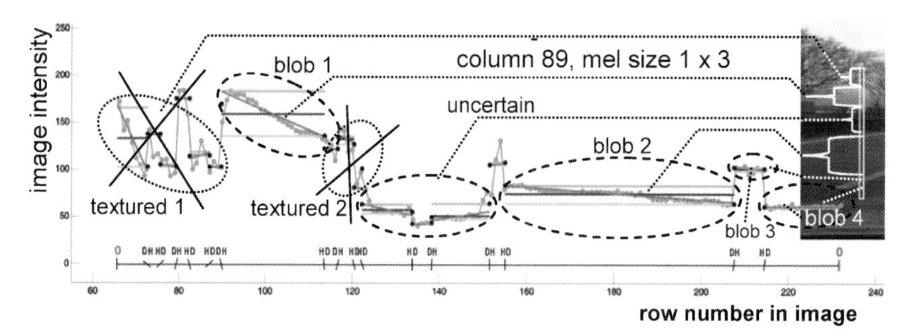

Figure 5.32. Intensity profile for one vertical image stripe, rows 60 to 200 (white rectangle in part of a video field, see insert at right boundary); sky (rows < 60) and motor hood (rows > 200) are excluded (basic figure from [Hofmann 2004], see text)

scale of analysis they may be considered as "textured". Since there will be very many of these local features, in general, they are not characteristic for understanding the scene observed. To concentrate on highly visible features that can be recognized and tracked well, the size of blob features should have an extension of a certain percentage of the whole image in at least one dimension, say, a few percent. Taking 500 pixels as a characteristic (average) image resolution in one dimension, the minimal segment length should be chosen as $5 < L_{segmin} < 25$ pixels, in general.

In Figure 5.32, blobs 1, 2, and 4 clearly satisfy this condition. If smaller regions show an image intensity level considerably different from their environment, they may qualify as *blob features* also, despite their small size. Maybe a characteristic number depending on the product of segment length and intensity difference should be chosen for decision-making. Lane markings on roads are typical examples; blob 3 (at the right-hand side), but also the spike left of blob 2 in Figure 5.32 are points in case. However, all four edges of these two lane markings are strong features by themselves and will be detected independently of the blob analysis, usually (see lower bar in the figure with characterization of edges as DH for dark-to-bright and HD for bright-to-dark transitions [Hofmann 2004]).

The basic mathematical tool used for the least squares fit of a straight line to segments with free floating boundary points at both ends in UBM is derived in Appendix C. This is an extension of the stripe-wise image interpretation underlying Figure 5.32. Figure C.3 in Appendix C shows a close-up view of blob 1 in the figure, indicating the improvements in line fit achievable. Therefore, after a recursive line fit in forward direction, the starting point should be checked against the variance of the fit achieved in the overall segment for possible improvements by dropping initial values (see Appendix C). This extension to a *floating fit* by checking the boundary points at both ends against the variance criterion has become the standard blob fit approach. Figure 5.33 shows some results for different vertical cross sections (columns) through an image (white lines in image top left with column numbers). The video field was compressed with $(m_c\ n_c) = (2\ 1)$; the threshold value for the variance allowed at the boundaries was VarLim = 7.5, and the minimal segment length accepted was $L_{seg} > L_{segmin} = 4$.

Several typical regions such as road, lane markings, dark shade underneath vehicles, and sky are designated; some correspondences are shown by dotted lines. In

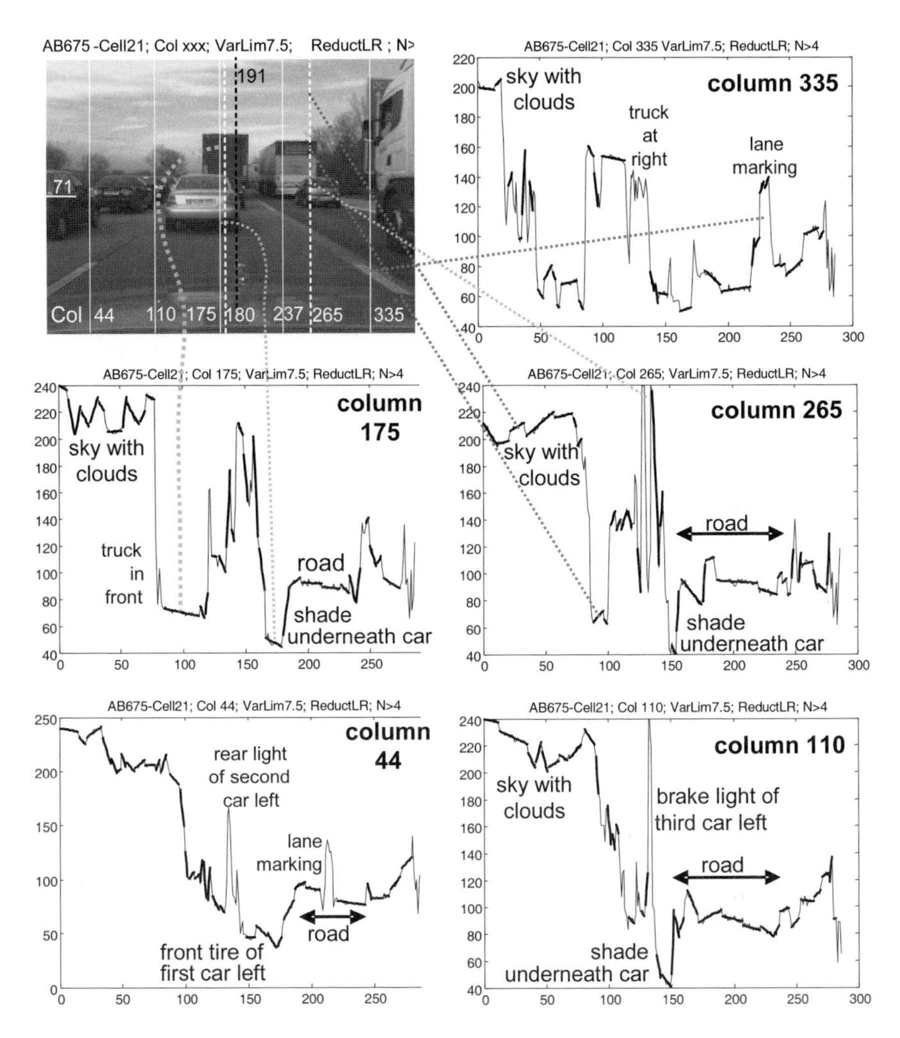

Figure 5.33. Intensity profiles through columns marked by white lines in image top left; the solid dark straight lines are the *1-D blobs* segmented

creasing L_{segmin} and possibly the threshold value VarLim to allow more generous grouping would lead to reduced numbers of *1-D blobs*; values for VarLim of up to 100 (ten intensity levels out of 256 typical for 8-bit intensity coding in video) yield acceptable results for many applications.

A good check for the usefulness of the blob concept is the image quality judged by humans, after the image has been reconstructed from its abstract blob representation by average gray values, shading parameters, and the locations of the segments. In [Hofmann 2004] this has been shown to work well for small masks, disregarding the special treatment of nonplanar regions and allowing arbitrarily small segment lengths. On the contrary, Figure 5.34 shows the reconstruction result from the first pyramid stage of a video field ($m_c = n_c = 2$) with mask elements $m = n = 2$

Figure 5.34. Image reconstruction for the parameter set (22.22) and vertical search direction, compressed horizontally 2:1 for display after processing. Scene recognition is no problem for the human observer

in column search (shorthand notation: 2222C). The resulting image has been compressed in row direction for standard viewing conditions. Recall that stepping of the mask in search direction is done by one-pixel-steps, while stripes are shifted by m = 2; after compression, this corresponds to one (pyramid-) pixel also.

The figure shows typical digitization effects, but object recognition is very well possible, even for an untrained person. Six vehicles can be distinguished clearly; four cars, one bus and one truck are easily distinguished. Lane markings and shadow boundaries look almost natural. The question is, how many of these details will be lost when image compression is done together with computing the pyramid image (i.e., $m_c\ n_c = 42$ instead of 22), when larger segments only are accepted for blob building, and when all nonlinear intensity regions (corner candidates and edges) are excluded beforehand.

The scene analyzed is only slightly different from the one before; the variance threshold VarLim has been set to 15 and the minimal segment length required is

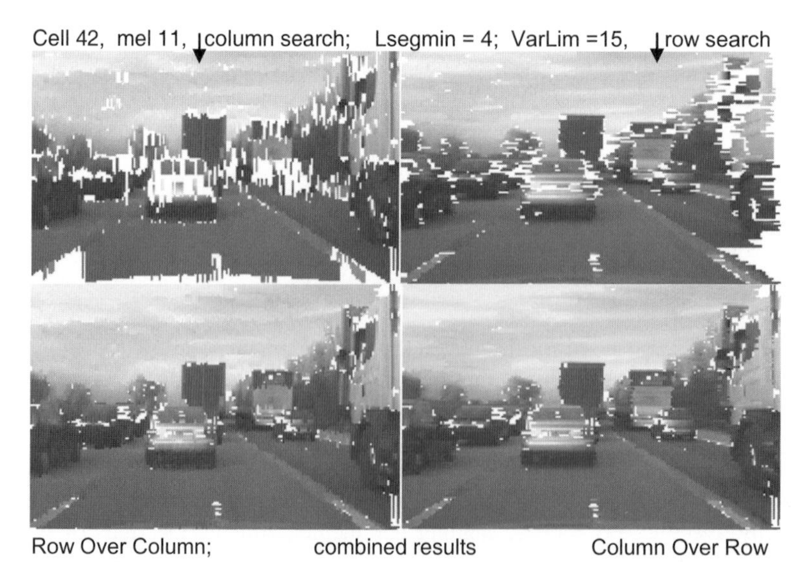

Figure 5.35. Reconstructed images from blob representations: Column search only (top left), row search only (top right), superimposed results (bottom); pixels missing have been filled by results from the alternate search direction

L_{segmin} = 4 pyramid-pixels or 16 original image pixels. Figure 5.35 shows the results for column search (top left), for row search (top right), and superimposed results, where pixels missing in one reconstructed image have been added from the other one, if available.

The number of blobs to be handled is at least one order of magnitude smaller than for the full representation underlying Figure 5.34. For a human observer, recognizing the road scene is not difficult despite the pixels missing. Since homogeneous regions in road scenes tend to be more extended horizontally, the superposition 'column over row' (bottom right) yields the more naturally looking results.

Note, however, that up to now no merging of blob results from one stripe to the next has been done by the program. When humans look at a scene, they cannot but do this unwillingly and apparently without special effort. For example, nobody will have trouble recognizing the road by its almost homogeneously shaded gray values. The transition from *1-D blobs* in separate stripes to *2-D blobs* in the image and to a 3-D surface in the outside world are the next steps of interpretation in machine vision.

5.3.2.5 Extended Shading Models in Image Regions

The *1-D blob* results from stripe analysis are stored in a list for each stripe, and are accumulated over the entire image. Each blob is characterized by

1. the image coordinates of its starting point (row respectively column number and its position j_{ref} in it),
2. its extension L_{seg} in search direction,
3. the average intensity I_c at its center, and
4. the average gradient components of the intensity a_u and a_v.

This allows easy merging of results of two neighboring stripes. Figure 5.36a shows the start of *1-D blob* merging when the *threshold conditions for merger* are satisfied in the region of overlap in adjacent stripes: (1) The amount of overlap should exceed a lower bound, say, two or three pixels. (2) The difference in image intensity at the center of overlap should be small. Since the *1-D blobs* are given by their cg-position ($u_{bi} = j_{ref} + L_{seg,i}/2$), their 'weights' (proportional to the segment length $L_{seg,i}$), and their intensity gradients, the intensities at the center of overlap can be computed in both stripes (I_{covl1}) and I_{covl2}) from the distance between the blob center and the center of overlap exploiting the gradient information. This yields the condition for acceptance

(a) Merging of first two *1-D blobs* to a *2-D blob*

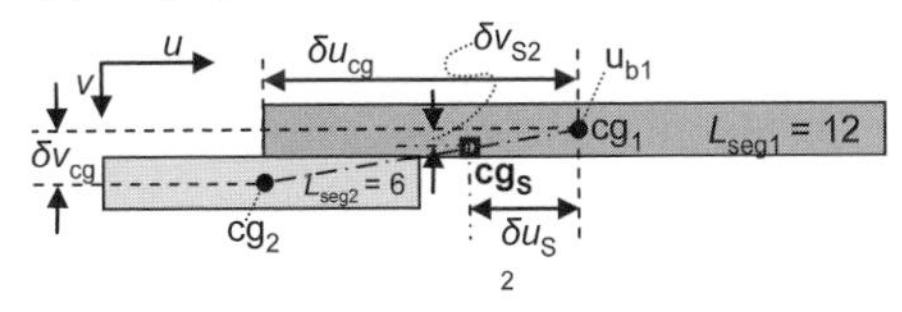

(b) Recursive merging of a *2-D blob* with an overlapping *1-D blob* to an extended *2-D blob*.

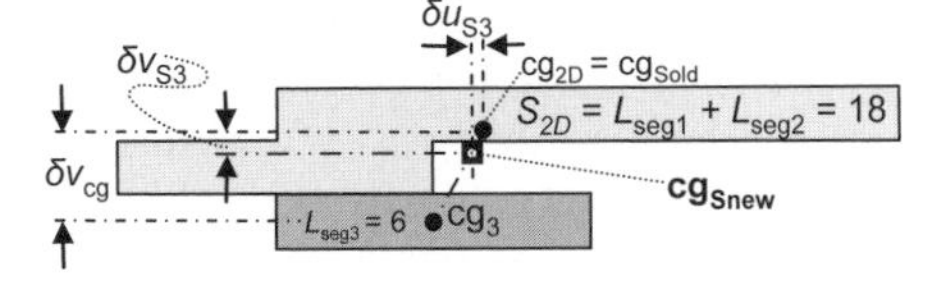

Figure 5.36. Merging of overlapping *1-D blobs* in adjacent stripes to a *2-D blob* when intensity and gradient components match within threshold limits

$$|I_{\mathrm{cov}l1} - I_{\mathrm{cov}l2}| < DelIthreshMerg . \qquad (5.37)$$

Condition (3) for merging is that the intensity gradients should also lie within small common bounds (difference < *DelSlopeThrsh*, see Table 5.1).

If these conditions are all satisfied, the position of the new cg after merger is computed from a balance of moments on the line connecting the cg's of the regions to be merged; the new cg of the combined areas S_{2D} thus has to lie on this line. This yields the equation (see Figure 5.36a)

$$\delta u_S \cdot L_{seg1} - (\delta u_{cg} - \delta u_S) \cdot L_{seg2} = 0, \qquad (5.38)$$

and, solved for the shift δu_S with $S_{2D} = L_{seg1} + L_{seg2}$, the relation

$$\delta u_S = L_{seg2} / (L_{seg1} + L_{seg2}) \cdot \delta u_{cg} = \delta u_{cg} \cdot L_{seg2} / S_{2D} \qquad (5.39)$$

is obtained. The same is true for the v-component

$$\delta v_S = L_{seg2} / (L_{seg1} + L_{seg2}) \cdot \delta v_{cg} = \delta v_{cg} \cdot L_{seg2} / S_{2D}. \qquad (5.40)$$

Figure 5.36b shows the same procedure for merging an existing *2-D blob*, given by its weight S_{2D}, the cg-position at cg_{2D}, and the segment boundaries in the last stripe. To have easy access to the latter data, the last stripe is kept in memory for one additional stripe evaluation loop even after the merger to *2-D blobs* has been finished. The equations for the shift in cg are identical to those above if L_{seg1} is replaced by S_{2Dold}. The case shown in Figure 5.36b demonstrates that the position of the cg is not necessarily inside the 2-D blob region.

A 2-D blob is finished when in the new stripe no area of overlap is found any more. The size S_{2D} of the 2-D blob is finally given by the sum of the L_{seg}-values of all stripes merged. The contour of the 2-D blob is given by the concatenated lower and upper bounds of the 1-D blobs merged. Minimum (u_{min}, v_{min}) and maximum values (u_{max}, v_{max}) of the coordinates yield the encasing box of area

$$A_{encbox} = (u_{max} - u_{min}) \cdot (v_{max} - v_{min}). \qquad (a) \qquad (5.41)$$

A measure of the compactness of a blob is the ratio

$$R_{compBlob} = S_{2D} / A_{encbox} . \qquad (b)$$

For close to rectangular shapes it is close to 1; for circles it is $\pi/4$, for a triangle it is 0.5, and for an oblique wide line it tends toward 0. The 2-D position of the blob is given by the coordinates of its center of gravity u_{cg} and v_{cg}. This robust feature makes highly visible blobs attractive for tracking.

5.3.2.6 Image Analysis on two Scales

Since coarse resolution may be sufficient for the near range and the sky, fine scale image analysis can be confined to that part of the image containing regions further away. After the road has been identified nearby, the boundaries of these image regions can be described easily around the subject's lane as looking like a "pencil tip" (possibly bent). Figure 5.37 shows results demonstrating that with highest resolution (within the white rectangles), almost no image details are lost both for the horizontal (left) and the vertical search (right).

The size and position of the white rectangle can be adjusted according to the actual situation, depending on the scene content analyzed by higher system levels. Conveniently, the upper left and lower right corners need to be given to define the

Figure 5.37. Foveal–peripheral differentiation of image analysis shown by the 'imagined scene' reconstructed from symbolic representations on different scales: Outer part 44.11, inner part 11.11 from video fields compressed 2:1 after processing; left: horizontal search, right: vertical search, with the Hofmann operator.

rectangle; the region of high resolution should be symmetrical around the horizon and around the center of the subject's lane at the look-ahead distance of interest, in general.

5.3.3 The Corner Detection Algorithm

Many different types of nonlinearities may occur on different scales. For a long time, so-called 2-D-features have been studied that allow avoiding the "aperture problem"; this problem occurs for features that are well defined only in one of the two degrees of freedom, like edges (sliding along the edge). Since general texture analysis requires significantly more computing power not yet available for real-time applications in the general case right now, we will also concentrate on those *points of interest* which allow reliable recognition and computation of feature flow [Moravec 1979; Harris, Stephens 1988; Tomasi, Kanade 1991; Haralick, Shapiro 1993].

5.3.3.1 Background for Corner Detection

Based on the references just mentioned, the following algorithm for corner detection fitting into the mask scheme for planar approximation of the intensity function has been derived and proven efficient. The structural matrix

$$N = \begin{pmatrix} (f_{r1N}^2 + f_{r2N}^2) & 2 \cdot f_{rN} \cdot f_{cN} \\ 2 \cdot f_{rN} \cdot f_{cN} & (f_{c1N}^2 + f_{c2N}^2) \end{pmatrix} = \begin{pmatrix} n_{11} & n_{12} \\ n_{21} & n_{22} \end{pmatrix} \tag{5.42}$$

has been defined with the terms from Equations 5.17 and 5.18. Note that compared to the terms used by previously named authors, the entries on the main diagonal are formed from local gradients (in and between half-stripes), while those on the cross-diagonal are twice the product of the gradient components of the mask (average of the local values). With Equation 5.18, this corresponds to half the sum of all four cross-products

$$n_{12} = n_{21} = 0.5 \cdot \sum_{i,j=1,2} (f_{riN} \cdot f_{cjN}) . \quad (5.43)$$

This selection yields proper tuning to separate corners from planar elements in all possible cases (see below). The determinant of the matrix is

$$w = \det N = n_{11} \cdot n_{22} - n_{12}^2 . \quad (5.44)$$

With the equations mentioned, this becomes

$$\det N = 0.75 \cdot n_{11} \cdot n_{22} - \\ -0.5 \cdot (n_{11} \cdot f_{c1} \cdot f_{c2} + n_{22} \cdot f_{r1} \cdot f_{r2}) - f_{r1} \cdot f_{r2} \cdot f_{c1} \cdot f_{c2} \cdot \quad (5.45)$$

Haralick calls $\det N = w$ the "Beaudet measure of cornerness", however, formed with a different term $n_{12} = \Sigma f_{ri} \cdot f_{ci}$. The eigenvalues λ of the structural matrix are obtained from

$$\begin{bmatrix} \lambda - n_{11} & -n_{12} \\ -n_{12} & \lambda - n_{22} \end{bmatrix} = (\lambda - n_{11})(\lambda - n_{22}) - n_{12}^2 = 0, \\ n_{11} \cdot n_{22} - \lambda(n_{11} + n_{22}) + \lambda^2 - n_{12}^2 = 0 . \quad (5.46)$$

With the quadratic enhancement term Q,

$$Q = (n_{11} + n_{22})/2 , \quad (5.47)$$

there follows for the two eigenvalues λ_1, λ_2,

$$\lambda_{1,2} = Q\left[1 \pm \sqrt{1 - \det N/Q^2}\right] . \quad (5.48)$$

Normalizing these with the larger eigenvalue λ_1 yields

$$\lambda_{1N} = 1 \quad ; \lambda_{2N} = \lambda_2 / \lambda_1 ; \\ \lambda_{2N} = \left(1 - \sqrt{1 - \det N / Q^2}\right) \Big/ \left(1 + \sqrt{1 - \det N / Q^2}\right) . \quad (5.49)$$

Haralick defines a measure of circularity q as

$$q = 1 - \left[\frac{\lambda_1 - \lambda_2}{\lambda_1 + \lambda_2}\right]^2 = \frac{4\lambda_1 \lambda_2}{(\lambda_1 + \lambda_2)^2} . \quad (5.50)$$

With Equation 5.48 this reduces to

$$q = \det N / Q^2 = 4 \cdot (n_{11} \cdot n_{22} - n_{12}^2)/(n_{11} + n_{22})^2 , \quad (5.51)$$

and in normalized terms (see Equation 5.49), there follows

$$q = 4 \cdot \lambda_{2N} / (1 + \lambda_{2N})^2 . \quad (5.52)$$

It can thus be seen that the normalized second eigenvalue λ_{2N} and circularity q are different expressions for the same property. In both terms, the absolute magnitudes of the eigenvalues are lost.

Threshold values for corner points are chosen as lower limits for the determinant $\det N = w$ and circularity q:

$$w > w_{min} \\ \text{and} \\ q > q_{min} . \quad (5.53)$$

In a post-processing step, within a user-defined window, only the maximal value of $w = w^*$ is selected.

Harris was the first to use the eigenvalues of the structural matrix for threshold definition. For each location in the image, he defined the performance value

$$R_H(y,z) = \det N - \alpha(trace\ N)^2 , \tag{5.54}$$

where $\det N = \lambda_1 \cdot \lambda_2$ and

$$trace\, N = \lambda_1 + \lambda_2 = 2Q , \tag{5.55}$$

yielding

$$R_H = \lambda_1\lambda_2 - \alpha(\lambda_1 + \lambda_2)^2 . \qquad \text{(a)} \tag{5.56}$$

With $\kappa = \lambda_2 / \lambda_1$ (λ_{2N}, see Equation 5.49), there follows

$$R_H = \lambda_1^2 \left[\kappa - \alpha(1+\kappa)^2 \right]. \qquad \text{(b)}$$

For $R_H \geq 0$ and $0 \leq \kappa \leq 1$, α has to be selected in the range,

$$0 \leq \alpha \leq \kappa / (1+\kappa)^2 \leq 0.25 . \tag{5.57}$$

Corner candidates are points for which $R_H \geq 0$ is valid; larger values of α yield fewer corners and *vice versa*. Values around α = 0.04 to 0.06 are recommended. This condition on R_H is equivalent to (from Equations 5.44, 5.53, and 5.54)

$$\det N > 4\alpha Q^2 . \tag{5.58}$$

Kanade et al. (1991) (KLT) use the following corner criterion: After a smoothing step, the gradients are computed over the region $D{\cdot}D$ ($2 \leq D \leq 10$ pixels). The reference frame for the structural matrix is rotated so that the larger eigenvalue λ_1 points in the direction of the steepest gradient in the region

$$\lambda_{1KLT} = \sqrt{f_r^2 + f_c^2} . \tag{5.59}$$

λ_1 is thus normal to a possible edge direction. A corner is assumed to exist if λ_2 is sufficiently large (above a threshold value λ_{2thr}). From the relation $\det N = \lambda_1 \cdot \lambda_2$, the corresponding value of λ_{2KLT} can be determined

$$\lambda_{2KLT} = \det N / \lambda_{1KLT} . \tag{5.60}$$

If

$$\lambda_{2KLT} > \lambda_{2thr} , \tag{5.61}$$

the corresponding image point is put in a candidate list. At the end, this list is sorted in decreasing order of λ_{2KLT}, and all points in the neighborhood with smaller λ_{2KLT} values are deleted. The threshold value has to be derived from a histogram of λ_2 by experience in the domain. For larger D, the corners tend to move away from the correct position.

5.3.3.2 Specific Items in Connection with Local Planar Intensity Models

Let us first have a look at the meaning of the threshold terms circularity (q in Equation 5.50) and trace N (Equation 5.55) as well as the normalized second eigenvalue (λ_{2N} in Equation 5.49) for the specific case of four symmetrical regions in a 2 × 2 mask, as given in Figure 5.20. Let the perfect rectangular corner in intensity distribution as in Figure 5.38b be given by local gradients $f_{r1} = f_{c1} = 0$ and $f_{r2} = f_{c2} = -K$. Then the global gradient components are $f_r = f_c = -K/2$. The determinant Equation

5.44 then has the value det $N = 3/4{\cdot}K^4$. The term Q (Equation 5.47) becomes $Q = K^2$, and the "circularity" q according to Equation 5.51 is

$$q = \det\ N / Q^2 = 4/3 = 0.75. \tag{5.62}$$

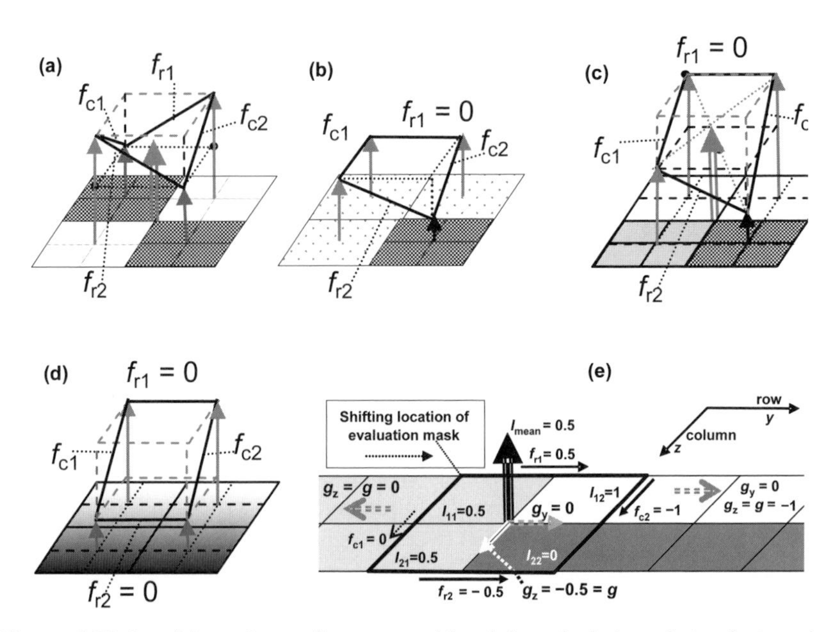

Figure 5.38. Local intensity gradients on mel-level for calculation of *circularity* q in corner selection: (a) Ideal checker-board corner: $q = 1$. (b) ideal single corner: $q = 0.75$; (c) slightly more general case (three intensity levels, closer to planar); (d) ideal shading, one direction only (linear case for interpolation, $q \approx 0$); (e) demanding (idealized) corner feature for extraction (see text).

The two eigenvalues of the structure matrix are $\lambda_1 = 1.5{\cdot}K^2$, and $\lambda_2 = 0.5{\cdot}K^2$ so that trace$N = 2Q$ is $4{\cdot}K^2$; this yields the normalized second eigenvalue as $\lambda_{2N} = 1/3$. Table 5.2 contains this case as the second row. Other special cases according to the intensity distributions given in Figure 5.38 are also shown. The maximum circularity of 1 occurs for the checkerboard corners in Figure 5.38a and row 1 in Table 5.2; the normalized second eigenvalue also assumes its maximal value of 1 in this case. The case Figure 5.38c (third row in the table) shows the more general situation with three different intensity levels in the mask region. Here, circularity is still close to 1 and λ_{2N} is above 0.8. The case in Figure 5.38e with constant average mask intensity in the stripe is shown in row 5 of Table 5.2: Circularity is rather high at $q = 8/9 \approx 0.89$ and $\lambda_{2N} = 0.5$. Note that from the intensity and gradient values of the whole mask this feature can only be detected by g_z (I_M and g_y) remain constant along the search path.

By setting the minimum required circularity q_{min} as the threshold value for acceptance to

$$q_{\min} \simeq 0.7\,, \tag{5.63}$$

all significant cases of intensity corners will be picked. Figure 5.38d shows an almost planar intensity surface with gradients $-K$ in the column direction and a very small gradient $\pm\,\varepsilon$ in the row direction ($K >> \varepsilon$). In this case all characteristic values: det N, circularity q, and the normalized second eigenvalue λ_{2N} all go to zero (row 4 in the table). The last case in Table 5.2 shows the special planar intensity distribution with the same value for all local and global gradients ($-K$); this corresponds to Fig 5.20c. It can be seen that circularity and λ_{2N} are zero; this nice feature for the general planar case is achieved through the factor 2 on the cross-diagonal of the structure matrix Equation 5.42.
When too many corner candidates are found, it is possible to reduce their number not by lifting q_{min} but by introducing another threshold value traceN_{min} that limits the sum of the two eigenvalues. According to the main diagonals of Equations 5.42 and 5.46, this means prescribing a minimal value for the sum of the squares of all local gradients in the mask.

Table 5.2. Some special cases for demonstrating the characteristic values of the structure matrix in corner selection as a function of a single gradient value K. Trace$N$ is twice the value of Q (column 4).

Example	Local gradient values	Det. N Equation 5.44	Term Q Equation 5.47	Circularity q	λ_1	$\lambda_{2N} = \lambda_2/\lambda_1$
Figure 5.38a	$+, -K$ (2 each)	$4\,K^4$	$2\,K^2$	1	$2\,K^2$	1
Figure 5.38b	$0, -K$ (2 each)	$\frac{3}{4}\,K^4$	K^2	0. 75	$1.5\,K^2$	0.3333
Figure 5.38c	$0, -K\ (f_{c1}, f_{r2}),\ -2K$	$5\,K^4$	$3\,K^2$	$5/9 = 0.556$	$5\,K^2$	0.2
Figure 5.38e	$f_{ri} = \pm\,K$ $f_{ci} = 0; -2K$	$8\,K^4$	$3\,K^2$	8/9	$4\,K^2$	0.5
Figure 5.38d	$f_{ri} = \pm\varepsilon\ (<< K)$ $f_{ci} = -K$	$4 * \varepsilon^2\,K^2 \approx 0$	$(\varepsilon^2+K^2) \approx K^2$	$\sim 4\,\varepsilon^2/K^2 \approx 0$	$\approx 2 * (K^2 - \varepsilon^2)$	$\approx 2\varepsilon^2/K^2 \approx 0$
Planar	$f_{i,j} = -K\ (4\times)$	0	$2\,K^2$	0	$4 * K^2$	0

This parameter depends on the absolute magnitude of the gradients and has thus to be adapted to the actual situation at hand. It is interesting to note that the planarity check (on 2-D curvatures in the intensity space) for interpolating a tangent plane to the actual intensity data has a similar effect as a low boundary of the threshold value, traceN_{min}.

5.3.4 Examples of Road Scenes

Figure 5.39 left shows the nonplanar regions found in horizontal search (white bars) with ErrMax = 3%. Of these, only those locations marked by cyan crosses have been found satisfying the corner condition $q_{min} = 0.6$ and trace$N_{min} = 0.11$. The figure on the right-hand side shows results with the same parameters except the reduction of the threshold value to trace$N_{min} = 0.09$, which leaves an increased

Figure 5.39. Corner candidates derived from regions with planar interpolation residues > 3% (white bars) with parameters (m, n, m_c, n_c = 3321). The circularity threshold q_{min} = 0.6 eliminates most of the candidates stemming from digitized edges (like lane markings). The number of corner candidates can be reduced by lifting the threshold on the sum of the eigenvalues $traceN_{min}$ from 0.09 (right: 103, 121) to 0.11 (left image: 63, 72 candidates); cyan = row search, red = column search.

number of corner candidates (over 60% more). Note that all oblique edges (showing minor corners from digitization), which were picked by the nonplanarity check, did not pass the corner test (no crosses in both figures). The crosses mark corner candidates; from neighboring candidates, the strongest yet has to be selected by comparing results from different scales. $m_c = 2$ and $n_c = 1$ means that two original pixels are averaged to a single cell value; nine of those form a mask element (18 pixels), so that the entire mask covers 18×4 = 72 original pixels.

Figure 5.40 demonstrates all results obtainable by the unified blob-edge-corner method (UBM) in a busy highway scene in one pass: The upper left subfigure shows the original full video image with shadows from the cars on the right-hand side. The image is analyzed on the pixel level with mask elements of size four pixels (total mask = 16 pixels). Recall that masks are shifted by steps of 1 in search direction and by steps of mel-size in stripe direction. About 10^5 masks result for evaluation of each image. The lower two subfigures show the small nonplanarity regions detected (about 1540), marked by white bars. In the left figure the edge elements extracted in row search (yellow, = 1000) and in column search (red, = 3214) are superimposed. Even the shadow boundaries of the vehicles and the reflections from the own motor hood (lower part) are picked. The circularity threshold of q_{min} = 0.6 and $traceN_{min}$ = 0.2 filter up to 58 corner candidates out of the 1540 nonplanar mask results; row and column search yield almost identical results (lower right). More candidates can be found by lowering ErrMax and $traceN_{min}$.

Combining edge elements to lines and smooth curves, and merging 1-D blobs to 2-D (regional) blobs will drastically reduce the number of features. These compound features are more easily tracked by prediction error feedback over time. Sets of features moving in conjunction, *e.g.* blobs with adjacent edges and corners, are indications of objects in the real world; for these objects, motion can be predicted and changes in feature appearance can be expected (see the following chapters). Computing power is becoming available lately for handling the features mentioned in several image streams in parallel. With these tools, machine vision is maturing for application to rather complex scenes with multiple moving objects. However, quite a bit of development work yet has to be done.

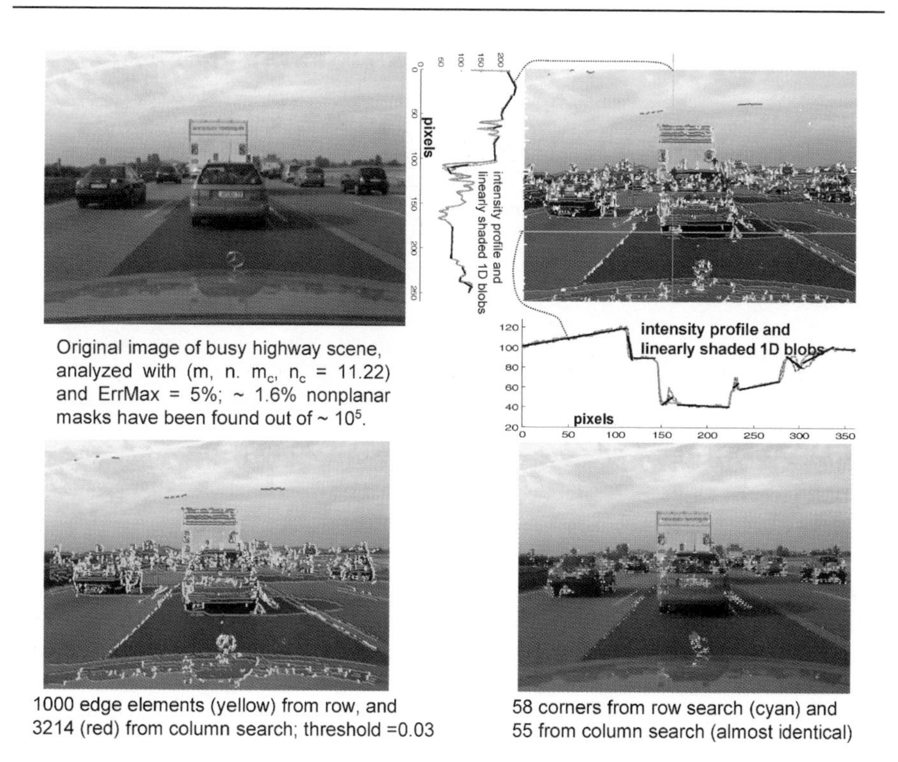

Figure 5.40. Features extracted with unified blob-edge-corner method (UBM): Bidirectionally nonplanar intensity distributions (white regions in lower two subfigures, ~ 1540), edge elements and corner candidates (column search in red), and linearly shaded blobs. One vertical and one horizontal example is shown (gray straight lines in upper right subfigure with dotted lines connecting to the intensity profiles between the images. Red and green are the intensity profiles in the two half-stripes used in UBM; about 4600 1-D blobs resulted, yielding an average of 15 blobs per stripe. The top right subfigure is reconstructed from symbolically represented features only (no original pixel values). Collections of features moving in conjunction designate objects in the world.

Conclusion of section 5.3 (UBM): Figure 5.41 shows a road scene with all features extractable by the unified blob-edge-corner method UBM superimposed. The image processing parameters were: MaxErr = 4%; $m = n = 3$, $m_c = 2$, $n_c = 1$ (33.21); anglefact = 0.8 and IntGradMin = 0.02 for edge detection; $q_{min} = 0.7$; trace$N_{min} = 0.06$ for corner detection and $Lseg_{min} = 4$, VarLim = 64 for shaded blobs. Features extracted were 130 corner candidates, 1078 nonplanar regions (1.7%), 4223 ~vertical edge elements, 5918 ~horizontal edge elements, 1492 linearly shaded intensity blobs (from row search) and 1869 from column search; the latter have been used only partially to fill gaps remaining from the row search. The nonplanar regions remaining are the white areas.

Only an image with several colors can convey the information contained to a human observer. The entire image is reconstructed from symbolic representations of the features stored. The combination of linearly shaded blobs with edges and corners alleviates the generation of good object hypotheses, especially when char-

Figure 5.41. "Imagined" feature set extracted with the unified blob-edge-corner method UBM: Linearly shaded blobs (gray areas), horizontally (green) and vertically extracted edges (red), corners (blue crosses) and nonhomogeneous regions (white).

acteristic sub-objects such as wheels can be recognized. With the background knowledge that wheels are circular (for smooth running on flat ground) with the center on a horizontal axis in 3-D space, the elliptical appearance in the image allows immediate determination of the aspect angle without any reference to the body on which it is mounted. Knowing some state variables such as the aspect angle reduces the search space for object instantiation in the beginning of the recognition process after detection.

5.4 Statistics of Photometric Properties of Images

According to the results of planar shading models (Section 5.3.2.4), a host of information is now available for analyzing the distribution of image intensities to adjust parameters for image processing to lighting conditions [Hofmann 2004]. For each image stripe, characteristic values are given with the parameters of the shading models of each segment. Let us assume that the intensity function of a stripe can be described by n_s segments. Then the average intensity b_s of the entire stripe over all segments i of length l_i and average local intensity b_i is given by

$$b_S = \sum_{i=1}^{n_s} (l_i \cdot b_i) / \sum_{i=1}^{n_s} l_i . \tag{5.64}$$

For a larger region G segmented into n_G image stripes, then follows

$$b_G = \sum_{j=1}^{n_G} \left[\sum_{i=1}^{n_{Sj}} (l_{ij} \cdot b_{ij}) \right] \Big/ \sum_{j=1}^{n_G} \left[\sum_{i=1}^{n_{Sj}} (l_{ij}) \right]. \tag{5.65}$$

The values of b_S and b_G are different from the mean value of the image intensity since this is given by

$$b_{MeanS} = \left(\sum_{i=1}^{n_S} b_i \right) \Big/ n_S \,,$$

resp., (5.66)

$$b_{MeanG} = \left[\sum_{j=1}^{n_G} \left(\sum_{i=1}^{n_{Sj}} b_{ij}) \right) \right] \Big/ \sum_{j=1}^{n_G} n_{Sj}$$

The absolute minimal and maximal value of all mel–intensities of a single stripe can be obtained by standard comparisons as $\text{Min}_{\min S}$ and $\text{Max}_{\max S}$; similarly, for a larger region, there follows

$$\text{Min}_{\min G} = \min_{j=1}^{n_G} \left[\min_{i=1}^{n_{Sj}} (\min_{ij}) \right]; \quad \text{Max}_{\max G} = \max_{j=1}^{n_G} \left[\max_{i=1}^{n_{Sj}} (\max_{ij}) \right]. \tag{5.67}$$

The difference between both expressions yields the dynamic range in intensity H_S within an image stripe, respectively, H_G an image region. The dynamic range in intensity of a single segment is given by $H_i = \max_i - \min_i$. The average dynamic range within a stripe, respectively in an image region, then follows as

$$H_{MeanS} = \left(\sum_{i=1}^{n_S} H_i \right) \Big/ n_S; \quad \text{resp.,} \quad H_{MeanG} = \left[\sum_{j=1}^{n_G} \left(\sum_{i=1}^{n_{Sj}} H_{ij} \right) \right] \Big/ \sum_{j=1}^{n_G} n_{Sj} \,. \tag{5.68}$$

If the maximal or minimal intensity value is to be less sensitive to single outliers in intensity, the maximal, respectively, minimal value over all average values b_i of all segments may be used:

$$b_{MinS} = \min_{i=1}^{n_S} (b_i) \,,$$

resp. (5.69)

$$b_{MaxS} = \max_{i=1}^{n_S} (b_i) \,;$$

similarly, for larger regions there follows

$$b_{MinG} = \min_{j=1}^{n_G} \left[\min_{i=1}^{n_{Sj}} (b_{ij}) \right] \quad ; \quad b_{MaxG} = \max_{j=1}^{n_G} \left[\max_{i=1}^{n_{Sj}} (b_{ij}) \right] . \tag{5.70}$$

Depending on whether the average value of the stripe is closer to the minimal or maximal value, the stripe will appear rather darker than brighter.

An interesting characteristic property of edges is the average intensity on both sides of the edge. This has been used for two decades in connection with the method CRONOS for the association of edges with objects. When using several cameras with independent apertures, gain factors, and shutter times, the ratio of these intensities varies least over time; absolute intensities are not that stable, generally. Statistics on local image areas, respectively, single stripes should always be judged in relation to similar statistics over larger regions. Aside from characteristics of image regions at the actual moment, systematic temporal changes should also be monitored, for example, by tracking the changes in average intensity values or in variances.

The next section describes a procedure for finding transformations between images of two cameras looking at the same region (as in stereovision) to alleviate joint image (stereo) interpretation.

5.4.1 Intensity Corrections for Image Pairs

This section uses some of the statistical values defined previously. Two images of a stereo camera pair are given that have different image intensity distributions due to slightly different apertures, gain values, and shutter times that are independently automatically controlled over time (Figure 5.42a, b). The cameras have approximately parallel optical axes and the same focal length; they look at the same scene. Therefore, it can be assumed that segmentation of image regions will yield similar results except for absolute image intensities. The histograms of image intensities are shown in the left-hand part of the figure. The right (lower) stereo image is darker than the left one (top). The challenge is to find a transformation procedure which allows comparing image intensities to both sides of edges all over the image.

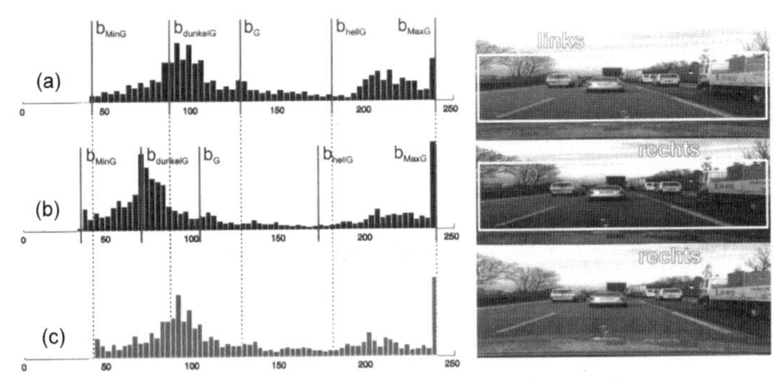

Figure 5.42. Images of different brightness of a stereo system with corresponding histograms of the intensity: (a) left image, (b) right-hand-side image, and (c) right-hand image adapted to the intensity distribution of the left-hand image after the intensity transformation described (see Figure 5.43, after [Hofmann 2004]).

The lower sub–figure (c) shows the final result. It will be discussed after the transformation procedure has been derived.

At first, the characteristic photometric properties of the image areas within the white rectangle are evaluated in both images by the stripe scheme described. The left and right bars in Figure 5.43a, b show the characteristic parameters considered. In the very bright areas of both images (top), saturation occurs; this harsh nonlinearity ruins the possibility of smooth transformation in these regions. The intensity transformation rule is to be derived using five support points: b_{MinG}, $b_{dunkelG}$, b_G , b_{hellG} and b_{MaxG} of the marked left and right image regions. The full functional relationship is approximated by interpolation of these values with a fourth-order polynomial. The central upper part of Figure 5.43 shows the resulting function as a curve; the lower part shows the scaling factors as a function of the intensity values. The support points are marked as dots. Figure 5.42c shows the adapted histogram on the left-hand side and the resulting image on the right-hand side. It can be seen that after the transformation, the intensity distribution in both images has become much more similar. Even though this transformation is only a coarse approximation, it shows that it can alleviate evaluation of image information and correspondence of features.

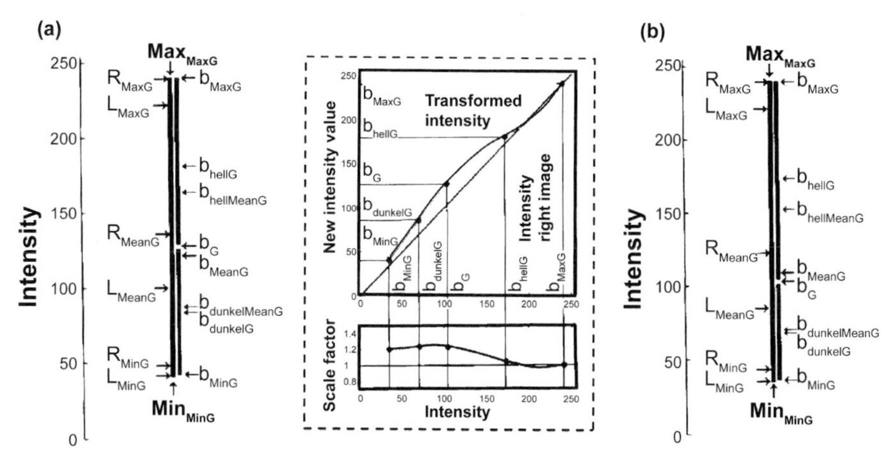

Figure 5.43. Statistical photometric characteristics of (a) the left-hand, and (b) the right-hand stereo image (Figure 5.42); the functional transformation of intensities shown in the center minimizes differences in the intensity histogram (after [Hofmann 2004]).

The largest deviations occur at high intensity values (right end of histogram); fortunately, this is irrelevant for road-scene interpretation since the corresponding regions belong to the sky.

To become less dependent on intensity values of single blob features in two or more images, the ratio of intensities of several blob features recognized with uncertainty and their relative location may often be advantageous to confirm an object hypothesis.

5.4.2 Finding Corresponding Features

The ultimate goal of feature extraction is to recognize and track objects in the real world, that is the scene observed. Simplifying feature extraction by reducing, at first search, spaces to image stripes (a 1-D task with only local lateral extent) generates a difficult second step of merging results from stripes into regional characteristics (2-D in the image plane). However, we are not so much interested in (virtual) objects in the image plane (as computational vision predominantly is) but in recognizing 3-D-objects moving in 3-D space over time! Therefore, all the knowledge about motion continuity in space and time in both translational and rotational degrees of freedom has to be brought to bear as early as possible. Self-occlusion and partial occlusion by other objects has to be taken into account and observed from the beginning. Perspective mapping is the link from spatio–temporal motion of 3-D objects in 3-D space to the motion of groups of features in images from different cameras.

So the difficult task after basic feature extraction is to find combinations of features belonging to the same object in the physical world and to recognize these features reliably in an image sequence from the same camera and/or in parallel images from several cameras covering the same region in the physical world, maybe under slightly different aspect conditions. Thus, finding corresponding features is a basic task for interpretation; the following are major challenges to be solved in dynamic scene understanding:

- Chaining of neighboring features to edges and merging local regions to homogeneous more global ones.
- Selecting the best suited feature from a group of candidates for prediction error feedback in recursive tracking (see below).
- Finding the corresponding features in sequences of images for determining *feature flow*, a powerful tool for motion understanding.
- Finding corresponding features in parallel images from different cameras for stereointerpretation to recover depth information lost in single images.

The rich information derived in previous sections from stripewise intensity approximation of one-dimensional segments alleviates the comparison necessary for establishing correspondence, that is, to quantify similarity. Depending on whether homogeneous segments or segment boundaries (edges) are treated, different criteria for quantifying similarity can be used.

For segment boundaries as features, the type of intensity change (bright to dark or *vice versa*), the position and the orientation of the edge as well as the ratio of average intensities on the right- (R) and left-hand (L) side are compared. Additionally, average intensities and segment lengths of adjacent segments may be checked for judging a feature in the context of neighboring features.

For homogeneous segments as features considered, average segment intensity, average gradient direction, segment length, and the type of transition at the boundaries (dark-to-bright or bright-to-dark) are compared. Since long segments in two neighboring image stripes may have been subdivided in one stripe but not in the other (due to effects of thresholding in the extraction procedure), chaining (concatenation) procedures should be able to recover from these arbitrary effects according to criteria to be specified. Similarly, chaining rules for directed edges are able to close gaps if necessary, that is, if the remaining parameters allow a consistent interpretation.

5.4.3 Grouping of Edge Features to Extended Edges

The stripewise evaluation of image features discussed in Section 5.3 yields (beside the nonplanar regions with potential corners) lists of corners, oriented edges, and homogeneously shaded segments. These lists together with the corresponding index vectors allow fast navigation in the feature database retaining neighborhood relationships. The index vectors contain for each search path the corresponding image row (respectively, column), the index of the first segment in the list of results, and the number of segments in each search path.

As an example, results of concatenation of edge elements (for column search) are shown in Figure 5.44, lower right (from [Hofmann 2004]); the steps required are discussed in the sequel. Analog to image evaluation, concatenation proceeds in search path direction [top-down, see narrow rectangle near (a)] and from left to right. The figure at the top shows the original video-field with the large white rectangle marking the part evaluated; the near sky and the motor hood are left off. All features from this region are stored in the feature database. The lower two subfigures are based on these data only.

At the lower left, a full reconstruction of image intensity is shown based on the complete set of shading models on a fine scale, disregarding nonlinear image intensity elements like corners and edges; these are taken here only for starting new blob segments. The number of segments becomes very large if the quality of the reconstructed image is requested to please human observers. However, if edges from neighboring stripes are grouped together, the resulting extended line features allow to reduce the number of shaded patches to satisfy a human observer and to appreciate the result of image understanding by machine vision.

Linear concatenation of directed 2-D edges: Starting from the first entry in the data structure for the edge feature, an entry into the neighboring search stripe is looked for, which approximately satisfies the colinearity condition with the stored edge direction at a distance corresponding to the width of the search stripe. To accept correspondence, the properties of a candidate edge point have to be similar to the average properties of the edge elements already accepted for chaining. To evaluate similarity, criteria like the *Mahalanobis*-distance may be computed, which allow weighting the contributions of different parameters taken into account. A threshold value then has to be satisfied to be accepted as sufficiently similar. An-

Figure 5.44. From features in stripes (here vertical) to feature aggregations over the 2-D image in direction of object recognition: (a) Reduced vertical range of interest (motor hood and sky skipped). (b) Image of the scene as internally represented by symbolically stored feature descriptions ('imagined' world). (c) Concatenated edge features which together with homogeneously shaded areas form the basis for object hypothesis generation and object tracking over time.

other approach is to define intervals of similarity as functions of the parameters to be compared; only those edge points are considered similar with respect to a yet concatenated set of points, whose parameters lie within the intervals.

If a similar edge point is found, the following items are computed as a measure of the quality of approximation to the interpolating straight line: The slope a, the average value b, and the variance Var of the differences between all edge points and the interpolating straight line. The edge point thus assigned is marked as "used" so that it will no longer be considered when chaining other edge candidates, thus saving computing time.

If no candidate in the interval of the search stripe considered qualifies for acceptance, a number of gaps up to a predefined limit may be bridged by the procedure until concatenation for the contour at hand is considered finished. Then a new contour element is started with the next yet unused edge point. The procedure ends when no more edge points are available for starting or as candidates for chaining. The result of concatenation is a list of linear edge elements described by the parameter set given in Table 5.3 [Hofmann 2004].

Linear concatenation admits edge points as candidates only when the orthogonal distance to the interpolated straight line is below a threshold value specified by the variance Var. If grouping of edge points along an arbitrary, smooth, continuous curve is desired, this procedure is not applicable. For example, for a constantly curved line, the method stops after reaching the *epsilon-tube* with a nominal curve. Therefore, the next section gives an extension of the concatenation procedure for grouping sets of points by local concatenation. By limiting the local extent of feature points already grouped, relative to which the new candidate point has to satisfy similarity conditions (local window), smoothly changing or constantly curved aggregated edges can be handled. With respect to the local window, the procedure is exactly the same as before.

Table 5.3. Data structure for results of concatenation of linear edge elements

uBegin, vBegin	Starting point in image coordinates (pixel)
uEnd, vEnd	End point in image coordinates
du, dv	Direction components in image coordinates
AngleEdge	Angle of edge direction in image plane
a	Slope in image coordinates
b	Reference point for the edge segment (average)
Len	Length of edge segment
MeanL	Average intensity value on left-hand side of the edge
MeanR	Average intensity value on right-hand side of the edge
MeanSegPos	Average segment length in direction of search path
MeanSegNeg	Average segment length in opposite direction of search path
NrPoints	Number of concatenated edge points
su, sv	sum of u-, resp., v-coordinate values of concatenated edge points
suu, svv	sum of the squares of u-, resp., v-coordinate values of concatenated edge points

When the new edge point satisfies the conditions for grouping, the local window is shifted so that the most distant concatenated edge point is dropped from the list for grouping. This grouping procedure terminates as soon as no new point for grouping can be found any longer. The parameters in Table 5.3 are determined for the total set of points grouped at the end. Since for termination of grouping the epsilon-tube is applied only to the local window, the variance of the deviations of all edge points grouped relative to the interpolated straight line may and will be larger than epsilon. Figure 5.44c (lower right) shows the results of this concatenation process. The image has been compressed after evaluation in row direction for standard viewing conditions).

The lane markings and the lower bounds of features originating from other vehicles are easily recognized after this grouping step. Handling the effects of shadows has to be done by information from higher interpretation levels.

5.5 Visual Features Characteristic of General Outdoor Situations

Due to diurnal and annual light intensity cycles and due to shading effects from trees, woods, and buildings, *etc.*, the conditions for visual scene recognition may vary to a large extent. To recognize weather conditions and the state of vegetation encountered in the environment, color recognition over large areas of the image may be necessary. Since these are slowly changing conditions, in general, the corresponding image analysis can be done at a much lower rate (*e.g.*, one to two orders of magnitude less than the standard video rate, *i.e.*, about every half second to once every few seconds).

One important item for efficient image processing is the adaptation of threshold values in the algorithms, depending on brightness and contrast in the image. Using a few image stripes distributed horizontally and vertically across the image and computing the statistic representation mentioned in Section 5.4 allows grasping the essential effects efficiently at a relatively high rate. If necessary, the entire image can be covered at different resolutions (depending on stripe width selected) at a lower rate.

During the initial summation for reducing the stripe to a single vector, statistics can be done yielding the brightest and the darkest pixel value encountered in each cross-section and in the overall stripe. If the brightest and the darkest pixel values in some cross-sections are far apart, the average values represented in the stripe vector may not be very meaningful, and adjustments should be made for the next round of evaluation.

On each pyramid level with spatial frequency reduced by a factor of 2, new mean intensities and spatial gradients (contrasts)of lower frequency are obtained. The maximum and minimum values on each level relative to those on other levels yield an indication of the distribution of light in spatial frequencies in the stripe. The top pixel of the pyramid represents the average image intensity (gray value) in the stripe. The ratio of minimum to maximum brightness on each level yields the maximum dynamic range of light intensity in the stripe. Evaluating these extreme and average intensity values relative to each other provides the background for threshold adaptation in the stripe region.

For example, if all maximum light intensity values are small, the scene is dark, and gradients should also be small (like when looking to the ground at dusk or dawn or moonlight). However, if all maximum light intensity values are large, the gradients may either be large or small (or in between). In the former case, there is good contrast in the image, while in the latter one, the stripe may be bright all over [like when looking to the sky (upper horizontal image stripe)], and contrast may be poor on a high intensity level. Vertical stripes in such an image may still yield good contrasts, as much so as to disallow image evaluation with standard threshold settings in the dark regions (near the ground). Looking almost horizontally toward a sunset is a typical example. The upper horizontal stripe may be almost saturated all over in light intensity; a few lower stripes covering the ground may have very low maximal values over all of the image columns (due to uniform automatic iris adjustment over the entire image. The vertical extension of the region with average intensities may be rather small. In this case, treating the lower part of the image with different threshold values in standard algorithms may lead to successful interpretation not achievable with homogeneous evaluations all over the entire image.

For this reason, cameras are often mounted looking slightly downward in order to avoid the bright regions in the sky. (Note that for the same reason most cars have shading devices at the top of the windshield to be adjusted by the human driver if necessary.)

Concentrating attention on the sky, even weather conditions may be recognizable in the long run. This field is wide open for future developments. Results of these evaluations of general situational aspects may be presented to the overall situation assessment system by a memory device similar to the scene tree for single objects. This part has been indicated in Figure 5.1 on the right-hand side. There is a 'specialist block' on level 2 (analogous to the object / subject-specialists labeled 3 and 4 to the left) which has to derive these non-object-oriented features which, nonetheless, contribute to the situation and have to be taken into account for decision-making.

6 Recursive State Estimation

Real-time vision is not perspective inversion of a sequence of images. Spatial *recognition* of objects, as the first syllable (*re-*) indicates, requires previous knowledge of structural elements of the 3-D shape of an object seen. Similarly, understanding of motion requires knowledge about some basic properties of temporal processes to grasp more deeply what can be observed over time. To achieve this deeper understanding while performing image evaluation, use will be made of the knowledge representation methods described in Chapters 2 and 3 (shape and motion). These models will be fitted to the data streams observed exploiting and extending least-squares approximation techniques [Gauss 1809] in the form of recursive estimation [Kalman 1960].

Gauss improved orbit determination from measurements of planet positions in the sky by introducing the theoretical structure of planetary orbits as cuts through cones. From Newton's gravitational theory, it was hypothesized that planets move in elliptical orbits. These ellipses have only a few parameters to completely describe the orbital plane and the trajectory in this plane. Gauss set up this description with the structure prespecified but all parameters of the solution open for adjustment depending on the measurement values, which were assumed to contain noise effects. The parameters now were to be determined such that the sum of all errors squared was minimal. This famous idea brought about an enormous increase in accuracy for orbit determination and has been applied to many identification tasks in the centuries following.

The model parameters could be adapted only after all measurements had been taken and the data had been introduced *en bloc* (so-called batch processing). It took almost two centuries to replace the batch processing method with sequential data processing. The need for this occurred with space flight, when trajectories had to be corrected after measuring the actual parameters achieved, which, usually, did not exactly correspond to those intended. Early correction is important for saving fuel and gaining payload. Digital computers started to provide the computing power needed for this corrective trajectory shaping. Therefore, it was in the 1960s that Kalman rephrased the least-squares approximation for evolving trajectories. Now, no longer could the model for the analytically solved trajectory be used (integrals of the equations of motion), but the differential equations describing the motion constraints had to be the starting point. Since the actually occurring disturbances are not known when a least-squares approximation is performed sequentially, the statistical distribution of errors has to be known or has to be estimated for formulating the algorithm. This step to "recursive estimation" opened up another wide field of applications over the last five decades, especially since the *extended Kalman filter* (EKF) was introduced for handling nonlinear systems when they were linearized around a nominal reference trajectory known beforehand [Gelb 1974; Maybeck 1979; Kailath *et al.* 2000].

This EKF method has also been applied to image sequence processing with arbitrary motion models in the image plane (for example, noise corrupted constant speed components), predominantly with little success in the general case. In the mid-1980s, it had a bad reputation in the vision community. The situation changed, when motion models according to the physical laws in 3-D space over time were introduced. Of course, now perspective projection from physical space into the image plane was part of the measurement process and had to be included in the measurement model of the EKF. This was introduced by the author's group in the first half of the 1980s. At that time, there was much discussion in the AI and vision communities about 2-D, 2.5-D, and 3-D approaches to visual perception of image sequences [Marr, Nishihara 1978; Ballard, Brown 1982; Marr 1982; Hanson, Riseman 1987; Kanade 1987].

Two major goals were attempted in our approach to visual perception by unlimited image sequences: (1) Avoid storing full images, if possible at all, even the last few, and (2) introduce continuity conditions over time right from the beginning and try to exploit knowledge on egomotion for depth understanding from image sequences. This joint use of knowledge on motion processes of objects in all four physical dimensions (3-D space and time) has led to the designation "4-D approach to dynamic vision" [Dickmanns 1987].

6.1 Introduction to the 4-D Approach for Spatiotemporal Perception

Since the late 1970s, observer techniques as developed in systems dynamics [Luenberger 1966] have been used at UniBwM in the field of motion control by computer vision [Meissner 1982; Meissner, Dickmanns 1983]. In the early 1980s, H.J. Wuensche did a thorough comparison between observer and Kalman filter realizations in recursive estimation applied to vision for the original task of balancing an inverted pendulum on an electro-cart by computer vision [Wuensche 1983]. Since then, refined versions of the extended Kalman filter (EKF) with numerical stabilization (UDU^T-factorization, square root formulation) and sequential updates after each new measurement have been applied as standard methods to all dynamic vision problems at UniBwM [Dickmanns, Wuensche 1999].

This approach has been developed based on years of experience gained from applications such as satellite docking [Wuensche 1986], road vehicle guidance, and on-board autonomous landing approaches of aircraft by machine vision. It was realized in the mid-1980s that the joint use of dynamic models and temporal predictions for several aspects of the overall problem *in parallel* was the key to achieving a quantum jump in the performance level of autonomous systems based on machine vision. Recursive state estimation has been introduced for the interpretation of 3-D motion of physical objects observed and for control computation based on these estimated states. It was the feedback of knowledge thus gained to image feature extraction and to the feature aggregation level, which allowed an increase in efficiency of image sequence evaluation of one to two orders of magnitude.

Figure 6.1 gives a graphical overview.

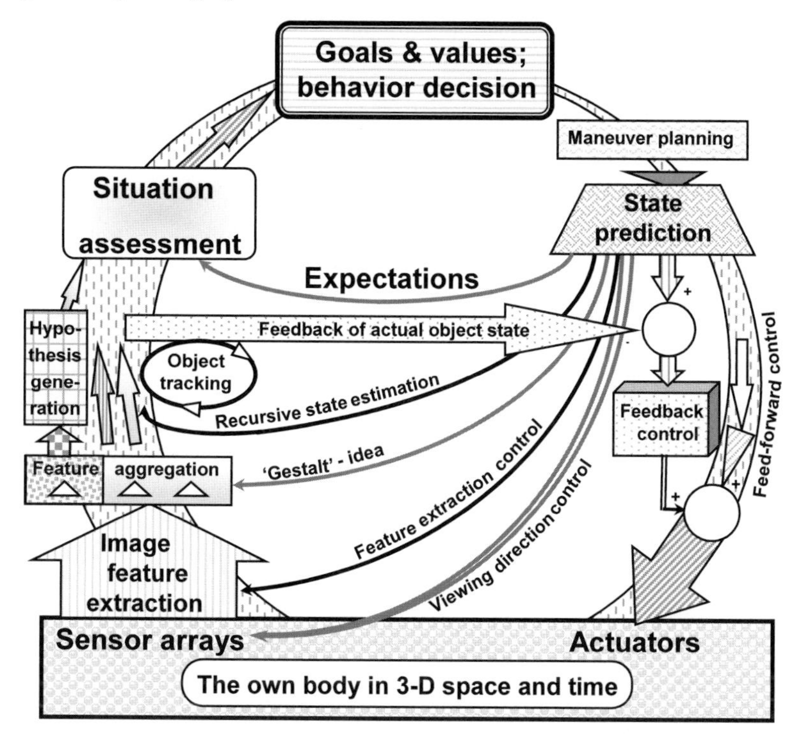

Figure 6.1. Multiple feedback loops on different space scales for efficient scene interpretation and behavior control: control of image acquisition and processing (lower left corner), 3-D "imagination" space in upper half; motion control (lower right corner).

Following state prediction, the shape and the measurement models were exploited for determining:

- viewing direction control by pointing the two-axis platform carrying the cameras with lenses of different focal lengths;
- locations in the image where information for the easiest, non-ambiguous and accurate state estimation could be found (feature selection),
- the orientation of edge features which allowed us to reduce the number of search masks and directions for robust yet efficient and precise edge localization,
- the length of the search path as a function of the actual measurement uncertainty,
- strategies for efficient feature aggregation guided by the idea of *gestalt* of objects, and
- the Jacobian matrices of first-order derivatives of feature positions relative to state components in the dynamic models that contain rich information for inter-

pretation of the motion process in a least-squares error sense, given the motion constraints, the features measured, and the statistical properties known.

This integral use of

1. dynamic models for motion of and around the center of gravity taking actual control outputs and time delays into account,
2. spatial (3-D) shape models for specifying visually measurable features,
3. the perspective mapping models, and
4. feedback of prediction-errors for estimating the object state in 3-D space and time simultaneously and in closed-loop form was termed the *4-D approach*.

It is far more than a recursive estimation algorithm based on some arbitrary model assumption in some arbitrary subspace or in the image plane. It is estimated from a scan of publications in the field of vision that even in the mid-1990s, most of the papers referring to Kalman filters did not take advantage of this integrated use of spatiotemporal models based on physical processes.

Initially, in our applications just the ego vehicle has been assumed to move on a smooth surface or trajectory, with the cameras fixed to the vehicle body. In the meantime, solutions of rather general scenarios are available with several cameras spatially arranged on a platform, which may be pointed by voluntary control relative to the vehicle body. These camera arrangements allow a wide simultaneous field of view, a central area for trinocular (skew) stereo interpretation, and a small area with high image resolution for "tele"-vision (see Chapter 11 below). The vehicle may move in full six degrees of freedom; while moving, several other objects may move independently in front of a stationary background. One of these objects may be "fixated" (tracked) by the pointing device using inertial and visual feedback signals to keep the object (almost) centered in the high-resolution image. A newly appearing object in the wide field of view may trigger a fast change in viewing direction such that this object can be analyzed in more detail by one of the telecameras. This corresponds to "saccadic" vision as known from vertebrates and allows very much reduced data rates for a complex sense of vision. It trades the need for time-sliced attention control and scene reconstruction based on sampled data (actual video image) for a *data rate reduction of one to two orders of magnitude* compared to full resolution in the entire simultaneous field of view.

The 4-D approach lends itself to this type of vision since both object-orientation and temporal ("dynamic") models are already available in the system. This complex system design for dynamic vision has been termed *EMS vision* (from expectation-based, multifocal, and saccadic vision). It has been implemented with an experimental set of up to five miniature TV cameras with different focal lengths and different spectral characteristics on a two-axis pointing platform named *multi-focal active/reactive vehicle eye* (*MarVEye*). Chapter 12 discusses the requirements leading to this design; some experimental results will be shown in Chapter 14.

For *subjects* (objects with the capability of information intake, behavior decision, and control output affecting future motion), knowledge required for motion understanding has to encompass typical time histories of control output to achieve some goal or state transition and the corresponding trajectories resulting. From the trajectories of subjects (or parts thereof) observed by vision, the goal is to recognize the maneuvers intended to gain reaction time for own behavior decision. In this closed-loop context, real-time vision means activation of animation capabilities, including the potential behavioral capabilities (maneuvers, trajectory control by feedback) of other subjects. As Figure 6.1 indicates, recursive estimation is not confined to perceiving simple physical motion processes of objects (proper) but allows recognizing diverse, complex motion processes of articulated bodies if the corresponding maneuvers (or trajectories resulting) are part of the knowledge base available. Even developments of situations can be tracked by observing these types of motion processes of several objects and subjects of interest. Predictions and expectations allow directing perceptual resources and attention to what is considered most important for behavior decision.

A large part of mental activities thus is an essential ingredient for understanding motion behavior of subjects. This field has hardly been covered in the past but will be important for future really intelligent autonomous systems. In the next section, a summary of the basic assumptions underlying the 4-D approach is given.

6.2 Basic Assumptions Underlying the 4-D Approach

It is the explicit goal of this approach to take advantage as much as possible of physical and mathematical models of processes happening in the real world. Models developed in the natural sciences and in engineering over the last centuries in simulation technology and in systems engineering (decision and control) over the last decades form the base for computer-internal representations of real-world processes (see also Chapters 1 to 3):

1. The (mesoscopic) world observed happens in *3-D space and time* as the four independent variables; non-relativistic (Newtonian) and non-quantum-mechanical models are sufficient for describing these processes.
2. All interactions with the real world happen *here and now*, at the location of the body carrying special input/output devices. Especially the locations of the sensors (for signal or data input) and of the actuators (for control output) as well as those body regions with strongest interaction with the world (as, for example, the wheels of ground vehicles) are of highest importance.
3. *Efficient interpretation of sensor signals* requires background knowledge about the *processes* observed and controlled, that is, both its spatial and temporal characteristics. Invariants for process understanding may be abstract model components not graspable at one time.
4. Similarly, *efficient computation of* (favorable or optimal) *control outputs* can be done only by taking complete (or partial) process models into account; control theory provides the methods for fast and stable reactions.
5. *Wise behavioral decisions* require knowledge about the longer term outcome of special feed-forward or feedback control modes in certain situations and environments; these results are obtained from integration of dynamic models. This may have been done beforehand and stored appropriately or may be done on the spot if analytical solutions are available or numerical ones can be derived in a small fraction of real time as becomes possible now with the increasing processing power at hand. Behaviors are realized by triggering the modes that are available from point 4 above.

6. *Situations* are made up of arrangements of objects, other active subjects, and of the goals pursued; therefore,
7. it is essential to recognize *single objects and subjects*, their relative state, and for the latter also, if possible, their intentions to make meaningful predictions about the future development of a situation (which are needed for successful behavioral decisions).
8. As the term **re-***cognition* tells us, in the usual case it is assumed that objects seen are (at least) generically known already. Only their appearance here (in the geometrical range of operation of the senses) and now is new; this allows a fast jump to an object hypothesis when first visual impressions arrive through sets of features. Exploiting background knowledge, the model based perception process has to be initiated. Free parameters in the generic object models may be determined efficiently by attention control and the use of special algorithms and behaviors.
9. To do step 8 efficiently, knowledge about "the world" has to be provided in the *context of task domains* in which likely co-occurrences are represented (see Chapters 4, 13, and 14). In addition, knowledge about discriminating features is essential for correct hypothesis generation (indexing into the object database).
10. Most efficient object (class) descriptions by *invariants* are usually done *in 3-D space* (for shape) *and time* (for motion constraints or stereotypical motion sequences). Modern microprocessors are sufficiently powerful for computing the visual appearance of an object under given aspect conditions in an image (in a single one, or even in several with different mapping parameters in parallel) at runtime. They are even powerful enough to numerically compute the elements of the Jacobian matrices for sensor/object pairs of features evaluated with respect to object state or parameter values (see Sections 2.1.2 and 2.4.2); this allows a very flexible general framework for recursive state and parameter estimation. The inversion of perspective projection is thus reduced to a least-squares model fit once the recursive process has been started. The underlying assumption here is that local linearization of the overall process is a sufficiently good representation of the nonlinear real process; for high evaluation rates like video frequency (25 or 30 Hz), this is usually the case.
11. In a running interpretation process of a dynamic scene, *newly appearing objects* will occur *in restricted areas* of the image such that bottom-up search processes may be confined to these areas. Passing cars, for example, always enter the field of view from the side just above the ground; a small class of features allows detecting them reliably.
12. *Subjects*, *i.e.*, objects with the capability of self-induced generation of control actuation, are characterized by typical (sometimes stereotypical, *i.e.*, predictive) motion behavior in certain situations. This may also be used for recognizing them (similar to shape in the spatial domain).
13. The same object/subject may be represented internally on *different scales with various degrees of detail*; this allows flexible and efficient use in changing contexts (*e.g.*, as a function of distance or degree of attention).

Since the use of the terms "state variables" and "dimensions" rather often are quite different in the AI/computer science communities, on one hand, and the natural sciences and engineering, on the other hand, a few sentences are spent here to avoid confusion in the sequel. In mesoscale physics of everyday life there are no more than four basic dimensions, three in space and time. In each spatial dimension there is one translational degree of freedom (d.o.f.) along the axis and one rotational d.o.f. around the axis, yielding six d.o.f. for rigid body motion, in total. Since Newtonian mechanics requires a second-order differential equation for properly describing physical motion constraints, a rigid body requires 12 *state variables* for full description of its motion; beside the 3 positions and 3 angles there are the corresponding temporal derivatives (speed components).

State variables in physical systems are defined as those variables that cannot be changed at one time, but have to evolve over time according to the differential equation constraints. A full set of state variables decouples the future evolution of a system from the past. By choosing these physical state variables of all objects of interest to represent scenes observed, there is no need to store previous images. The past of all objects (relevant for future motion) is captured in the best estimates for the actual state of the objects, at least in theory for objects known with certainty. Uncertainty in state estimation is reflected into the covariance matrix which is part of the estimation process to be discussed. However, since object hypothesis generation from sparse data is a weak point, some feature data have to be stored over a few cycles for possible revisions needed later on.

Contrary to common practice in natural sciences and engineering, in computer science, state variables change their values at one time (update of sampled data) and then remain constant over the cycle time for computing. Since this fate is the same for any type of variable data, the distinct property of *state variables in physics* tends to be overlooked and the term *state variable* (or in short *state*) tends to be used abundantly for any type of variable, *e.g.*, an acceleration as well as for a control variable or an output variable (computed possibly from a collection of both state and control variables).

Another point of possible misunderstanding with respect to the term "dimension" stems from discretizing a state variable according to thresholds of resolution. A total length may be subdivided into a sequence of discrete "states" (to avoid exceedingly high memory loads and search times); each of these new states is often called a *dimension in search space*. Dealing with 2-D regions or 3-D volumes, this discretization introduces strong increases in "problem dimensions" by the second or third power of the subdivisions. Contrary to this approach often selected, for example, in [Albus, Meystel 2001], here the object state in one entire degree of freedom is precisely specified (to any resolution desired) by just two state variables: position (pose or angle) and corresponding speed component. Therefore, in our approach, a rigid body does not need more than 12 state variables to describe its actual state (at time "now") in all three spatial dimensions.

Note that the motion constraints through dynamic models prohibit large search spaces in the 4-D approach once the pose of an object/subject has been perceived correctly. Easily scalable homogeneous coordinates for describing relative positions/orientations are the second item guaranteeing efficiency.

6.3 Structural Survey of the 4-D Approach

Figure 6.2 shows the main three activities running in parallel in an advanced version of the 4-D approach:

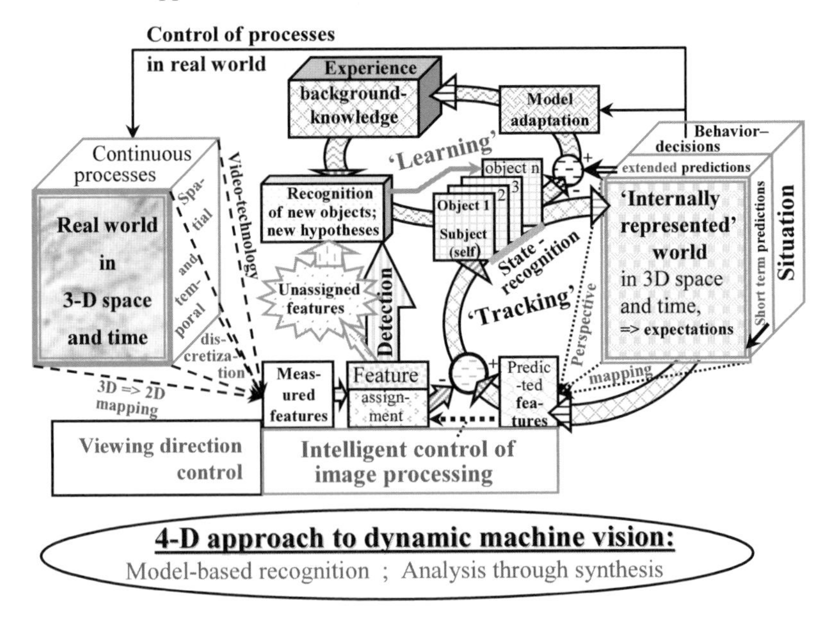

Figure 6.2. Survey of the 4-D approach to dynamic machine vision with three major areas of activity: Object detection and recognition (central arrow upward), object tracking and state estimation (recursive loop at lower right), and learning (loop at center top); the latter two are driven by prediction-error feedback.

1. **Detection of objects** from typical collections of features not yet assigned to some object already tracked (center, arrow upward). When these feature collections are stable over several frames, an object hypothesis is formed, and the new object is added to the list of objects regularly tracked (arrow in upper part of center to the right).
2. **Tracking of objects and state estimation** is shown in the loop at the lower right in Figure 6.2; first, with the control output chosen, a single step prediction is done in 3-D space and time, the "imagined real world." This new state is taken for projecting features to be tracked into the individual image planes for each camera involved. Prediction-error feedback is then used for improving state and parameter estimation based on the rich first-order derivative information contained in the Jacobian matrices for each camera-object pair. This chapter is devoted mainly to this topic.
3. **Learning from observation** is done with the same data as for tracking; however, this is not a single step loop but rather a low frequency estimation component concentrating on "constant" parameters, or it is even an off-line component with batch processing of stored data. This area is still under development

at present; it will open up the architecture to becoming more autonomous in knowledge acquisition in new task domains as experience with the system grows. Both dynamical models (for the "where"-part) and shape models (for the "what"-part) shall be learnable in the future.

Note that in the 4-D approach, system behavior is always considered in a closed, real-time loop; both perception and behavior control are derived from spatio-temporal models. The method has become mature and has remained stable for visual relative state estimation over two decades by now.

First, one proven implementation of the extended Kalman filter (EKF) method for recursive estimation by vision will be discussed; based on the insight gained, steps toward initialization of the tracking method will be mentioned.

6.4 Recursive Estimation Techniques for Dynamic Vision

As mentioned in the introduction to this chapter, the basic method is a recursive reformulation of the Gaussian least squares approach to fitting a solution curve of known structure to a set of measurement data. Many variants have been derived in the meantime; they will not be discussed here. There is a rich literature on this topic available; even Web-search machines turn up a host of references and reviews. Some standard references are [Maybeck 1979, 1990; Bar Shalom, Li 1998; Kailath *et al.* 2000]. A recent book dedicated to "probabilistic robotics" is [Thrun *et al.* 2005]. As a basic reference to probability theory, the book [Jaynes 2003] is recommended. The reader not acquainted with Kalman filtering should refer to one of these references if the description given here misses depth.

In view of these facts, here only the approach as developed at UniBwM will be retraced. It has allowed achieving many "firsts" in visual guidance for autonomous vehicles. It has been appreciated from the beginning that the linearization of perspective mapping, taking the prediction of motion components into account, would yield excellent first-order approximations compensating for the loss of depth information in a single snapshot image. This turned out to be true and immediately allowed motion stereointerpretation of dynamic scenes observed.

6.4.1 Introduction to Recursive Estimation

The *n*-vector of state variables $x(t)$ of a dynamical system defined by the differential equation

$$\dot{x}(t) = f[x(t), u(t), z(t), p_s], \tag{6.1}$$

with $u(t)$ an *r*-vector of control variables, $z(t)$ an n-vector of disturbances, and p_s a vector of parameters of the system, usually, cannot be measured directly but only through an m-vector $y(t_k)$ of output variables at discrete points t_k in time (index *k*, for short: at time *k*) spoiled by some superimposed measurement noise $w'(t)$, whose statistics are assumed to be known sufficiently well.

The continuous, usually nonlinear dynamic model Equation 6.1 is transformed into a discrete linear state transition model for sampled data with cycle time T by one of the standard methods in systems theory (see Sections 2.2.3, 3.4.1 and, for example, [Kailath 1980]). The standard form then is (Equation 3.7)

$$x_k = A \cdot x_{k-1} + Bu_{k-1} + v_k \tag{6.2}$$

with A the $n{\cdot}n$ state transition matrix from $(k\text{-}1)T$ to kT, B the $n{\cdot}r$ control effect matrix, and $v(z(t),\ T)$ the discrete noise term. For linearized systems, all these terms may depend on the nominal state and control variables x_N and u_N, in general.
The observable output variables y(k) may also depend nonlinearly on the state variables, and on some known or unknown (measurement) parameters p_m, like in perspective mapping for vision

$$y(k) \ = \ h\ [x(k),\ p_m\ ] \ + \ w(k)\,. \tag{6.3}$$

Therefore, the state variables have to be reconstructed from the output variables by exploiting knowledge about the dynamical process itself and about the measurement process h.

In vision, the measurement process is perspective projection, assumed here to be sufficiently well described by a pinhole camera model (straight lines of light rays through a pinhole). This is a nonlinear mathematical operation with no direct inversion available. As will be detailed later, the nonlinear measurement equation is linearized around the predicted nominal state x_N of the process and the nominal parameter set p_N (q values) yielding (without the noise term)

$$\begin{aligned} y^*(k) &= y_N(k) + \delta y(k) \\ &= h(x_N(k), p_N, k) + C_x(k)\cdot \delta x(k) + C_p(k)\cdot \delta p \end{aligned} \tag{6.4}$$

where $C_x = \partial h/\partial x|_N$ and $C_p = \partial h/\partial p|_N$ are the Jacobian matrices of perspective mapping with respect to the state components and the parameters involved. Since the first terms on the right-hand side of the equality sign are equal by definition, Equation 6.4 may be used to determine δx and δp as deviations from the nominal values x_N and p_N in a least-squares sense from δy, the measured prediction errors:

$$\delta y(k) = C_x(k)\cdot \delta x(k) + C_p(k)\cdot \delta p\ . \tag{6.5}$$

To achieve this goal under the side constraints of Equations 6.2 and 6.4, the actual error covariance matrix P of the overall system and the covariance matrices Q of the dynamic system as well as R of the measurement model have to be known; observability is assumed to be given. This is the core of recursive estimation. The basic mathematical relationships will be given in the next section.

Note that through Equation 6.2 and the covariance matrix Q, temporal continuity with some room for stochastic adjustment is introduced into the interpretation process. The challenge is to find out how prediction errors δy should contribute to improving the various components of δx and δp.

6.4.2 General Procedure

The reader is assumed to have some understanding of Kalman filtering as a recursive estimation process; therefore, a somewhat unusual approach is taken here. If difficulties in understanding should occur, the reader is recommended to one of the basic textbooks to close the gap. The description is rather pragmatic and completely oriented towards real-time vision. For this specific application, the treatment is rather exhaustive, so that the reader should be able to gain full understanding of the basic ideas and application details.

6.4.2.1 Sequence of Steps

First, starting values $x^*(0)$ for the iteration have to be found before the recursive loop can be initiated.

- If not available otherwise, find a good guess for the n initial state components $x^*(0)$ to start with (initial hypothesis).
- Find an estimate for the probability distribution of this initial state in terms of the first and second moments of a Gaussian distribution (mean value $\hat{x}_0$ and the (n·n) error covariance matrix P_0. The mean value is set to the estimated initial state: $\hat{x}_0 = x^*(0)$, which is believed to represent the best knowledge available, but of course, is not the real value of x at $k = 0$. With the covariance matrix P_0, the Gaussian probability distribution from the second moment by definition is the following expression:
$$p(\hat{x}_0) = \det(2\pi P_0)^{-1/2} \exp\left[-\tfrac{1}{2}(x-\hat{x}_0)^T P_0^{-1}(x-\hat{x}_0)\right]. \qquad (6.6)$$
- Assuming that the state components are independent of each other, the off-diagonal terms in P_0 are zero, and the diagonal terms are the variance components σ_i^2 of the states. This gives hints for finding good guesses for these terms.
- Now the time index is incremented ($k-1 = k_{old}$; $k = k_{old}+1$) and the state variables are predicted over one computing cycle exploiting the dynamic model of the plant (Equation 6.2 with transition matrix A and control effect matrix B, but without the noise term) => $x^*(k)$. For nonlinear systems, this may be done either with an analytical solution of the linearized state equations or by numerical integration using the nonlinear model.
- From the measurement model (Equation 6.3), compute the predicted nominal measurement values y^* (no noise)
$$y^*(k) = h[x^*(k), p_m]. \qquad (6.7)$$
The usual noise model assumed is
$$E\{w\} = 0$$
and (6.8)
$$E[w^T w] = R \quad (\textit{white noise}).$$
- Take the actual measurement values y(k) and compute the prediction errors:
$$\delta y(k) = y(k) - y^*(k), \qquad (6.9)$$
for driving the iteration. Deeper understanding of visual interpretation is achieved through the linear approximation around the expected point according to Equation 6.5. Therefore, compute the Jacobian matrices C_i (see Section 2.1.2). [To simplify notation from here on, new state variables x_{n+1} to x_{n+q} are defined with derivatives = 0 (constants) and driven by a noise term (for ad-

justment). By increasing the order of the system from n_{old} to $n = n_{old} + q$ and by proper adjustment of all matrices involved, there need not be a distinction any longer between states and parameters. The C-matrix now has dimension $m \cdot (n_{old} + q)$. For efficient coding, state variables and parameters should be treated separately, however.]

- Improve the state estimate by the "innovation step" exploiting a (yet open) gain matrix $K(k)$ with the assumption that changes are small so that the linear terms capture most of the functional relationships around $x^*(k)$:

$$\hat{x}(k) = x^*(k) + K(k) \cdot \delta y^*(k) \tag{6.10}$$

This new estimate for the state variables is the base for the next recursion loop.

6.4.2.2 Selection of the Matrix for Prediction-Error Feedback

Depending on the computational scheme for determining $K(k)$, different methods are distinguished:

deterministic scheme (no noise modeled) ⇒ *Luenberger observer*
stochastic scheme ⇒ *Kalman filter.*

Here, only the Kalman filter will be discussed, though in the first several years of our research into recursive vision, the Luenberger observer has shown very good performance, too [Meissner 1982]. In contrast to the (deterministic) observer, the so-called "Kalman-gain" matrix $K(k)$ is obtained from stochastic reasoning. Let the dynamic system (plant) be described by Equation 6.2 with noise property

$$E[v] = 0, \qquad E[v^T v] = Q. \tag{6.11}$$

The covariance matrices $Q(k)$ for the motion prediction step and $R(k)$ for the measurement noise have to be determined from existing knowledge or heuristic reasoning. Since they affect the convergence process in recursive estimation, special care has to be taken when specifying these values. This process is known as filter tuning, and the topic has to be revisited below. $x^*(0)$, $v(0)$ and $w(0)$ are assumed to be uncorrelated.

Start of recursive loop: Increment time index $k = k+1$; predict expected values of state variables (Equation 6.2 without noise v) and the expected error covariance matrix (both marked by * as expected values);

$$\begin{aligned} x^*_k &= A(k-1) \cdot \hat{x}_{k-1} + B(k-1) \cdot u_{k-1}, \\ P^*(k) &= A(k-1) \cdot P(k-1) \cdot A^T(k-1) + Q(k-1). \end{aligned} \tag{6.12}$$

The challenge now is to find a gain matrix $K(k)$ such that the error covariance after inclusion of the new measurement values is minimized [matrix $P(k)$]. The prediction error $\delta\hat{x}(k)$ will be the unknown real state $x(k)$ minus the best estimate $\hat{x}(k)$ yet to be found:

$$\delta\hat{x}(k) = x(k) - \hat{x}(k). \tag{6.13}$$

The covariance matrix of the estimation error $P = E[\delta\hat{x} \cdot \delta\hat{x}^T]$ follows with Equation 6.10 for $\hat{x}(k)$ as

$$P = E\{[x(k) - \hat{x}(k)] \cdot [x(k) - \hat{x}(k)]^T\}$$
$$= E\{[x - x^* - K \cdot C \cdot \delta x^*] \cdot [x - x^* - K \cdot C \cdot \delta x^*]^T\}. \quad (6.14)$$

The output error covariance with $\delta y = C\delta x^* + v$ and Equation 6.11 is

$$E\{\delta y \cdot \delta y^T\} = E\{(C\delta x^* + v) \cdot (C\delta x^* + v)^T\} = C \cdot P^* \cdot C^T + R. \quad (6.15)$$

Finding that matrix K that minimizes P under the given side constraints, yields

$$K = P^* \cdot C^T \{C \cdot P^* \cdot C^T + R\}^{-1}. \quad (6.16)$$

With this result Equation 6.14 reduces to the well-known form for updating the error covariance matrix

$$P(k) = P^*(k) - K(k) \cdot C(k) \cdot P^*(k). \quad (6.17)$$

6.4.2.3 Complete Recursion Loop in Kalman Filtering

These results are summarized in the following table as algorithmic steps for the basic version of the extended Kalman filter for real-time vision (4-D approach):

1. Find a good guess for the n initial state components $x^*(0)$ to start with [initial hypothesis $k = 0$, $\hat{x}_0 = x^*(0)$].
2. Find an estimate for the probability distribution of this initial state in terms of the first and second moments of a Gaussian distribution (mean value $\hat{x}_0$ and the $(n \cdot n)$ error covariance matrix P_0). The diagonal terms are the components σ_i^2 of the variance.
3. Find an estimate for the covariance matrices $Q = \mathrm{E}\{v^\mathrm{T} v\}$ of system noise v and $R = \mathrm{E}\{w^\mathrm{T} w\}$ of measurement noise w.

→ *Entry point for recursively running loop*

4. Increment time index $k = k+1$;
5. Compute expected values for state variables at time k+1 (state prediction $x^*(k)$): $x^*_k = A(k-1) \cdot \hat{x}_{k-1} + B(k-1) \cdot u_{k-1}$.
6. Predict expected error covariance matrix P^*(k) (components: state prediction and noise corruption): $P^*(k) = A(k-1) \cdot P(k-1) \cdot A^T(k-1) + Q(k-1)$.
7. Compute the expected measurement values $y^*(k) = h[x^*(k), p_m]$ and the (total) Jacobian matrix $C = \partial y^* / \partial x|_N$ as first-order approximations around this point.
8. Compute the gain matrix for prediction error feedback: $K = P^* \cdot C^T \cdot \{C \cdot P^* \cdot C^T + R\}^{-1}$.
9. Update the state variables (innovation) to the +best estimates, including the last measurement values: $\hat{x}(k) = x^*(k) + K(k) \cdot [y(k) - y^*(k)]$.
10. Update the error covariance matrix (innovation of statistical properties): $P(k) = P^*(k) - K(k) \cdot C(k) \cdot P^*(k)$.

Go back to step 4 for next loop.

Steps for monitoring convergence and progress in the vision process will be discussed in connection with applications.

6.4.3 The Stabilized Kalman Filter

Equation 6.17 is not well conditioned for use on a computer with limited word length. Canceling of significant digits may lead to asymmetries not allowed by the definition of P^*. Numerically better conditioned is the following equation (I = identity matrix):

$$P = (I - K \cdot C) \cdot P^* \cdot (I - K \cdot C)^T + K \cdot R \cdot K^T, \qquad (6.18)$$

which results from using the reformulated Equation 6.16 (multiplying by the {}-term)

$$KCP^* C^T + KR = P^* C^T \rightarrow \; K \cdot R = (I - K \cdot C) P^* \cdot C^T. \qquad (6.19)$$

Multiplying the right-hand form by K^T from the right and shifting terms to the left side of the equality sign yields

$$-(I - K \cdot C) \cdot P^* \cdot C^T \cdot K^T + K \cdot R \cdot K^T = 0. \qquad (6.20)$$

Adding this (0) to Equation 6.17 in the form $P = (I - K\,C)\,P^*$ yields Equation 6.18.

6.4.4 Remarks on Kalman Filtering

Filter tuning, that is, selecting proper parameter settings for the recursive algorithm, is essential for good performance of the method. A few points of view will be discussed in this section.

6.4.4.1 Influence of the Covariance Matrices Q and R on the Gain Matrix K

The basic effects of the influence of the covariance matrices Q and R on the gain matrix K can be seen from the following simple scalar example: Consider the case of just one state variable which may be measured directly as the output variable r:

$$\begin{aligned} x_k &= a\, x_{k\text{-}1} + b\, u_{k\text{-}1} + v_{k\text{-}1} = r_{k\text{-}1} + 0 + s_{x,\,k\text{-}1}, \\ y_k &= c\, x_k + w_k = r_{m,k} + s_{y,k}, \\ Q(k) &= E\{s_x^2\} = \sigma_x^2; \qquad R(k) = E\{s_y^2\} = \sigma_y^2. \end{aligned} \qquad (6.21)$$

The variance of the prediction error then is

$$P^*(k) = p^*(k) = p(k\text{-}1) + \sigma_x^2, \qquad (6.22)$$

and for the scalar gain factor K, one obtains

$$K(k) = [p(k-1) + \sigma_x^2\,]/[p(k-1) + \sigma_x^2 + \sigma_y^2]. \qquad (6.23)$$

Two limiting cases for the noise terms σ_x^2 and σ_y^2 show the effect of different values of variances on the progress of the iteration. For $\sigma_y^2 << \sigma_x^2$, *i.e.*, very good measurements, a K value just below 1 results; for example,

$$\hat{r}(k) = r^*(k) + 0.95\{r_m(k) - r^*(k)\} = 0.05\, r^*(k) + 0.95\, r_m(k). \qquad (6.24)$$

This tells us for example, that when initial conditions are poor guesses and measurement data are rather reliable, the R elements should be much smaller than the Q-

elements. But when the dynamic models are good and measurements are rather noisy, they should be selected the other way round. For $\sigma_y^2 >> \sigma_x^2$, *i.e.*, for very poor measurement data, a small filter gain factor K results; for example,

$$\hat{r}(k) = r^*(k) + 0.1[r_m(k) - r^*(k)] = 0.9\, r^*(k) + 0.1\, r_m(k). \qquad (6.25)$$

With poor measurements, also the variance of the estimation error will be improved only a little:

$$p(k) = p^*(k) - K \cdot p^*(k) = (1-K)\, p^*(k)\,.$$
$$e.g., \quad K = 0.1 \quad => \quad p(k) = 0.9 \cdot p^*(k). \qquad (6.26)$$

These considerations carry over to the multivariate case.

6.4.4.2 Kalman Filter Design

For the observer (not treated here), the specification of the filter gain matrix is achieved by assigning desired eigenvalues to the matrix (pole positions). In the Kalman filter, the gain matrix $K(k)$ is obtained automatically from the covariance matrices $Q(k)$ and $R(k)$, assumed to be known for all t_k, as well as from the error covariance matrix P^*_0.

$K(k)$ results from P^*_0, R(k) and Q(k): P^*_0 may be obtained by estimating the quality of in-advance-knowledge about x_0 (how good/certain is x_0?). $R(k)$ may eventually be derived from an analysis of the measurement principle used. On the contrary, finding good information about $Q(k)$ is often hard: Both the inaccuracies of the dynamic model and perturbations on the process are not known, usually.

In practice, often *heuristic methods* are used to determine $K(k)$: Initially, change $Q(k)$ only, for fixed $R(k)$, based on engineering judgment. Then, after achieving useful results, also change $R(k)$ as long as the filter does not show the desired performance level (filter tuning).

The initial transient behavior is essentially determined by the choice of P^*_0: The closer P^*_0 is chosen with respect to (the unknown) $P^*(k)$, the better the convergence will be. However, the larger initial covariance values P^*_0 selected, the more will measured values be taken into account. (Compare the single variable case: Large initial values of $p^*(k)$ resulted in K-values close to 1, => strong influence of the measured values.) But note that this can be afforded only if measurement noise R is not too large; otherwise, there may be no convergence at all. Filter tuning is often referred to as an art because of the difficulties in correctly grasping all the complex interrelationships.

6.4.4.3 Computational Load

For m measurement values, an $(m \times m)$ matrix has to be inverted to compute the filter gain matrix $K(k)$. In the numerical approach, first a transformation into triangular form via the Householder transformation is done; then recursive elimination is performed.

Especially in the stabilized form (Equation 6.18, called the "Joseph"-form), computing load is considerable. This has led to alternative methods: Briefly discussed will be *sequential Kalman filtering* and *UDU^T-factored Kalman filtering*.

Gauss-Markov estimators with postfiltering are not discussed here; in this approach, in which more measurement equations than state variables have to be available, the noise corrupted state variables arrived at by a Gauss-Markov estimator are filtered in a second step. Its advantage lies in the fact that for each state variable a filter factor between 0 and 1 can be set separately. Details on several recursive filter types may be found in the reference [Thrun *et al.* 2005].

6.4.5 Kalman Filter with Sequential Innovation

At time t_k, the m_k measurement values are obtained:

$$y(k) = h[x(k)] + w(k); \qquad E\{w\} = 0; \qquad E\{w^T w\} = R. \tag{6.27}$$

For innovation of the state variables (see Section 6.4.2)

$$\hat{x}(k) = x^*(k) + K(k) \cdot [y(k) - y^*(k)]$$

the $n \times n$ dimensional filter gain matrix

$$K(k) = P^*(k) \cdot C^T(k)[C(k) \cdot P^*(k) \cdot C^T(k) \ + \ R(k)]^{-1}$$

is obtained by inversion of a matrix of dimension $m_k \times m_k$. The central statement for the so-called "sequential innovation" is

Under certain conditions, an m_k-dimensional measurement vector can always be treated like m_k scalar single measurements that are exploited sequentially; this reduces the matrix inversion to a sequence of scalar divisions.

6.4.5.1 Preconditions

The m_k measurement values have to be uncorrelated, that means $R(k)$ has to be (essentially) a diagonal matrix. If the covariance matrix $R(k)$ is blockwise diagonal, the corresponding subvectors may be treated sequentially. Correlated measurements may always be transformed into uncorrelated pseudo-measurement data which then may be treated sequentially [Maybeck 1979, p. 375].

In image processing, different features derived from video signals can be assumed to be uncorrelated if the attributes of such features are gained by different processors and with different algorithms. Attributes of a single feature (*e.g.*, the *y*- and *z*- components in the image plane) may well be correlated.

6.4.5.2 Algorithm for Sequential Innovation

The first scalar measurement at a sampling point t_k,

$$y_1(k) = g_1[x(k)] + w_1(k), \quad E\{w_1\} = 0, \quad E\{w_1^2\} = s_1^2, \tag{6.28}$$

leads to the first partial innovation [in the sequel, the index k will be given only when needed for clarity; k_i in the following is a column-vector of Kalman gains]:

$$d\hat{x}_1 = k_1(y_1 - y^*_1). \tag{6.29}$$

The $(n \times 1)$-dimensional filter vector k_1 is computed by

$$k_1 = P_0 \, c_1^T (c_1 \, P_0 \, c_1^T + s_1^T)^{-1}, \qquad P_0 = P^*(k). \tag{6.30}$$

Here, c_1 is the first row (vector) of the Jacobian matrix $C(k)$ corresponding to y_1. After this first partial innovation, the following terms may be computed:

$$\hat{x}_1(k) = x^*(k) + d\hat{x}_1(k), \text{ (improved estimate for the state)}$$
$$P_1 = P_0 - k_1 c_1 P_0 \quad \text{(improved error covariance matrix).} \tag{6.31}$$

For the I–th partial innovation

$$d\hat{x}_I = d\hat{x}_{I-1} + k_I(y_I - y^*_I), \qquad I = 1, \ldots\ldots, m_k, \qquad d\hat{x}_0 \tag{6.32}$$

momentarily the best estimate for y_I has to be inserted. This estimate is based on the improved estimated value for x:

$$x^*_I(k) = \hat{x}_{I-1}(k) = x^*(k) + d\hat{x}_{I-1}\ ; \tag{6.33}$$

it may be computed either via the complete perspective mapping equation

$$y_I(k) = h_I[x^*_I(k)], \tag{6.34}$$

or, to save computer time, by correction of the predicted measurement value (at t_{k-1} for t_k) $y^*_{I,0}$:

$$y^*_I(k) = y^*_{I,0}(k) + c_I \cdot d\hat{x}_{I-1}\ ; \text{ (simplification: } c_I \sim \text{const for } t = t_k). \tag{6.35}$$

With this, the complete algorithm for sequential innovation is

Initialization:

$$\begin{aligned}
& y^*_{I,0} = g_I\,[\,x^*(k)\,], && I = 1, \ldots\ldots, m_k, \\
& s_I^2 = E\{w_I^2\}, && y_I = g\,[x_I] + w_I, \\
& c_I = dy^*_{I,0}/dx, && I = 1, \ldots\ldots, m_k, \\
& d\hat{x}_0 = 0, && P_0 = P^*(k).
\end{aligned} \tag{6.36}$$

Recursion for I = 1, … , m_k:

$$\begin{aligned}
& y^*_I = y^*_{I,0} + c_I \cdot d\hat{x}_{I-1}, \text{ (simplification: } c_I \sim \text{const for } t = t_k) \\
& k_I = P_{I-1}\, c_I^T / (c_I\, P_{I-1}\, c_I^T + s_I^2), \\
& d\hat{x}_I = d\hat{x}_{I-1} + k_I(\,y_I - y^*_I), \\
& P_I = P_{I-1} - k_I\, c_I\, P_{I-1}, \\
& \text{Final step at } t_k: \quad \hat{x}(k) = x^*(k) + d\hat{x}_{m_k}, \qquad P(k) = P_{m_k}.
\end{aligned} \tag{6.37}$$

6.4.6 Square Root Filters

Appreciable numerical problems may be encountered even in the so-called stabilized form, especially with computers of limited word length (such as in navigation computers onboard vehicles of the not too far past): Negative eigenvalues of P(k) for poorly conditioned models could occur numerically, for example,

- very good measurements (eigenvalues of R small compared to those of P^*), amplified, for example, by large eigenvalues of P_0;
- large differences in the observability of single state variables, *i.e.* large differences in the eigenvalues of $P(k)$.

These problems have led to the development of so-called square root filters [Potter 1964] for use in the Apollo onboard computer. The equations for recursive computation of P are substituted by equations for a recursion in the square root matrix $P^{1/2}$. The filter gain matrix K then is directly computed from $P^{1/2}$. This has the advantage that the eigenvalues allowable due to the limited word length of the covariance matrices P and R are increased considerably. For example, let the variance of the state variable "position of the space vehicle" be $s_x^2 = 10^6$ m^2 and the variance of the measurement value "angle": $s_y^2 = 10^{-4}$ rad^2. The numerical range for P , R then is 10^{10}, while the numerical range for $P^{1/2}$, $R^{1/2}$ is 10^5.

For modern general-purpose microprocessors, this aspect may no longer be of concern. However, specialized hardware could still gain from exploiting this problem formulation.

What is the *"square roots of a matrix"*? For each positive semidefinite matrix P there exist multiple roots $P^{1/2}$ with the property $P^{1/2}\,P^{1/2} = P$. Of especial interest for numerical computations are *triangular decompositions*. Cholesky decompositions use the "lower" triangular matrices L: $P = L\,L^T$. With the so-called Carlson filter, "upper" triangular matrices U' are being used: $P = U'\,U'^T$. Their disadvantage is the relative costly computation of at least n scalar square roots for each recursion step. This may be bypassed by using triangular matrices with 1s on the diagonal (unitarian upper triangular matrices U) and a diagonal matrix D that may be treated as a vector.

6.4.6.1 Kalman Filter with UDU^T-Factorized Covariance Matrix

This method has been developed by Thornton and Bierman [Thornton, Bierman 1980]. It can be summarized by the following bullets:

- Use the decomposition of the covariance matrix

$$P = U\,D\,U^T. \tag{6.38}$$

Due to the property $U\,D^{1/2} = U'$, this filter is considered to belong to the class of square root filters.

- In the recursion equations, the following replacement has to be done:

$$P^* \quad \text{by} \quad U^* \cdot D^* \cdot U^{*T}. \tag{6.39}$$

Modified innovation equations

- Starting from the sequential formulation of the innovation equations (with k_i as column vectors of the Kalman gain matrix and c_i as row vectors of the Jacobian matrix)

$$k_i = P_{i\text{-}1}\,c_i^T\,/(c_i\,P_{i\text{-}1}\,c_i^T + s_i^2), \tag{6.40}$$

$$d\hat{x}_i = d\hat{x}_{i\text{-}1} + k_i(\,y_i - y^*_i), \tag{6.41}$$

$$P_i = P_{i\text{-}1} - k_i\,c_i\,P_{i\text{-}1}. \tag{6.42}$$

Equation 6.40 is introduced into Equation 6.42; by substituting

$$P_{i\text{-}1} \quad \text{by} \quad U_-\,D_-\,U_-^T \qquad (P_0 = P^* = U^*\,D^*\,U^{*T}),$$

$$P_i \quad \text{by} \quad U_+\,D_+\,U_+^T \qquad (P_{m_k} = P = U\,D\,U^T),$$

there follows

$$\begin{aligned} U_+D_+U_+^T &= U_-D_-U_-^T - (U_-D_-U_-^T c_i^T c_i U_-D_-U_-^T)/(c_i U_-D_-U_-^T c_i^T + s_i^T), \\ &= U_-\{D_- - [(D_-U_-^T c_i^T)\,(D_-U_-^T c_i^T)^T]/(\,c_i U_-D_-U_-^T c_i^T + s_i^2)\}U_-^T. \end{aligned} \tag{6.43}$$

- With the definition of two column vectors with n components each

$$f =: U_-^T\;c_i^T, \qquad\qquad e = D_-f, \;, \tag{6.44}$$

a scalar q, called the innovation covariance,

$$q = c_i U\_D\_U\_^T c_i^T + s_i^2 = s_i^2 + f^T \cdot e \equiv D_{jj} \cdot f_j, \tag{6.45}$$

and

$$U_+ D_+ U_+^T = U\_(D\_ - e \cdot e^T / q) U\_^T. \tag{6.46}$$

the innovation of U and D is reduced to a decomposition of the symmetric matrix $(D\_ - e \cdot e^T / q)$:

$(D\_ - e \cdot e^T / q) = U^0 \; D^0 \; U^{0T}$. There follows:

$$U_+ D_+ U_+^T = (U\_ \; U^0) \, D^0 \; (U^{0T} \; U\_^T), \tag{6.47}$$

that is, $U_+ = (U\_ \; U^0) \; D_+ = D^0$.

- Computation of the $n \cdot 1$ filter vector k_i follows directly from Equation 6.40:

$$k_i = U\_ \, e / q. \tag{6.48}$$

- The recursive innovation for each of the m_k measurement values then is
  1. $f_i = U_{j-}^T \; c_i^T$.
  2. $q_i = s_i^2 + D_{i-,jj} \cdot f_{i,j}; \quad j = 1, \; ...., \; n$.
  3. $e_i =: D_{i-} \, f_i$.
  4. $k_i = U_{i-} \; e_i / q$.
  5. recursive computation of U_i^0 and D_i^0 from $(D_{i-} - e_i \cdot e_i^T / q)$.
  6. $U_{i+} = U_{i-} \; U_i^0$.
  7. $D_{i+} = D_i^0$.
  8. Set $U_{i+1,-} = U_{i+}$ and $D_{i+1,-} = D_{i+}$ for the next measurement value.
  9. Repeat steps 1 to 8 for $i = 1, \; .... , \; m_k$.
  10. Set $U = U_{m_k +}$ and $D = D_{m_k +}$ as starting values for the prediction step.

Extrapolation of the covariance matrix in UDU^T-factorized form:

The somewhat involved derivation is based on a Gram-Schmidt vector orthogonalization and may be found in [Maybeck 1979, p. 396] or [Thornton, Bierman 1977].

- The starting point is the prediction of the covariance matrix:

$$P^*(k) = A(k-1) \, P(k-1) \, A^T(k-1) + Q(k-1), ,$$

with the decompositions,

$$P = U \, D \, U^{\mathrm{T}} \quad \text{and} \quad Q = U_q \; D_q \; U_q^T. \quad (U_q = I \text{ if } Q \text{ is diagonal}). \tag{6.49}$$

$$P_{i-1} \quad \text{by} \quad U\_ \; D\_ \; U\_^T \qquad (P_0 = P^* = U^* \; D^* \; U^{*T}),$$

$$P_i \quad \text{yb} \quad U_+ \; D_+ \; U_+^T \qquad (P_{m_k} = P^* = U^* \; D^* \; U^{*T}).$$

- U^* and D^* are to be found such that

$$\begin{aligned} U_+ D_+ U_+^T &= \underbrace{A \, U}_{} \, D \underbrace{U^T A^T}_{} + U_q \, D_q \, U_q^T \\ &= W \;\; D \;\; W^{\mathrm{T}} + U_q \, D_q \, U_q^T. \end{aligned} \tag{6.50}$$

For this purpose, the following matrices are defined:

$$W = (A \cdot U, \; U_q) \quad \text{dimension: } n \times 2n, \tag{6.51}$$

$$D = \begin{bmatrix} D & 0 \\ 0 & D_q \end{bmatrix} \qquad \text{dimension: } 2n \times 2n \,. \tag{6.52}$$

- The recursion obtained then is

Initialization: $[\, a_1,\, a_2,\, \ldots\ldots,\, a_n \,] \equiv W^T$ (column vectors a_i of the transition matrix A are rows of W). *Recursion backwards* for $\kappa = n, n\text{-}1, \ldots., 1$.

1. $h_\kappa = D\_\, a_\kappa;\qquad (h_{\kappa j} = D_{-jj}\, a_{\kappa j},\, j = 1,\, 2,\, \ldots.,\, n;$ $\dim\{h_\kappa\}:\ 2n).$ (6.53)

2. $D_{\kappa\kappa} = a_\kappa^T\, h_\kappa$ (scalar). (6.54)

3. $d_\kappa = h_\kappa / D_{\kappa\kappa}$ (normalizes diagonal terms to 1). (6.55)

4a. For $j = 1, \ldots., \kappa\text{-}1:\ U_{j\kappa} = a_j^T \cdot d_\kappa$. (a)

4b. Replace a_j by $(a_j - U_{j\kappa} \cdot a_\kappa)$. (b) (6.56)

6.4.6.2 Computational Aspects

The matrices U and D are stored as compact vectors. The recursion may be further speeded up for sparse transition matrices $A(k)$. In this case, index vectors for the rows of A indicate at what column index the nonzero elements start and end; only these elements contribute to the vector products. In an example with $n = 8$ state variables and $m_k = 6$ measurement values, by using this approach, half the computing time was needed with simultaneously improved stability compared to a Kalman filter with sequential innovation.

6.4.6.3 General Remarks

With the UDU^T-factored Kalman filter, an innovation, that is, an improvement of the existing state estimation, may be performed with just one measurement value (observability given). The innovation covariance in Equation 6.45

$$\begin{aligned} q = c_i U\_ D\_ U\_^T c_i^T + s_i^2 = s_i^2 + D_{jj} f_j, \qquad j = 1, \ldots. n, \\ f = U^T c_i^T, \end{aligned} \tag{6.57}$$

is a measure for the momentary estimation quality; it provides information about an error zone around the predicted measurement value y_i* that may be used to judge the quality of the arriving measurement values. If they are too far off, it may be wise to discard this measured value all together.

6.4.7 Conclusion of Recursive Estimation for Dynamic Vision

The background and general theory of recursive estimation underlying the 4-D approach to dynamic vision have been presented in this chapter. Before the overall integration for complex applications is discussed in detail, a closer look at major components including the specific initialization requirements is in order: This will be done for road detection and tracking in Chapters 7 to 10; for vehicle detection and tracking, it is discussed in Chapter 11. The more complex task with many vehicles on roads will be treated in Chapter 14 after the resulting system architecture for integration of the diverse types of knowledge needed has been looked at in Chapter 13.

The sequence of increasing complexity will start here with an ideally flat road with no perturbations in pitch on the vehicle and the camera mounted directly onto the vehicle body. Horizontal road curvature and self-localization on the road are to be recognized; [this has been dubbed SLAM (self-localization and mapping) since the late 1990s]. As next steps, systematic variations in road (lane) width and recognition of vertical curvature are investigated. These items have to be studied in conjunction to disambiguate the image inversion problem in 3-D space over time. There also is a cross-connection between the pitching motion of the vehicle and estimation of lane or road width. It even turned out that temporal frequencies have to be separated when precise state estimation is looked for. Depending on the loading conditions of cars, their stationary pitch angle will be different; due to gas consumption over longer periods of driving, this quasi-stationary pitch angle will slowly change over time. To adjust this parameter that is important for visual range estimation correspondingly, two separate estimation loops with different time constants and specific state variables are necessary (actual *dynamic pitch angle* θ_V, and the *quasi-stationary bias angle* θ_b. This shows that in the 4-D approach, even subtle points in understanding visual perception can be implemented straightforwardly.

Recursive estimation is not just a mathematical tool that can be applied in an arbitrary way (without having to bear the consequences), but the models both for motion in the real world and for perspective mapping (including motion blur!) have to be kept in mind when designing a high-performance dynamic vision system.

Provisions to be implemented for intelligent control of feature extraction in the task context will be given for these application domains. As mentioned before, most gain in efficiency is achieved by looking at the perception and control process in closed-loop fashion and by exploiting the same spatiotemporal models for all subtasks involved.

The following scheme summarizes the recursive estimation loop with sequential innovation and UDU^T-factorization as it has been used in standard form with minor modifications over almost two decades.

Complete scheme with recursion loops in sequential Kalman filtering and UDU^T-factorization.

1. Find a good guess for the n initial state components $x^*(0)$ to start with (initial hypothesis $k = 0$, $\hat{x}_0 = x^*(0)$).
2. Find an estimate for the probability distribution of this initial state in terms of the first and second moments of a Gaussian distribution (mean value $\hat{x}_0$ and the $(n{\times}n)$ error covariance matrix P_0 in factorized form $U_0 D_0 U_0^T$). The terms on the main diagonal of matrix D_0 now are the variance components σ_i^2.
3. If the plant noise covariance matrix Q is diagonal, the starting value for U_q is the identity matrix I, and Q is $Q = U_q\, D_q\, U_q^T$ with D_q the (guessed or approximately known)

values on the diagonal of the covariance matrix $E\{v^T v\}$. In the sequential formulation, the measurement covariance matrix $R = E\{w^T w\}$ is replaced by the diagonal terms $s_i^2 = E\{w_i^2\}$ (Equation 6.28)

→ *Entry point for recursively running main loop (over time)*

4. Increment time index $k = k+1$.
5. Compute expected values for state variables at time $k+1$ [state prediction $x^*(k)$]: $x^*_k = A(k-1)\cdot \hat{x}_{k-1} + B(k-1)\cdot u_{k-1}$.
6. Compute the expected measurement values $y^*(k) = h[x^*(k), p_m]$ and the (total) Jacobian matrix $C = \partial y^* / \partial x|_N$ as first-order approximations around this point.
7. Predict expected error covariance matrix $P^*(k)$ in factorized form (Equations 6.43 and 6.49):

7.1 Initialize matrices $W = [A\cdot U, \ U_q]$ (dimension: $n \cdot 2n$; Equation 6.51)

and $D = \begin{bmatrix} D & 0 \\ 0 & D_q \end{bmatrix}$ (dimension: $2n \cdot 2n$, Equation 6.52).

→ *Entry point for sequential computation of expected error covariance matrix*

7.2 Recursion backward for $\kappa = n, n-1, \ldots, 1$ (Equations. 6.53 – 6.56)

$h_\kappa = D\_a_\kappa$; $(h_{\kappa j} = D_{-jj}\ a_{\kappa j}$, $j = 1, 2, \ldots, n$; $\dim\{h_\kappa\}: 2n+1$);

scalar $D_{\kappa\kappa} = a_\kappa^T \cdot h_\kappa$;vector $d_\kappa = h_\kappa / D_{\kappa\kappa}$;

(for $j = 1, \ldots, \kappa-1$: $U_{j\kappa} = a_j^T \cdot d_\kappa$); replace a_j by $(a_j - U_{j\kappa}\ a_\kappa)$.

go back to step 7.2 for prediction of error covariance matrix

8. Read new measurement values y(k); m_k in size (may vary with time k).
9. Recursive innovation for each of the m_k measurement values: Initialize with $i = 1$.

→ *Entry point for sequential innovation (Equations 6.40 – 6.48)*

9.1: $f_i = U_{j-}^T\ c_i^T$., 9.2: $q_i = s_i^2 + D_{i-,jj}\cdot f_{i,j}$, $j = 1, \ldots, n$.

9.3: $e_i =: D_{i-} f_i$, 9.4: $k_i = U_{i-}\ e_i / q$,

9.5: recursive computation of U_i^0 and D_i^0 from $(D_{i-} - e_i\ e_i^T / q)$,

9.6: $U_{i+} = U_{i-}\ U_i^0$, 9.7: $D_{i+} = D_i^0$,

9.8: Set $U_{i+1,-} = U_{i+}$ and $D_{i+1,-} = D_{i+}$ for the next measurement value.

Increment i by 1

go back to 9.1 for next inner innovation loop as long as $i \leq m_k$.

(Loop 9 yields the matrices $U = U_{m_k+}$ and $D = D_{m_k+}$ for prediction step 7.)

10. Monitor progress of perception process;

go back to step 4 for next time step k = k+1.

This loop may run indefinitely, controlled by higher perception levels, possibly triggered by step 10.

7 Beginnings of Spatiotemporal Road and Ego-state Recognition

In the previous chapters, we have discussed basic elements of the 4-D approach to dynamic vision. After an introduction to the general way of thinking (in Chapter 1), the basic general relations between the real world in 3-D space and time, on one hand, and features in 2-D images, on the other, have been discussed in Chapter 2. Chapter 5 was devoted to the basic techniques for image feature extraction. In Chapter 6, the elementary parts of recursive estimation developed for dynamic vision have been introduced. They avoid the need for storing image sequences by combining 3-D shape models and 3-D motion models of objects with the theory of perspective mapping (measurement models) and feature extraction methods. All this together was shown to constitute a very powerful general approach for dynamic scene understanding by research groups at UniBwM [Dickmanns, Graefe 1988; Dickmanns, Wünsche 1999].

Recognition of well-structured roads and the egostate of the vehicle carrying the camera relative to the road was one of the first very successful application domains. This method has been extensively used long before the SLAM–acronym (simultaneous localization and mapping) became fashionable for more complex scenes since the 1990s [Moutarlier, Chatila 1989; Leonhard, Durrant-White 1991; Thrun *et al.* 2005]. This chapter shows the beginnings of spatiotemporal road recognition in the early 1980s after the initial simulation work of H.G. Meissner between 1977 and 1982 [Meissner 1982], starting with observer methods [Luenberger 1964].

Since road recognition plays an essential role in a large application domain, both for driver assistance systems and for autonomously driving systems, the next four chapters are entirely devoted to this problem class. Over 50 million road vehicles are presently built every year worldwide. With further progress in imaging sensor systems and microprocessor technology including storage devices and high-speed data communication, these vehicles would gain considerably in safety if they were able to perceive the environment on their own. The automotive industry worldwide has started developments in this direction after the "Pro-Art" activities of the European EUREKA-project "PROMETHEUS" (1987–1994) set the pace [Braess, Reichart 1995a, b]. Since 1992, there is a yearly *International Symposium on Intelligent Vehicles* [Masaki 1992++] devoted to this topic. The interested reader is referred to these proceedings to gain a more detailed understanding of the developments since then.

This chapter starts with assembling and arranging components for spatiotemporal recognition of road borderlines while driving on one side of the road. It begins with the historic formulation of the dynamic vision process in the mid-

1980s which allowed driving at speeds up to ~ 100 km/h on a free stretch of Autobahn in 1987 with a half dozen 16-bit Intel 8086 microprocessors and 1 PC on board, while other groups using AI approaches without temporal models, but using way more computing power, drove barely one-tenth that speed. Remember that typical clock rates of general purpose microprocessors at that time were around 10 MHz and the computational power of a microprocessor was ~ 10^{-4} (0.1 per mille) of what it is today. These tests proved the efficiency of the 4-D approach which afterward was accepted as a standard (without adopting the name); it is considered just another extension of Kalman filtering today.

The task of robust lane and road perception is investigated in Section 7.4 with a special approach for handling the discontinuities in the clothoid parameter C_1. Chapter 8 discusses the demanding initialization problem for getting started from scratch. The recursive estimation procedure (Chapter 9) treats the simple planar problem first (Section 9.1) without perturbations in pitch, before the more general cases are discussed in some detail. Extending the approach to nonplanar roads (hilly terrain) is done in Section 9.2, while handling larger perturbations in pitch is treated in Section 9.3. The perception of crossroads is introduced in Chapter 10.

Contrary to the approaches selected by the AI community in the early 1980s [Davis *et al.* 1986; Hanson, Riseman 1987; Kuan *et al.* 1987; Pomerleau 1989, Thorpe *et al.* 1987; Turk *et al.* 1987; Wallace *et al.* 1986], which start from quasi-static single image interpretation, the 4-D approach to real-time vision uses both shape and motion models over time to ease the transition from measurement data (image features) to internal representations of the *motion process* observed.

The motion process is given by a vehicle driving on a road of unknown shape; due to gravity and vehicle structure (Ackermann steering, see Figure 3.10), the vehicle is assumed to roll on the ground with specific constraints (known in AI as "nonholonomic") and the camera at a constant elevation H_c above the ground. Under normal driving conditions (no free rotations), vehicle rotation around the vertical axis (ψ) is geared to speed V and steering input λ, while pitch (θ) and roll (bank) angles (Φ) are assumed to be small; while pitch angles may vary slightly but cannot be neglected for large look-ahead ranges, the bank is taken as zero in the average here.

7.1 Road Model

The top equation in Figure 3.10 indicates that for small steering angles, curvature C of the trajectory driven is proportional to the steering angle λ. Driving at constant speed $V = dl/dt$ and with a constant steering rate $d\lambda/dt$ as control input, this means that the resulting trajectory has a linear change of curvature over the arc length. This class of curves is called *clothoids* and has become the basic element for the design of high-speed roads. Starting the description from an initial curvature C_0 the general model for the road element *clothoid* can then be written

$$C = C_0 + C_1 \cdot l, \qquad (7.1)$$

with C_1 as the clothoid parameter fixing the rate of change of the inverse of the radius R of curvature ($C = 1/R$). This differential geometry model has only two parameters and does not fix the location x, y and the heading χ of the trajectory. With the definition of curvature as the rate of change of heading angle over arc length l ($C = d\chi/dl$), there follows as first integral

$$\chi = \int C \cdot dl = \chi_0 + C_0 \cdot l + C_1 \cdot l^2 / 2 . \qquad (7.2)$$

Heading direction is seen to enter as integration constant χ_0. The transition to Cartesian coordinates x (in the direction of the reference line for χ) and y (normal to it) is achieved by the sine and cosine component of the length increments l. For small heading changes ($\leq 15°$), the sine can be approximated by its argument and the cosine by 1, yielding (with index h for horizontal curvature):

$$x = \int \cos\chi \cdot dl = x_0 + \Delta x, \qquad \text{(a)}$$

$$y = \int \sin\chi \cdot dl = \int (\chi_0 + C_{0h} \cdot l + C_{1h} \cdot l^2 / 2) \cdot dl \qquad (7.3)$$

$$= y_0 + \chi_0 l + C_{0h} \cdot l^2 / 2 + C_{1h} \cdot l^3 / 6 = y_0 + \chi_0 l + \Delta y_C . \qquad \text{(b)}$$

The integration constants x_0 and y_0 of this second integral specify the location of the trajectory element. For local perception of the road by vision, all three integration constants are of no concern; setting them to zero means that the remaining description is oriented relative to the tangent to the trajectory at the position of the vehicle. Only the last two terms in Equation 7.3b remain as curvature-dependent lateral offset Δy_C. This description is both very compact and convenient. Figure 7.1 shows the local trajectory with heading change $\Delta\chi_C$ and lateral offset Δy_C due to curvature.

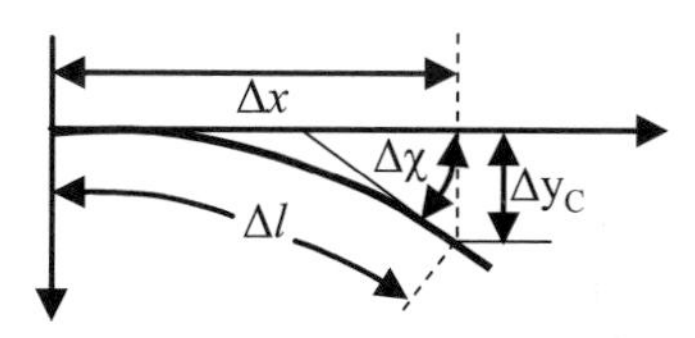

Figure 7.1. From differential geometry to Cartesian coordinates

Roads are pieced together from clothoid elements with different parameters C_0 and C_1, but without jumps in curvature at the points of transition (segment boundaries). Figure 7.2 shows in the top part a road built according to this rule; the lower part shows the corresponding curvature over the arc length. For local vehicle guidance on flat ground, it is sufficient to perceive the two curvature parameters of the road C_{0h} and C_{1h} as well as the actual state of the vehicle relative to the road: lateral offset y_0, lateral velocity v, and heading angle ψ [Dickmanns, Zapp, 1986].

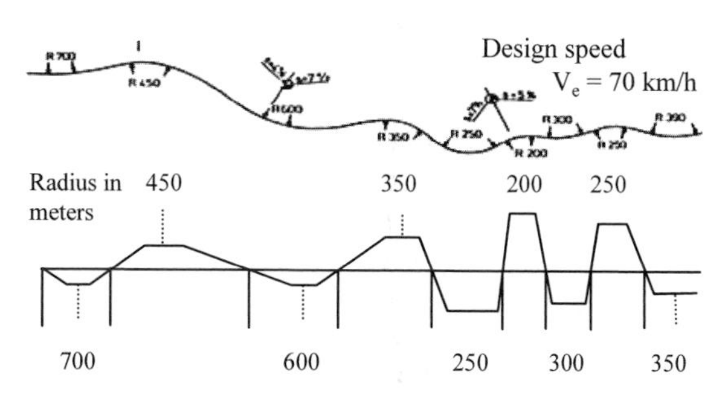

Figure 7.2. Road (upper curve in a bird's-eye view) as a sequence of clothoid elements with continuity in position, heading, and curvature; the curvature change rate C_1 is a sequence of step functions. The curvature over the arc length is a polygon (lower curve).

Note that in the general case, there may be a slight discrepancy between the tangent to the trajectory and the vehicle heading angle ψ, namely, the slip angle β, due to the type of axles with front wheel steering and due to tire softness.

7.2 Simple Lateral Motion Model for Road Vehicles

Figure 7.3 shows simplified models based on the more complete one of Chapter 3. Figure 3.10 described the simple model for road vehicle motion (see also Section 3.4.5.2). The lateral extension of the vehicle is idealized to zero, the socalled bicycle model (with no bank angle degree of freedom!). Equation 3.37 and the block diagram Figure 3.24 show the resulting fifth order dynamic model. Neglecting the dynamic effects on yaw acceleration $\ddot{\psi}_{abs}$ and slip angle rate $\dot{\beta}$ (valid for small speeds, T_ψ and $T_\beta \rightarrow 0$, Equation 3.38), a lateral motion model of third order according to block diagram Figure 7.3a results.

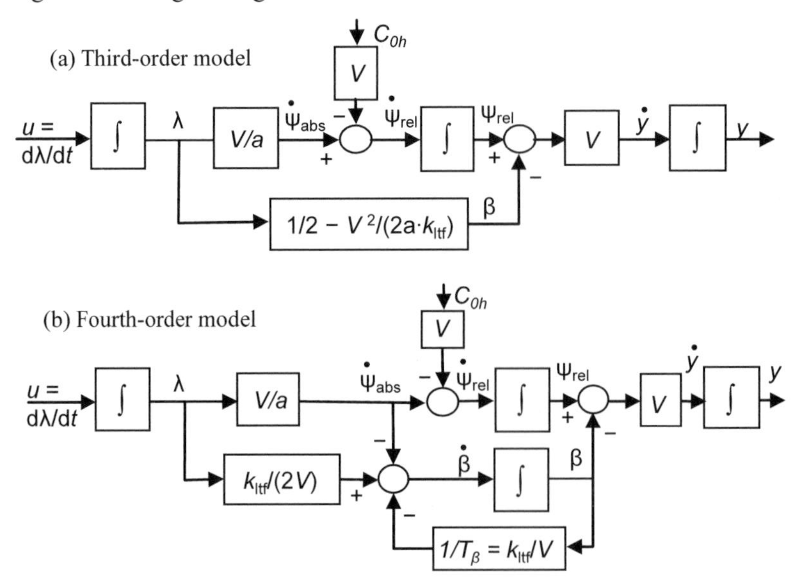

Figure 7.3. Block diagram for lateral control of road vehicles taking rotational dynamics around the vertical axis only partially into account. (a) Infinite tire stiffness yields simplest model. (b) Taking tire force build-up for slip angle into account (not for yaw rates) improves observation results.

Neglecting the dynamic term only in the differential equation for yaw acceleration, the turn rate becomes directly proportional to steering angle $(d\psi_{abs}/dt) = V{\cdot}\lambda/a$, and the dynamic model reduces to order four. The split in omitting tire softness effects for the yaw rate but not for the slip angle β can be justified as follows: The yaw angle is the next integral of the yaw rate and does not influence forces or moments acting on the vehicle; in the estimation process, it will finally be approximated by the correct value due to direct visual feedback. On the contrary, the slip angle β directly influences tire forces but cannot be measured directly.

Steering rate is the control input $u(t)$, acting through a gain factor k_λ; the steering angle directly determines the absolute yaw rate in the fourth-order model. By subtracting the temporal heading rate of the road $V{\cdot}C_{0h}$ from this value, the heading rate $\dot{\psi}_{rel}$ relative to the road results. After integrating, subtracting the slip angle, and multiplying the sum by speed V, the lateral speed relative to the road $v = dy_V/dt$ is obtained. The set of differential equations for the reduced state variables then is

$$\begin{pmatrix} \dot{\lambda} \\ \dot{\beta} \\ \dot{\psi}_{rel} \\ \dot{y}_V \end{pmatrix} = \begin{pmatrix} 0 & 0 & 0 & 0 \\ a_{12} & -1/T_\beta & 0 & 0 \\ V/a & 0 & 0 & 0 \\ 0 & -V & V & 0 \end{pmatrix} \begin{pmatrix} \lambda \\ \beta \\ \psi_{rel} \\ y_V \end{pmatrix} + \begin{pmatrix} k_\lambda \\ 0 \\ 0 \\ 0 \end{pmatrix} \cdot u(t) + \begin{pmatrix} 0 \\ 0 \\ -V \\ 0 \end{pmatrix} \cdot C_{0h}, \quad (7.4)$$

with $a_{12} = 1/(2T_\beta) - V/a; T_\beta = V/k_{ltf}$; a = axle distance; $k_{ltf} = F_y/(m_{WL} \cdot \alpha_f)$ (in $m/s^2/rad$) (lateral acceleration as a linear function of angle of attack between the tire and the ground, dubbed *tire stiffness*, see Section 3.4.5.2. For the UniBwM test van VaMoRs, the actual numbers are a = 3.5 m, $k_{ltf} \approx 75$ $m/s^2/rad$, and $T_\beta = 0.0133 \cdot V\ s^{-1}$. For a speed of 10 m/s (36 km/h), the eigenvalue of the dynamic tire mode is $1/T_\beta = 7.5\ s^{-1}$ (= 0.3 times video rate of 25 Hz); this can hardly be neglected for the design of a good controller since it is in the perceptual range of human observers. For higher speeds, this eigenvalue becomes even smaller. The k_{ltf} – value corresponds to an average lateral acceleration buildup of 1.3 m/s^2 per degree angle of attack. [This is a rather crude model of the vehicle characteristics assuming the *cg* at the center between the two axles and the same tire stiffness at the front and rear axle; the precise data are as follows: The center of gravity is 1.5 m in front of the rear axle and 2 m behind the front axle; the resulting axle loads are rear 2286 and front 1714 kg (instead of 2000 each). Due to the twin wheels on each side of the rear axle, the tire stiffness at the axles differs by almost a factor of 2. The values for the simple model selected yield sufficiently good predictions for visual guidance; this resulted from tests relative to the more precise model.]

7.3 Mapping of Planar Road Boundary into an Image

This section has been included for historical reasons only; it is not recommended for actual use with the computing power available today. The simple mapping model given here allowed the first autonomous high-speed driving in 1987 with very little computing power needed (a few 16-bit microprocessors with clock rates of order of magnitude 10 MHz). Only intelligently selected subregions of images (256×256 pixels) could be evaluated at a rate of 12.5 Hz. Moving at speed V along the road, the curvature changes $C(l)$ appear in the image plane of the camera as time-varying curves. Therefore, exploiting Equation 7.1 and the relationship $V = dl/dt$, a simple dynamic model for the temporal change of the image of the road while driving along the road can be derived.

7.3.1 Simple Beginnings in the Early 1980s

For an eye–point at elevation h above the road and at the lane center, a line element of the borderline at the look-ahead distance L is mapped onto the image plane y_B, z_B according to the laws of perspective projection (see Figure 7.4).

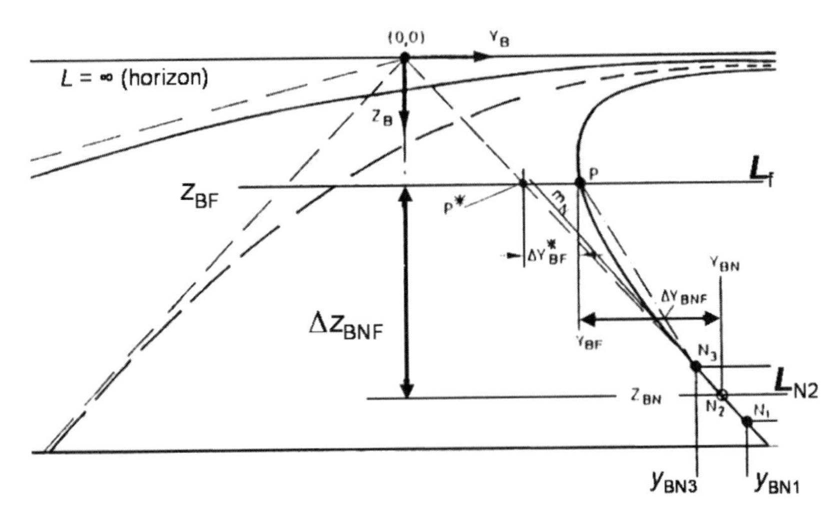

Figure 7.4. Symbols for planar curvature analysis, look-ahead range from nearby (index N) to L_f (row z_{BF} in image)

P* would be the image point for a straight planar road and the camera looking parallel to the road. Road curvature according to a clothoid element of the form Equation 7.1 would yield the lateral offset Δy^*_{BF} corresponding to Δy_C given by Equation 7.3b at the look-ahead distance $l = L_f$ between points P* and P.

Figure 7.5 shows the more general situation in a top-down view where both a lateral offset y_V of the camera from the lane center line ($b/2 - y_v$) and a viewing direction ψ_K not tangential to the road at the vehicle (camera) position have been allowed. At the look-ahead distance L_f, ψ_K yields an offset $L_f \cdot \psi_K$ in addition to y_V. Both contribute to shifts of image points P and P*.

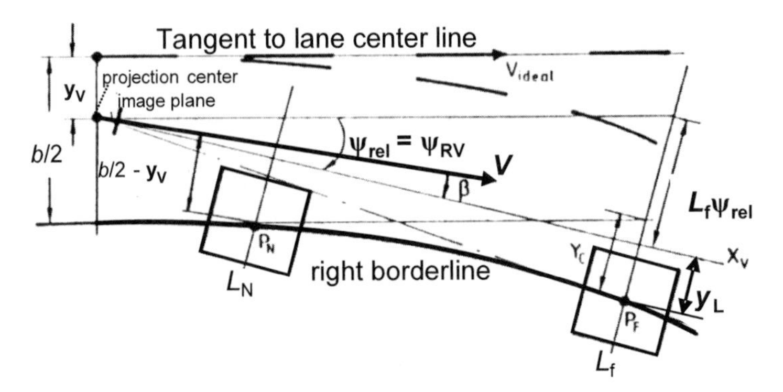

Figure 7.5. Top view of general imaging situation for a borderline with lateral offset y_V, nontangential viewing direction ψ_K, and road curvature; two preview ranges L_N and L_f

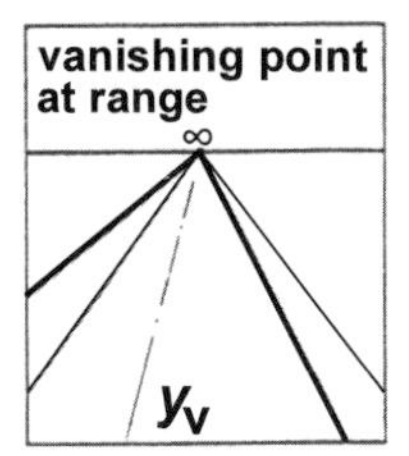

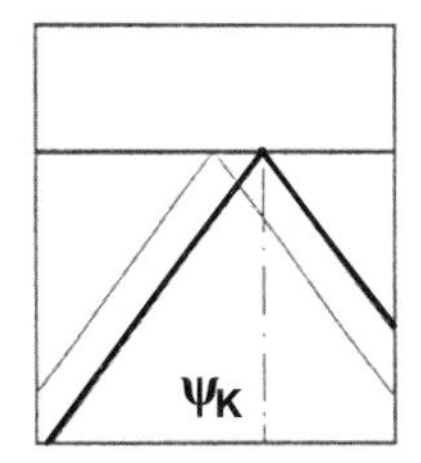

Figure 7.6. Effects of lateral offset y_v (left) and camera pointing direction ψ_{rel} (right) on image of a straight road (reference: thin lines). Positive y_v turns lines to left around the "vanishing" point at ∞; negative ψ_{rel} shifts the entire image horizontally to the right.

The effects of small offsets on P* can be seen from Figure 7.6. Lateral offsets of the camera to the road or lane center turn the road image around the vanishing point at $L = \infty$ (horizon) and show up in the lower part of the image essentially (near ranges), while gaze shifts translate the mapped object laterally in the image.

Here, the camera is assumed to be fixed directly to the vehicle body in the x_V-direction which forms an angle ψ_{rel} with the road tangent (Figure 7.5). The effects on a point on the curved borderline of the road are more involved and require perspective mapping. Using the pinhole model (see Figure 2.4, Equation 2.4), perspective projection onto a focal plane at distance f from the projection center (on left side in Figure7.5 at lateral position y_V from the lane center line) yields for small angles and perturbations (cos ≈ 1, sine ≈ argument) the approximate image coordinates

$$\begin{aligned} y_B &= f \cdot (b/2 - y_V + y_C - L\psi_{rel})/L \\ z_B &= fh/L. \end{aligned} \tag{7.5}$$

Let us introduce for convenience the back-projected lateral offset y_B in the image to the look-ahead distance L as y_L. Then with the first of Equations 7.5 multiplied by L/f,

$$y_L = b/2 - y_V + y_C - L\psi_{rel} = Ly_B / f\,. \tag{7.6}$$

A dynamic model for road curvature determination is obtained by taking the time derivative (constant lane width and look-ahead distance are assumed)

$$\dot{y}_L = -\dot{y}_V + \dot{y}_C - L\dot{\psi}_{rel}\,. \tag{7.7}$$

Now some substitutions are introduced: The lateral speed $\dot{y}_V$ relative to the road may be expressed by the relative path angle $\Delta\chi$ between the vehicle velocity vector and the road tangent. With the viewing direction fixed to the car body axis x_V, the sideslip angle β (see Figure 7.5) between the velocity vector V and ψ_{rel} has to be taken into account yielding

$$\dot{y}_V = V \cdot \Delta\chi = V(\psi_{rel} - \beta)\,. \tag{7.8}$$

This equation represents the right-hand part of Figure 7.3. For the term $\dot{y}_C$ regarding the curvature effects from the position of the vehicle ($l = 0$) to the look-ahead distance L, by applying the chain rule, one obtains

$$dy_C / dt = dy_C / dl \cdot dl / dt = dy_C / dl \cdot V\,. \tag{7.9}$$

From Equation 7.3b, there follows for constant values C_{0h}, C_{1h},

$$dy_C / dl = C_{0h} \cdot L + C_{1h} \cdot L^2 / 2\,. \tag{7.10}$$

Similarly, there follows for the change with time of the curvature C_{0h} at the location of the vehicle with Equation 7.1,

$$dC_{0h}/dt = dC_{0h}/dl \cdot dl/dt = (C_{1h} + l \cdot \Delta C_{1i} \cdot \delta(l - l_{Ci})) \cdot V . \quad (7.11)$$

C_{1h} is piecewise constant; the stepwise change at transition points l_{Ci} means a Dirac impulse $\Delta C_{1i} \cdot \delta(l - l_{Ci})$ in the derivative dC_{1h}/dl. For practical purposes, the approximation of the last term in Equation 7.11 by Gaussian random noise $w(t)$ driving the time derivative of C_{1h} has proven to work sufficiently well with proper tuning of the recursive filter (at least for these early trials, see Section 7.4).

The term $\dot{\psi}_{rel}$ in Equation 7.7 is the inertial yaw rate of the vehicle (= camera) minus the heading change of the road over time, which is the curvature at the vehicle location times speed

$$\dot{\psi}_{rel} = \dot{\psi}_{abs} - C_{0h}V = (V/a) \cdot \lambda - C_{0h}V . \quad (7.12)$$

Equations 7.7 – 7.12 yield the state space model for estimating the curvature of the right-hand road boundary; since the effects of a lateral offset y_V are contained in the variable y_L and since dy_L/dt does not depend on it, for curvature estimation (as a separately coded routine), this variable can be deleted from the state vector yielding the third-order dynamic model,

$$\begin{pmatrix} \dot{y}_L \\ \dot{C}_{0h} \\ \dot{C}_{1h} \end{pmatrix} = \begin{pmatrix} 0 & 2LV & L^2V/2 \\ 0 & 0 & V \\ 0 & 0 & 0 \end{pmatrix} \begin{pmatrix} y_L \\ C_{0h} \\ C_{1h} \end{pmatrix} + \begin{pmatrix} -LV/a & V & -V \\ 0 & 0 & 0 \\ 0 & 0 & 0 \end{pmatrix} \begin{pmatrix} \lambda \\ \beta \\ \psi_{rel} \end{pmatrix} + \begin{pmatrix} 0 \\ 0 \\ w(t) \end{pmatrix} . \quad (7.13)$$

The second block gives the coupling of vehicle motion into the process of road observation. The fourth state variable y_V of the vehicle has been omitted here since its coefficients in the vehicle part (right) are all zero; the observation part is included in y_L. This yields the historic formulation for road perception which looks a bit unsystematic nowadays. For the near range, the effects of C_{1h} have been neglected and the four vehicle states have been used to compute from Equation 7.5 the predicted distance y_{BN} in the image for any look-ahead distance L_N

$$y_{BN} = f \cdot \left[\{ b/2 - y_V + y_C(L_N) \} / L_N - \psi_{rel} \right] . \quad (7.14)$$

With Equation 7.3b for the curvature effects, the combined (total) state vector (index t) becomes from Equations 7.4 and 7.13

$$x_t^T = (\lambda, \beta, \psi_{rel}, y_V, y_L, C_{0h}, C_{1h}) . \quad (7.15)$$

With this terminology, the measurement equations for image interpretation from the two edge positions y_{BN} and y_{BF} in the near and far range are

$$\begin{pmatrix} y_{BN} \\ y_{BF} \end{pmatrix} = \left(\begin{array}{cccc|ccc} 0 & f & -f & -f/L_N & 0 & f \cdot L_N/2 & 0 \\ 0 & 0 & 0 & 0 & f/L_F & 0 & 0 \end{array} \right) \cdot x_t + \begin{pmatrix} b \cdot f/(2L_N) \\ 0 \end{pmatrix} . \quad (7.16)$$

Note that the introduction of the new state variable y_L for visual measurements in the far range (Equation 7.6) contains the effects of the goal distance $b/2$ from the road boundary for the trajectory to be driven, of the lateral offset y_V from this line and of the curvature y_C; for this reason, all entries in the lower row of Equation 7.16 are zero except for the position corresponding to y_L. However, y_L also contains a term proportional to the heading direction ψ; since this angle determines future lateral offsets. Its feedback to corrective control in steering automatically yields a lead term, which is known to be beneficial for damping oscillations [Zapp 1988].

7.3.2 Overall Early Model for Spatiotemporal Road Perception

Since the images are taken cyclically as temporal samples with evaluation time twice the video cycle time (T_B = 80 ms), the linear differential Equations 7.4 and 7.13 are transformed into difference equations by one of the standard methods in sampled data theory. The complete set of differential equations to be transformed into algebraic transition equations in matrix form for the time step from kT_B to $(k+1)\, T_B$ is

$$\dot{x}_t = F \cdot x_t + g \cdot u(t) + g_n \cdot n(t), \quad \text{where} \tag{7.17}$$

$$F = \left(\begin{array}{cccc|ccc} 0 & 0 & 0 & 0 & 0 & 0 & 0 \\ f_{12} & -1/T_\beta & 0 & 0 & 0 & 0 & 0 \\ V/a & 0 & 0 & 0 & 0 & -V & 0 \\ 0 & -V & V & 0 & 0 & 0 & 0 \\ \hline (-LV/a) & V & -V & 0 & 0 & 2LV & L^2V/2 \\ 0 & 0 & 0 & 0 & 0 & 0 & V \\ 0 & 0 & 0 & 0 & 0 & 0 & 0 \end{array}\right); x_t = \begin{pmatrix} \lambda \\ \beta \\ \psi_{rel} \\ y_V \\ \hline y_L \\ C_{0h} \\ C_{1h} \end{pmatrix}; g_n = \begin{pmatrix} k_\lambda \\ 0 \\ 0 \\ 0 \\ \hline 0 \\ 0 \\ 0 \end{pmatrix};$$

$$g_n^T = (0, 0, 0, 0, 0, 0, n_{C_1})$$

with $f_{12} = 1/(2T_\beta) - V/a$; $\quad T_\beta = V/k_{ltf}$; $\quad$ Equation 3.30.

This equation shows the internal dependences among the variables. The steering angle λ depends only on the control input $u(t)$ = steering rate. The lateral position of the vehicle y_V as well as the variable y_L does not affect vehicle dynamics directly (columns 4, 5 in F). Vehicle dynamics (upper left block) depends on steering angle (first column), slip and heading angle as well as curvature input (sixth column). [Equation 3.37 shows that in the fifth-order model, with an additional time constant in yaw rate buildup, this yaw rate also plays a role (*e.g.*, for higher speeds). To gain experience, it has sometimes been included in the model; note that for five state variables, the number of entries in the upper left block of the F-matrix goes up by 56 % compared to just four.]

The two models for vehicle state estimation (upper left) and for road perception (lower right) have been coded as separate fourth-order and third-order models with cross-feeds of $V \cdot C_{0h}$ into the yaw equation, and of the first three vehicle states into the y_L equation $[-\lambda \cdot LV/a + V(\beta - \psi_{rel})]$. Instead of 49 elements of the full matrix F, the sum of elements of the decoupled matrices is just 25. This rigorous minimization may have lost importance with the processors available nowadays.

The clothoid parameter *curvature change with arc length,* C_{1h} is driven by a noise input $n_{C1}(t)$ here; since C_{1h} affects C_{0h} via integrals, introducing a properly chosen noise level for this variable directly may help improve convergence; similarly, introducing some noise input into the equation for ψ_{rel} and y_L may compensate for neglecting dynamic effects in the real world. The best parameters have to be found by experimentation with the real setup (filter tuning).

The big advantage of this formulation of the visual perception process as a prediction-error feedback task is automatic adjustment of the variables in a least squares sense to the models chosen. The resulting linear measurement model Equation 7.16 yields the Jacobian matrix directly for prediction-error feedback in recursive estimation.

A perception model like that given (including the yaw acceleration equation) was used for the first high-speed tests with VaMoRs [Dickmanns, Zapp 1987] when autonomous driving elsewhere was at least one order of magnitude slower.

7.3.3 Some Experimental Results

This setup has allowed driving behavior up to higher speeds; details may be found in [Zapp 1988]. Two consequential tests will be shown here since they had an impact on the development of vision for road vehicle guidance in Europe. Figure 7.7 shows a demonstration setup in the skidpan of the Daimler-Benz AG (now part of DaimlerChrysler) in November/December 1986.

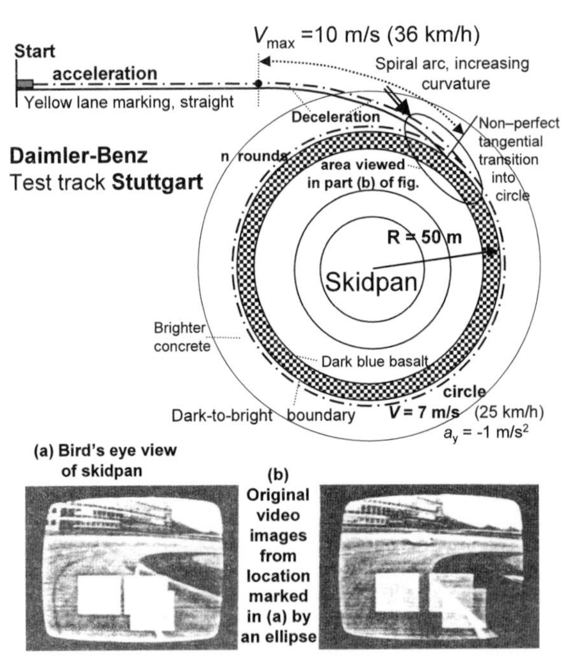

Figure 7.7. Historic first demonstration of visual road vehicle guidance to Daimler-Benz AG in their skidpan in Stuttgart: The test vehicle VaMoRs performed autonomous longitudinal and lateral guidance. It initially accelerated along a straight line which then turned into a spiral with increasing curvature that ended at a transition boundary between circular areas of concrete and basalt in the skid pan. Speed had to be adjusted automatically such that a lateral acceleration of 0.1 g (~ 1 m/s^2) was not exceeded (see text).

The skidpan consists of circular areas with different road surface materials. This is normally used for testing reactions to different friction coefficients between tire and ground. For our purposes, the differences in brightness between concrete and basalt were essential. After accelerating along a straight line (*upper left*), a tightening spiral arc followed, leading to the natural boundary between concrete and basalt (*area in ellipse, upper right*). The vehicle then had to pick up the edge in brightness between these materials (no additional line markers; see the two video images in lower part). It can be seen that the transition was not exactly tangential (as asked for) but showed a slight corner (to the left), which turned out to pose no problem to perception; the low-pass filtering component of the algorithm just smoothed it away. (In crisp detailed perception, there should have been a small section with negative curvature.) The downward dip in the lower part of Figure 7.8 around 100 m was all that resulted.

The task then was to follow the circular edge at a speed such that the lateral acceleration was 0.1 g (~ 1 m/s^2). The top curve in Figure 7.8 shows on the right-hand side that speed varied around V = 7.3 m/s, while the estimated curvature oscillated between $1/R = C = 0.019$ and 0.025 m^{-1}. These demonstrations encouraged Daimler-Benz to embark on a joint project with UniBwM named *Autonomous Mobile Systems* that was supported by the German Ministry of Research and Technology (BMFT). During this project, a major achievement in 1987 was the demonstration of the capability of visual autonomous driving at high speed on a free stretch of Autobahn near Munich.

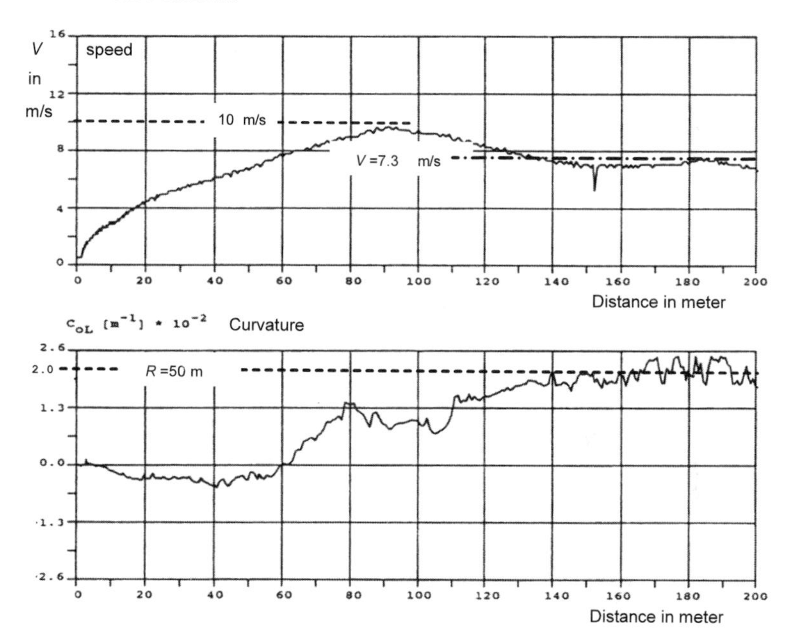

Figure 7.8. Skidpan test results (1986) for speed driven (top) and curvature estimated (bottom). (See text for a more detailed explanation)

Figure 7.9 shows the speed history (top) over a stretch of 13 kilometers and the estimated road curvature (bottom). It can be seen that speed is around 25 m/s (90 km/h) for about ten kilometers. The dips with speed reduction are due to erroneous feature selection and stronger perturbations in the estimation process. The vehicle then reduces speed which, usually, means fewer hazards and more time for reaction; when the correct features are recovered after a few evaluation cycles of 80 ms, the vehicle accelerates again. Top speed achieved at around 9 km was 96 km/h, limited by the engine power of the 5-ton van (actual weight was 4 metric tons).

The far look-ahead range was a little more than 20 m with a distribution of three windows for image evaluation similar to Figure 7.7b. Due to the limited computing power of the (16-bit) Intel 8086 microprocessors used for image processing, only

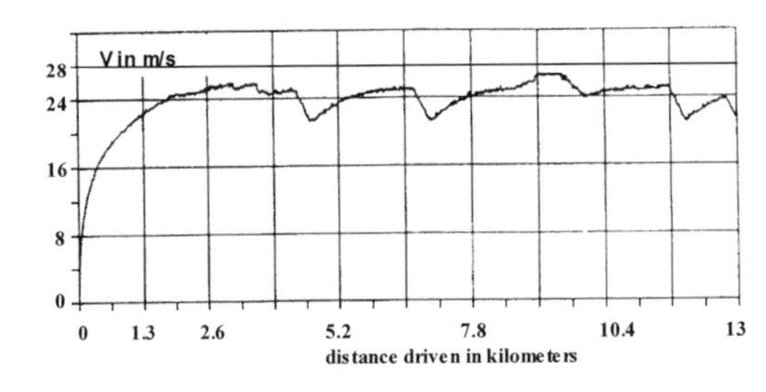

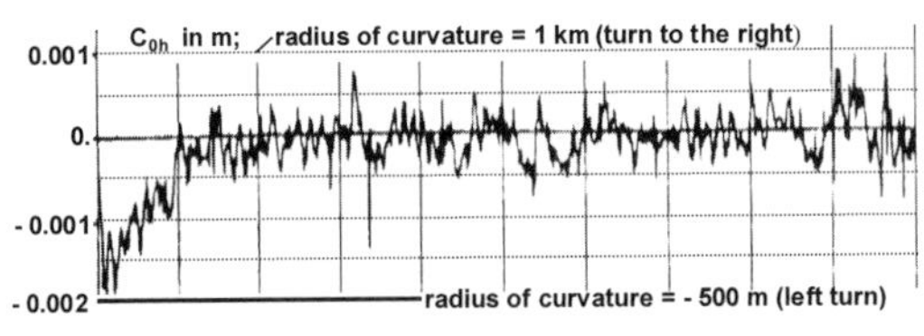

Figure 7.9. Historic autonomous high-speed driving with VaMoRs on a free stretch of Autobahn near Munich in 1987: *Top:* Speed over distance driven (V_{max} = 26.7 m/s = 96 km/h ≈ 60 mph). *Bottom:* Curvature as estimated (note initial transient when starting from straight-road assumption $C = 0$)

the video signals within these windows were grabbed by one processor each. The location of the windows in the image could be shifted from one frame to the next. Up to three search paths could be set up within each window, but only the best looking one would be used for interpretation (see further down below). The left-hand window was used for picking up the left lane or road boundary when higher curvatures were met (mainly on minor roads also investigated).

The relatively short look-ahead range is the reason for the long transient in averaging out the correct road curvature (*bottom graph, left side*). In almost flat countryside, a "post-WW2 Autobahn" in Germany should have radii of curvature $R \geq 1$ km (C_{0h} = 0.001 m-1); it is seen that the estimated curvature, starting from the initial value zero, increases in magnitude to almost 0.002 (R = 500 m) before leveling out in the correct range $|C_{0h}| \leq 0.001$ m^{-1} only after about 1 km. Even though curvature estimation looks rather jagged, the lateral offset from the ideal centerline averaged about 15 cm with occasional deviations up to 30 cm. This has to be judged against a usual lane width of 3.75 m provided for vehicles less than ~2 m wide; thus, the initial autonomous results were better than those normally shown by an average human driver.

With the advent of higher performance microprocessors, more robust road and lane recognition using multiple windows on lane markings or border lines on both sides of the vehicle has been achieved with more advanced perception models [Mysliwetz 1990]. The early results shown are considered of special importance since they proved the high potential of computer vision for lateral guidance of road vehicles versus buried cables as initially intended in the EUREKA-project "Prometheus" in 1987. This has led, within a few years, about a dozen car manufacturing companies and about four dozen university institutes to start initiatives in visual road vehicle guidance in the framework of this large EUREKA-project running from 1987 till the end of 1994. As a consequence of the results demonstrated since 1986 both in the U.S. and in Europe, Japanese activities were initiated in 1987 with the *Personal Vehicle System* (PVS) [Tsugawa, Sadayuki 1994]; these activities were followed soon by several other car companies.

Though *artificial neural net* approaches (ANN) have been investigated for road vehicle guidance on all continents, the derivatives of the method shown above finally outperformed the ANN-approaches; nowadays, spatiotemporal modeling and recursive estimation predominate all applications for vehicle guidance around the globe.

7.3.4 A Look at Vertical Mapping Conditions

Vertical mapping geometry is shown in Figure 7.10. The elevation of the camera above the ground strongly affects the depth–resolution in the viewing direction. Since the sky is of little interest, in general, but may introduce high image brightness in the upper part of the image (sky–region), cameras in vehicles are mounted with a downward looking pitch angle, usually.

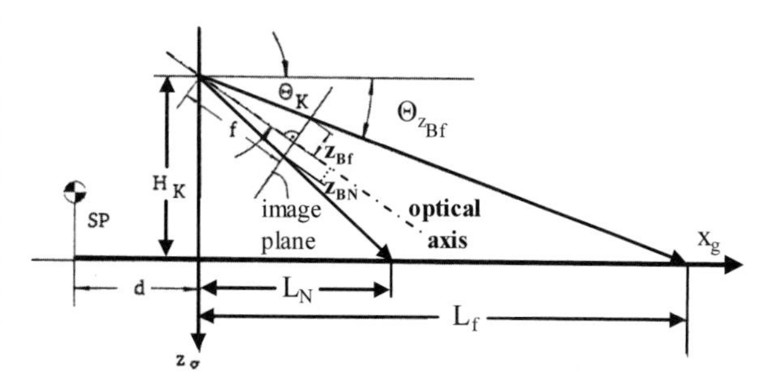

Figure 7.10. Vertical perspective mapping geometry with downward looking camera (θ_K) for better imaging in near range (index N)

The slope of the mapping ray through points at the far look-ahead range L_f relative to the road tangent in the vertical plane according to the figure is

$$\theta_{z_{Bf}} = \theta_K + \arctan(z_{Bf} /(f \cdot k_z)$$

with (7. 18)

$$\theta_{z_{Bf}} = \arctan(H_K / L_f).$$

Equating the tan of these expressions and exploiting the addition theorem for tan $(\alpha + \beta) = (\tan \alpha + \tan \beta)/(1 - \tan \alpha \cdot \tan \beta)$ leads to

$$H_K / L_f = \left[\tan\theta_K + z_{Bf} /(f \cdot k_z)\right] / \left[1 - \tan\theta_K \cdot z_{Bf} /(f \cdot k_z)\right]. \tag{7.19}$$

Solving for the vertical image coordinate z_{Bf} as a function of the look-ahead range L_f yields

$$z_{Bf} = f \cdot k_z \left(H_K - L_f \tan\theta_K\right) / \left(L_f + H_K \tan\theta_K\right). \tag{7.20}$$

For $\theta_K = 0$, this reduces to the known simple pinhole model of perspective mapping. Inverting Equation 7.19 and taking the derivative with respect to the camera pitch angle shows the sensitivity of the look-ahead range L_f with respect to changes in pitch angle θ (shorthand: $z'_B = z_{Bf} / f \cdot k_z$):

$$\frac{d}{d\theta}\left(\frac{L_f}{H_K}\right) = \frac{d}{d\theta}\left(\frac{1 - z'_B \tan\theta}{\tan\theta + z'_B}\right) = -\frac{1 + (z'_B)^2}{(\sin\theta + z'_B \cos\theta)^2}, \tag{7.21}$$

which for the optical axis $z_{Bf} = 0$ reduces to

$$d/d\theta \left(L_f\right)_{|z_B = 0} = -H_K /(\sin\theta)^2. \tag{7.22}$$

For the van *VaMoRs*, the camera elevation H_K above the ground is 1.8 to 2 m; for the sedan *VaMP*, it is ~ 1.3 m. Table 7.1 gives some typical numbers for the gaze angles relative to the horizontal (downward negative) and for the sensitivity of the look-ahead range to small changes in pitch angles for these elevations. Columns 3, 4 and 7, 8 show range changes in meters per image row (pixel) for two different focal lengths realizing resolutions of 8, respectively, 40 pixels per degree; columns 3 and 7 are typical for images with 240 lines and a 30° vertical field of view which has often been used as a standard lens, while columns 4 and 8 correspond to a tele–lens with an 8° vertical field of view.

Table 7.1. Pitch angles θ of optical axes for certain look-ahead ranges L, for two camera elevations H_K above a planar ground, and for two focal lengths (resolutions)

0	1	2	3	4	5	6	7	8
vehicle	VaMoRs				VaMP			
H_K in m	camera elevation = 1.8 m				= 1.3 m		above ground	
resolut.			8 pel/°	30 pel/°			8 pel/°	30 pel/°
$L_{opt\ axis}$ in m	θ in degrees	dL/dθ m/deg	dL/dpel m/pixel	dL/dpel m/pixel	θ in degrees	dL/dθ m/deg	dL/dpel m/pixel	dL/dpel m/pixel
100	-1.03	97	12.1	3.2	-0.745	134	16.7	4.47
60	-1.72	19.4	2.4	0.65	-1.24	48	6	1.6
40	-2.58	15.5	1.9	0.52	-1.86	21.5	2.7	0.72
20	-5.14	3.91	0.5	0.13	-3.72	5.4	0.68	0.18
15	-6.84	2.2	0.275	0.073	-4.95	3.04	0.38	0.10
10	-10.2	1.00	0.125	0.033	-7.4	1.36	0.17	0.045
5	-19.8	0.274	0.034	0.009	-14.6	0.358	0.045	0.012
1.8	-45	0.063	0.008	0.002	(-35.8)	(0.066)	(0.008)	(0,002)

1° change in pitch angle at a look-ahead range of 20 m leads to look-ahead changes of 3.9 m (~ 20 %, column 2) for the van and 5.4 m (27 %, column 6) for the car; this corresponds to a half meter range change for the standard focal length in the van and 68 cm in the car. For the telelens, the range change is reduced to about 10 cm per pixel for the van and 14 cm for the car. However, at a look-ahead range of 100 m, even the telelens experiences a 2.4 m range change per pixel (= 1/40 = 0.025° pitch angle) in the van and 3.35 m in the car.

A look at these values for larger look-ahead ranges shows that the pitch angle cannot be neglected as a state variable for these cases, since minor nonplanarity of the ground may lead to pitch perturbations in the range of a few tenths of 1°. This problem area will be discussed in Section 9.3 after some experience has been gained with the basic estimation process for perceiving road parameters.

7.4 Multiple Edge Measurements for Road Recognition

Since disturbances in the visual perception process in natural environments are more the rule than an exception, multiply redundant measurements are mandatory for robust road or lane recognition; the least-squares estimation process handles this case without additional problems. However, it has proven advantageous to take a special approach for dealing with the discontinuity in C_1 over arc the length.

7.4.1 Spreading the Discontinuity of the Clothoid Model

Figure 7.11 shows a piece of a road skeleton line consisting of several clothoid arcs: (1, *see bottom of figure*) straight line with $C_{0h} = C_{1h} = 0$; (2): clothoid (proper) with increasing curvature C; (3) circular arc (positive curvature = turning to the right); (4 and 5) clothoid with decreasing curvature (inflection point at the transition from 4 to 5); (6) negative circular arc (turning left); (7) clothoid back to straight driving. [Note that this maneuver driven by a vehicle on a wide road would mean a lane change when the parameters are properly chosen. The steering rate as control input for constant speed driven would have to make jumps like C_{1h} (*second curve from top*); this of course is an idealization.] The location and magnitude of these jumps is hard to measure from perspective image data.

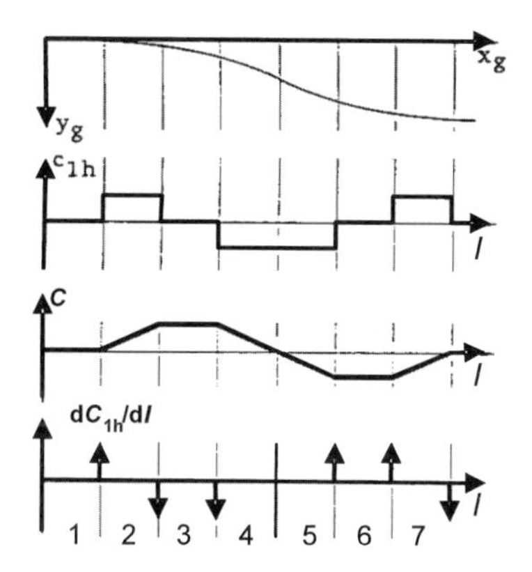

Figure 7.11. Clothoid road section: Top in Cartesian coordinates x, y; second row: clothoid parameters C_{1h} = dC/dl; third row: curvature C over arc length; fourth row: dC_{1h}/dl as Dirac impulses (step jumps in C_{1h})

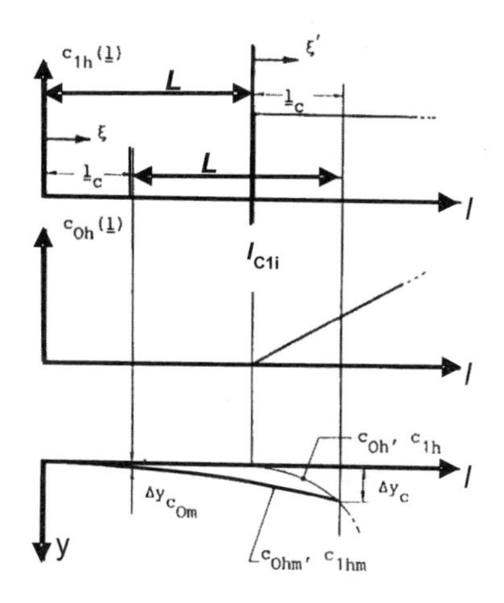

Figure 7.12. Handling a discontinuity in the clothoid parameter C_1 (top) by introducing an *averaging* model (index m) for the curvature parameters (bottom)

Figure 7.12 displays the basic idea of the approach for avoiding the ideal Dirac impulse and approximating it by a substitute system (index m); this should yield the same lateral offset Δy_C at the far end of the look-ahead distance. In principle, it starts working when the look-ahead region up to the distance L in front of the vehicle just touches the location of the impulse l_{C1i}, which is of course unknown. Continuing egomotion, an increasing part of the road with $C_{1h} \neq 0$ moves into the field of view; its extension is measured by the running variable l_c starting from zero at the location $l = l_{C1i} - L$. To obtain a non-dimensional description of the process, the variable $\xi = l_c/L$ is introduced (l_c is a fraction of the look-ahead distance). The idea is to obtain "spread-out" variables as new curvature parameters C_{0hm} and C_{1hm} of a dynamic system that generate the same lateral offset Δy_C of the curved line at the look-ahead distance L at any moment as the real road curvature (see lower right corner of Figure 7.12).

The assumption for computing the offset Δy_C of the real clothoid by the spread-out system is that C_{0hm} and C_{1hm} of this virtual system are constants with respect to integration over the look-ahead distance (at each moment) even though they are dynamic variables evolving with penetration of the look-ahead range into the region with $C_{1h} \neq 0$.

The real lateral offset as function of penetration l_c for $0 \leq l_c \leq L$ and $C_{0h} = 0$, is

$$\Delta y_{Cr}(l_c) = C_{1h} l_c^3 / 6 . \tag{7.23}$$

For the spread-out system, we write

$$\Delta y_{Cm}(l_c) = \Delta y_{C0hm} + \Delta y_{C1hm} ;$$

where

$$\Delta y_{C1hm} = C_{1hm}(l_c) \cdot L^3 / 6 \tag{7.24}$$

$$\Delta y_{C0hm} = C_{0hm}(l_c) \cdot L^2 / 2 = \int_0^{l_c} C_{1hm}(\lambda)\, d\lambda \cdot L^2 / 2.$$

Setting Equations 7.23 and 7.24 equal yields

$$\int_0^{l_c} C_{1hm}(\lambda)\, d\lambda \cdot L^2 / 2 + C_{1hm}(l_c) \cdot L^3 / 6 = C_{1h} \cdot l_c^3 / 6 . \tag{7.25}$$

By dividing by $L^2/6$, differentiating with respect to l_c, and introducing $\xi = l_c/L$, the following differential equation is obtained:

$$dC_{1hm} / d\xi + 3 \cdot C_{1hm}(\xi) = 3 \cdot C_{1h} \cdot \xi^2 . \tag{7.26}$$

Solving by one of the standard methods (*e.g.* Laplace transform) finally yields:

$$C_{1hm}(\xi) = C_{1hm}(0) \cdot \exp(-3\xi) + C_{1h} \{\xi^2 - 2/3 \cdot \xi + 2/9[1 - \exp(-3\xi)]\}. \tag{7.27}$$

With $C_{1hm}(0) = 0$, one obtains the value $C_{1hm}(1) = 0.5445\, C_{1h}$ at the location of the jump in C_{1h} ($\xi = 1$). The magnitude of C_{1h} is yet unknown. Due to the discontinuity in C_{1h}, a new descriptive equation has to be chosen for $\xi > 1$. Selecting $\xi' = (\xi - 1)$ and the final value from the previous part ($C_{1hm}(l_{C1i}) = 0.5445\, C_{1h}$) as the initial value for the following one, again the resulting lateral offset Δy_{Cm} should be equal to the real one. The vehicle is now moving in the region where C_{0h} is increasing linearly with C_{1h}. Therefore, instead of Equation 7.24, we now have

$$C_{0h}(l_c) \cdot L^2 / 2 + C_{1h} \cdot L^3 / 6 = C_{0hm}(l_c) \cdot L^2 / 2 + C_{1hm}(l_c) \cdot L^3 / 6 . \tag{7.28}$$

Dividing again by $L^2/6$, changing to ξ' as a dimensionless variable, and taking the derivative with respect to ξ' now yields Equation 7.29. Since $dC_{0h}/d\xi' = C_{1h}$, $dC_{1h}/d\xi' = 0$ and $dC_{0hm}/d\xi' = C_{1hm}$, there follows the differential equation for the part after passing the discontinuity:

$$dC_{1hm} / d\xi' + 3 \cdot C_{1hm}(\xi') = 3 \cdot C_{1h} . \tag{7.29}$$

It has the solution

$$\begin{aligned} C_{1hm}(\xi') &= C_{1hm}(\xi' = 0) \cdot \exp(-3\xi') + C_{1h}[1 - \exp(-3\xi)] \\ &= C_{1h}[1 - 0.4555 \cdot \exp(-3\xi)]. \end{aligned} \tag{7.30}$$

Figure 7.13 shows these functions scaled by C_{1h}. It can be seen that the (theoretical) impulse at $\xi = 1$ is now spread out in the range 0 to ≈ 2.5; it has a maximum value of 1,3665 at $\xi = 1$. At this point, the variable C_{1hm} is about 54 % of C_{1h} and continues to grow smoothly toward this value (1).

The differential Equations 7.26 and 7.29 can be written as time derivatives with $d(\cdot)/dl = d(\cdot)/dt/dl/dt$ and $dl/dt = V$ in the form

$$\dot{C}_{1hm} = 3V/L \cdot \{C_{1h} \cdot [f(l_c/L)] - C_{1hm}\},$$

where

$$f(l_c/L) = (l_c/L)^2 \quad \text{for} \quad l_c < L \tag{7.31}$$

or

$$f(l_c/L) = 1$$

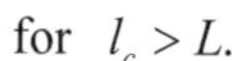

for $l_c > L$.

Since the locations l_{C1i} of steps in C_{1h} are unknown, the term in square brackets is always set equal to 1. C_{1h} is driven by noise in the recursive estimation process. This corresponds to neglecting the contributions of C_{1h} to curvature as long as the vehicle has not passed the point of the occurrence; in Figure 7.13 this means neglecting the shaded half plane on the left. For this reason, the starting value for C_{1hm} now has to be zero at $\xi' = 0$ (instead of 0.5445 C_{1h}); it can be seen from the dashed-double-dotted line how fast the new variable C_{1hm} approaches the actual value C_{1h}. After driving the distance of one look-ahead range L, the 5 % limit of $(C_{1hm}(0) - C_{1h})$ is reached. In [Mysliwetz 1990; Behringer 1996] it was shown that this simple, pragmatic approach yields very good estimation results compared with both known solutions in simulation studies and real experiments. Neglecting the part $\xi < 1$ has the additional advantage that for multiple observation windows along the look-ahead range, no special management is necessary for checking which of those are within or outside the $C_{1h} \neq 0$ region.

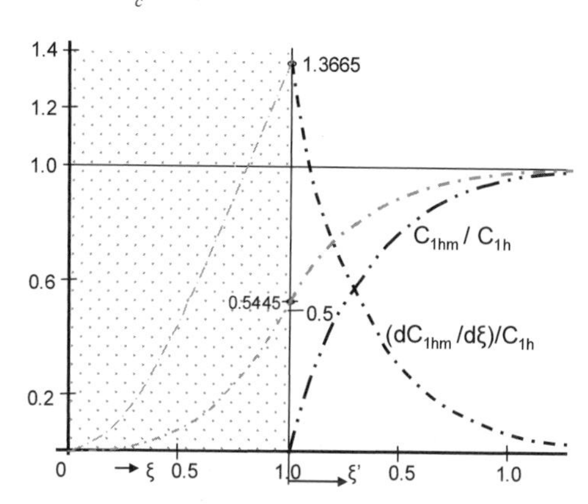

Figure 7.13. Approximation of the Dirac impulse in curvature change rate $dC_{1h}/d\xi$ by a finite model for improved recursive estimation with the 4-D approach. For simplicity, only the non-shaded part ($\xi' > 0$) is used in the method (see text); convergence is given by the *dash-double-dotted curve* starting from zero for ξ'.

The model for the new spread-out curvature variables is supposed to be valid in the entire look-ahead range; all windows at different observation ranges contribute to the same model via prediction-error feedback, so that an averaged model will result.

7.4.2 Window Placing and Edge Mapping

Contrary to Figure 7.4, where the goal was to drive in the center of an imagined lane of width b by observing just the right-hand borderline, here, we assume to be driving on a road with lane markings. Lane width b and the subject's lateral position in the lane y_V off the lane center have to be determined visually by exploiting many edge measurements for improved robustness and accuracy. In each of the horizontal stripes in Figure 7.14, edges (and eventually bright lines) have to be found that mark the lanes for driving. Two basically different approaches may be chosen: One can first look for lines that constitute lane markings and determine their shape over the arc length. Second-order or third-order parabolas have been used frequently; in a second step, the lane is determined from two neighboring lane markings. This approach, looking straightforward at a first glance, is not taken here.

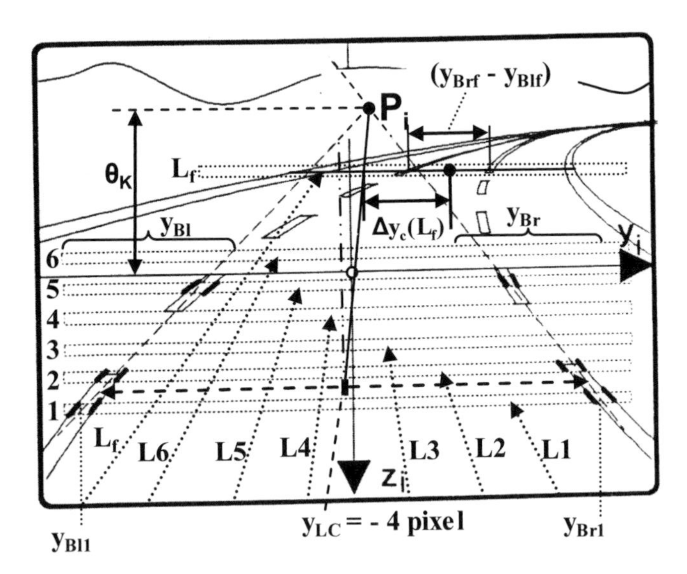

Figure 7.14. Multiple horizontal stripes with measurement windows for lane recognition by edge detection and model based grouping. Image rows correspond to look-ahead ranges L_i; $y_{Br} - y_{Bl} = b_B(L_i)$ shows lane width. Features are used for derivation of initial values to start recursive estimation (see text).

The main reason is that the vehicle is supposed to drive in the center of the lane, so the clothoid model should be valid there; the lane markings then are normal to this skeleton line at distances $\pm\, b/2$. Lane width should be observed steadily while driving; this model allows width to be one (smoothly varying) parameter in the recursive estimation process. Lane observation can thus be continued in the case that one marking line is strongly perturbed or even missing. Using the curvature model of the previous section for the skeleton line, it can be described at the vertical center of each horizontal stripe by

$$y_{csk_i} = C_{0hm} L_i^2 / 2 + C_{1hm} L_i^3 / 6 . \tag{7.32}$$

Assuming small heading changes of the road in the look-ahead region, the lateral positions of the left and right lane markings then are at

$$y_{cl_i} = y_{csk_i} - b/2; \qquad y_{cr_i} = y_{csk_i} + b/2. \tag{7.33}$$

To obtain their mapping into the image, a more detailed overall model of the vehicle–sensor arrangement has to be taken into account. First, the vehicle center of gravity (cg for which the lateral offset y_V is defined) is not identical with the location of the camera; in the test vehicle VaMoRs, there is a distance $d \approx 2$ m in the longitudinal direction (see Figure 7.15). Similar to Figure 7.5, one can now write for the lateral distance y_L on the left-hand road boundary with $-b/2$ (index l) or on the right-hand side with $+ b/2$ (index r),

$$y_{L_{l,r}} = \pm b/2 + y_{C_{l,r}} - (y_V + y_{\psi V} + y_{\psi KV}), \tag{7.34}$$

where $y_{\psi V} = L \cdot \psi_V$ and $y_{\psi KV} = L \cdot \psi_{KV}$ are the shifts due to vehicle and camera heading angles. With a pinhole camera model and the scaling factor k_y for the camera, one obtains for the image coordinates $y_{Bl,r}$ in the row direction

$$y_{Bi_{l,r}} = f \cdot k_y \cdot y_{Li_{l,r}} / L_i \,. \tag{7.35}$$

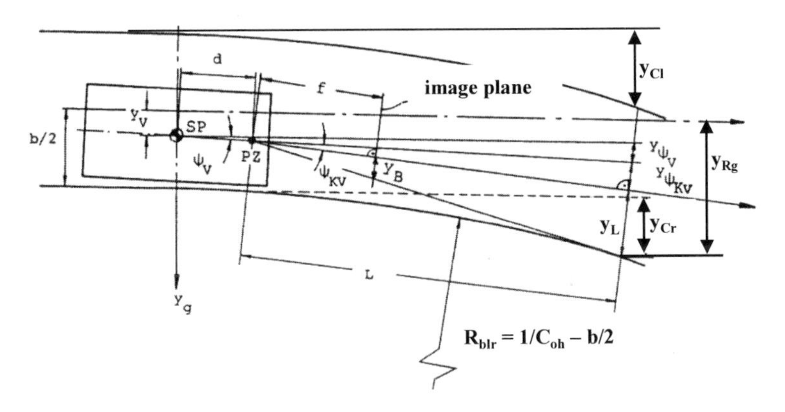

Figure 7.15. Precise model for recognition of lane and relative egostate: exact camera location d and gaze angle ψ_{KV} relative to the vehicle body. (Focal length f is extremely exaggerated for better visibility)

For the vertical image coordinates z_{Bi}, Figure 7.10 and Equation 7.20 are valid here, too. Subtracting the left-hand value y_{Bil} from the right-hand one y_{Bir} for the same z_{Bi}-coordinate yields the imaged lane width $b_B(L_i)$. However, in the 4-D approach, this need not be done since b is a parameter in the measurement (imaging) model (Equation 7.35 with 7.34); via the Jacobian matrix, this parameter is iterated simultaneously for all horizontal measurement stripes L_i together, which makes it more stable and precise. Real lane width b is obtained together with the curvature parameters.

7.4.3 Resulting Measurement Model

The spread-out dynamic clothoid model for the skeleton line of the lane can be written from Equations 7.24 – 7.27 in the time domain as

$$\dot{C}_{0hm} = C_{1hm} \cdot V; \quad \dot{C}_{1hm} = (3V/L) \cdot (C_{1h} - C_{1hm}); \quad \dot{C}_{1h} = n_{C_{1h}}; \tag{7.36}$$

or in matrix form

$$\dot{x}_{c_h} = A_{c_h} x_{c_h} + n_{c_{h1}} = \begin{pmatrix} 0 & V & 0 \\ 0 & -3V/L & 3V/L \\ 0 & 0 & 0 \end{pmatrix} \begin{pmatrix} C_{ohm} \\ C_{1hm} \\ C_{1h} \end{pmatrix} + \begin{pmatrix} 0 \\ 0 \\ n_{c_{h1}} \end{pmatrix}. \tag{7.36 (a)}$$

Equation 7.35 can be written with Equations 7.32 – 7.34 [two positions for lane markings (left and right) with two edges each (dark-to-bright and bright-to-dark) in each search stripe at z_{Bi}, yield the look-ahead distance L_i, see Figure 7.14]:

$$y_{Bi_{l,r}} = f \cdot k_y \left[\frac{\pm \frac{b}{2} - y_V}{L_i} - \psi_V - \psi_{VK} + \frac{L_i}{2} C_{0hm} + \frac{L_i^2}{6} C_{1hm} \right]. \tag{7.37}$$

This measurement equation clearly indicates the dependency of edge features on the variables and parameters involved: Lane width b and lateral position y_v can best be measured in the near range (when L_i is small); measurement of the heading angles is not affected by range (see Figure 7.6). Curvature parameters can best be measured at large ranges (higher up in the image at smaller row indices z_{Bi}). Since all variables for a given range L_i enter Equation 7.37 linearly, the coefficients accompanying the variables are the elements in the corresponding columns of the Jacobian matrix. (The rows of the Jacobian correspond to each feature measured.)

Since all expressions in Equation 7.37 except the sign of the $b/2$ term are the same for the left- (y_{Bil}) and the right-hand boundary lines (y_{Bir}), their difference directly yields the lane (road) width b_i

$$b_i = L_i \cdot (y_{Bi_r} - y_{Bi_l}) / (f \cdot k_y). \tag{7.38}$$

The center of the lane in the image is at $y_{BLCi} = (y_{Bil} + y_{Bir})/2$. The difference between the y coordinate of the vanishing point y_{BP} and y_{BLCi} is a measure of the lateral offset y_V. Analogous to Equation 7.35, one can write

$$y_{Vi} = L_i \cdot (y_{BP}(\infty) - y_{BLCi}) / (f \cdot k_y). \tag{7.39}$$

The dynamic model for the variables y_V and ψ_V (describing vehicle state relative to the road) may be selected as fourth order, as in Section 7.2, or as fifth order (Equation 3.37) depending on the application. With small perturbations in pitch from roughness of the ground onto the vehicle, look-ahead distance L_i is directly linked to the vertical image coordinate z_{Bi} (Equation 7.19); with θ_K representing the pitch angle between the optical axis of the camera and a horizontal line, this equation in generalized form is valid here:

$$z_{Bi} = f \cdot k_z (H_K - L_i \tan\theta_K) / (L_i + H_K \tan\theta_K). \tag{7.40}$$

The gaze angle ψ_{KV} of a camera as well as the camera pitch angle θ_{KV} relative to the vehicle can usually be measured with sufficient accuracy directly in the vehicle. The case that vehicle pitch also changes dynamically will be treated in Section 9.3. For driving on (smooth) rural dirt roads, the concept of this section is also valid.

Figure 7.16 shows such a case. Only feature extraction has to be adjusted: Since the road boundaries are not crisp, large masks with several zeros at the center in the feature extractor CRONOS are advantageous in the near range; the mask shown in Figure 5.10b [$(n_d, n_0, n_d) = 7, 3, 7$] yielded good performance (see also Figure 5.11, center top). With the method UBM, it is advisable to work on higher pyramid levels or with larger sizes for mask elements (values m and n). To avoid large disturbances in the far range, no edges but only the approximate centers of image regions signifying "road" by their brightness are determined there (in Figure 7.16; other characteristics may be searched for otherwise). Search stripes may be selected orthogonal to the expected road direction (windows 7 to 10). The y_B and z_B coordinates of the road center point in the stripe determine the curvature offset and the range of the center on the skeleton line.

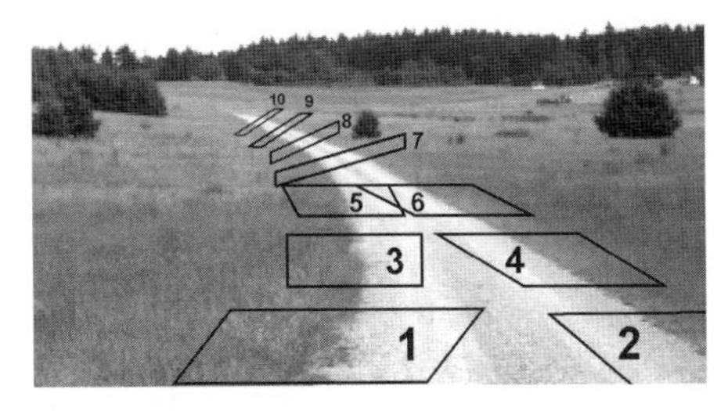

Figure 7.16. Recognition of the skeleton line of a dirt road by edges in near range (with large masks) and by the center of brighter regions in ranges further away for improved robustness. Road width is determined only in the near range.

7.4.4 Experimental Results

In this section, early results (1986) in robust road recognition with multiple redundant feature extraction in eight windows are shown. In these windows, displayed in Figure 7.17, one microprocessor Intel 8086 each extracted several edge candidates for the lane boundaries (see figure). On the left-hand side of the lane, the tar filling in the gaps between the plates of concrete that form the road surface, gave a crisp edge; however, disturbances from cracks and dirt on the road were encountered. On the right-hand side, the road boundary changed from elevated curbstones to a flat transition on grass expanding onto the road.

Features accepted for representing the road boundary had to satisfy continuity conditions in curvature (heading change over arc length) and colinearity. Deviations from expected positions according to spatio-temporal prediction also play a role:

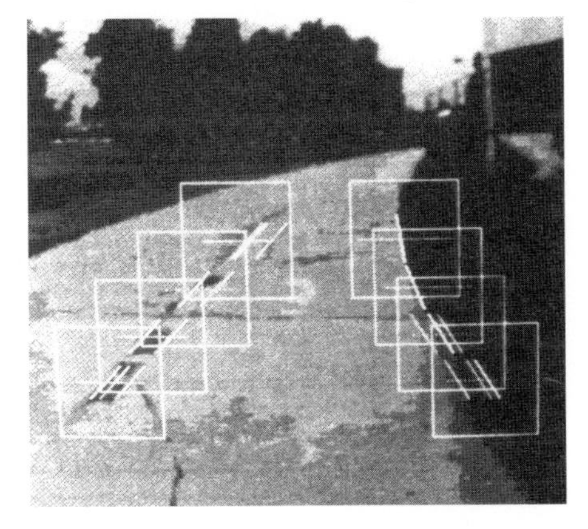

Figure 7.17. Multiple oriented edge extraction in eight windows with first-generation, real-time image processing system BVV2 [Mysliwetz 1990]

Features with offsets larger than 3σ from the expected value, were discarded altogether; the standard deviation σ is obtained from the error covariance matrix of the estimation process. This conservative approach stabilizes interpretation; however, one has to take caution that unexpected real changes can be handled. Especially in the beginning of the estimation process, expectations can be quite uncertain or even dead wrong, depending on the initially generated hypothesis. In these cases it is good to have additional potential interpretations of the feature arrangement available to start alternative hypotheses. At the time of the experiments described here, just one hypothesis could be started at a time due to missing computing power; today (four orders of magnitude in processing power per microprocessor later!), several likely hypotheses can be started in parallel.

In the experiments performed on a campus road, the radius of curvature of about 140 m was soon recognized. This (low-speed) road was not designed as a clothoid; the estimated C_{1hm} parameter even changed sign (*dotted curve* in Figure 7.18a around the 80 m mark). The heading angle of the vehicle relative to the road tangent stayed below 1° (Figure 7.18b) and the maximum lateral offset y_V was always less than 25 cm (Figure 5.18c). The steering angle (Figure 5.18d) corresponds directly to road curvature with a bit of a lead due to the look-ahead range and feed-forward control.

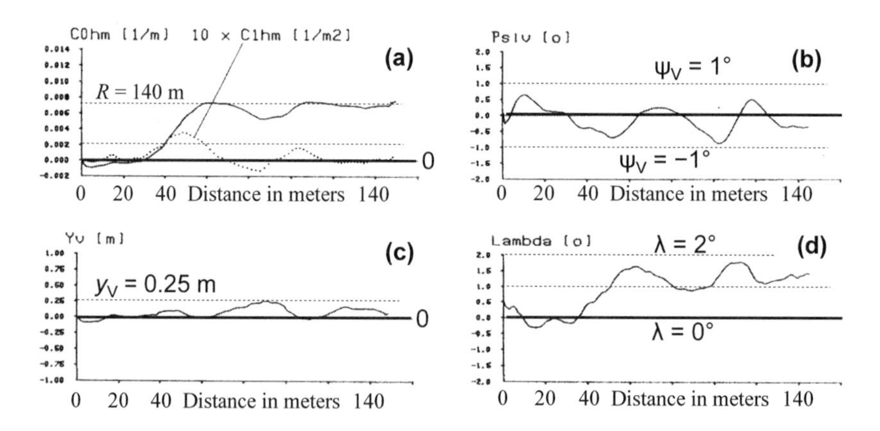

Figure 7.18. Test results in autonomous driving on unmarked campus–road: Transition from straight to radius of curvature ≈ 140 m. (a) Curvature parameters, (b) vehicle heading relative to the road (<~ 0.9°), (c) lateral offset (< 25 cm), (d) steering angle (time integral of control input). Speed was ~ 30 km/h.

8 Initialization in Dynamic Scene Understanding

Two very different situations have to be distinguished for initialization in road scenes: (1) The vehicle is being driven by a human operator when the visual perception mode is switched on, and (2) the vehicle is at rest somewhere near the road and has to find the road on its own. In the latter, much more difficult case, it has sufficient time to apply more involved methods of static scene recognition. This latter case will just be touched upon here; it is wide open for future developments.

It is claimed here that 3-D road recognition while driving along a road is easier than with a static camera if some knowledge about the motion behavior of the vehicle carrying the camera is given. In the present case, it is assumed that the egovehicle is an ordinary car with front wheel steering, driving on ordinary roads. Taking the known locomotion measured by odometer or speedometer into account, integration of measurements over time from a single, passive, monocular, 2-D imaging sensor allows motion stereointerpretation in a straightforward and computationally very efficient way.

With orientation toward general road networks, the types of scenes investigated are the human-built infrastructure “roads” which is standardized to some extent but is otherwise quasi-natural with respect to environmental conditions such as lighting including shadows, such as weather, and possible objects on the road; here, we confine ourselves just to road recognition. The bootstrap problem discussed here is the most difficult part and is far from being solved at present for the general case (all possible lighting and weather conditions). At the very first start of the vision process, alleviation for the task, of course, is the fact that during this self-orientation phase *no real-time control activity* has to be done. Several approaches may be tried in sequence; during development phases, there is an operator checking the results of recognition trials independently. Solution times may lie in the several-second range instead of tens of milliseconds.

8.1 Introduction to Visual Integration for Road Recognition

Some aspects of this topic have already been mentioned in previous chapters. Here, the focus will be on the overall interpretation aspects of roads and how to get started. For dynamic scene understanding based on edge and stripe features, the spatial distribution of recognizable features has to be combined with translational and rotational motion prediction and with the laws of central projection for mapping spatial features into the image plane. The recursive visual measurement process fits the best possible parameters and *spatial state time histories* to the data measured.

These estimates satisfy the motion model in the sense of least-squares errors taking the specified (assumed) noise characteristics into account. Once started, direct nonlinear, perspective inversion is bypassed by prediction-error feedback. To get started, however, either an initial perspective inversion has to be done or an intuitive jump to sufficiently good starting values has to be performed somehow, from which the system will converge to a stable interpretation condition. On standard roads in normal driving situations, the latter procedure often works well.

For hypothesis generation, corresponding object databases containing both motion characteristics and all aspects geared to visual feature recognition are key elements of this approach. Tapping into these databases triggered by the set of features actually measured is necessary for deriving sufficiently good initial values for the state variables and other parameters involved to get started. This is the task of hypothesis generation to be discussed here.

When applying these methods to complex scenes, simple rigid implementation will not be sufficient. Some features may have become occluded by another object moving into the space between the camera and the object observed. In these cases, the interpretation process must come up with proper hypotheses and adjustments in the control parameters for the interpretation system so that feature matching and interpretation continues to correspond to the actual process happening in the scene observed. In the case of occlusion by other objects/subjects, an information exchange with higher interpretation levels (for situation assessment) has to be organized over time (see Chapter 13).

The task of object recognition can be achieved neither fully bottom-up nor fully top-down exclusively, in general, but requires joint efforts from both directions to be efficient and reliable. In Section 5.5, some of the bottom-up aspects have already been touched upon. In this section, purely visual integration aspects will be discussed, especially the richness in representation obtained by exploiting the first-order derivative matrix of the connection between state variables in 3-D space and features in the image (the *Jacobian matrix*; see Sections 2.1.2 and 2.4.2). This will be done here for the example of recognizing roads with lanes. Since motion control affects conditions for visual observation and is part of autonomous system design in closed-loop form, the motion control inputs are assumed to be measured and available to the interpretation system. All effects of active motion control on visual appearance of the scene are predicted as expectations and taken into account before data interpretation.

8.2 Road Recognition and Hypothesis Generation

The presence of objects has to be hypothesized from feature aggregations that may have been collected in a systematic search covering extended regions of the image. For roads, the coexistence of left- and right-hand side boundaries in a narrow range of meaningful distances (say, 2 to 15 m, depending on the type of road) and with low curvatures are the guidelines for a systematic search. From the known elevation of the camera above the ground, the angle of the (possibly curved) "pencil tip" in the image representing the lane or road can be determined as a function of lane or road width. Initially, only internal hypotheses are formed by the specialist algorithm for road recognition and are compared over a few interpretation cycles taking the conventionally measured egomotion into account (distance traveled and steering angle achieved); the tracking mode is switched on, but results are published to the rest of the system only after a somewhat stable interpretation has been found. The degree of confidence in visual interpretation is also communicated to inform the other perception and decision routines (agents).

8.2.1 Starting from Zero Curvature for Near Range

Figure 7.14 showed some results with a search region of six horizontal stripes. Realistic lane widths are known to be in the range of 2 to 4.5 m. Note that in stripes 3 and 4, no edge features have been found due to broken lines as lane markings (indicating that lane changes are allowed). To determine road direction nearby robustly, approximations of tangents to the lane borderlines are derived from features in well separated stripes (1, 2, and 5 here). The least-squares fit on each side (*dashed lines*) yields the location of the vanishing point (designated P_i here). If the camera is looking in the direction of the longitudinal axis of the vehicle ($\psi_{KV} = 0$), the offset of P_i from the vertical centerline in the image represents directly the scaled heading angle of the vehicle (Figure 7.6). Similarly, if a horizonline is clearly visible, the offset of P_i from the horizon is a measure of the pitch angle of the camera θ_K. Assuming that θ_K is 0 (negligibly small), Equation 7.40 and its derivative with respect to range L_i can be written

$$z_{Bi}(L_i) = f \cdot k_z H_K / L_i; \qquad dz_{Bi} / dL_i = -f \cdot k_z H_K / L_i^2 . \tag{8.1}$$

Similarly, for zero curvature, the lateral image coordinate as a function of range L_i and its derivative become from Equation 7.37,

$$\begin{aligned} y_{Bi_{l,r}}(L_i) &= f \cdot k_y \left[(\pm b/2 - y_V)/L_i - \psi_V - \psi_{VK}\right]; \\ dy_{Bi_{l,r}} / dL_i &= -f \cdot k_y \cdot (\pm b/2 - y_V)/L_i^2 . \end{aligned} \tag{8.2}$$

Dividing the derivative in Equation 8.2 by that in Equation 8.1 yields the expressions for the image of the straight left (+b) and right (−b) boundary lines;

$$\begin{aligned} dy_{Bl} / dz_B &= -(b/2 - y_V) \cdot (k_y / k_z) / H_K; \\ dy_{Br} / dz_B &= (b/2 - y_V) \cdot (k_y / k_z) / H_K . \end{aligned} \tag{8.3}$$

Both slopes in the image are constant and independent of the yaw angles ψ (see Figure 7.6). Since z is defined positive downward, the right-hand boundary–coordinate increases with decreasing range as long as the vehicle offset is smaller than half the lane width; at the vanishing point $L_i = \infty$, the vertical coordinate z_{Bi} is zero for $\theta_K = 0$. The vehicle is at the center of the lane when the left and right boundary lines are mirror images relative to the vertical line through the vanishing point.

Assuming constant road (lane) width on a planar surface and knowing the camera elevation above the ground, perspective inversion for the ranges L_i can be done in a straightforward manner from Equation 8.1 (left);

$$L_i = f \cdot k_z H_k / z_{Bi} . \tag{8.4}$$

Equation 8.2 immediately yields the lumped yaw angle ψ for $L_i \rightarrow \infty$ as

$$\psi = \psi_V + \psi_{VK} = -y_B(L_i = \infty) / f \cdot k_y \,. \qquad (8.5)$$

These linear approximations of road boundaries usually yield sufficiently accurate values of the unknown state variables (y_V and ψ_V) as well as the parameter lane width b for starting the recursive estimation process; it can then be extended to further distances by adding further search stripes at smaller values z_{Bi} (higher up in the image). The recursive estimation process by itself has a certain range of convergence to the proper solution, so that a rough approximate initialization is sufficient, mostly. The curvature parameters may all be set to zero initially for the recursive estimation process when look-ahead distances are small. A numerical example will be shown at the end of the next section.

8.2.2 Road Curvature from Look-ahead Regions Further Away

Depending on the type of road, the boundaries to be found may be smooth (*e.g.*, lane markings) or jagged [*e.g.*, grass on the shoulder (Figure 7.16) or dirt on the roadside]. Since road size in the image decreases with range, various properly sized edge masks (templates) are well suited for recognizing these different boundary types reliably with the method CRONOS (see Section 5.2). Since in the near range on roads, some *a priori* knowledge is given, usually, the feature extraction methods can be parameterized reasonably well. When more distant regions are observed, working with multiple scales and possibly orientations is recommended; a versatile recognition system should have these at its disposal. Using different mask sizes and/or sub–sampling of pixels as an inverse function of distance (row position in the vertical direction of the image) may be a good compromise with respect to efficiency if pixel noise is low. When applying direction-sensitive edge extractors like UBM (see Section 5.3), starting from the second or third pyramid level at the bottom of the image is advisable.

Once an edge element has been found, it is advisable for efficient search to continue along the same boundary in adjacent regions under colinearity assumptions; this reduces search intervals for mask orientations and search lengths. Since lanes and (two-lane) roads are between 2 and 7 m wide and do have parallel boundaries, in general, this gestalt knowledge may be exploited to find the adjacent lane marking or road boundary in the image; mask parameters and search regions have to be adjusted correspondingly, taking perspective mapping into account. Looking almost parallel to the road surface, the road is mapped into the image as a triangular shape, whose tip may bend smoothly to the left (Figure 7.16) or right (Figure 7.17) depending on its curvature.

As a first step, a straight road is interpreted into the image from the results of edge finding in several stripes nearby, as discussed in the previous section; in Figure 7.14 the dashed lines with the intersection point P_i result. From the average of the first two pairs of lane markings, lane width and the center of the lane y_{LC} are determined. The line between this center point and P_i (shown solid) is the reference line for determining the curvature offset Δy_c at any point along the road. Further lane markings are searched in stripes higher up in the image at increasingly further distances. Since Equation 7.37 indicates that curvature can best be determined from look-ahead regions far away, this process is continued as long as lane markings are highly visible. During this process, search parameters may be adapted to the results found in the previous stripe. Let us assume that this search is stopped at the far look-ahead distance L_f.

Now the center point of the lane at L_f is determined from the two positions of the lane markings y_{Brf} and y_{Blf}. The difference of these values yields the lane width in the image b_{Bf} at L_f (Equation 7.38). The point where the centerline of this search stripe hits the centerline of the virtual straight road is the reference for determining the offset due to road curvature $\Delta y_c(L_f)$ (see distance marked in Figure 7.14). Assuming that the contribution of C_{1hm} is negligible against that of C_{0hm}, from Equation 7.24, $\Delta y_c(L_f) = C_{0hm} \cdot L_f^2/2$. With Equation 7.37 and the effects of y_V and ψ taken care of by the line of reference, the curvature parameter can be estimated from

$$\Delta y_{Cf} = \Delta y_{CBf} \cdot L_f / (f \cdot k_y) = C_{0hm} \cdot L_f^2 / 2$$

as (8.6)

$$C_{0hm} = 2 \cdot \Delta y_{Cf} / L_f^2 = \Delta y_{CBf} \cdot 2 / L_f \cdot f \cdot k_y \ .$$

On minor roads with good contrast in intensity (as in Figure 7.16), the center of the road far away may be determined better by region-based methods like UBM.

8.2.3 Simple Numerical Example of Initialization

Since initialization in Figure 7.16 is much more involved due to hilly terrain and varying road width, this will be discussed in later chapters. The relatively easy initialization procedure for a highway scene while driving is discussed with the help of Figure 7.14. The following parameters are typical for the test vehicle VaMoRs and one of its cameras around 1990: Focal length $f \approx 8$ mm; scaling factors for the imaging sensor: $k_z \approx 50$ pixels/mm and $k_y \approx 40$ pixels/mm; elevation of camera above the ground $H_K = 1.9$ m. The origin of the y_B, z_B image coordinates is selected here at the center of the image.

By averaging the results from stripes 1 and 2 for noise reduction, the lane width measured in the image is obtained as 280 pixels; its center lies at $y_{LC} = -4$ pixels and $z_{LC} = 65$ pixels (average of measured values). The vanishing point P_i, found by intersecting the two virtual boundary lines through the lane markings nearby and in stripe 5 (for higher robustness), has the image coordinates $y_{BP} = 11$ and $z_{BP} = -88$ pixels. With Equation 8.5, this yields a yaw angle of $\psi \approx 2°$ and with Equation 7.40 for $L_f \rightarrow \infty$, a pitch angle of $\theta_K \approx -12°$. The latter value specifies with Equation 7.40 that the optical axis ($z_B = 0$) looks at a look-ahead range L_{oa} (distance to the point mapped into the image center) of

$$L_{oa} = H_K / \tan\theta_K = 1.9 / 0.22 = 8.6 \,\text{m} \ . \qquad (8.7)$$

For the far look-ahead range L_f at the measured vertical coordinate $z_{Bf} = -55$ pixel, the same equation yields with $F = z_{Bf}/(f \cdot k_z) = -0.1375$

$$L_f = H_K(1 - F \cdot \tan\theta_K)/(F + \tan\theta_K) = 22.3 \,\text{m} \ . \qquad (8.8)$$

With this distance now the curvature parameter can be determined from Equation 8.6. To do this, the center of the lane at distance L_f has to be determined. From the measured values $y_{Brf} = 80$ and $y_{Blf} = 34$ pixels the center of the lane is found at $y_{BLCf} = 57$ pixels; Equations 7.38 and 8.8 yield an estimated lane width of $b_f = 3.2$ m. The intersection point at L_f with the reference line for the center of the virtual straight road is found at $y_{BISP} = 8$ pixel. The difference $\Delta y_{CBf} = y_{BLCf} - y_{BISP} = 49$ pixels according to Equation 8.6, corresponds directly to the curvature parameter C_{0hm} yielding

$$C_{0hm} = \Delta y_{CBf} \cdot 2/(L_f \cdot f \cdot k_y) = 0.0137\ \text{m}^{-1}, \tag{8.9}$$

or a radius of curvature of $R \approx 73$ m. The heading change of the road over the look-ahead distance is $\Delta\chi(L_f) \approx C_{0hm} \cdot L_f = 17.5°$. Approximating the cosine for an angle of this magnitude by 1 yields an error of almost 5%. This indicates that to determine lane or road width at greater distances, the row direction is a poor approximation. Distances in the row direction are enlarged by a factor of ~ $1/\cos\,[\Delta\chi(L_f)]$.

Since the results of (crude) perspective inversion are the starting point for recursive estimation by prediction-error feedback, high precision is not important and simplifications leading to errors in the few percent range are tolerable; this allows rather simple equations and generous assumptions for inversion of perspective projection. Table 8.1 shows the collected results from such an initialization based on Figure 7.14.

Table 8.1. Numerical values for initialization of the recursive estimation process derived from Figure 7.14, respectively, assumed or actually measured

Name of variable	Symbol (dimension)	Numerical value	Equation
Gaze angle in yaw	ψ (degrees)	2	8.5
Gaze angle in pitch	θ (degrees)	−12	7.40
Look-ahead range (max)	L_f (meter)	22.3	8.8
Lane width	b (meter)	3.35	7.38
Lateral offset from lane center	y_V (meter)	0.17	7.39
Road curvature parameter	C_{0hm} (meter^{-1})	$0.0137 \approx 1/73$	8.9
Slip angle	β (degrees)	unknown, set to 0	
	C_{1hm} (meter^{-2})	unknown, set to 0	
	C_{1h} (meter^{-2})	unknown, set to 0	
Steering angle	λ	actually measured	
Vehicle speed	V (m/s)	actually measured	

Figure 8.1 shows a demanding initialization process with the vehicle VaMoRs at rest but in almost normal driving conditions on a campus road of UniBwM near Munich without special lane markings [Mysliwetz 1990]. On the right-hand side, there are curbstones with several edge features, and the left lane limit is a very narrow, but very visible tar-filled gap between the plates of concrete forming the lane surface. The shadow boundaries of the trees are much more pronounced in intensity difference than the road boundary; however, the hypothesis that the shadow of the tree is a lane can be discarded immediately because of the wrong dimensions in lateral extent and the jumps in the heading direction.

Without the gestalt idea of a smoothly curved continuous road, mapped by perspective projection, recognition would have been impossible. Finding and checking single lines, which have to be interpreted later on as lane or road boundaries in a separate step, is much more difficult than introducing essential shape parameters of the object lane or road from the beginning at the interpretation level for single edge features.

Figure 8.1. Initialization of road recognition; example of a successful instantiation of a road model with edge elements yielding smoothly curved or straight boundary lines and regions in between with perturbed homogeneous intensity distributions. Small local deviations from average intensity are tolerated (dark or bright patches). The long white lines in the right image represent the lane boundaries for the road model accepted as valid.

For verification of the hypothesis "road," a region-based intensity or texture analysis in the hypothesized road area should be run. For humans, the evenly arranged objects (trees and bushes) along the road and knowledge about shadows from a deep-standing sun may provide the best support for a road hypothesis. In the long run, machine vision should be able to exploit this knowledge as well.

8.3 Selection of Tuning Parameters for Recursive Estimation

Beside the initial values for the state variables and the parameters involved, the values describing the statistical properties of the dynamic process observed and of the measurement process installed for the purpose of this observation also have to be initialized by some suitable starting values. The recursive estimation procedure of the extended Kalman filter (EKF) relies on the first two moments of the stochastic process assumed to be Gaussian for improving the estimated state after each measurement input in an optimal way. Thus, both the initial values of the error covariance matrix P_0 and the entries in the covariance matrices Q for system perturbations as well as R for measurement perturbations have to be specified. These data describe the knowledge one has about uncertainties of the process of perception.

In Chapter 6, Section 6.4.4.1, it was shown in a simple scalar example that choosing the relative magnitude of the elements of R and Q determines whether the update for the best estimate can trust the actual state x and its development over time (relatively small values for the variance σ_x^2) more than the measurements y

(smaller values for the variance of the measurements σ_y^2). Because of the complexity of interdependence between all factors involved in somewhat more complex systems, this so called "filter tuning" is considered more an art than a science. Vision from a moving platform in natural environments is very complex, and quite some experience is needed to achieve good behavior under changing conditions.

8.3.1 Elements of the Measurement Covariance Matrix *R*

The steering angle λ is the only conventionally measured variable beside image evaluation; in the latter measurement process, lateral positions of inclined edges are measured in image rows. All these measurements are considered unrelated so that only the diagonal terms are nonzero. The measurement resolution of the digitized steering angle for the test vehicle VaMoRs was 0.24° or 0.0042 rad. Choosing about one-quarter of this value as standard deviation ($\sigma_\lambda = 0.001$ rad), or the variance as $\sigma_\lambda^2 = 10^{-6}$, showed good convergence properties in estimation.

Static edge extraction to subpixel accuracy in images with smooth edges has standard deviations of considerably less than 1 pixel. However, when the vehicle drives on slightly uneven ground, minor body motion in both pitch and roll occurs around the static reference value. Due to active road following based on noisy data in lateral offset and heading angle, the yaw angle also shows changes not modeled, since loop closure has a total lumped delay time of several tenths of a second. To allow a good balance between taking previous smoothed measurements into account and getting sufficiently good input on changing environmental conditions, an average pixel variance of $\sigma_{yBi}^2 = 5$ pixel^2 in the relatively short look-ahead range of up to about 25 m showed good results, corresponding to a standard deviation of 2.24 pixels. According to Table 7.1 (columns 2 and 3 for $L \approx 20$ m) and assuming the slope of the boundaries in the image to be close to ±45° (tan ≈ 1), this corresponds to pitch fluctuations of about one-quarter of 1°; this seems quite reasonable.

It maybe surprising that body motion is considered measurement noise; however, there are good reasons for doing this. First, pitching motion has not been considered at all up to now and does not affect motion in lateral degrees of freedom; it comes into play only through the optical measurement process. Second, even though the optical signal path is not directly affected, the noise in the sensor pose relative to the ground is what matters. But this motion is not purely noise, since eigen-motion of the vehicle in pitch exists that shows typical oscillations with respect to frequency and damping. This will be treated in Section 9.3.

8.3.2 Elements of the System State Covariance Matrix *Q*

Here again it is generally assumed that the state variations are uncoupled and thus only the diagonal elements are nonzero. The values found to yield good results for the van VaMoRs by iterative experimental filter tuning according to [Maybeck 1979; Mysliwetz 1990] are (for the corresponding state vector see Equation 9.17)

$$Q = \mathrm{Diag}(10^{-7}, 10^{-5}, 10^{-7}, 10^{-4},$$
$$10^{-9}, 10^{-11}, 10^{-10}, 10^{-9}, 10^{-10}). \quad (8.10)$$

These values have been determined for the favorable observation conditions of cameras relatively high above the ground in a van (H_K = 1.8 m). For the sedan VaMP with H_K = 1.3 m (see Table 7.1) but much smoother driving behavior, the variance of the heading angle has been 2 to 3 · 10^{-6} (instead of 10^{-7}), while the variance for the lateral offset y_V was twice as high (~ $2{\cdot}10^{-4}$) in normal driving; it went up by another factor of almost 2 during lane changes [Behringer 1996]. Working with Q as a constant diagonal matrix with average values for the individual variances usually suffices. One should always keep in mind that all these models are approximations (more or less valid) and that the processes are certainly not perfectly Gaussian and decoupled. The essential fact is that convergence occurs to reasonable values for the process to be controlled in common sense judgment of practical engineering; whether this is optimal or not is secondary. If problems occur, it is necessary to go back and check the validity of all models involved.

8.3.3 Initial Values of the Error Covariance Matrix P_0

In contrast to the covariance matrices Q and R which represent long-term statistical behavior of the processes involved, the initial error covariance matrix P_0 determines the transient behavior after starting recursion. The less certain the initially guessed values for the state variables, the larger the corresponding entries on the diagonal of P_0 should be. On the other hand, when relatively direct measurements of some states are not available but reasonable initial values can be estimated by engineering judgment, the corresponding entries in P_0 should be small (or even zero) so that these components start being changed by the estimation process only after a few iterations.

One practically proven set of initial values for road estimation with VaMoRs is

$$\begin{array}{llllllllll} P_0 = \mathrm{Diag}(0.1, & 0, & 0.1, & 0.1, & 0.1, & 0, & 0, & 0.1, & 0), \\ \text{corresp. to } \lambda & \beta & \psi_{rel} & y_V & C_{0hm} & C_{1hm} & C_{1h} & C_{0vm} & C_{1vm}. \end{array} \quad (8.11)$$

This means that the initial values estimated by the user (zeros) for the slip angle β and the C_1 parameters in all curvature terms are trusted more than those derived from the first measurements. Within the first iterations all values will be affected by the transition matrix A and the covariance matrices Q and R (according to the basic relations given in Equations 6.12, 6.16, and 6.17 or the corresponding expressions in sequential filtering).

From Figure 7.9, it can be seen, that during the initial acceleration phase (changing V in upper graph) horizontal curvature estimation is rather poor; only after about 1 km distance traveled are the horizontal curvature parameters estimated in a reasonable range ($R \geq 1$ km according to [RAS-L-1 1984]). This is partly due to using only 2 windows on one side, relatively close together like shown in Figure 7.7. The detailed analysis of this nonstationary estimation process is rather involved and not discussed here. It is one example of the fact that clean academic conditions are hardly found in steadily changing natural environments. However, the idealized methods (properly handled) may, nonetheless, be sufficient for achieving useful results, partly thanks to steady feedback control in a closed-loop action-perception cycle, which prohibits short-term divergence.

8.4 First Recursive Trials and Monitoring of Convergence

Depending on the quality of the initial hypothesis, short-term divergence may occur in complex scenes; once a somewhat stable interpretation has been achieved, the process is likely to continue smoothly. In order not to disturb the rest of the overall system, the perception process for a certain object class should look at the convergence behavior internally before the new hypothesis is made public. For example, in Figure 8.2, the long shadow from a tree in winter may be interpreted as a road boundary in one hypothesis.

Figure 8.2. Ambiguous situation for generating a good road hypothesis without additional measurement data and higher level reasoning or other input

Starting with the vehicle at rest and assuming that the vehicle is oriented approximately in the road direction near the center of the lane, knowledge about usual lane widths leads to the hypothesis marked by white line segments. When nothing can be assumed about the subject's position, other road hypotheses cannot be excluded completely, *e.g.*, one road side following the shadow of the tree in front of the vehicle. The road then could be as well on the left-hand as on the right-hand side of the shadow from the tree; for the resulting feature positions expected from this hypothesis, new measurement tasks have to be ordered. Features found or not found then have to be judged in conjunction with the hypothesis.

Starting initialization under these conditions while the vehicle is being driven by a human, a few most likely hypotheses should be set up in parallel and started internally (without giving results to the rest of the system or to the operator). Assuming that the human driver guides the vehicle correctly, the system can prune away the road models not corresponding to the path driven by observing the convergence process of the other (parallel) hypotheses; while driving, all features from stationary, vertically extended objects in the environment of the road will move in a predictable way corresponding to ego speed. Those hypothesized objects, having large prediction errors, a high percentage of rejected features, or that show divergence, will be eliminated. This can be considered as some kind of learning in the actual situation.

In the long run, road recognition systems do not have to recognize just the road surface, possibly with lanes, but a much larger diversity of objects forming the infrastructure of different types of roads. During the search for initial scene recognition, all objects of potential interest for vehicle guidance should be detected and correctly perceived. In Section 8.5, the road elements to be initialized in more advanced vision systems will be discussed; here we look first at some steps for testing hypotheses instantiated.

8.4.1 Jacobian Elements and Hypothesis Checking

The Jacobian matrix contains all essential first-order information, how feature positions in an image (or derived quantities thereof) depend on the states or parameters of the object descriptions involved. Analyzing the magnitude of these entries yields valuable information on the mapping and interpretation process, which may be exploited to control and adapt the process of hypothesis improvement.

8.4.1.1 Task Adjustment and Feature Selection for Observability

The basic underlying equation links the *m* vector *dy* of measurements (and thus the prediction errors) to the *n* vector of optimal increments for the state variables and parameters *dx* to be iterated:

$$dy = C \cdot dx \tag{8.13}$$

If one column (index c) of the C matrix is zero (or its entries are very small in magnitude), this means that all measured features do not (or hardly) depend on the variable corresponding to index c (see Figure 8.3); therefore one should not try to determine an update for this component dx_c from the actual measurements.

In the case shown, features 1, 3, 6, and $m = 8$ depend very little on state component x_c (fourth column) and the other ones not at all; if prediction error dy_3 is large, this is

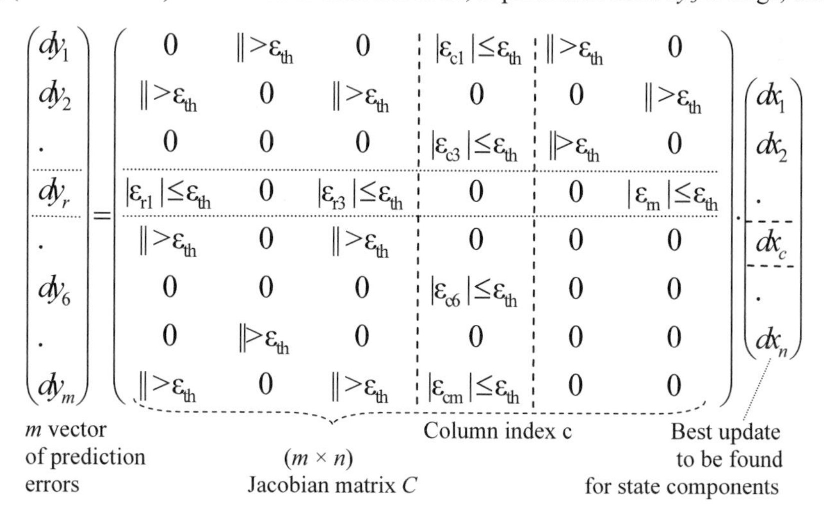

Figure 8.3. Example of a degenerate Jacobian matrix with respect to measurement values y_r, y_6 and state component x_c (each of the absolute values of ε_{ri} and ε_{ci} is below a specific threshold, see text; || means magnitude of the corresponding value)

reflected in changes of dx_c and especially in the state component (c + 1), here 5, since this influences y_3 by a larger factor. However, if prediction error dy_6 is large, the only way of changing it within the feature set selected is by adjusting x_c; since the partial derivative J_{6c} is small, large changes in x_c will achieve some effect in error reduction. This often has detrimental effects on overall convergence. Very often, it is better to look for other features with larger values of their Jacobian elements, at least in column c, to substitute them for y_6.

However, since we are dealing with dynamic systems for which dynamic links between the state variables may be known (integral relationships and cross-feeds), the system may be observable even though entire columns of C are zero. To check observability of all n components of a dynamic system, a different test has to be performed. For systems with single eigenvalues (the case of multiple eigenvalues is more involved), observability according to Kalman may be checked by using the matrix of right eigenvectors V (defined by $A \cdot V = V \cdot \Lambda$, where Λ is the diagonal matrix of eigenvalues of the transition matrix A of the dynamic system). By performing the linear similarity transformation

$$x = V \cdot x' \tag{8.14}$$

the linearized measurement equation becomes

$$dy = C \cdot V \cdot dx' = C' \cdot dx \qquad \text{with} \qquad C' = C \cdot V\,. \tag{8.15}$$

If all elements of a column in the C' matrix are small or zero, this means that all features positions measured do not depend on this characteristic combination of state variables or parameters. Therefore, no meaningful innovation update can be done for this component of the eigenmotions. Maybe, due to other conditions in the future, some new features will occur which may allow also to have an innovation update for this component.

The case of almost vanishing columns or rows of the matrix C is shown in Figure 8.3 for *row indices r and 6* and for *column index c*. Here, the $\varepsilon_{i,j}$ mean numerically small values in magnitude compared to the noise level in the system. If all entries in a matrix row are small or zero, this means that the position in the image of the corresponding feature does not depend on any of the state variables or parameters of the object represented by the C matrix, at least under the current conditions. Therefore, the feature may be discarded altogether without affecting the update results. This reduces the workload for the numerical computation and may help in stabilizing the inversion process buried in the recursion algorithm. Due to noise, a prediction error is likely to occur which will then be interpreted as a large change in the state vector because of the smallness of the C elements; this will be avoided by removing this measurement value altogether, if the row vector in C has only very small entries.

The judgment of smallness, of course, has to be made with reference to the dimension of the corresponding state variable. An angle in radians determining a feature position at the end of a long lever (say several meters) has to be weighted correspondingly differently compared to another variable whose the dimension has been chosen in centimeter (for whatsoever reason). So-called “balancing” by proper weighting factors for each variable may bring the numerical values into the same numerical range required for meaningful selection of thresholds for the ε_{jl} to be considered effectively zero.

To obtain good results, the thresholds of the C element values should be chosen generously. This is good advice if it can be expected that with time the aspect conditions will become more favorable for determining the corresponding variable and if the prediction model is already sufficiently good for handling the observation task meanwhile. In summary, checking the conditioning of the Jacobian matrix and adapting the observation process before things go wrong, is a better way to go than waiting until interpretation has already gone wrong.

8.4.1.2 Feature Selection for Optimal Estimation Results

If computing power available is limited, however, and the number of features "*s*" measurable is large, the question arises which features to choose for tracking to get best state estimation results. Say, the processor capability allows m features to be extracted, with $m < s$. This problem has been addressed in [Wuensche 1986, 1987]; the following is a condensed version of this solution. Since feature position does not depend on the speed components of the state vector, this vector is reduced x_R (all corresponding expressions will be designated by the index R).

To have all state variables in the same numerical range (*e.g.*, angles in radians and distances in meters), they will be balanced by a properly chosen diagonal scaling matrix S. The equation of the Gauss-Markov estimator for the properly scaled reduced state vector (of dimension n_R) is

$$\delta\hat{x}_R = S \cdot \delta\hat{x}_{NR} = S \cdot \{(C_R \cdot S)^T \cdot R^{-1} \cdot (C_R \cdot S)^T\}^{-1} (C_R \cdot S)^T \cdot R^{-1} \cdot (y - y^*), \tag{8.16}$$

where C_R is the Jacobian matrix, with the zero-columns for speed components removed; R is the diagonal measurement covariance matrix, and S is the scaling matrix for balancing the state variables. With the shorthand notation

$$C_N = R^{-1/2} \cdot C_R \cdot S, \tag{8.17}$$

the performance index $J = | C_N^T \, C_N |$ is chosen with the recursive formulation

$$J = \sum_i |C_i^T \cdot C_i|, \tag{8.18}$$

where C_i is a $2 \times n_R$ matrix from C_N corresponding to feature i with two positional components each in the image plane:

$$J = \begin{vmatrix} A_1 & A_2 \\ A_2^T & A_3 \end{vmatrix} = | A_1 | \cdot | A_3 - A_2^T \cdot A_1^{-1} \cdot A_2 | = | A_1 | \cdot | A_4 | . \tag{8.19}$$

A_1 is diagonal and A_4 is symmetrical. This expression may be evaluated with 16 multiplications, 2 divisions, and $9\,(m-1) + 7$ additions. For initialization, a complete search with (s over m) possible feature combinations is performed. Later on, in the real-time phase, a suboptimal but efficient procedure is to find for each feature tracked another one out of those not tracked that gives the largest increase in J. Since one cycle is lost by the replacement, which leads to an overall deterioration, the better feature is accepted only if it yields a new performance measure sensibly larger than the old one:

$$J_{new} > \alpha \cdot J_{old}, \quad \text{with} \quad \alpha = 1 + \varepsilon . \tag{8.20}$$

The best suited value of ε is of course problem dependent. Figure 8.4 shows an experimental result from a different application area: Satellite docking emulated with an air-cushion vehicle in the laboratory. Three corner features of a known polyhedral object being docked to are to be selected such that the performance index for recognizing the relative egostate is maximized. The maneuver consists of an initial approach from the right-hand side relative to the plane containing the docking device (first 40 seconds), an intermediate circumnavigation phase to the left with view fixation by rotating around the vertical axis, and a final phase of homing in till mechanical docking by a rod is achieved [Wuensche 1986]. At point 2 (~ 57 s) a self-occlusion of the side to be docked to starts disappearing on the left-hand side of the body, leading to small spacing between features in the image and possibly to confusion; therefore, these features should no longer be tracked (lower left corner starts being discarded). Only at point 4a are the outermost left features well separable again. In region 5, the features on the right-hand side of the body are very close to each other, leading again to poor measurement results (confusion in feature correlation). The performance index improves considerably when at around 100 s both upper outer corners of the body become very discernible.

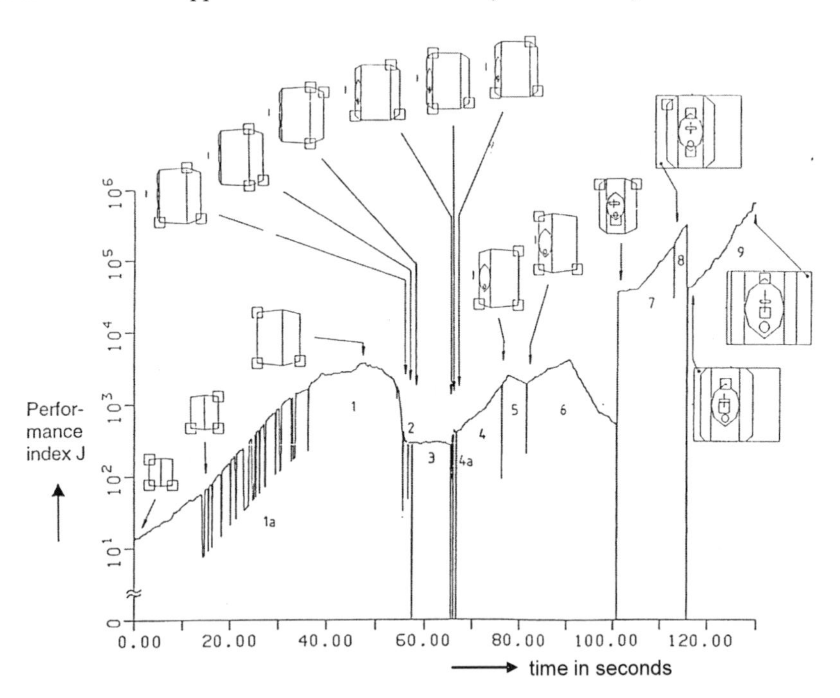

Figure 8.4. Selection of optimal feature combinations of an autonomous air-cushion vehicle with the sense of vision: Approach (first 40 seconds), circumnavigation of the polyhedral object with view fixation, and homing in for docking (last 20 seconds) [Wuensche 1986]

In road traffic, similar but – due to rounded corners and unknown dimensions of cars – more complicated situations occur for changing the aspect conditions of another vehicle. Of course, looking straight from behind, the length of the vehicle body cannot be determined; looking straight from the side, body width is non-observable. For aspect conditions in between, the size of the body diagonal can be estimated rather well, while width and length separately are rather unstable due to poor visibility of features on the rounded edges and corners of the body inside the outer contour [Schmid 1995]. This means that both shape parameters to be estimated and promising features for recognition and tracking have to be selected in dependence on the aspect conditions.

8.4.2 Monitoring Residues

A very powerful tool for hypothesis checking in the 4-D approach consists of watching and judging residues of predicted features. If no features corresponding to predicted ones can be found, apparently the hypothesis is no good. Since it is normal that a certain percentage of predicted features cannot be found, the number of features tried should be two to three times the minimum number required for complete state estimation. More features make the measurement and interpretation process more robust, in general.

If the percentage of features not found (or found with too large prediction errors) rises above a certain threshold value (*e.g.*, 30%), the hypothesis should be adjusted or dropped. The big advantage of an early jump to an object hypothesis is the fact that now there are many more features available for testing the correctness. Early starts with several (the most likely) different hypotheses, usually, will lead to earlier success in finding a fitting one than by trying to accumulate confidence through elaborate feature combinations (danger of combinatorial explosion).

8.5 Road Elements To Be Initialized

In the long run, road recognition systems do not have to recognize just the road surface, possibly with lanes, but a much larger diversity of objects forming the infrastructure of different types of roads (see Table 8.2 below).

Lane markings do not just define the lanes for driving. Information about maneuvering allowed or forbidden is also given by the type of marking: Solid markings indicate that this line is not to be crossed by vehicles in normal situations; in case of a failure, these lines may be crossed, *e.g.*, for parking the vehicle on the road shoulder. Dashed markings allow lane changes in both directions. A solid and a dashed line beside each other indicate that crossing this marking is allowed only from the dashed side while forbidden from the solid side. Two solid lines beside each other should never be crossed. In some countries, these lines may be painted in yellow color regularly while in others yellow colors are spared for construction sites.

Within lanes, arrows may be painted on the surface indicating the allowed driving directions for vehicles in the lane [Baghdassarian et al. 1994]. Other more complex markings on roads can be found when the number of lanes changes, either upward or downward; typical examples are exits or entries on high-speed roads or turnoffs to the opposite direction of driving (left in right-hand traffic and *vice versa*).

Vertical poles at regular spacing along the road (25 or 50 m), sometimes with reflecting glass inserts, are supposed to mark the road when the surface is covered with snow, dirt, or leaves in the fall. The reflecting glass may have different colors (*e.g.*, in Austria) or shapes (in Germany) to allow easier discrimination of left and right road boundaries further away. Availability of this background knowledge in vision systems makes road recognition easier.

Guide rails along the road also help affirming a road hypothesis. They have nice features, usually, with elongated horizontal edges and homogeneous areas (bands of similar color or texture) in between. Since they stand up about a half meter above the road surface and at a certain distance to the side, they are separated from the road surface by irregular texture from vegetation, usually. At sharper bends in the road, especially at the end of a straight high-speed section, often red and white arrow heads are painted onto their surface, signaling the driver to slow down. When no guide rails are there, special signs at the usual location of guide rails with arrows in the direction of the upcoming curve, or triangular traffic signs with bent arrows in elevated positions usual for traffic signs may be found for alerting the driver. Machine vision allows autonomous vehicles to take advantage of all this information, put up initially for the human driver, without additional cost.

Table 8.2. Objects of road infrastructure to be recognized on or in the vicinity of a road

Name of object (class)	Location relative to road	Characteristic shape (2-D, 3-D	Colors
Road markings	On the surface	Lines (0.1 to 0.5 m wide; solid, dashed - single or in pairs, rectangles, arrows	Chite, yellow
Vertical poles	Beside road	Round, triangular cylinders (spacing every 25 or 50 m)	Black, white orange
Traffic regulation signs	Beside or above	Round, triangular, square, rectangular, octagonal	White, red and black, blue
Arrows	On guide rails, posts	Solid arrows, diagonal and cross-diagonal stripes	White and red, black
Traffic lights	Beside or above	Round, rectangles (with bars and arrows)	Red, yellow, green
Signs for navigation	Beside or above	Large rectangular	White, blue, green
Signs for city limits	Beside	Large rectangular (some with black diagonal bar)	Yellow, black, white
Signs for construction sites	On, beside or above	Round, triangular, rectangular with diagonal stripes	Yellow, black, red
Guide rails	Above and beside road	Curved steel bands	Metal, white and red
Reflecting nails	On surface	Shallow rounded pyramid	Yellow, red
Tar patches	On surface	Any	Black
Turnoff lane	At either side	Widening lane with arrow	White, yellow
Road fork	Splitting road	Y-shape	Different shades or textures

Traffic sign recognition has been studied for a long time by several groups, *e.g.*, [Estable *et al.* 1994; Priese *et al.* 1995; Ritter 1997]. The special challenge is, on one hand, separating these traffic signs from other postings, and on the other, recognizing the signs under partial occlusion (from branches in summer and snow in winter). Recognition of traffic and navigational signs will not be followed here since it can be decoupled from road recognition proper. What cannot be decoupled from normal road recognition is detecting "reflecting nails" in the road surface and recognizing that certain patches in the road surface with different visual appearance are nonetheless smooth surfaces, and it does no harm driving over them. Stereovision or other range sensing devices will have advantages in these cases.

On the contrary, driving on poorly kept roads requires recognizing potholes and avoiding them. This is one of the most difficult tasks similar to driving cross-country. On unmarked roads of low order with almost no restrictions on road curvature and surface quality, visual road (track) recognition of autonomous vehicles has barely been touched. Mastering this challenge will be required for serious military and agricultural applications.

8.6 Exploiting the Idea of Gestalt

Studies of the nature of human perception support the conclusion that perception is not just to reflect the world in a simple manner. Perceived size is not the same as physical size, perceived brightness is not the same as physical intensity, perceived velocity is not physical velocity, and so on for many other perceptual attributes. Moreover, the perception of composite stimuli often elicits interpretations which are not present when the components are perceived separately. Or in other words, "The whole is different from the sum of its parts". The gestalt laws deal with this aspect in greater detail. Some remarks on history:

Ernst Mach (1838–1916) introduced the concepts of *space forms* and *time forms*. We see a square as a square, whether it is large or small, red or blue, in outline or as textured region. This is space form. Likewise, we hear a melody as recognizable, even if we alter the key in such a way that none of the notes are the same. Motion processes are recognized by just looking at dots on the joints of articulated bodies, everything else being dark [Johansson 1973].

Christian von Ehrenfels (1859–1932) is the actual originator of the term *gestalt* as the gestalt psychologists were to use it. In 1890, he wrote a book *On Gestalt Qualities*. One of his students was Max Wertheimer to whom *Gestalt Psychology* is largely attributed.

Wolfgang Köhler (1887–1967) received his PhD in 1908 from the University of Berlin. He then became an assistant at the Psychological Institute in Frankfurt, where he met and worked with Max Wertheimer. In 1922, he became the chair and director of the psychology lab at the University of Berlin, where he stayed until 1935. During that time, in 1929, he wrote *Gestalt Psychology*. The original observation was Wertheimer's, when he noted that we perceive motion where there is nothing more than a rapid sequence of individual sensory events. This is what he saw in a toy stroboscope he bought by chance at a train station and what he saw in his laboratory when he experimented with lights flashing in rapid succession (like fancy neon signs in advertising that seem to move). The effect is called the *phi phenomenon*, and it is actually the basic principle of motion pictures!

If we see what is not there, what is it that we are seeing? One could call it an illusion, but it is not a hallucination. Wertheimer explained that we are seeing an effect of the whole event, not contained in the sum of the parts. We see a coursing string of lights, even though only one light lights up at a time, because the whole event contains relationships among the individual lights that we experience as well. This is exploited in modern traffic at construction sites to vividly convey the (unexpected) trajectory to be driven.

In addition, say the gestalt psychologists, we are built to experience the structured whole as well as the individual sensations. And not only do we have the ability to do so, we have a strong tendency to do so. We even add structure to events which do not have gestalt structural qualities.

In perception, there are many organizing principles called *gestalt laws*. The most general version is called the *law of Praegnanz*. It is supposed to suggest being pregnant with meaning. This law says that we are innately driven to experience things in as good a gestalt as possible. “Good” can mean many things here, such as regular, orderly, simplicity, symmetry, and so on, which then refer to specific gestalt laws.

For example, a set of dots outlining the shape of an object is likely to be perceived as the object, not as a set of dots. We tend to complete the figure, make it the way it ‘should’ be, and finish it in the context of the domain perceived. Typical in road scenes is the recognition of *triangular* or *circular* traffic signs even though parts of them are obscured by leaves from trees or by snow sticking to them.

Gestalt psychology made important contributions to the study of visual perception and problem solving. The approach of gestalt psychology has been extended to research in areas such as thinking, memory, and the nature of aesthetics. The Gestalt approach emphasizes that we perceive objects as well-organized patterns rather than an aggregation of separate parts. According to this approach, when we open our eyes, we do not see fractional particles in disorder. Instead, we notice larger areas with defined shapes and patterns. The "whole" that we see is something that is more structured and cohesive than a group of separate particles. That is to say, humans tend to make an early jump to object hypotheses when they see parts fitting that hypothesis.

In visual perception, a simple notion would be that to perceive is only to mirror the objects in the world such that the physical properties of these objects are reflected in the mind. But is this really the case? Do we “measure” the scene we watch? The following examples show that perception is different from this simple notion and that it is more constructive. The nature of perception fits more with the notion *to provide* ***a useful description*** *of objects* in the outside world instead of being an accurate mirror image of the physical world. This description has to represent features that are *relevant to our behavior*.

The focal point of gestalt theory is the idea of "grouping", or how humans tend to interpret a visual field or problem in a certain way. The main factors that determine grouping are

- *proximity* - how elements tend to be grouped together depending on their spatial closeness;

- *similarity* - items that are similar in some way tend to be grouped together;
- *closure* - items are grouped together if they tend to complete a shape or pattern;
- *simplicity* - organization into figures according to symmetry, regularity, and smoothness.

In psychology, these factors are called the *laws of grouping* in the context of perception. Gestalt grouping laws do not seem to act independently. Instead, they appear to influence each other, so that the final perception is a combination of the entire gestalt grouping laws acting together. Gestalt theory applies to all aspects of human learning, although it applies most directly to perception and problem-solving.

8.6.1 The Extended Gestalt Idea for Dynamic Machine Vision

Not only is spatial appearance of importance but also temporal gestalt (patterns over time, oscillations, optical flow): Objects are perceived within an environment according to all of their elements taken together as a global construct. This gestalt or "whole form" approach has tried to isolate *principles of perception*: seemingly innate mental laws that determine the way in which objects are perceived.

This capability of humans, in general, has been exploited in designing roads and their infrastructure for humans steering vehicles.

Road curvature is recognized from smooth bending of the "pencil tip", also over time, as when a road appears in video sequences taken by a camera in a vehicle guided along the road. The steering angle needed to stay at the center of the lane (road) is directly proportional to curvature; steering rate is thus linked to speed driven (see Equations 3.10/3.11). However, continuously seeing the road is not necessary for perceiving a smoothly curved road. When snow covers both road and shoulders and there is no vertical surface profile perpendicular to the road (everything is entirely flat so that there are no surface cues for recognizing the road), humans are, nonetheless, able to perceive a smoothly curved road from poles regularly spaced along the side of the road. Introduced standards of spacing are 25 m (on state roads) or every 50 m (on freeways designed for higher speeds). While driving continuously at constant speed, each pole generates a smoothly curved trajectory in the image; the totality of impressions from all poles seen thus induces in the human observer the percept of the smoothly curved road. Technical systems can duplicate this capability by application of sampled data theory in connection with a proper road model, which is standard state of the art.

Similarly, guide rails or trees along the road can serve the same purpose. Guide rails usually are elevated above the ground (~ 0.5 m) and allow recognizing curve initiation at long distances, especially when marked with arrowheads (usually black or red and white); while driving in a curve, the arrowheads give a nice (optical) feature flow field which the unstructured continuous guide rail is not able to provide.

In *Gestalt psychology*, the mechanism behind these types of percepts is labeled *Principle of Totality*. This is to say that conscious experience must be considered globally (by taking into account all the physical and mental aspects of the perceiving individual simultaneously) because the nature of the mind demands that each component be considered as part of a system of dynamic relationships in a task context. This is sometimes called the *Phenomenon*; in experimental analysis, in relation to the *Totality Principle*, any psychological research should take phenomena as a starting point and not be solely focused on sensory qualities.

In terms of the 4-D approach to dynamic machine vision, this means that early jumps to higher level hypotheses for perception may be *enlightening*, shedding new light (richer empirical background knowledge) for solving the vision task based on impoverished (by perspective mapping) image data.

8.6.1.1 Spatial Components: Shape Characteristics

The semantics of perception are components not directly derived from bottom-up signal processing; they originate from internal top-down association of relational structures with derived image structures. Most reliable structures are least varying based on a wealth of experience. It is claimed that 3-D space in connection with central projection has to offer many advantages over 2-D image space with respect to least variations (idealized as "invariance") in the real world. This is the reason why in the 4-D approach all internal representations are done in 3-D space directly (beside continuous temporal embedding to be discussed later). From a collection of features in the image plane, an immediate jump is made to an object hypothesis in 3-D space for a physical object observed under certain spatial aspect conditions with the imaging process governed by straight light ray propagation. This may include mirroring.

Note that in this approach, the (invariant) 3-D shape always has to be associated with the actual aspect conditions to arrive at its visual appearance; because of the inertia of objects, these objects tend to move smoothly. Knowing 3-D shape and motion continuity eliminates so called *catastrophic events* when in 2-D projection entire faces appear or disappear from frame to frame.

From the task context under scrutiny, usually, there follows a strong reduction in meaningful hypotheses for *objects under certain aspect conditions* from a given collection of features observed. In order not to be too abstract, the example of scene interpretation in road traffic will be discussed. Even here, it makes a large difference whether one looks at and talks about freeway traffic, cross-country traffic or an urban scene with many different traffic participants likely to appear. As usual in the interpretation of real world dynamic scenes, two different modes of operation can be distinguished: (1) The initial orientation phase, where the system has to recognize the situation it is in (discussed here) and (2) the continuous tracking and control phase, where the system can exploit knowledge on temporal processes for single objects to constrain the range of interpretations by prediction-error feedback (4-D part proper to be discussed later).

Certain *Gestalt laws* have been formulated by psychologists since humans cannot but perceive groups of features in a preferred way: *The law of proximity* states that objects near each other tend to be seen as a unit. A useful example in road marking is the following: Parallel, tightly spaced double lines, one solid and the other broken, are a single perceptual unit indicating that crossing this line is allowed only from the side of the broken line and not from the other. Two solid lines mean that crossing is never allowed, not even in critical situations.

The *law of closure* says that, if something is missing in an otherwise complete figure, we will tend to add it. A circle or a rectangle, for example, with small parts of their edges missing, will still be seen as a circle or a rectangle, maybe described by the words "drawn with interrupted lines"; if these lines are interrupted regularly, a specification such as "dashed" or "dash-dotted" is immediately understood in a conversation between individuals if they have the geometric concept of circle or rectangle at their disposal. For the outer rectangle in Figure 8.5, the gap at the lower left corner will be "closed" (maybe with a footnote like "one corner open".

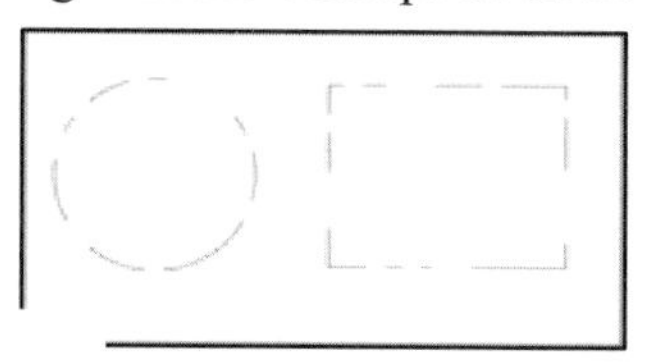

Figure 8.5. The *law of closure* in Gestalt psychology for completing basic shapes (under perturbations?)

The *law of similarity* says that we will tend to group similar items together, to see them as forming a gestalt, within a larger form. An example in road traffic may be recognizing wheels and vehicle bodies. Recognizing the separate subparts may be ambiguous. When they are assembled in standard form, tapping knowledge about form and function can help finding the correct hypotheses more easily. An idealized example is given in Figure 8.6. Covering up the rest of the figure except the upper left part (a), the graph could be read as the digit zero or the letter 'o' written in a special form for usage in posters. If the additional hint is given that the figure shows a 3-D body, most of us (with a multitude of wheeled vehicles around in everyday life) would tend to see an axially symmetrical ring (tube for a wheel?) under oblique viewing conditions; whether it is seen from the left or from the right side cannot be decided from the simple graphic display.

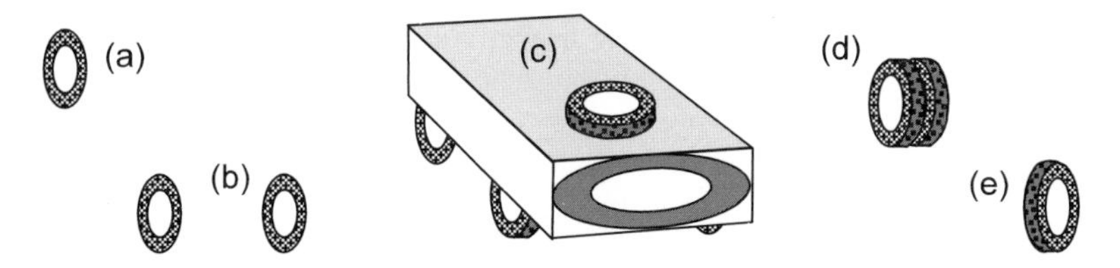

Figure 8.6. Percepts in road traffic: (a) An elliptically shaped, ring-like figure (could be an axially symmetrical tube for tires under an oblique angle, seen either from the left or the right). (b) A pair of shapes like (a) (see text); (c) adding a simple vehicle body obscuring most of the three wheels, the arrangement of (a) and (b) turns into the percept of a wheeled vehicle clearly seen from the rear left side (or the front right side!). (d) and (e), adding minor details to (a) resolves ambiguity by spatial interpretation: d) is a twin-wheel seen from the rear left while (e) is viewed from the rear right. [There is more background knowledge available affirming the interpretations in (c) to (e), see text.]

Part (b) shows two such objects side by side; they are perceived as separate units. Only when in part (c) a very simple box-like objects is added, covering most of the three objects arranged as in parts a) and b), the percept immediately is that of a cart seen from above, the left and the rear. The visible parts of the formerly separate three objects now fit the two hypotheses of a vehicle with four wheels, seen either (1) from the rear left or (2) from the front right; one wheel is self-occluded. In case (1) it is the front right wheel, while in case (2) it is the rear right one. There is no doubt that the wheels are both seen from the same side; this is due to the interpretation derived from the rectangular box. In parts d) and e) of the figure, the object from (a) has been supplemented with a differently looking surface representing the contact area of the wheel to the ground. From this additional cue it is apparent that the "tires" in (d) are seen from the rear left, while the one in (e) is seen from the rear right (or *vice versa*, depending on the direction of travelling that is needed for disambiguation). The twin-tires in (d) immediately induce the percept of a truck since cars do not have this arrangement of wheels, in general.

The object on the upper surface of the vehicle body is also hypothesized as a wheel in the perception process because it fits reasonably to the other wheels (size and appearance). If not mounted on an axle, the pose shown is the most stable one for isolated wheels. As a spare part, it also fits the functional aspect of standard usage. Quite contrary, the ellipses seen on the rear (front) side of the vehicle will not be perceived as a wheel by humans, normally; at best, it is painted there for whatever reason. Note that except size, color, and pose it is identical to the leftmost wheel of the vehicle seen only partially (the contact area of the tire to the ground is not shown in both cases); nonetheless, the human percept is unambiguous. It is this rich background of knowledge on form and function of objects that allows intelligent systems easy visual perception.

Next, there is the *law of symmetry*. We tend to perceive objects as symmetrical in 3-D space even though they appear asymmetrical in perspective projection. Other vehicles tend to be seen as symmetrical since we are biased by previous experience; only when seen straight from the back or the front are they close to symmetrical, usually. Traffic signs of any shape (rectangular, circular, and triangular) are perceived as symmetrical under any aspect conditions; knowledge about perspective distortion forces us to infer the aspect conditions from the distortions actually seen.

The law of good continuation states that objects arranged in either a straight line or a smooth curve tend to be seen as a unit. In Figure 8.7, we distinguish two lines, one from A to B and another from C to D, even though this graphic could represent another set of lines, one from A to D and the other from C to B. Nevertheless, we are more likely to identify line A to B, which has better continuation than the line from A to D, which has an obvious discontinuity in direction at the corner. If the context in which this arrangement appeared would have been objects with corners of similar appearance, continuity over time could invoke the other interpretation as more likely. Bias from context is known to have considerable influence.

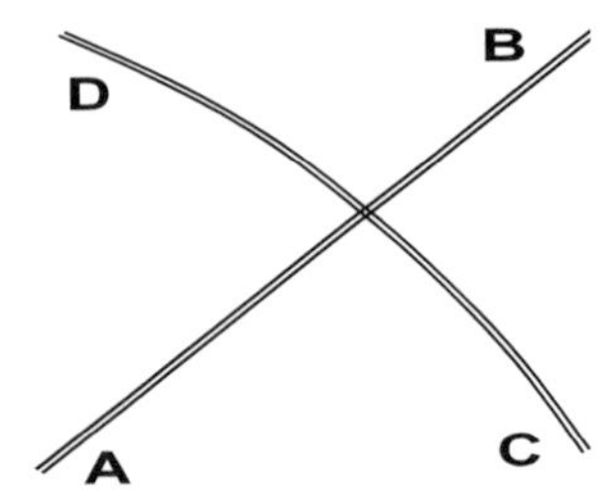

Figure 8.7. The law of *good continuation* tends to favor 'two lines crossing' as percept (A-B crossing C-D) and not two corners touching each other (AD touches BC) or (AC touches BD)

The idea behind these examples, and much of the gestalt explanation of things, is that the world of our experiencing is meaningfully organized, to one degree or another; we receive a better payoff taking these (assumed) facts into account.

The *gestalt effect* refers to the form–forming capability of our senses, particularly with respect to the visual recognition of figures and whole forms instead of just a collection of simple lines and curves. Key properties of gestalt systems are *emergence*, *reification*, *multistability*, and *invariance*.

Emergence: Some parts are seen only after the whole has been hypothesized. *Reification* is the *constructive* or *generative* aspect of perception, by which the experienced percept contains more explicit spatial information than the sensory stimulus on which it is based. For instance, based on the fact that straight lines can be fit between the centers of three black circles, the human observer will see a ("Kaniza") triangle in Figure 8.8, although no triangle has actually been drawn.

Figure 8.8. The percept of a white triangle, based on an obscuration hypothesis for three separate dark circular disks in triangular arrangement

We have noticed our vehicle perception system in a highway scene overlooking two vehicles driving side by side in front at almost constant speed, but claiming a vehicle (for several cycles) in the space between the vehicles; only after several cycles has this misinterpretation been resolved by the system through newly hypothesized vehicles at the correct location without intervention of an operator.

Multi-stability (or *Multi-stable perception*) is the tendency of ambiguous perceptual experiences to pop back and forth unstably between two or more alternative interpretations. This is seen for example in the well-known *Necker cube* shown in Figure 8.9. It is even claimed that one cannot suppress this alternation by conscious attention [Pöppel *et al.* 1991]; this indicates that hypothesis generation is not fully dependent on one's will.

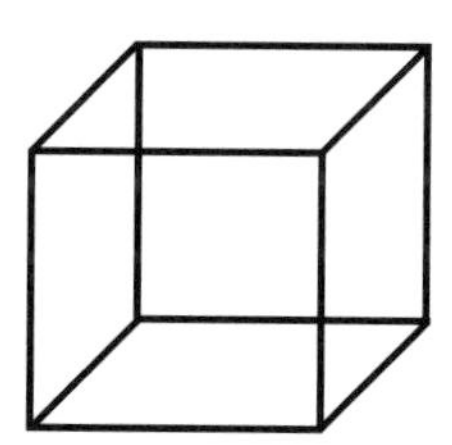

Figure 8.9. The Necker cube (wire frame) showing two stable 3-D interpretations: Either from top right above or from left below the cube.

This phenomenon of multiple interpretations may not only happen in pictorial (snapshot) image interpretation as shown, but also in image sequences with moving light dots. As psychologists have shown, individuals tend to interpret some of these sequences differently, depending on the context discussed previously or on the personal background of experience. A difficult problem arises in interpreting a snapshot from an action by just one single frame; the actual pose may occur in several different action patterns. It is again likely to find the correct interpretation only by referring to the context which may be inferred from some other components of the image or which has to be known from a different source; this problem is often encountered in interpreting pieces of art. Sometimes this type of ambiguity is introduced purposely to stimulate discussion.

Invariance is the property of perception whereby simple geometric objects are recognized independently of rotation, translation, and scale, as well as several other variations such as elastic deformations, different lighting, and different component features. For example, all wheels are immediately recognized as having the same basic shape. They are recognized despite perspective and elastic deformations as well as partial occlusion (see Figure 8.4).

Emergence, reification, multistability, and invariance are not separable modules to be modeled individually, but they are different aspects of a single unified dynamic mechanism. These examples show that image interpretation often needs embedding in a temporal and situational framework to yield unique results. For machine vision, the 4-D approach is capable of providing this background in a natural way.

8.6.1.2 Temporal Components, Process Characteristics

The law of common fate states that when objects or groups of features move in the same direction, we tend to see them as a unit. This is the most common reason for hypothesizing other vehicles or moving subjects in traffic. Biological subjects using legs for locomotion have swinging leg (and possibly arm) movements superimposed on body motion. The location of joints and their motion is characteristic of the type of living being. Humans moving their arms, possibly with objects in their hands, can be part of organized traffic signaling in unusual traffic situations.

Depth separation for traffic signs located in front of background texture has already been mentioned. Vehicles oscillating with decreasing amplitude after a perturbation in pitch or bank angle generate typical percepts of temporal gestalt: Oscillation frequency and damping ratio.

Blinking lights, when one-sided, signal the intention for lane change or turnoff of a vehicle, or when both-sided, signal “danger”; blinking lights can only be recognized after a few cycles as “blinking.” These are percepts of relevance for decision–making in traffic. Another use of blinking lights, but now sequentially of many properly spaced ones on a curved line, can be found at construction sites; the consecutive flashing of the next light in the row induces in the observer the percept of an object moving on the curve. This vividly tells the driver the type of steering he has to do (curvature ≈ steering angle) to follow the trajectory indicated.

The phenomenon of multiple possibilities for interpretation may not only happen in pictorial (snapshot) image interpretation but also in image sequences with moving light dots. As psychologists have shown, individuals tend to interpret some of these sequences differently, depending on the context discussed previously or on the personal background of experience [Johannson 1973]. These examples show that image interpretation often needs embedding in a temporal and situational framework to yield unique results.

A ball bouncing on the road with children playing at the side allows deriving certain expectations which will require attention focusing for the vision system (ball and/or children). The trajectory of the ball, when up in the air, is expected to be close to a sequence of parabolic arcs with decreasing energy due to friction and with points of reflection at approximately the same angle relative to the surface normal each time it hits the ground or another surface. Since humans have been playing with the ball, usually, it can be expected that someone may run after the ball or that another may come from a different direction to catch the ball.

A bicyclist in right-hand traffic lifting his left arm (or his right arm in left-hand traffic) can be expected to cross the lane driven to turn off to the side of the arm lifted. The temporal sequence of arm lifting and lane crossing can be considered a temporal gestalt, which in common language we call a *maneuver*.

If the process observed involves a subject capable of (intelligent) motion control aimed at some goal, similar predictions for motion processes may be possible by recognizing stereotypical control sequences, as for lane change maneuvers in road vehicle guidance. This gestalt idea extended into the temporal domain for typical motion processes again allows us to concentrate limited computing power onto image regions and to confine parameter variations of the feature extraction methods to ranges likely to yield most of the information necessary for efficient process tracking and control. Seen from this point of view, the knowledge of real-world spatiotemporal processes represented in dynamic models is a powerful means for reducing the possible danger of combinatorial explosion in feature aggregation compared to ignoring these temporal constraints.

Audio signals of certain temporal shapes may indicate an ambulance or a fire brigade vehicle approaching.

8.6.2 Traffic Circle as an Example of Gestalt Perception

Nowadays, a spreading example in traffic of spatiotemporal gestalt perception is the "roundabout" (traffic circle, in some countries exploited to the extreme as a circular area with a central dot marked by color). It consists of an extended, very often partially obscured circular area distorted by perspective projection. Several roads are linked to the outer lane of the circle, in some regions yielding a star-like appearance when seen in a bird's-eye view, due to the density of connections. Vehicles in the circle usually have the right-of-way; their direction signaling by blinking eases the decision of entering vehicles whether or not to proceed. Viewing patterns and the capability of developing proper expectations for the behavior of other vehicles (maybe not according to the valid rules) are important for handling this complex traffic situation.

Here an aphorism attributed to the painter Salvador Dali may be most valid:

> *"To see is to think"*.

8.7 Default Procedure for Objects of Unknown Classes

When none of the established object or subject classes is able to provide a sufficiently good model for a detected object since too many predicted features have not been found, residues are too large, or no convergence occurred, the following approach for learning more about the new object is advisable:

1. Reduce the object to the center of gravity of jointly moving features and a simple shape encasing all these features.
2. The envelope for the shape may be composed of a few simple shapes connected together if the area in the initial convex hull not containing features is too large (*e.g.*, choose a long cylinder of small diameter on a larger rectangular body, *etc.*).
3. Assume a standard Newtonian model for motion (see Section 2.2.5.3, translation separate from rotation); shrinking or expansion of the feature set may allow depth (range) perception.
4. A rotational model is meaningful only if collections of features can be recognized moving relative to each other on top of translation. Appearance and disappearance of certain features on different sides give an indication of the direction of rotation of the body carrying the features.
5. Store statistics on motion behavior and other parameters observed in a list of "observed but unknown" objects (extended data logging in a DOB).

This may be the starting point for learning about hitherto unknown objects or subjects. Recognizing this type of object/subject at a later time and continuing collection of information is the next step toward intelligence. In the long run, expanding the knowledge base by posing meaningful questions to humans about this object/subject, maybe together with a replay of the corresponding video sequences stored, may lead to real machine intelligence. This, however, seems to be off in the far future right now.

9 Recursive Estimation of Road Parameters and Ego State while Cruising

The goal of real-time vision for road vehicles is to understand the environment with as little time delay as possible including all relevant processes that are happening there. For vehicles with large extensions in space (roughly 2 × 2 × 5 m for a van, ~ 1.8 × 1.5 × 4.5 m for an average car, and much larger dimensions for a truck), there are different points of reference for certain subtasks. Cameras and their projection centers should be elevated as high up as possible above the ground for best imaging conditions of the surface they are driving on. There has to be a coordinate system associated with visual mapping with its origin at the projection center of the camera. On the other hand, motion perception and control are most easily described for the center of gravity (cg); therefore, the relevant coordinate system has its origin at the cg. In a closed-loop application like autonomous vehicle guidance by vision, these different reference systems have to be described in a coherent way and set into relation.

The very general approach presented in Chapter 2 via homogeneous coordinates will be used in later sections for more complex vision systems. Here, to demonstrate the early beginnings, the model according to Figure 9.1 is used.

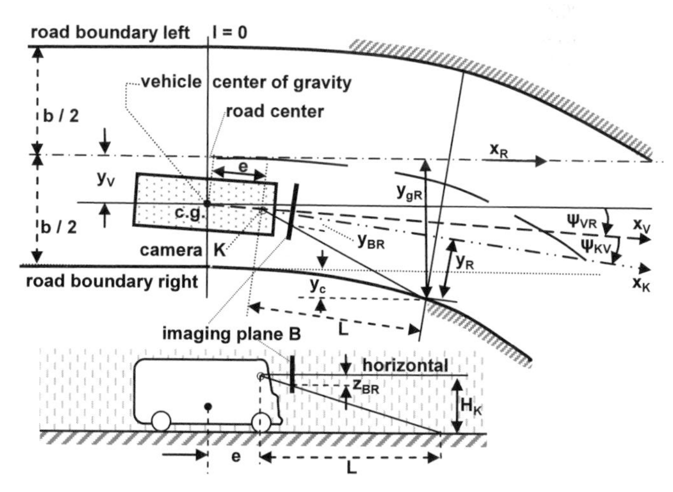

Figure 9.1. More detailed process model than Figure 7.5 for visual road vehicle guidance. Top: Bird's-eye view with variables and parameters for lateral guidance; bottom: Vertical mapping conditions for vision and center of gravity

Table 9.1. Collection of results for recursive estimation of road parameters from Chapter 7: Dynamic models (to be turned into discrete form for sampled data videointerpretation), measurement model, and relations for determining initial values for starting recursive estimation.

Simple **model for lateral dynamics** of a road vehicle [Section 7.1.2, Equation 7.4] (bicycle model, planar, no pitch and bank angle changes, linear system):

$$\begin{pmatrix} \dot\lambda \\ \dot\beta \\ \dot\psi_{rel} \\ \dot y_V \end{pmatrix} = \begin{pmatrix} 0 & 0 & 0 & 0 \\ a_{12} & -1/T_\beta & 0 & 0 \\ V/a & 0 & 0 & 0 \\ 0 & V & V & 0 \end{pmatrix} \begin{pmatrix} \lambda \\ \beta \\ \psi_{rel} \\ y_V \end{pmatrix} + \begin{pmatrix} \mathrm{k}_\lambda \\ 0 \\ 0 \\ 0 \end{pmatrix} \cdot u(t) + \begin{pmatrix} 0 \\ 0 \\ -V \\ 0 \end{pmatrix} \cdot C_{0h}$$

with $a_{12} = 1/(2T_\beta) - V/a$; $T_\beta = \mathrm{V}/\mathrm{k}_{\mathrm{ltf}}$; (Equation 3.30).

Spread-out **clothoid model** [Section 7.1.4]:

$$\dot x_{C_h} = A_{C_h} x_{C_h} + n_{C_{h1}} = \begin{pmatrix} 0 & V & 0 \\ 0 & -3V/L & 3V/L \\ 0 & 0 & 0 \end{pmatrix} \begin{pmatrix} C_{ohm} \\ C_{1hm} \\ C_{1h} \end{pmatrix} + \begin{pmatrix} 0 \\ 0 \\ n_{C_{h1}} \end{pmatrix}. \tag{7.36a}$$

Measurement model for lane (simple road) recognition [Section 7.1.4.3]:
Edge features (left and right) as a function of (road or lane) width b, lateral offset y_V, camera heading $\psi = \psi_V + \psi_{KV}$, C_{0hm} and C_{1hm}.

$$y_{Bi_{l,r}} = f \cdot k_y \left[\frac{\pm\frac{b}{2} - y_V}{L_i} - \psi_V - \psi_{VK} + \frac{L_i}{2} C_{0hm} + \frac{L_i^3}{6} C_{1hm} \right], \tag{7.37}$$

$$z_{Bi} = f \cdot k_z \left(H_K - L_i \tan\theta_K\right) / \left(L_i + H_K \tan\theta_K\right). \tag{7.40}$$

From initialization procedure [Section 8.2]:
Camera gaze direction relative to vanishing point for road nearby (tangents to borderlines intersecting at $L_i = \infty$):

$$\psi = \psi_V + \psi_{VK} = -y_B(L_i = \infty) / f \cdot k_y. \tag{8.5}$$

For $L_i = \infty$ [vanishing point $P(y_{BP}, z_{BP})$]

$$\tan\theta_K = z_{BP} / f \cdot k_z, \quad \text{or} \quad \theta_K = \arctan(z_{BP} / f \cdot k_z). \tag{7.18}$$

Look-ahead ranges:

$$L_i = H_K \left[1 - \tan\theta_K \cdot z_{Bi} / (f \cdot k_z)\right] / \left[z_{Bi} / (f \cdot k_z) + \tan\theta_K\right]. \tag{8.8}$$

Lane (road) **width:** $b_i = L_i \cdot (y_{Bi_r} - y_{Bi_l}) / (f \cdot k_y)$. (7.38)

Lateral offset y_V from lane center: $y_V = L_i \cdot (y_{BP} - y_{BLCi}) / (f \cdot k_y)$. (7.39)

Road curvature parameter:

$$C_{ohm} = 2 \cdot \Delta y_{Cf} / L_f^2 = \Delta y_{CBf} \cdot 2 / (L_f \cdot f \cdot k_y). \tag{8.9}$$

They were shown sufficient for VaMoRs in the late 1980s. The differential geometry model for road representation of Section 7.4.3 now allows interpretation of the spatial continuity conditions for the road as temporal continuity constraints in the form of difference equations for the estimation process while the vehicle moves along the road.

By this choice, the task of recursive estimation of road parameters and relative egostate can be transformed into a conventional online estimation task with two cooperating dynamic submodels. A simple set of equations for planar, undisturbed motion has been given in Chapter 7. In Chapter 8, the initialization problem has been discussed. The results for all elements needed for starting recursive estimation are collected in Table 9.1.

Numerical values for the example in Figure 7.14 extracted from image data have been given in Table 8.1. The steering angle λ and vehicle speed V are taken from conventional measurements assumed to be correct. The slip angle β cannot be determined from single image interpretation and is initialized with zero. An alternative would be to resort to the very simple dynamic model of third order in Figure 7.3a and determine the idealized value for infinite tire stiffness, as indicated in the lower feed-forward loop of the system:

$$\beta = [1/2\text{-}V^2/(2a \cdot k_{ltf})] \cdot \lambda. \qquad (9.1)$$

The estimation process with all these models is the subject of the next section.

9.1 Planar Roads with Minor Perturbations in Pitch

When the ground is planar and the vehicle hardly pitches up during acceleration or pitches down during braking (deceleration), there is no need to explicitly consider the pitching motion of the vehicle (damped second-order oscillations in the vertical plane) since the measurement process is affected only a little. However, in the real world, there almost always are pitch effects on various timescales involved. Acceleration and decelerations, usually, do affect the pitch angle time history, but also the consumption of fuel leads to (very slow) pitch angle changes. Loading conditions, of course, also have an effect on pitch angle as well as uneven surfaces or a flat tire. So, there is no way around taking the pitch degree of freedom into account for precise practical applications.

However, the basic properties of vision as a perception process based on cooperating spatiotemporal models can be shown for a simple example most easily: (almost) unperturbed planar environments. The influence of adding other effects incrementally can be understood much more readily once the basic understanding of recursive estimation for vision has been developed.

9.1.1 Discrete Models

The dynamic models described in previous sections and summarized in Table 9.1 (page 254) have been given in the form of differential equations describing constraints for the further evolution of state variables. They represent in a very efficient way general knowledge about the world as an evolving process that we want to use to understand the actual environment observed under noisy conditions and for decision-making in vehicle guidance.

First, the dynamic model has to be adapted to sampled data measurement by transforming it into a state transition matrix A and the control input matrix B (see Equation 3.7) for the specific cycle time used in imaging (40 ms for CCIR and 33 1/3 for NTSC). Since speed V enters the elemental expressions at several places, the elements of the transition and control input matrices have to be computed anew every cycle. To reduce computing time, the terms have been evaluated analytically via Laplace transform (see Appendix B.1) and can be computed efficiently at runtime [Mysliwetz 1990].

The *measurement model* is given by Equations 7.20 and 7.37:

$$z_{Bi} = f \cdot k_z \left(H_K - L_i \tan\theta_K\right) / \left(L_i + H_K \tan\theta_K\right), \qquad (9.2)$$

$$y_{Bi_{l,r}} = f \cdot k_y \left[\frac{\pm\frac{b}{2} - y_V}{L_i} - \psi_V - \psi_{VK} + \frac{L_i}{2} C_{0hm} + \frac{L_i^2}{6} C_{1hm} \right], \qquad (9.3)$$

with θ_K as angle of the optical axis relative to the horizontal. None of the lateral state variables enters the first equation. However, if there are changes in θ_K due to vehicle pitch, with image row z_{Bi} evaluated kept constant, the look-ahead distance L_i will change. Since L_i enters the imaging model for lateral state variables (lower equation), these lateral measurement values y_{Bi} will be affected by changes in pitch angle, especially the lateral offset and the curvature parameters. The same is true for the road parameter 'lane or road width b' at certain look-ahead ranges L_i (Equation 7.38):

$$b_i = L_i \cdot (y_{Bi_r} - y_{Bi_l}) / (f \cdot k_y). \qquad (9.4)$$

Since b depends on the difference of two measurements in the same row, it scales linearly with look-ahead range L_i and all other sensitivities cancel out. Note however, that according to Table 7.1, the effects of changes in the look-ahead range due to pitch are small in the near range and large further away.

Introducing b as an additional state variable (constant parameter with $db/dt = 0$) the state vector to be estimated by visual observation can be written

$$x^T = (\lambda,\ \beta,\ \psi,\ y_V,\ C_{0hm},\ C_{1hm},\ C_{1h},\ b). \qquad (9.5)$$

Note that these variables are those we humans consider the most compact set to describ a given simple situation in a road scene. Derivation of control terms for guiding the vehicle efficiently on the road also uses exactly these variables; they constitute the set of variables that by multiplication with the feedback gain matrix yields optimal control for linear systems. There simply is no more efficient cycle for perception and action in closed-loop form.

9.1.2 Elements of the Jacobian Matrix

These elements are the most important parameters from which the 4-D approach to dynamic vision gains its superiority over other methods in computational vision. The prediction component integrates temporal aspects through continuity conditions of the physical process into 3-D spatial interpretation, including sudden changes in one's own control behavior. The first-order relationship between states and parameters included as augmented states of the model, on one hand, and feature positions in the image, on the other, contains rich information for scene understanding according to the model instantiated; this relationship is represented by the elements of the Jacobian matrix (partial derivatives). Note that depth is part of the measurement model through the look-ahead ranges L_i which are geared to image rows by Equation 9.2 for given pitch angle and camera elevation.

Thus, measuring in image stripes around certain rows directly yields road parameters in coordinates of 3-D space. Since the vehicle moves through this space and knows about the shift in location from measurements of odometry and steering angle, motion stereointerpretation results as a byproduct.

The *i*th row of the Jacobian matrix C (valid for the *i*th measurement value y_{Bi}) then has the elements

$$c_i = \partial y_{Bi} / \partial x|_{x^*} = f \cdot k_y \left[0 \,|\, 0 \,|\, -1 \,|\, \frac{-1}{L_i} \,|\, \frac{L_i}{2} \,|\, \frac{L_i^2}{6} \,|\, 0 \,|\, \frac{\pm 1}{2L_i} \right], \tag{9.6}$$

where $+1/(2L_i)$ is valid for edges on the right-hand and $-1/(2L_i)$ for those on the left-hand border of the lane or road. The zeros indicate that the measurements do not depend on the steering and the slip angle as well as on the driving term C_1 for curvature changes. *Lateral offset* y_V (fourth component of the state vector) and *lane* or *road width b* (last component) go with 1/(range L_i) indicating that *measurements nearby* are best suited for their update; *curvature parameters* go with range (C_1 even with range squared) telling us that *measurements far away* are best suited for iteration of these terms.

With no perturbations in pitch assumed, the Jacobian elements regarding z_{Bi} are all zero. (The small perturbations in pitch actually occurring are reflected into the noise term of the measurement process by increasing the variance for measuring lateral positions of edges.)

9.1.3 Data Fusion by Recursive Estimation

The matrix R (see Section 8.3.1) is assumed to be diagonal; this means that the individual measurements are considered independent, which of course is not exactly true but has turned out sufficiently good for real-time vision:

$$R = \mathrm{Diag}(r_i) = \mathrm{Diag}(\sigma_\lambda^2, \sigma_{y_{B1}}^2, \sigma_{y_{B2}}^2, \sigma_{y_{B3}}^2, \ldots\ldots, \sigma_{y_{B8}}^2). \tag{9.7}$$

Standard deviations (and variances) for different measurement processes have been discussed briefly in Section 8.3.1. The type of measurement does not show up in Equation 9.7; here, only the first component is a conventionally measured quantity; all others come from image processing with complex computations for interpretation. What finally matters in trusting these measurements in EKF processing is just their standard deviation. (For highly dynamic processes, the delay time incurred in processing may also play a role; this will be discussed later when inertial and visual data have to be fused for large perturbations from a rough surface; angular motion then leads to motion blur in vision.)

For high-quality lane markings and stabilized gaze in pitch, much smaller values are more reasonable than the value of standard deviation $\sigma = 2.24$ pixels selected here for tolerating small pitch angle variations not modeled. This is acceptable only for these short look-ahead ranges on smooth roads; for the influence of larger pitch angle perturbations, see Section 9.3.

If conventional measurements can yield precise data with little noise for some state variables, these variables should not be determined from vision; a typical example is measurement of one's own speed (*e.g.*, by optical flow) when odometry and solid ground contact are available. Visual interpretation typically has a few tenths of a second delay time, while conventional measurements are close to instantaneous.

9.1.4 Experimental Results

The stabilizing and smoothing effects of recursive estimation including feature selection in the case of rather noisy and ambivalent measurements, as in the lower right window of Figure 7.17 marked as a white square, can be shown by looking at some details of the time history of the feature data and of the estimated states in Figure 9.2.

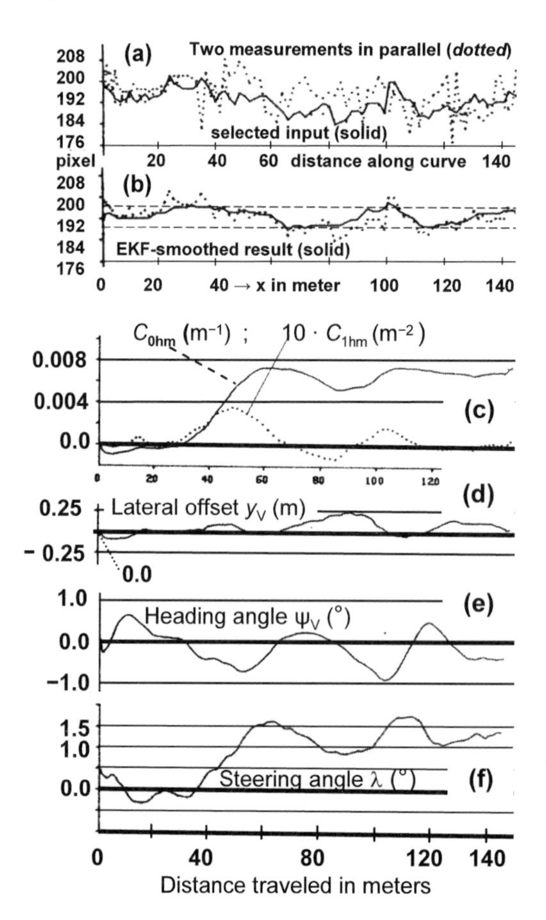

Figure 9.2. From noise corrupted measurements of edge feature positions (*dots, top graph* (a) via pre-selection through expectations from the process model [*dots in graph* (b) = solid curve in (a)] to symbolic representations through high-level percepts: (c) road curvature at cg location of vehicle C_{0hm} and smoothed (averaged) derivative term C_{1hm} (multiplied by a factor of 10 for better visibility). (d) Lateral offset y_V of vehicle from lane center; lane = right half part of road surface (no lane markings except a tar-filled gap between plates of concrete forming the road surface). Note that errors were less than 25 cm. (e) Heading angle ψ_V of vehicle relative to road tangent direction ($|\psi_V| < 1°$). The bottom graph (f) shows the control input generated in the closed-loop action–perception cycle: a turn to the right with a turn radius R of about 160 m (= $1/C_{0hm}$) and a steering angle between ~1 and 1.7°, starting at ~ 40 m distance traveled.

The measured pixel positions of edge candidates vary by ~ 16 (in extreme cases up to almost 30) pixels (*dotted curve in top part*). Up to four edge candidates per window are considered; only the one fitting the predicted location best is selected and fed into the recursive estimation process if it is within the expected range of tolerance (~ 3 σ) given by the innovation variance according to the denominator in the second of Equations 6.37.

The solid line in the top curve of Figure 9.2 represents the input into the recursive estimation process [repeated as dotted line in (b)-part of the figure]. The solid line there shows the result of the smoothing process in the extended Kalman filter; this curve has some resemblance to the lateral offset time history y_V in Figure 7.18, right (repeated here as Figure 9.2c–f for direct comparison with the original data). The dynamic model underlying the estimation process and the characteristics of the car by Ackermann–steering allow a least-squares error interpretation that distributes the measurement variations into combinations of road curvature changes (c), yaw angles relative to the road over time (e), and the lateral offset (d), based also on the steering rate output [= control time history (f)] in this closed-loop perception–action cycle. The finite nonzero value of the steering angle in the right-hand part of the bottom Figure 9.2f confirms that a curve is being driven.

It would be very hard to derive this insight from temporal reasoning in the quasi-static approaches initially favored by the AI community in the 1980s. In the next two sections, this approach will be extended to driving on roads in hilly terrain, exploiting the full 4-D capabilities, and to driving on uneven ground with stronger perturbations in pitch.

9.2 Hilly Terrain, 3-D Road Recognition

The basic appearance of vertically curved straight roads in images differs from flat ones in that both boundary lines at constant road width lie either below (for downward curvature) or above (for upward curvature) the typical triangle for planar roads (see Figure 9.3).

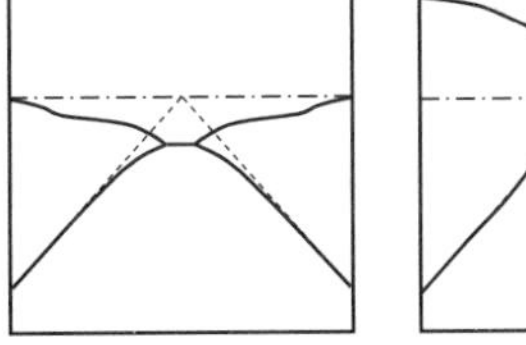

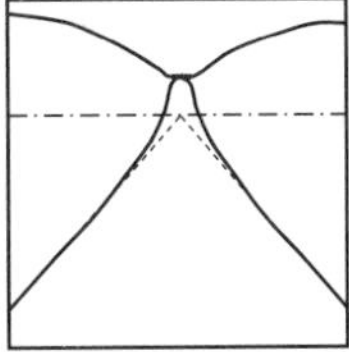

Figure 9.3. Basic appearance of roads with vertical curvature: Left: Curved downward (negative); right: curved upward (positive curvature)

From Figure 9.4, it can be seen that upward vertical curvature shortens the look-ahead range for the same image line and camera angle from L_0 down to L_{cv}, depending on the elevation of the curved ground above the tangent plane at the location of the vehicle (flat ground).

Similar to the initial model for horizontal curvature, assuming constant vertical curvature C_{0v} , driven by a noise term on its derivative C_{1v} as a model, has turned out to allow sufficiently good road perception, usually:

$$C_v = C_{0v} + C_{1v} \cdot l,$$
$$dC_{0v} / dl = C_{1v}, \qquad dC_{1v} / dl = n_{cv}(l). \tag{9.8}$$

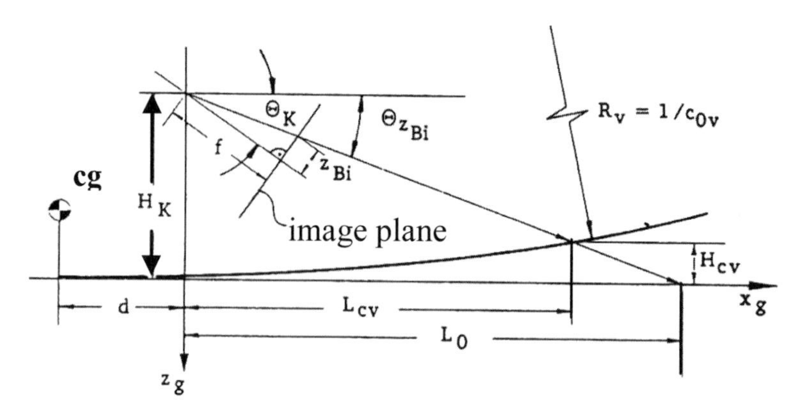

Figure 9.4. Definition of terms for vertical curvature analysis (vertical cut seen from right-hand side). Note that positive pitch angle θ and positive curvature C_v are upward, but positive z is downward.

9.2.1 Superposition of Differential Geometry Models

The success in handling vertical curvature independent of the horizontal is due to the fact that both are dependent on arc length in this differential geometry description. Vertical curvature always takes place in a plane orthogonal to the horizontal one. Thus, the vertical plane valid in Equation 9.8 is not constant but changing with the tangent to the curve projected into the horizontal plane. Arc length is measured on the spatial curve. However, because of the small slope angles on normal roads, the cosine is approximated by 1, and arc length becomes thus identical to the horizontal one. Lateral inclination of the road surface is neglected here, so that this model will not be sufficient for driving in mountainous areas with (usually inclined) switchback curves (also called 'hair pine' curves). Road surface torsion with small inclination angles has been included in some trials, but the improvements turned out to be hardly worth the effort.

A new phenomenon occurs for strong downward curvature (see Figure 9.5) of the road. The actual look-ahead range is now larger than the corresponding planar one L_0. At the point where the road surface becomes tangential to the vision ray (at L_{cv} in the figure), self-occlusion starts for all regions of the road further away. Note that this look-ahead range for self-occlusion is not well defined because of the tangency condition; small changes in surface inclination may lead to large changes in look-ahead distance. For this reason, the model will not be applied to image regions close to the cusp which is usually very visible as a horizontal edge (*e.g.*, Figure 9.3).

9.2.2 Vertical Mapping Geometry

According to Figure 9.4, vertical mapping geometry is determined mainly by the camera elevation H_K above the local tangential plane, the radius of curvature $R_v = 1/C_{0v}$ and the pitch angle θ_K. The longitudinal axis of the vehicle is assumed to be always tangential to the road at the vehicle cg, which means that high-frequency pitch disturbances are neglected. This has proven realistic for stationary driving states on 'standard,' *i.e.*, smoothly curved and well-kept roads.

The additional terms used in the vertical mapping geometry are collected in the following list:

- k_z camera scaling factor, vertical (pixels/mm)
- H_K elevation of the camera above the tangential plane at cg (m)
- θ_K camera pitch angle relative to vehicle pitch axis (rad)
- z_B vertical image coordinate (pixels)
- L_0 look-ahead distance for planar case (m)
- L_{cv} look-ahead distance with vertical curvature (m)
- H_{cv} elevation change due to vertical curvature (m)
- C_{0v} average vertical curvature of road (1/m)
- C_{1v} average vertical curvature rate of road (1/m^2).

To each scan line at row z_{Bi} in the image, there corresponds a pitch angle relative to the local tangential plane of

$$\theta_{z_{Bi}} = \theta_K + \arctan[z_{Bi} / (f \cdot k_z)]. \tag{9.9}$$

From this angle, the planar look-ahead distance determined by z_{Bi} is obtained as

$$L_{0i} = H_K / \tan(\theta_{z_{Bi}}). \tag{9.10}$$

Analogous to Equation 7.3, the elevation change due to the vertical curvature terms at the distance $L_{cv} + d$ relative to the vehicle cg (see Figure 9.4) is

$$H_{cv} = C_{0v} \cdot (L_{cv} + d)^2 / 2 + C_{1v} \cdot (L_{cv} + d)^3 / 6. \tag{9.11}$$

From Figure 9.4, the following relationship can be read immediately:

$$H_{cv} = H_K - L_{cv} \cdot \tan(\theta_{z_{Bi}}). \tag{9.12}$$

Combining this with Equation 9.11 yields the following third-order polynomial for determining the look-ahead distance L_{cv} with vertical curvature included:

$$a_3 L_{cv}^3 + a_2 L_{cv}^2 + a_1 L_{cv} + a_0 = 0,$$

where

$$a_3 = C_{1v} / 6; \qquad a_1 = d \cdot (C_{0v} + d \cdot C_{1v} / 2) + \tan(\theta_{z_{Bi}}); \tag{9.13}$$

$$a_2 = (C_{0v} + d \cdot C_{1v}) / 2; \qquad a_0 = d^2 \cdot C_{0v} / 2 + d^3 \cdot C_{1v} / 6) - H_K.$$

This equation is solved numerically via Newton iteration with the nominal curvature parameters of the last cycle; taking the solution of the previous iteration or the planar solution according to Equation 9.10 as a starting value, the iteration typically converges in two or three steps, which means only small computational expense. Neglecting the a_3 term in Equation 9.13 or the influence of C_{1v} on the look-ahead range entirely would lead to a second-order equation that is easily solvable analytically. Disregarding the C_{1v} term altogether resulted in errors in the look-ahead range when entering a segment with a change in vertical curvature and led to wrong predictions in road width. The lateral tracking behavior of the feature extraction windows with respect to changes in road width resulting from vertical curvature could be improved considerably by explicitly taking the C_{1v} term into account (see below). (There is, of course, an analytical solution available for a third-order equation; however, the iteration is more efficient computationally since there is little change over time from k to $k + 1$. In addition, this avoids the need for selecting one out of three solutions of the third-order equation).

Beyond a certain combination of look-ahead distance and negative (downward) vertical curvature, it may happen that the road image is self-occluded. Proceeding from near to far, this means that the image row z_{Bi} chosen for evaluation should no longer decrease with range (lie above the previous nearer one) but start increasing again; there is no extractable road boundary element above the tangent line to the cusp of the road (shown by the x_K vector in Figure 9.5). The curvature for the limiting case, in which the ray through z_{Bi} is tangential to the road surface at that distance (and beyond which self-occlusion occurs), can be determined approximately by the second-order polynomial which results from neglecting the C_{1v} influence as mentioned above. In addition, neglecting the $d \cdot C_{0v}$ terms, the approximate solution for L_{cv} becomes

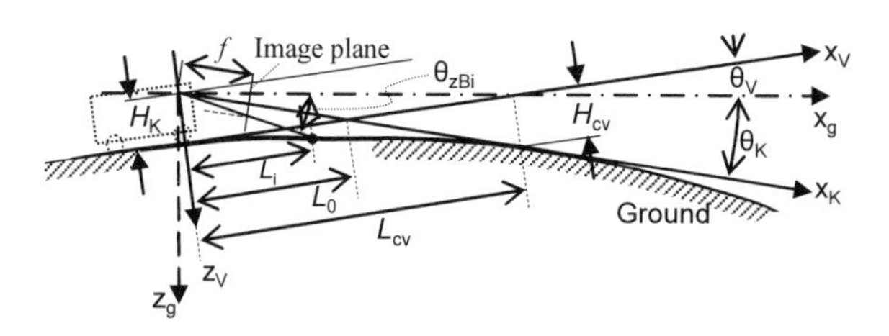

Figure 9.5. Negative vertical curvature analysis including cusp at L_{cv} with self-occlusion; magnitude of L_{cv} is ill defined due to the tangency condition of the mapping ray

$$L_{cv} \approx \frac{\tan(\theta_{z_{Bi}})}{-C_{0v}} \left[1 - \sqrt{1 + 2 \frac{H_K C_{0v}}{\tan^2(\theta_{z_{Bi}})}} \right]. \tag{9.14}$$

The limiting tangent for maximal negative curvature is reached when the radicand becomes zero, yielding

$$C_{0v,\lim}(z_{B_i}) = -\tan^2(\theta_{z_{Bi}})/(2 \cdot H_K). \tag{9.15}$$

Because of the neglected terms, a small “safety margin” ΔC may be added. If the actually estimated vertical curvature C_{0v} is smaller than the limiting case corresponding to Equation 9.15 (including the safety margin of, say, $\Delta C = 0.0005$), no look-ahead distance will be computed, and the corresponding features will be eliminated from the measurement vector.

9.2.3 The Overall 3-D Perception Model for Roads

The dynamic models for vehicle motion (Equation 7.4) and for horizontal curvature perception (Equations 7.36 and 7.37) remain unchanged except that in the latter the look-ahead distance L_{cv} is now determined from Equation 9.13 which includes the effects of the best estimates of vertical curvature parameters.

With $dC_v/dt = dC_v/dl \cdot dl/dt$ and Equation 9.8, the following additional dynamic model for the development of vertical curvature over time is obtained, which is completely separate from the other two:

$$\frac{d}{dt}\begin{pmatrix} C_{0v} \\ C_{1v} \end{pmatrix} = \begin{pmatrix} 0 & V \\ 0 & 0 \end{pmatrix}\begin{pmatrix} C_{0v} \\ C_{1v} \end{pmatrix} + \begin{pmatrix} 0 \\ n_{c1v} \end{pmatrix}. \tag{9.16}$$

There are now four state variables for the vehicle, three for the horizontal, and two for the vertical curvature parameters, in total nine without the road width, which is assumed to be constant here, for simplicity. Allowing arbitrary changes of road width and vertical curvature may lead to observability problems to be discussed later. The state vector for 3-D road observation is

$$x^T_{t3D} = (\lambda, \beta, \psi_{rel}, y_V \mid C_{0hm}, C_{1hm}, C_{1h} \mid C_{0v}, C_{1v}), \tag{9.17}$$

which together with Equations 7.4 and 7.34 (see top of table 9.1 or Equation B.1 in Appendix B) yields the overall dynamic model with a 9 × 9 matrix F

$$F = \left(\begin{array}{cccc|ccc|cc} 0 & 0 & 0 & 0 & 0 & 0 & 0 & 0 & 0 \\ a_{12} & -1/T_\beta & 0 & 0 & 0 & 0 & 0 & 0 & 0 \\ V/a & 0 & 0 & 0 & -V & 0 & 0 & 0 & 0 \\ 0 & V & V & 0 & 0 & 0 & 0 & 0 & 0 \\ \hline 0 & 0 & 0 & 0 & 0 & V & 0 & 0 & 0 \\ 0 & 0 & 0 & 0 & 0 & -3V/L & 3V/L & 0 & 0 \\ 0 & 0 & 0 & 0 & 0 & 0 & 0 & 0 & 0 \\ \hline 0 & 0 & 0 & 0 & 0 & 0 & 0 & 0 & V \\ 0 & 0 & 0 & 0 & 0 & 0 & 0 & 0 & 0 \end{array}\right), \tag{9.18 (a)}$$

the input vector g, and the noise vector n(t);

$$\begin{aligned} g^T &= (k_\lambda, 0, 0, 0, \mid 0, 0, 0, \mid 0, 0) \\ n^T(t) &= [0, 0, 0, 0, \mid 0, 0, n_{c1h}(t), \mid 0, n_{c1v}(t)]. \end{aligned} \tag{b}$$

This analogue dynamic model,

$$\dot{x}(t) = F \cdot x(t) + g \cdot u(t) + n(t), \tag{9.18}$$

has to be transformed into a discrete model with the proper sampling period T according to the video standard used (see Equation B.4). All coefficients of the A matrix given there remain the same; dropping road width b again, two rows and columns have to be added now with zero entries in the first seven places of rows and columns, since vertical curvature does not affect the other state components. The 2×2 matrix in the lower right corner has a '1' on the main diagonal, a '0' in the lower left corner, and the coefficient a_{89} is just $a_{89} = VT$.

9.2.4 Experimental Results

The spatiotemporal perception process based on two superimposed differential geometry models for 3-D roads has been tested in two steps: First, in a simulation loop where the correct results are precisely known, and second, on real roads with the test vehicle VaMoRs. These tests were so successful that vertical curvature estimation in the meantime has become a standard component for all road vehicles. Especially for longer look-ahead ranges, it has proven very beneficial with respect to robustness of perception.

9.2.4.1 Simulation Results for 3-D Roads

Figures 9.6 and 9.7 show results from a hardware-in-the-loop simulation with video-projected computer-generated imagery interpreted with the advanced first-generation real-time vision system BVV2 of UniBwM [Graefe 1984]. This setup has the advantage over field tests that the solution is known to high accuracy beforehand. Figure 9.6 is a perspective display of the tested road segment with both horizontal and vertical curvature. Figure 9.7 shows the corresponding curvatures recovered by the estimation process described (*solid*) as compared with those used for image generation (*dashed*).

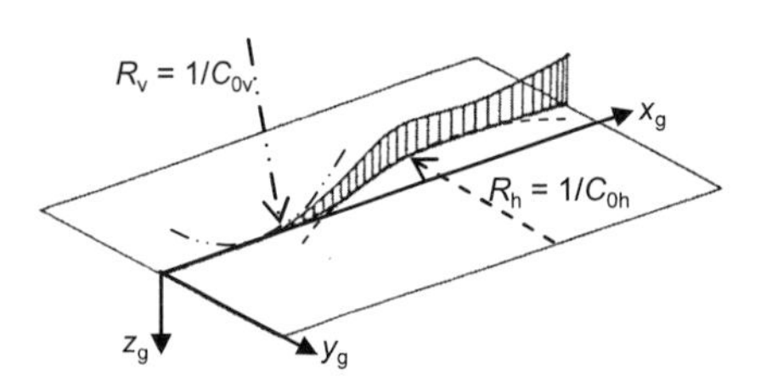

Figure 9.6. Simulated spatial road segment with 3-D horizontal and vertical curvature

Figure 9.7 (top) displays the good correspondence between the horizontal curvature components (C_{0hm} , as input: *dashed*, and as recovered: *solid line*); the dashed polygon for simulation contains four clothoid elements and two circular arcs with a radius of 200 m ($C_{0h} = \pm\ 1/200 = \pm\ 0.005$). Even though the C_{1hm} curve is relatively smooth and differs strongly from the series of step functions as derivatives of the dashed polygon (not shown), C_{0h} and C_{0hm} as integrals are close together.

Under cooperative conditions in the simulation loop, vertical radii of curvature of ~ 1000 m have been recognized reliably with a look-ahead range of ~ 20 m. The relatively strong deviation at 360 m in Figure 9.7 bottom is due to a pole close to the road (with very high contrast), which has been mistaken as part of the road boundary. The system recovered from this misinterpretation all on its own when the

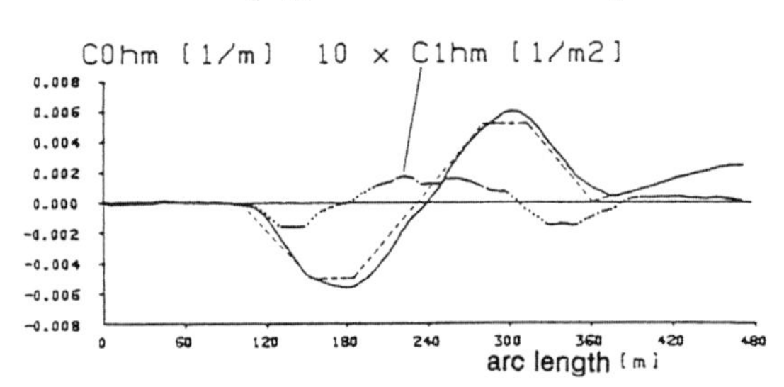

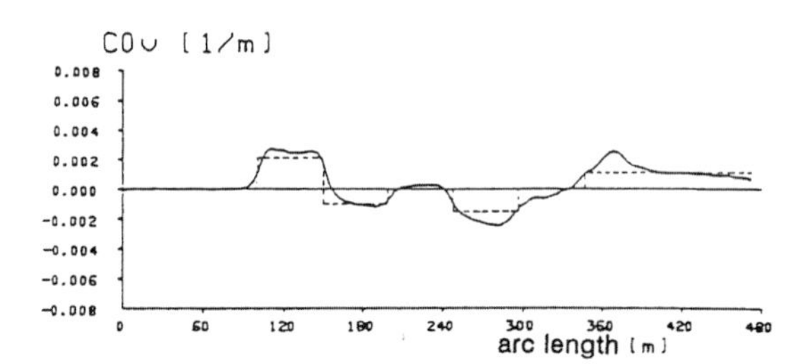

Figure 9.7. Simulation results comparing input model (*dashed*) with curvatures recovered from real-time vision (*solid lines*). Top: horizontal curvature parameters; bottom: Vertical curvature.

pole was approached; the local fit with high vertical curvature became increasingly contradictory to new measurement data of road boundary candidates in the far look-ahead range. The parameter C_{0v} then converged back to the value known to be correct. Since this approach is often questioned as to whether it yields good results under stronger perturbations and noise conditions, a few remarks on this point seem in order. It is readily understood that the interpretation is most reliable when it is concerned with regions close to the vehicle for several reasons:

1. The resolution in the image is very high; therefore, there are many pixels per unit area in the real world from which to extract information; this allows achieving relatively high estimation accuracies for lane (road) width and lateral position of the camera on the road.
2. The elevation above the road surface is well known, and the vehicle is assumed to remain in contact with the road surface due to Earth gravity; because of surface roughness or acceleration/deceleration, there may be a pitching motion, whose influence on feature position in the image is, again, smallest nearby. Therefore, predictions through the dynamic model are trusted most in those regions of the image corresponding to a region spatially close to the camera in the real world; measured features at positions outside the estimated 3σ range from the predicted value are discarded (σ is the standard deviation determinable from the covariance matrix, which in turn is a by-product of recursive estimation).
3. Features actually close to the vehicle have been observed over some period of time while the vehicle moved through its look-ahead range; this range has been increased with experience and time up to 40 to 70 m. For a speed of 30 m/s (108 km/h), this corresponds to about 2 s or 50 frames traveling time (at 40 ms interpretation cycle time). If there are some problems with data interpretation in the far range, the vehicle will have slowed down, yielding more time (number of frames) for analysis when the trouble area is approached.
4. The gestalt idea of a low curvature road under perspective projection, and the ego- motion (under normal driving conditions, no skidding) in combination with the dynamic model for the vehicle including control input yield strong expectations that allow selection of those feature combinations that best fit the generic road (lane) model, even if their correlation value from oriented edge templates is only locally but not globally maximal in the confined search space. In situations like that shown in Figure 8.2, this is more the rule than the exception.

In the general case of varying road width, an essential gestalt parameter is left open and has to be determined in addition to the other ones from the same measurements; in this case, the discriminative power of the method is much reduced. It is easy to imagine that any image from road boundaries of a hilly road can also be generated by a flat road of varying width (at least in theory and for one snapshot). Taking temporal invariance of road shape into account and making reasonable assumptions about road width variations, this problem also is resolvable, usually, at least for the region nearby, when it has been under observation for some time (*i.e.*, due to further extended look-ahead ranges). Due to limitations in image resolution at a far look-ahead distance and in computing power available, this problem had not been tackled initially; it will be discussed in connection with pitch perturbations in Section 9.3.

Note that the only low-level image operations used are correlations with local edge templates of various orientations (covering the full circle at discrete values. *e.g.*, every 11°). Therefore, there is no problem of prespecifying other feature operators. Those to be applied are selected by the higher system levels depending on the context. To exploit continuity conditions of real-world roads, sequences of feature candidates to be measured in the image are defined from near to far (bottom up in the image plane), taking conditions for adjacency and neighboring orientation into account.

9.2.4.2 Real-world Experiments

Figure 9.8 shows a narrow, sealed rural road with a cusp in a light curve to the left followed by an extended positively curved section that has been interpreted while driving on it with VaMoRs (bottom part). For vertical curvature estimation, road width is assumed to be constant. Ill-defined or irregular road boundaries as well as vehicle oscillations in pitch affect the estimation quality correspondingly.

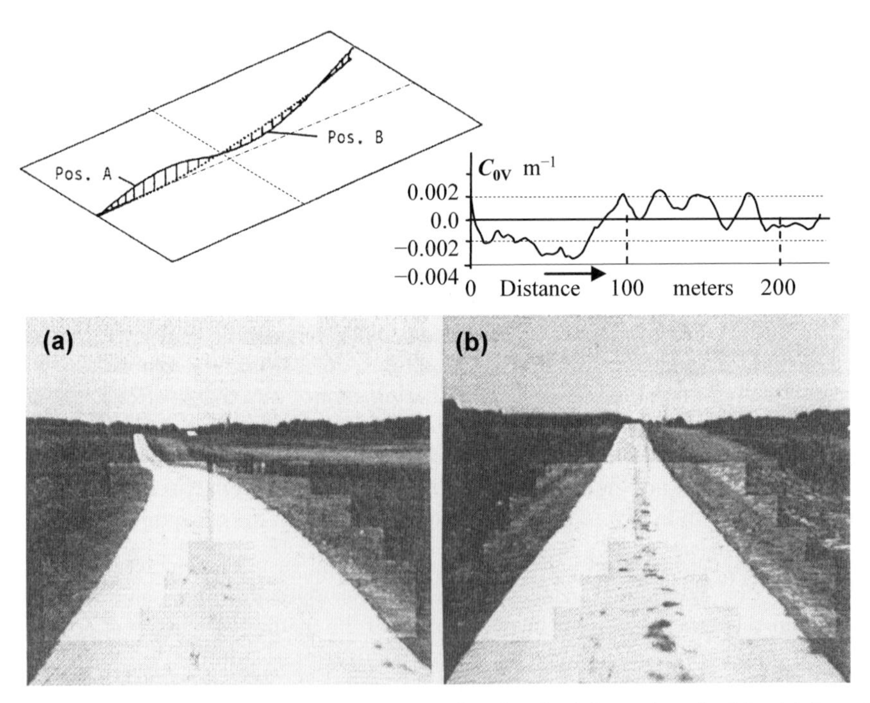

Figure 9.8. Differential geometry parameter estimation for 3-D rural road while driving on it with VaMoRs in the late 1980s: Top left: Superimposed horizontal and vertical curvature derived from recursive estimation. Top right: Estimated vertical curvature C_{0v} (1/m) over run length in meters. Bottom (a): View from position A marked in top left-hand subfigure; (b) view from position B (bottom of the dip, after [Mysliwetz 1990]).

These effects are considered the main causes for the fluctuations in the estimates of the vertical curvature in the top right part. To improve these results in the framework of the 4-D approach geared to dynamic models of physical objects for the representation of knowledge about the world it is felt that the pitching motion of the vehicle has to be taken into account. There are several ways of doing this:

1. The viewing direction of the camera may be stabilized by inertial angular rate feedback. This well-known method has the advantage of reducing motion blur. There are, however, drift problems if there is no position feedback. Therefore, the feedback of easily discriminated visual features yields nice complementary signals for object fixation.
2. The motion in pitch of the egovehicle is internally represented by another dynamic model of second order around the pitch axis. Tracking horizontal features far away (like the horizon) vertically allows estimating pitch rate and angular position of the vehicle recursively by prediction-error feedback. Again, knowledge about the dynamics in the pitch degree of freedom of the massive inertial body is exploited for measurement interpretation. Picking features near the longitudinal axis of the body at large ranges, so that the heave component (in the z-direction) is hardly affected, decouples this motion component from other ones.
3. Purely visual fixation (image registration from frame to frame) may be implemented. This approach has been developed by Franke (1992).

The first two have been investigated by members of our group; they will be discussed in the next section. The third one has been studied elsewhere, *e.g.*, [Bergen 1990, Pele, Rom 1990].

Tests to recognize vertical curvatures of unsealed roads with jagged boundaries of grass spreading onto the road failed with only intensity edges as features and with small look-ahead ranges. This was one of the reasons to proceed toward multi focal camera sets and area-based features.

Conclusion: The 4-D approach to real-time 3-D visual scene understanding allows spatial interpretation of both horizontally and vertically curved roads while driving. By exploiting recursive estimation techniques that have been well developed in the engineering sciences, this can be achieved at a high evaluation rate of 25 Hz with a rather small set of conventional microprocessors. If road width is completely unconstrained, ill-conditioned situations may occur. In the standard case of parallel road boundaries, even low curvatures may be recovered reliably with modest look-ahead ranges.

Adding temporal continuity to the spatial invariance of object shapes allows reducing image processing requirements by orders of magnitude. Taking physical objects as units for representing knowledge about the world results in a spatiotemporal internal representation of situations in which the object state is continuously servoadapted according to the visual input, taking perspective projection and motion constraints into account for the changing aspect conditions. The object road is recognized and tracked reliably by exploiting the gestalt idea of feature grouping. Critical tests have to be performed to avoid "seeing what you want to see." This problem is far from being solved; much more computing power is needed to handle more complex situations with several objects in the scene that introduce fast changing occlusions over time.

9.3 Perturbations in Pitch and Changing Lane Widths

As mentioned in previous sections several times, knowledge of the actual camera pitch angle $\theta_K(t)$ can improve the robustness of state estimation. The question in image interpretation is: What are well-suited cues for pitch angle estimation? For the special case of a completely flat road in a wide plane, a simple cue is the vertical position of the horizon in the image z_{hor} (in pixel units; see Sections 7.3.4 and 9.2.2). Knowing the vertical calibration factor k_z and the focal length f, one can calculate the camera pitch angle θ_K according to Equation 7.18. The pitch angle θ_K is defined positive upward (Figure 9.4).

In most real situations, however, the horizon is rarely visible. Frequently, there are obstacles occluding the horizon like forests, buildings, or mountains, and often the road is inclined, making it impossible to identify the horizon line. The only cue in the image that is always visible while driving is the road itself. Edges of dashed lane markings are good indicators for a pitch movement, if the speed of the vehicle is known. But unfortunately, a road sometimes has sections where both lane markers are solid lines. In this case, only then the mapped lane width b_i at a certain vertical scan line z is a measure for the pitch angle θ_K, if the lane width b is known.

9.3.1 Mapping of Lane Width and Pitch Angle

In Figure 9.4, the relevant geometric variables for mapping a road into a camera with pitch angle θ_K are illustrated. Since the magnitude of the pitch angle usually is less than 15°, the following approximations can be made:

$$\sin\theta_K \cong \theta_K; \qquad \cos\theta_K \cong 1. \tag{9.19}$$

As discussed in Section 9.2.2, the camera pitch angle θ_K determines the look-ahead distance L_i on a flat road, when a fixed vertical scan line z_i is considered; with Equation 9.19, the approximate relation is:

$$L_i \cong \frac{k_z f + z_i\theta_K}{z_i - k_z f\theta_K} \cdot H_K . \tag{9.20}$$

This yields the following mapping of lane width b_{lane} (assumed constant) into the image lane width b_i

$$b_i \cong \frac{k_y f b_{lane}}{H_K}\left(\frac{z_i}{k_z f} - \theta_K\right). \tag{9.21}$$

If only measurements in one image row per frame are taken, the influence of variations in pitch angle and of changing road width cannot be distinguished by measuring the mapped lane width. But, if more than one measurement b_i is obtained from measuring in different image rows z_i, the effects can be separated already in a single static image, as can be seen from Figure 9.9 (left).

The relations are valid only for a flat road without any vertical curvature. Both pitch and elevation (heave) do change the road image nearby; the elevation effect vanishes with look-ahead range, while the pitch effect is independent of range. In the presence of vertical road curvature, the look-ahead distance L_i has to be modified according to the clothoid model, as discussed in Section 9.2. Figure 9.3 had shown that curvature effects leave the image of the road nearby unchanged; deviations show up only at larger look-ahead distances.

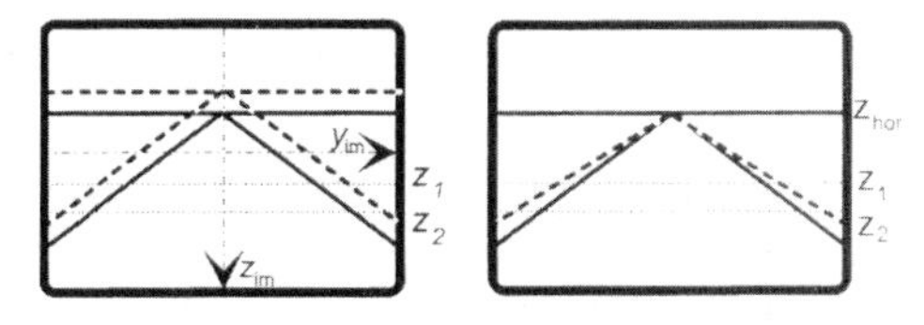

Figure 9.9. Effects of camera pitch angle (left) and elevation above ground (right) on the image of a straight road in a plane

Usually, the camera is not mounted at the center of gravity of the vehicle (cg), but at some point shifted from the cg by a vector $\underline{\Delta x}$; in vehicles with both cg and camera laterally in the vertical center plane of the vehicle, only the longitudinal shift d and elevation Δh are nonzero. The axis of vehicle pitch movements goes through the cg. Therefore, a body pitch angle yields not only the same shift for the camera pitch angle, but also a new height (elevation) above the ground H_K. This vertical shift is only a few centimeters and will be neglected here.

From Figure 9.9, it can be seen that the vertical coordinate z_{hor} of the vanishing point depends only on the camera pitch angle θ_K (Figure 9.9, left), and not on camera elevation or lane width (right). Accordingly, the pitch angle θ_K can be computed directly from the image row with $z_{Bi} = 0$ (optical axis) and the location of the vanishing point. Once the pitch angle is known, the width of the lane b_{lane} can be computed by Equation 9.21. This approach can be applied to a single image, which makes it suitable for initialization (see Chapter 8).

In curves, the road appears to widen when looking at a constant image line (z_{Bi}-coordinate). This effect is reduced when the camera pan angle ψ_K is turned in the direction of the curve. The widening effect without camera pan adjustment depends on the heading (azimuth) angle $\chi(l)$ of the road. The real lane width is smaller than the measured one in a single image line approximately by a factor $\cos(\chi)$. Assuming that the road follows a clothoid model, the heading angle χ of the road as the first integral of the curvature function is given by Equation 7.2. The effect of χ is reduced by the camera pan angle ψ_K (see Figure 9.10). Thus, the entire set of measurement equations for lane width estimation is (for each window pair and row z_i)

$$y_i = b \cdot k_y f / [L_i \cdot \cos(\chi - \psi_K)] \,. \tag{9.22}$$

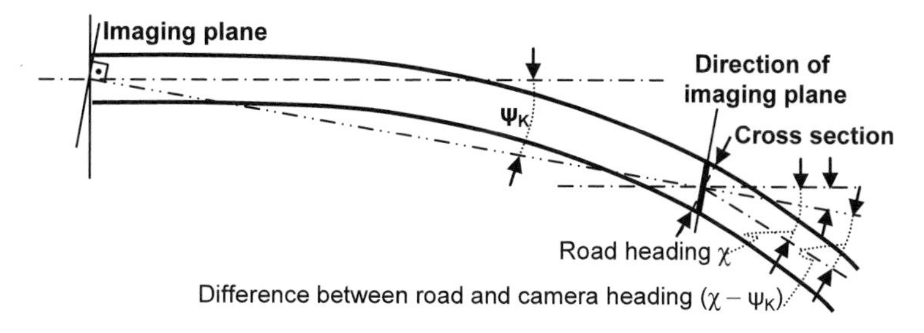

Figure 9.10. Mapping with active control gaze of road width for a curved road

9.3.2 Ambiguity of Road Width in 3-D Interpretation

Problems in interpreting an image occur when the image shows a lane or road with width changing over a section of run length. Figure 9.11 illustrates this problem for an ideally simple case with two parameter sets for the road in front of or behind the (unrealistically harsh) transition at line L_{ch}.

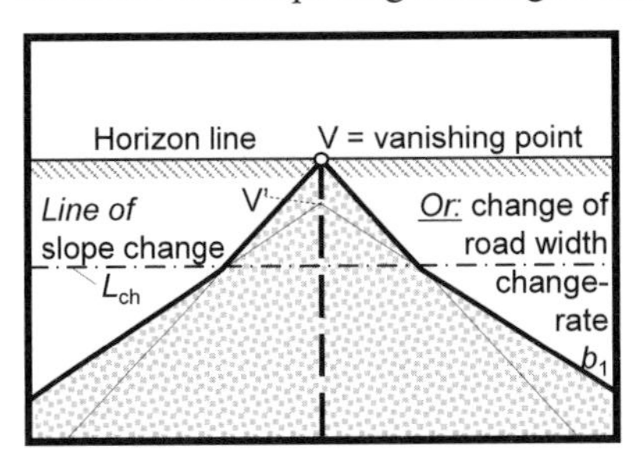

Figure 9.11. Ambiguous road image if slope and road width are open parameters; a change in slope at line L_{ch} has the same effect as a change in the parameter linear change rate b_l of road width (in each single image)

This isolated (single) image is completely ambiguous: First it can be interpreted as a road of constant width with a downward slope in the near range (going downhill, since V' lies below the vanishing point V on the horizon) switching to a flat horizontal road at line L_{ch}; second, it fits an interpretation as a road in the horizontal plane, of which the near part has linearly decreasing width with change rate b_1 according to the model ($b_1 < 0$)

$$b = b_0 + b_1 \cdot l \,. \tag{9.23}$$

Of course, an infinite number of interpretations with different combinations of values for slope γ and change rate of width (b_1) is possible.

These ambiguities cannot be resolved from one single image. However, the interpretation of sequences of images allows separating these effects. This can most easily be achieved if dynamic models are used for the estimation of these mapping effects. Consider hypothesis A in Figure 9.12 top right, solid lines (two inclined planes meeting at distance L_{ch}). If the road width is assumed constant as b_∞, for a given elevation H_K of the camera above the ground, the road width in the near range is given by the vanishing point V' and the slopes of the left and right boundary lines.

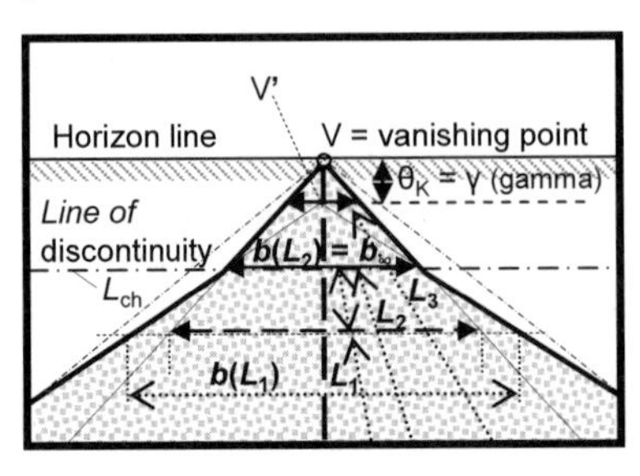

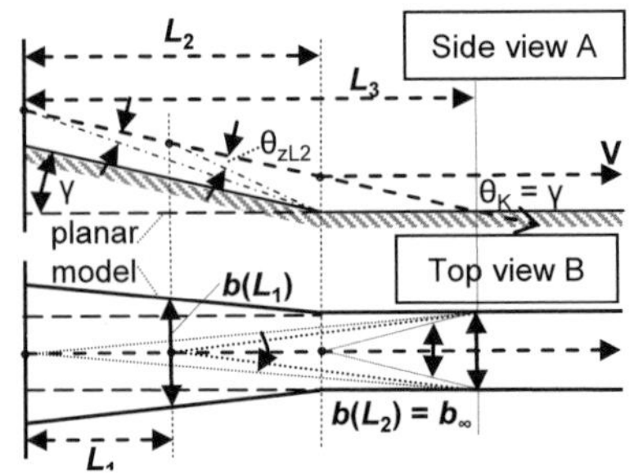

Figure 9.12. Two specific interpretations of the image in Figure 9.11: Hypothesis A: Constant lane width b_∞ and jump in slope angle γ at $L_2 = L_{ch}$ (top right). Hypothesis B: Flat ground and linearly decreasing road width for $l < L_2$: Parameter $b_1 = (b(L_2) - b(L_1))/(L_2 - L_1)$ [*dashed road boundary* bottom right]

However, the elevation of the camera above the plane of the far road depends linearly on the distance to the line of discontinuity according to the slope angle between the two planes; when line L_{ch} is approached, the borderlines of the far road have to widen toward the dash-dotted lines indicated in Figure 9.12, left. Right on line L_{ch}, the vehicle will experience a step input in pitch angle (smoothed by the axle distance), shifting the vanishing point from V' to V. Thus, the distance between V' and V corresponds to the slope angle γ of the near plane if the far plane is horizontal. So the conclusion has to be the following.

The discontinuity in slope angles of the borderlines in the image of a road of constant width can stem only from an inclined near plane if the borderlines in the near range stay constant while the angle between the borderlines in the far range increases during the approach to L_{ch}. These lines move toward those formed by the two rays from the vanishing point V through the points of intersection of the borderlines in the near range with the image boundaries. This optical flow component thus tells us something about the shape of the vertical terrain. In the 4-D approach, this shows up in the elements of the Jacobian matrix and allows hypothesis testing while driving.

Hypothesis B, assuming one single plane for both road sections, yields constant borderlines in the far range (beyond L_{ch}) and shrinking lane widths at a constant look-ahead range L_i while driving (lower right, *solid lines* in Figure 9.12). Here, the (lateral) optical flow of the borderlines occurs in the near range (shrinking toward the prolongation of the far-range boundary lines; see look-ahead distance L_1); again, this hypothesis is reflected in the values of the Jacobian matrix elements for given look-ahead ranges. These partial derivatives are the basis for monocular motion stereo in the 4-D approach.

The feedback of prediction errors in feature position from both the near and the far range thus allows adapting the hypotheses simultaneously. Since the effects are rather small and in real situations the geometry is much more involved with unknown radii of curvature (both horizontal and vertical in the worst case), stable convergence can hardly be expected for all combinations. Luckily, with change rate constraints for curvature on standard high-speed roads, sufficiently correct shape recognition can be expected in most of these standard cases, even with pitch angle perturbations. Pitch perturbation shows up as a vertical shift of the constant road shape in the image.

9.3.3 Dynamics of Pitch Movements: Damped Oscillations

The suspension system of a vehicle in pitch can be approximated by the model of a one-dimensional damped harmonic oscillator, which is excited by perturbations (rough road or accelerations, decelerations). The differential equation for an oscillating pitch angle θ_v of the vehicle is

$$\ddot{\theta}_v + 1/\tau \cdot \dot{\theta}_v + \omega_0^2 \cdot \theta_v = n(t), \tag{9.24}$$

with τ = relaxation time, ω_0 = characteristic frequency, and $n(t)$ = perturbation (noise term). The parameters τ and ω_0 can be derived from vehicle mass, its inertial momentum around the y-axis, the axle distance, and suspension characteristics. The solution of the homogeneous part of the differential equation is:

$$\theta_v(t) = \theta_{v0} \cdot \exp(-t/(2\tau)) \cdot \sin(\omega t),$$
$$\text{with} \quad \omega = \sqrt{\omega_0^2 - 1/(4\tau^2)} \qquad (\omega_0 \tau^2 > ½). \tag{9.25}$$

9.3.3.1 Offset-free Motion in Pitch

Equation 9.24 is transformed into the standard form of a linear, second-order system by introducing the additional state component $q_\theta = \dot{\theta}_v$. The transition to time-discrete form with cycle time T yields a set of two difference equations as motion constraints for the state vector $x_{\theta_v}^T = [\theta_v, q_\theta]$. This dynamic model represents knowledge about the pitching characteristics of the vehicle; in discrete form, it can be written as

$$x_k = \Phi_{\theta,k-1}(T) \cdot x_{k-1} + b_{k-1} \cdot u_{k-1} + g_{k-1} \cdot n_{k-1}. \tag{9.26}$$

This model describes the unperturbed vehicle pitch angle θ_v as an oscillation around zero. The control input u may come from acceleration or deceleration and acts on the pitching motion through the gain vector b, while noise (*e.g.*, from rough ground) acts through the coefficient vector g; both may depend on speed V actually driven.

The transition matrix Φ_θ for the discrete second-order model in pitch with sampling time T can be obtained by standard methods from Equation 9.24 as

$$\begin{aligned}
\phi_{\theta,11}(T) &= \exp(-\frac{T}{2\tau}) \cdot (\cos \omega T + \frac{1}{2\omega\tau} \sin \omega T),\\
\phi_{\theta,12}(T) &= \frac{1}{\omega} \exp(-\frac{T}{2\tau}) \cdot \sin \omega T,\\
\phi_{\theta,21}(T) &= -\omega_0^2 \cdot \phi_{\theta,12},\\
\phi_{\theta,22}(T) &= \exp(-\frac{T}{2\tau}) \cdot (\cos \omega T - \frac{1}{2\omega\tau} \sin \omega T).
\end{aligned} \tag{9.27}$$

Control input effects are not considered here; usually, an acceleration yields some pitch up response while downward pitching is typical of deceleration. The latter may be strong and amount to several degrees, depending on the softness of the suspension system. With vanishing control input, the pitch angle goes back to zero.

9.3.3.2 Constant or Slowly Varying Pitch Offset

For cameras mounted directly on the vehicle body, the camera pitch angle θ_K is the same as the vehicle pitch angle θ_v described above, if the optical axis of the camera is aligned with the body axis. But mostly, the camera is mounted with a fixed pitch angle offset θ_{offs} toward the ground. Therefore, the dynamic vehicle pitch angle θ_v can be written as

$$\theta_v = \theta_K - \theta_{offs}. \tag{9.28}$$

Usually, this offset is constant and can be measured by calibration procedures in a well-defined environment. But it may happen that due to mechanical perturbations the offset is shifted during operation. In this case, an incorrectly assumed offset value yields erroneous results for the estimated state variables q_θ and θ_v. To achieve self-correction, the pitch offset from the horizontal reference is introduced as a new state variable and is also estimated in each cycle. The complete state vector for the pitch angle system then can be written as

$$\mathrm{x}_\theta^T = [\theta_K, q_\theta, \theta_{offs}]. \tag{9.29}$$

If the vehicle pitch angle is not known, this formulation allows an even more versatile interpretation. Due to fuel consumption, the pitch angle of the vehicle may change slowly by a small amount. Similarly, shifting loads from front to rear or *vice versa* in the vehicle may also change the static equilibrium pitch angle. To separate the slowly varying (long-term) pitch offset from the dynamic pitching motion (usually in the range of about 1 Hz eigenfrequency), it is necessary to choose the process noise $Q_{\theta\mathrm{offs}}$ of this variable small compared to the process noise of θ_K. This method of estimating the camera pitch offset provides a further step toward a fully autonomous, self-gauging system. The system transition matrix Φ_θ now has five more elements referring to the additional state variable θ_{offs}:

$$\begin{aligned} \phi_{\theta,13}(T) &= 1 - \phi_{\theta,11} \quad ; \quad \phi_{\theta,23}(T) = -\phi_{\theta,21} \\ \phi_{\theta,31}(T) &= \phi_{\theta,32}(T) = 0 \quad ; \quad \phi_{\theta,33}(T) = 1. \end{aligned} \tag{9.30}$$

The full discrete model for motion in pitch with Equation 9.27 then has the form

$$\begin{pmatrix} \theta_K \\ q_\theta \\ \theta_{offs} \end{pmatrix}_{k+1} = \begin{pmatrix} \phi_{\theta,11} & \phi_{\theta,12} & 1-\phi_{\theta,11} \\ \phi_{\theta,21} & \phi_{\theta,22} & -\phi_{\theta,21} \\ 0 & 0 & 1 \end{pmatrix}_k \cdot \begin{pmatrix} \theta_K \\ q_\theta \\ \theta_{offs} \end{pmatrix}_k + \begin{pmatrix} g_1 \\ g_2 \\ 0 \end{pmatrix}_k a_x + \begin{pmatrix} 0 \\ n_\theta \\ n_{offs} \end{pmatrix}_k . \tag{9.31}$$

The longitudinal acceleration/deceleration a_x acts as excitation mainly on q_θ, but in the sampled data structure, due to integration over cycle time T, there is also a small effect on the pitch angle itself. The systems dynamics approach quite naturally yields a better approximation here than quasi-steady AI approaches. The offset angle is driven by an appropriately chosen discrete noise term $n_{\mathrm{offs}}(k)$.

9.3.4 Dynamic Model for Changes in Lane Width

Usually, a road consists of long sections with constant lane width according to standard road design rules [RAS-L-1 1984]. Between these sections, there is a short section where the lane width changes from the initial to the final value. In this transition section, the dynamics of lane width can be modeled under the assumption of piecewise constant linear change along the run length l according to Equation 9.23, where b_0 denotes the lane width at the point of observation (current position of the vehicle) and b_1 the lane width change rate ($\mathrm{d}b/\mathrm{d}l$).

A problem arises at the point of transition between sections of different values of b_1, where a jump in b_1 occurs. To obtain a robust model, the unsteady transitions between segments of b_1 should be approximated by steady functions. Moreover, the precise location where the lane width begins to change cannot be determined exactly from a single image. A possible way of approximation along the path similar to that for road curvature may be chosen (Section 7.4). For each moment, the real road curvature is approximated by a clothoid that yields the same lateral offset relative to the current vehicle position.

This dynamic approach, when applied to lane width, leads to a straight line approximation for the lane markings at each moment. In sections of changing lane width, the lane markings are approximated by straight lines on each side of the lane that produce the same widening or tapering effect as the real lines. Furthermore, it must be assumed that within the local look-ahead distance L of the camera, only one change in lane width change rate b_1 occurs. With b_e as the lane width at the end of the look-ahead distance L, this assumption leads to

$$b_e = b_{0m} + L \cdot b_{1m}, \tag{9.32}$$

where b_{om} is the approximate lane width at the current vehicle location and b_{1m} = db_{om}/dl the approximate linear change rate of the lane width within the look-ahead distance. Also, b_{om} can be expressed as

$$b_{0m} = b_0 + \int_0^{l_i} b_{1m} dl, \tag{9.33}$$

l_i denotes the distance covered after the b_1-change began to appear within L; b_0 is the lane width just before that moment. Thus, the lane width at the end of the look-ahead distance can be expressed as

$$b_e = b_0 + \int_0^{l_i} b_{1m} dl + b_{1m} L. \tag{9.34}$$

On the other hand, the lane width at the end of L is given by the real lines for lane marking. Admitting one change of b_0 within L results in

$$b_e = b_0 + b_{11} L + b_{12} l_i, \tag{9.35}$$

where b_{11}, b_{12} are the change rates in lane width before and after the step jump, respectively. Combining Equations 9.34 and 9.35 yields

$$b_{11} L + b_{12} l_i = \int_0^{l_i} b_{1m} dl + b_{1m} L. \tag{9.36}$$

Differentiating by dl_i and replacing dl_i by $V \cdot dt$ leads to

$$\frac{db_{1m}}{dt} = b_2 \frac{V}{L} - b_{1m} \frac{V}{L}. \tag{9.37}$$

As stipulated above, $b_{1m} = db_{0m}/dt$. By replacing dl by $V \cdot dt$ this yields

$$db_{0m}/dt = v \cdot b_{1m}. \tag{9.38}$$

Thus, the complete system of differential equations is

$$\begin{aligned} \dot{b}_{0m} &= V b_{1m} \\ \dot{b}_{1m} &= b_{12} \frac{V}{L} - b_{1m} \frac{V}{L}. \end{aligned} \tag{9.39}$$

The complete state vector using averaged values b_{0m}, b_{1m} can be written as

$$x_b^T = [b_{0m}, b_{1m}, b_{12}]. \tag{9.40}$$

The state variable b_{12} is not deterministic, as the change of lane width along the path is not predictable without prior knowledge of the overall road course. For the purpose of Kalman filtering, $\dot{b}_{12}$ is assumed to have Gaussian noise characteristics

$$\dot{b}_{12} = n_b(t) . \tag{9.41}$$

Equations 9.39 and 9.41 constitute the third-order analogue model for observing lane width changes over time while driving at speed V. For the discrete formulation, the transition matrix $\Phi_B(T)$ has to be determined; it is the solution of the system of homogenous differential equations for time step T. The result is as follows:

$$\begin{aligned}
&\Phi_{11}(T) = \Phi_{33}(T) = 1; \qquad \Phi_{21}(T) = \Phi_{31}(T) = \Phi_{32}(T) = 0;\\
&\Phi_{12}(T) = L\left(1 - \exp(-\frac{VT}{L})\right); \qquad \Phi_{22}(T) = \exp(-\frac{VT}{L});\\
&\Phi_{13}(T) = L\left(\exp(-\frac{VT}{L}) + \frac{VT}{L} - 1\right); \; \Phi_{23}(T) = 1 - \exp(-\frac{VT}{L}).
\end{aligned} \tag{9.42}$$

The discrete dynamic model for lane or road width observation, which is completely separate from the other variables, finally is

$$x_{b,k+1} = \begin{pmatrix} b_{0m} \\ b_{1m} \\ b_{12} \end{pmatrix}_{k+1} = \begin{pmatrix} 1 & \Phi_{12} & \Phi_{13} \\ 0 & \Phi_{22} & \Phi_{23} \\ 0 & 0 & 1 \end{pmatrix}_k \cdot \begin{pmatrix} b_{0m} \\ b_{1m} \\ b_{12} \end{pmatrix}_k + \begin{pmatrix} 0 \\ 0 \\ n_B \end{pmatrix}_k . \tag{9.43}$$

9.3.5 Measurement Model Including Pitch Angle and Width Changes

The dependence of look-ahead range L on pitch angle θ_K (Equation 9.20) and the dependence of both curvature and lane (or road) width on look-ahead range L has to be captured in the measurement model if these variables have to be estimated. Therefore, Equations 9.22 and 9.23 have to be combined with Equation 9.20, finally yielding the mapping equation

$$y_i = \frac{k_y f}{\cos(\chi - \psi_K)} \cdot \left[\frac{b_0}{H}\left(\frac{z_i}{k_z f} - \theta_K\right) + b_1\left((\frac{z_i \theta_K}{k_z f} + 1) \right) \right] . \tag{9.44}$$

The elements of rows of the Jacobian matrix C corresponding to this measurement equation for several image rows z_i are obtained as partial derivatives with respect to all state variables or parameters to be iterated. With the computing power available today, these terms are most easily obtained by numerical differencing of results with slightly perturbed variables and division by the perturbation applied. To obtain the width of the lane (or road) in the row direction from the video image, the horizontal positions y_i of the right and left lane markings (boundary lines) in each row are subtracted from each other.

9.4 Experimental Results

Validation of the approach has been achieved in two steps: First, the quality of estimation was checked by simulations. Second, experimental results with real test vehicles in the real world have been performed and some are discussed here.

9.4.1 Simulations with Ground Truth Available

In simulations, ground truth is known from simulation input; by varying noise parameters, some experience can be gained on convergence behavior and favorable parameter selection.

9.4.1.1 Validation of Pitch Estimation

Estimation of the pitch angle state vector has been validated by a simple test with simulated measurements based on damped oscillations:

Damped negative sine wave, initial amplitude of $\theta_{K0} = 1°$;
undamped natural frequency of 1 Hz, that is, $\omega_0 = 2\pi = 6.28$ rad/s;
relaxation time $\tau = 1$ s (damping coefficient of $\sigma = 1/(2\tau) = 0.5\ \text{s}^{-1}$);
pitch offset $\theta_{offs} = -5°$. (9.45)

Random noise has been added to the simulated measurement values. The initial state vector for the estimation process has been chosen as zero in all components. The starting value for lane width was the same as the simulated one: 3.75 m and constant. In Figure 9.13, the estimated pitch angle (*crosses*) is compared to the simulated one (*solid line*). In addition, at each time frame, the vertical position of the vanishing point ('horizon') has been computed from the intersection of the borderlines; for comparison, the pitch angle was extracted according to Equation 7.39. From the faint dotted graph in Figure 9.13, it can be seen clearly that this pitch angle is much more noise-corrupted than the pitch angle estimated by EKF.

The variance vector Q of the process has been set such that the estimated actual pitch angle θ_K follows the real pitch movement quickly. On the other hand, pitch

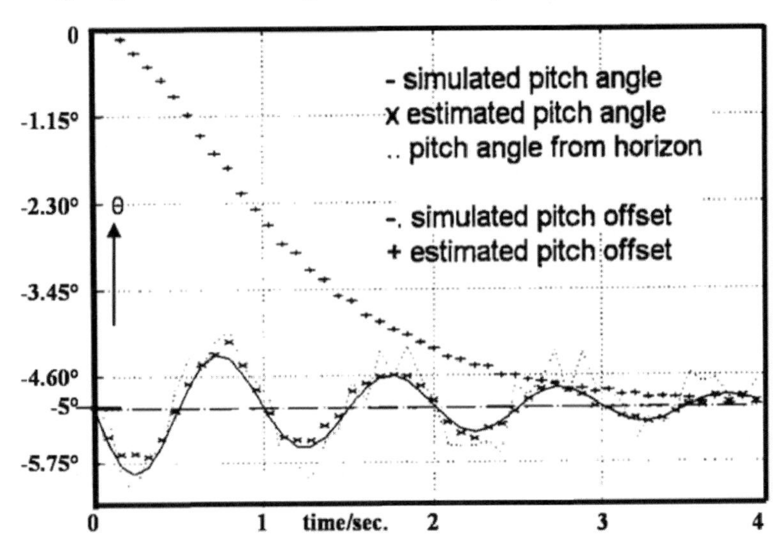

Figure 9.13. Estimation of pitch angle in simulation: Damped pitch oscillation (1 Hz, 1°) and offset (−5°). Convergence from $\theta_{offs} = 0$ occurred in ~ 4 s.

offset estimation from the wrong initial value converges much more slowly toward the real offset. This is based on the assumption that the offset, usually, is a slowly variable quantity or a constant after the camera has been mounted in a fixed pose (angle and position). However, a mechanical shock may also have perturbed the intended viewing direction. Therefore, a compromise has to be found between a sufficiently quick reaction of the filter and robustness to measurement noise. Of course, the filter can be tuned such that sudden jumps in θ_{offs} can also be estimated, but then there will be a conflict with respect to which part of the measured discrepancy should be assigned to which variable. Separating low- and high-frequency components is the most stable way to go: stepwise changes in pitch offset angle (when somebody inadvertently hits the camera mount) will be discovered in due time since the new orientation remains fixed again and will show up according to the time constant chosen for the low-frequency part of the estimation process.

Figure 9.14 shows the pitch rate of the test run for motion estimation from a simulated damped oscillation with superimposed noise. The difference between the incorrectly assumed pitch offset (0°) and the real one (−5°, see Figure 9.13) together with the increase in negative magnitude from prediction over one cycle time of 80 ms in the example (negative sine wave) first gives rise to positive pitch rates (crosses in Figure 9.14) according to the internal model with large positive prediction errors in this variable. It can be seen that only after the estimated pitch offset approaches its real value (Figure 9.13, at ~ 1.5 s) the estimated pitch rate q (crosses in 9.14) comes closer to the simulated value (to the right of the dash-dotted line).

During the initial transient phase of the filter, the estimated pitch rate should not be used for control purposes. General good advice with respect to the measured pitch rate is to use inertial angular rate sensors which are inexpensive and yield good (high-frequency) results with almost no delay time; results from visual interpretation, usually, have much larger time delays (2 to 4 video cycles) in addition to the uncertainty from object hypotheses chosen. On the other hand, inertial data tend to drift slowly over time; this can be easily counteracted by joint visual/inertial interpretation.

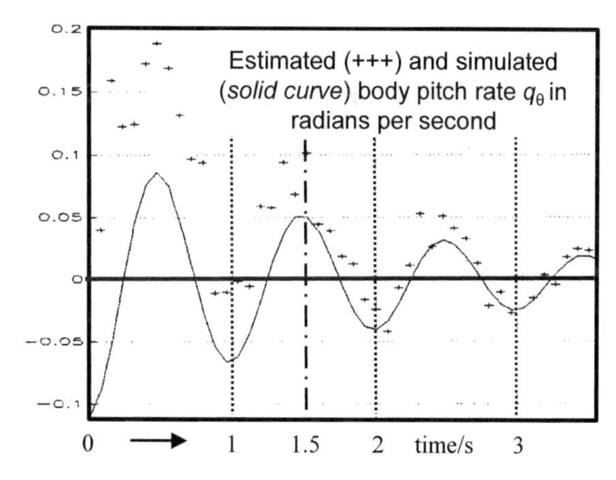

Figure 9.14. Damped pitch oscillation and initial transients from visual perception: it takes about 1½ seconds (about 20 recursive estimation cycles of 80 ms) until the perceived motion corresponds closely to the simulated one (initial pitch rate estimated excluding offset angle ≈ +10°/s)

9.4.1.2 Validation of Lane Width Estimation

The dynamic system for lane width estimation has also been validated by simulation; Figure 9.15 shows an example of an arbitrarily changed lane width from 4 m in both directions (−5 % short perturbation and +11 % step input) with some disturbances added (*solid line*). Lag effects and dynamic overshoot are apparent (*dotted line*) in acceptable limits; the high-frequency perturbations are completely smoothed out. The absolute error of the estimated width (after transient decay) is only a few centimeters.

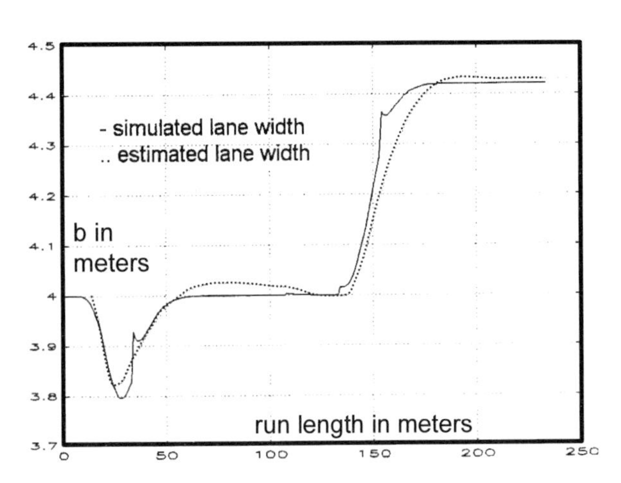

Figure 9.15. Simulation test for road width estimation: Imperfect software for lane width simulation generated small peaks and jumps (*solid curve*), which have not been removed on purpose to see the response of the estimation process (*dotted curve*). Nice smoothing and some overshoot occurred.

9.4.2 Evaluation of Video Scenes

The video scenes were taken from a CCD camera mounted behind the front window of test vehicle VaMP (sedan Mercedes 500 SEL). The data recorded on a VCR were played into the transputer image processing system in the laboratory for evaluation in real time.

9.4.2.1 Results from Test Track

Lane width on high-speed road section: Some video scenes were taken on the test track of UniBwM, the taxiway of the former airport Neubiberg, which is a straight concrete road 2 km long with standard boundary and lane markings for a German Autobahn; it includes sections with narrowing lane width. Figure 9.16 shows the results of simultaneous estimation of pitch angle and lane width at the end of the test track, where the lane width changes first from 3.75 m to 3.5 m (at around 9 seconds on the timescale) and down to 3.2 m (at 23 seconds). The error of lane width estimation is less than 5 %. The estimated (negative) pitch angle shows a slight increase as the lane becomes smaller. This is due to deceleration with the vehicle slightly pitching down because the lane actually ends there. Figure 3.20 visualizes the forces and moments acting that lead to the downward pitch in a brake maneuver; the induced damped small oscillations can be recognized in the

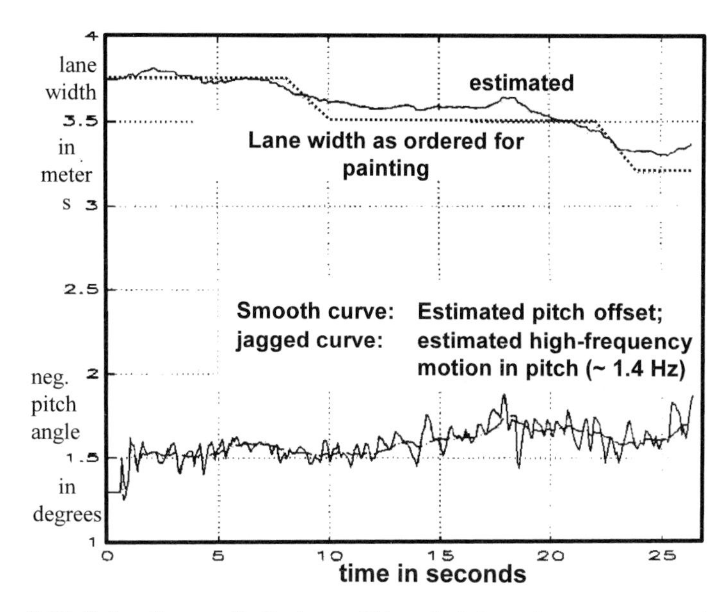

Figure 9.16. Estimation results for lane width and pitch angle components: The pitch angle is split into a dynamic part (of approximately 1.4 Hz eigenfrequency) and a 'quasi-static offset' part

graph. Measuring deceleration directly by inertial sensors and feedback would improve visual perception.

Horizontal curvature on test course: Taking perturbations in pitch angle into account, good estimation results for curvature over arc length can be obtained. Figure 9.17 shows, on the left side, the test course designed for developing and testing autonomous vehicles on the former airport Neubiberg. In the upper part, two circular arcs are directly adjoined to straight sections; this is the way road networks were built for horse carts. In the lower part, transient clothoid arcs as introduced for high-speed roads are implemented (dotted polygons). Here, curves are driven with constant steering rate.

It can be seen that abrupt steps in designed curvature (top right) lead to oscillations. This is due to the fact that the vehicle cannot perform jumps in steering angle, and that the harsh reaction in steering rate leads also to oscillations in roll and yaw of the autonomously controlled vehicle. The time delay in the visual control loop – from image taking to front wheel turn based on this information – was about a half second; this is similar to normal human performance. This closed-loop perception-and-action cycle leads to almost undamped oscillations of small amplitude in the two curves; the average radii of curvature are slightly smaller than the design values of 50 and 40 m (curvatures are slightly higher).

The trajectories perceived as really driven are the solid curves; it can be seen that these curves are closer to clothoids in the transition phases than to the idealized straight line to circular-arc junctions (step input for curvature). Due to the look-ahead range, curve steering is started ahead of the step input on the trajectory. Driving on the clothoid arcs (lower right), deviations between designed and actual trajectories are smaller. There is a strange kink in the first down-going arc between R3 and R4 (dashed ellipse). To check its validity, this part has been analyzed especially carefully by driving in both directions. The kink turned out to be stable in interpretation. Talking with the people who realized the lane markings, the following background was discovered: This clothoid arc had been painted starting from both sides. When the people approached each other, they noticed the discrepancy and realized a smooth transition by a short almost circular arc. This is exactly what the automatic vision system interpreted without knowing the background; it concluded that the radius is ~ 100 m ($C_h = 0.01$).

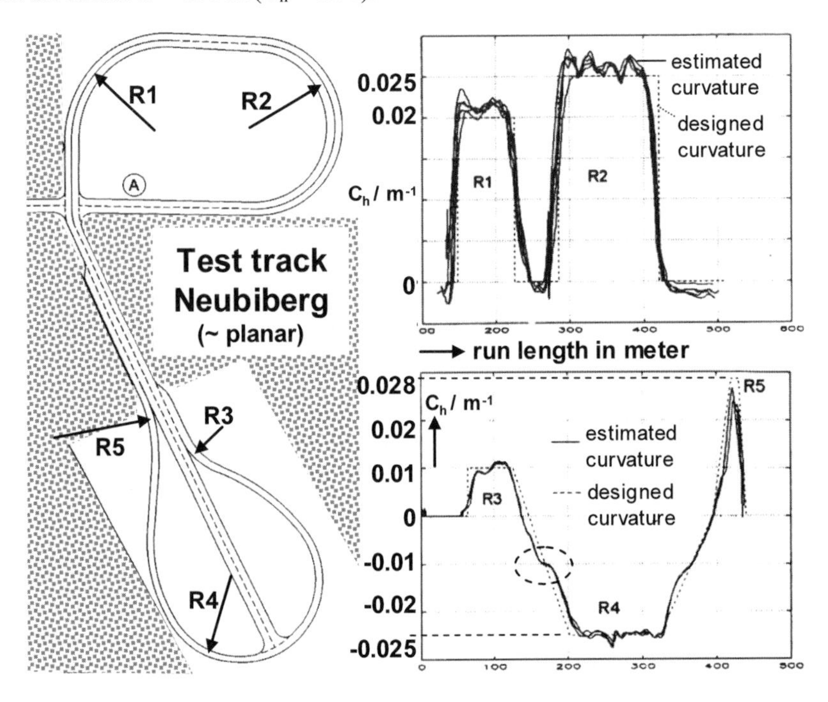

Figure 9.17. Test track for autonomous driving at UniBwM Neubiberg, designed with different radii of curvature R_i, which are connected with and without clothoids (left: bird's-eye view). The right-hand part shows the design parameters (*dotted*) and the recovered curvatures from vision with test vehicle VaMoRs.

9.4.2.2 Results from Rides on Public Roads

Inclusion of pitch angle estimation has led to an improvement in the robustness of the autonomous overall system, especially for driving at higher speeds with increased look-ahead ranges. Monocular distance estimation to other vehicles ahead showed best performance increases due to more stable look-ahead range assignment (see Table 7.1).

Figure 9.18 shows the results of pitch angle estimation from a ride on Autobahn A8 (Munich – Salzburg near the intersection *Munich-South*). Velocity was $V = 120$ km/h, focal length used was $f = 24$ mm, and the farthest look-ahead range was $L_{max} = 70$ m. Pitch angle perturbations are in the range of ± 0.2°, and pitch rates are less than 1.5 °/s (top curve with large peaks). According to column 8 in Table 7.1, row 2, a pitch angle change of 0.2° corresponds to a shift of 8 pixels for the test data with VaMP. At a look-ahead range of 60 m (row 5), the correction in look-ahead range for this pitch angle is 8 pixels · 1.2 m/pixel = 9.6 m, a change of 16%.

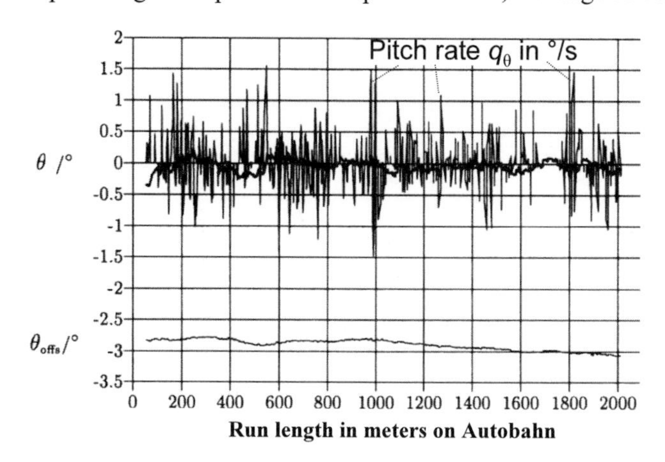

Figure 9.18. Pitch angle estimation ($|\theta| \leq 0.2°$) during a smooth Autobahn ride at $V = 120$ km/h and look-ahead ranges up to 70 m; θ_{offs} of camera estimated from local horizontal as lower graph

These numbers make immediately clear that even a small motion in pitch plays a role for large look-ahead ranges. As can be seen from Equation 9.44, the pitch angle enters into many elements of the Jacobian matrix and thus has an effect in interpreting the measured image features and their contribution to state updates. Because of the low sensitivity for nearby regions, the effects on interpreting images of wide-angle cameras are small; considering these effects as noise and adjusting the Q elements correspondingly has worked well (Chapter 7).

However, to interpret tele–images correctly, the pitch effects on the look-ahead range should be included in any case where even minor perturbations from surface roughness are present; a fraction of a degree in pitch cannot be neglected if precise results are looked for. Figure 9.19 shows a result which at first glance looks surprising. Displayed are results for lateral state estimation, especially the heading angle of the vehicle relative to the road, in real time from stored video sequences with (*solid line*) and without (*dotted line*) taking pitch effects into account for interpretation. The video data base thus was absolutely identical. During this test on the Autobahn, the pitch angle varied in the range $|\theta| \leq 0.5°$ depending on the road surface state.

If both lane markings are well visible, the effects of pitch changes cancel out; this is the case up to a run length of about 600 m. However, if lane markings are poorly detectable on one side and very visible on the other side, changes in pitch angle have the edge features glide along the well visible borderline also in a lateral direction not counteracted by the features on the other side. This is partially interpreted as yaw motion. From 600 to 1200 m in Figure 9.19, this leads to differences in yaw angles estimated at more than 1°. During this stretch, some edge feature measurements on one side of the lane were not successful, so that more emphasis was put on the data from the well visible lane marking. Vehicle pitch due to perturbations not modeled was interpreted partially as measured yaw angle.

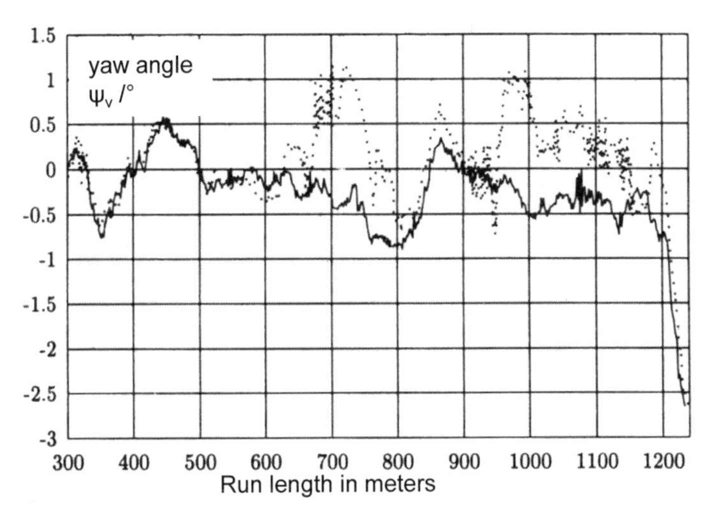

Figure 9.19. The effects of pitch motion modeled (*solid line*) or neglected (*dotted line* for θ = constant) on other variables not directly involved in this degree of freedom; here: yaw angle estimation when quality of lane markings on each side differs (see text, after [Behringer 1996])

Toward the end (run length > 1200 m), the vehicle decelerated and started a lane change to the left (yaw angle ψ down to about −3°).

9.4.2.3 Estimation of Lane Width

State road with T-junction on left-hand side: The pitch angle can be estimated correctly based on lane width only when the real lane width is known. On German 'Autobahnen' for a long time, the standard lane width was defined as b = 3.75 m. On older sections and in mountainous regions, it may be less. The assumption of a wrong and constant lane width leads to errors in pitch angle estimation, since the measured residues effect an innovation of the pitch angle even though this pitch angle adaptation cannot make prediction-errors vanish. In any case, changes of lane width on normal sections of freeways are relatively small, usually. Strong variations in lane width occur on construction sites.

Since changes in lane width on standard roads are much more pronounced on state roads or minor highways, one such case has been taken as a test. Large changes in road width are sometimes accepted for state roads with low traffic density when a side road connects to the state road by a T-junction; vehicles turning off onto the side road move to the side of the connection and slow down, while through traffic may continue to the right at normal or slightly reduced speed. In Figure 9.20, widening of the lane over ~ 100 m (220 to 320, top curve) is about 40 % (1.5 m on top of ~ 3.3 m), built up over a distance of about 60 m. Both integrated terms b_1 and b_0 are estimated as smooth curves; the smoothing effect of the artificially introduced variable b_1 is clearly visible. The closed-loop action-perception cycle yields an evolving symbolic representation hard to achieve with quasi-static AI approaches.

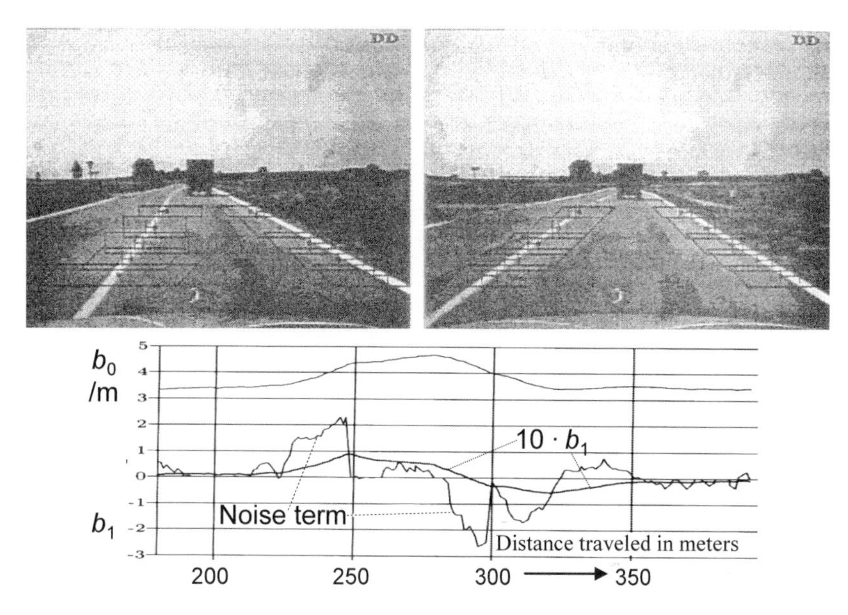

Figure 9.20. Lane widening at a left-turn road junction (in right-hand traffic) on a state road (top). Two snapshots at different distances from the T-junction: Left: further away; right: closer to junction seen at left. Bottom: Estimated parameters of lane width model.

With lane width estimation active but pitch estimation inactive, prediction errors relative to edge positions for constant pitch angle are interpreted as changes in lane width. With both models active, the same edge features drive both models simultaneously. With pitch estimation inactive, two types of errors occur: (1) lane width estimated is erroneous if pitch angle changes occur, and (2) actually occurring changes in look-ahead distances are not recognized. To study these effects, test data from the same run on a stretch of Autobahn have been analyzed with lane width estimation active twice, once with pitch estimation inactive and once with pitch estimation active.

Effects of pitching motion on estimation of lane width: Figure 9.21 shows results from [Behringer 1996] nicely indicating the improvement in stability when motion in pitch is estimated simultaneously with lane width; the small change in lane width occurs at around a run length of 150 m. If pitch estimation is not active but pitching motion occurs, the changes in road width in a certain image row due to actually occurring pitching motion are interpreted as changes in width. Estimation of lane width on a 1 km stretch of Autobahn with (*solid curve*) and without (*dotted curve*) simultaneous pitch angle estimation is shown.Through the corresponding Jacobian elements and the updates in the feedback gain for the prediction errors in width, these changes due to pitching motion are interpreted as changes in real road width. Misinterpretations of up to ~ 40 cm (10 % of lane width) occur around the real value of 3.75 m.

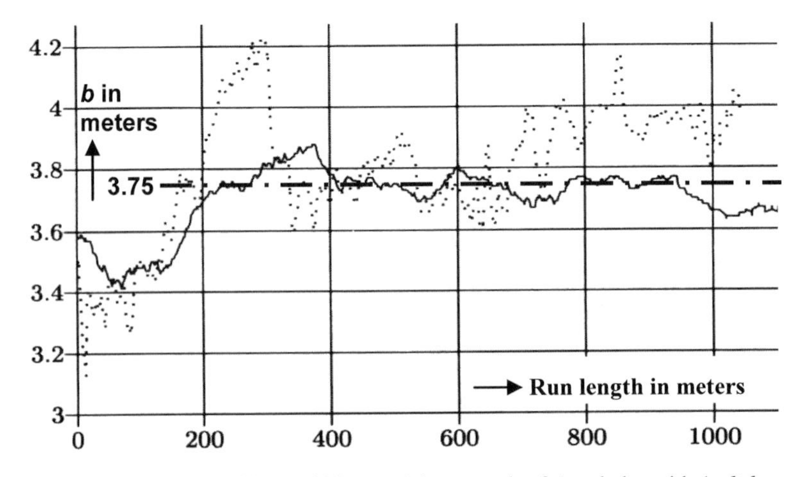

Figure 9.21. Estimation of lane width on a 1-km stretch of Autobahn with (*solid curve*) and without (*dotted curve*) simultaneous pitch angle estimation

Without pitch angle estimation, lane width varies with a standard deviation of ~ 0.1 to 0.3 m, depending on the smoothness of the road surface; with simultaneous pitch angle estimation, this value is reduced to 0.05 m (5 cm) and less. By adapting elements of the covariance matrix Q correspondingly, estimation quality is improved further.

9.4.2.4 Estimation of Small Vertical Curvatures

Surprising results have been achieved with respect to motion stereointerpretation of the height profile of a stretch of Autobahn close to UniBwM, north of the Autobahn intersection Munich South. Figure 9.22 shows five snapshots of the scene and the vertical curvature profile recovered. The upper left image shows an underpass under a bridge; the increasing slope behind the bridge is clearly visible from this image. The center image in the top row was taken shortly before passing the bridge, whose shadow is seen in the near range. The remaining three images show the vehicle going uphill (top right), in front of the cusp (lower left), and back to level driving behind the underpass (lower center). At the lower right, the vertical curvature profile recovered while traveling this stretch over a distance of a half kilometer is shown. Note that the unit for curvature is 100 km! The peak value of curvature of $7.2 \cdot 10^{-5}$ [1/m] corresponds to a radius of curvature of about 14 km. This means that for 100 m driven the slope change is ~ 0.4°.

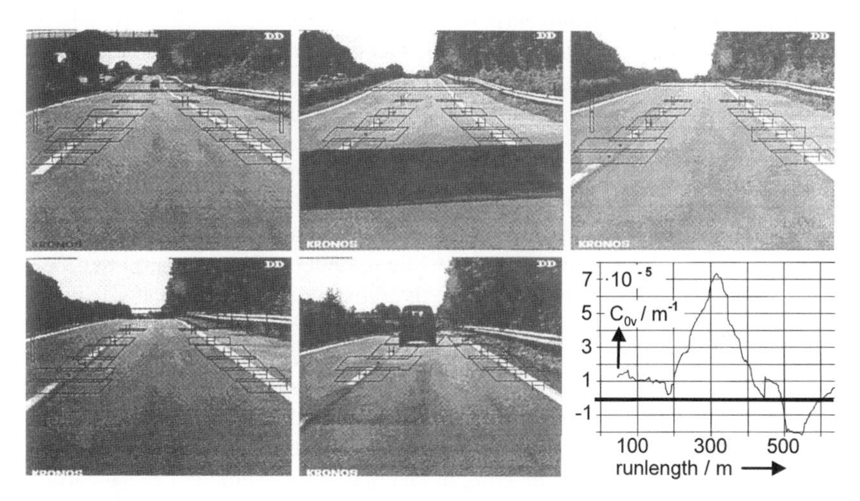

Figure 9.22. Precise estimation of vertical curvature with simultaneous pitch estimation on underpass of a bridge across Autobahn A8 (Munich – Salzburg) north of Autobahn crossing Munich-south: Top left: Bridge and bottom of underpass can be recognized; top center: vehicles shortly before underpass, shadow of bridge is highly visible. Top right and bottom left: Cusp after underpass is approached; bottom center: leaving the underpass area. Bottom right: Estimated vertical curvature over distance driven: The peak value corresponds to ~ a half degree change in slope over 100 m.

Of course, this information has been collected over a distance driven of several hundred meters; this shows that motion stereo with the 4-D approach in this case cannot be beaten by any kind of multiocular stereo. Note, however, that the stereo base length (a somewhat strange term in this connection) is measured by the odometer with rather high precision. Without the smoothing effects of the EKF (double integration over distance driven) this would not be achievable.

The search windows for edge detection are marked in black as parallelepipeds in the snapshots. When lane markings are found, their lateral positions are marked with dark-to-bright and bright-to-dark transitions by three black lines looking like a compressed letter capital H. If no line is found, a black dot marks the predicted position for the center of the line. When these missing edges occur regularly, the lane marking is recognized as a broken line (allowing lane changes).

9.4.2.5 Long-distance Test Run

Till the mid-1990s most of the test runs served one or a few specific purposes to demonstrate that these tasks could be done by machine vision in the future. Road types investigated were freeways (German Autobahn and French Autoroute), state roads of all types, and minor roads with and without surface sealing as well as with and without lane markings. Since 1992, first VaMoRs and since 1994 also VaMP continuously did many test runs in public traffic. Both the road with lane markings (if present) and other vehicles of relevance had to be detected and tracked; the latter will be discussed in Chapter 11.

After all basic challenges of autonomous driving had been investigated to some degree for single specific tasks, the next step planned was designing a vision based system in which all the separate capabilities were integrated into a unified approach. To improve the solidity of the database on which the design was to be founded, a long-distance test run with careful monitoring of failures, and the reasons for failures, had been planned. Earlier in 1995, CMU had performed a similar a test run from the East to the West Coast of the United States; however, only lateral guidance was done autonomously, while a human driver actuated the longitudinal controls. The human was in charge of adjusting speed to curvature and keeping a proper distance from vehicles in front. Our goal was to see how poor or well our system would do in fully autonomously performing a normal task of long-distance driving on high-speed roads, mainly (but not exclusively) on Autobahnen.

This also required using the speed ranges typical on German freeways which go up to and beyond 200 km/h. Figure 9.23 shows a section of about 38 minutes of this trip to a European project meeting in Denmark in November 1995 according to [Behringer 1996; Maurer 2000]. The safety driver, always sitting in the driver's seat,

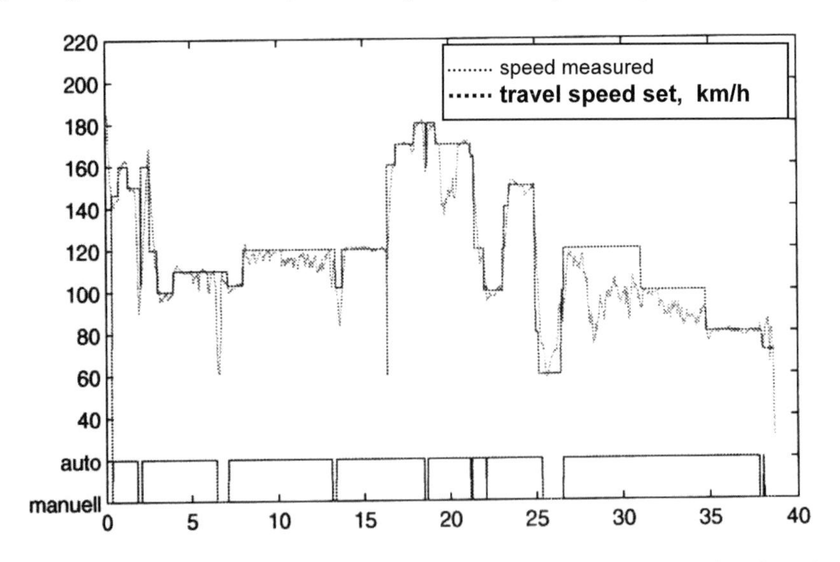

Figure 9.23. Speed profile of a section of the long-distance trip (over time in minutes)

or the operator of the computer and vision system selected and prescribed a desired speed according to regulations by traffic signs or according to their personal interpretation of the situation. The stepwise function in the figure shows this input. Deviations to lower speeds occurred when there were slower vehicles in front and lane changes were not possible. It can be seen that three times the vehicle had to decelerate down to about 60 km/h. At around 7 minutes, the safety driver decided to take over control (see gap in lower part of the figure), while at around 17 minutes the vehicle performed this maneuver fully autonomously (apparently to the satisfaction of the safety driver). The third event at around 25 minutes again had the safety driver intervene. Top speed driven at around 18 minutes was 180 km/h (50 m/s or 2 m per video cycle of 25 Hz). Two things have to be noted here: (1) With a look-ahead range of about 80 m, the perception system can observe each specific section of lane markings up to 36 times before losing sight nearby ($L_{\min}$ ~ 6 m), and (2) stopping distance at 0.8 g (-8 m/s^2) deceleration is ~ 150 m (without delay time in reaction); this means that these higher speeds could be driven autonomously only with the human safety driver assuring that the highway was free of vehicles and obstacles for at least ~ 200 m.

Figure 9.24 gives some statistical data on accuracy and reliability during this trip. Part (a) (left) shows distances driven autonomously without interruption (on a logarithmic scale in kilometers); the longest of these phases was about 160 km. Almost all of the short sequences (≤ 5 km) were either due to construction sites (lowest of three rows top left), or could be handled by an automatic reset (top row); only one required a manual reset (at ~ 0.7 km).

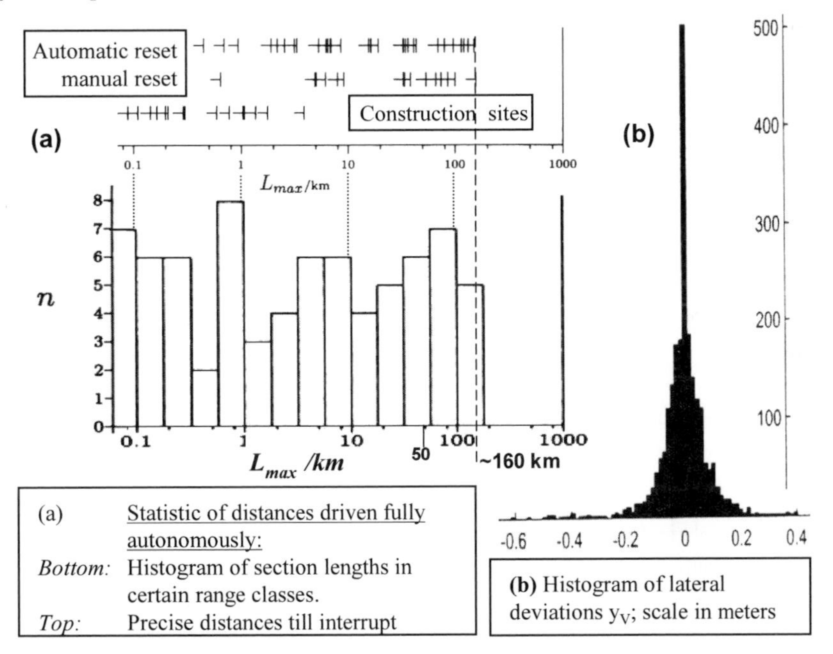

Figure 9.24. Some statistical data of the long-distance test drive with VaMP (Mercedes 500 SEL) from Munich to Odense, Denmark, in November 1995. Total distance driven autonomously was 1678 km (~ 95 % of system in operation).

This figure clearly shows that robustness in perception has to be increased significantly over this level, which has been achieved with black-and-white images from which only edges had been extracted as features. Region–based features in gray scale and color images as well as textured areas with precisely determinable corners would improve robustness considerably. The computing power in microprocessors is available nowadays to tackle this performance improvement. The figure also indicates that an autonomous system should be able to recognize and handle construction sites with colored and nonstandard lane markings (or even without any) if the system is to be of practical use.

Performance in lane keeping is sufficient for most cases; the bulk of lateral offsets are in the range ± 0.2 m (Figure 9.24b, lower right). Taking into account that normal lane width on a standard Autobahn (3.75 m) is almost twice as large as vehicle width, lateral guidance is more than adequate; with humans driving, deviations tend to be less strictly observed every now and then. At construction sites, however, lane widths of down to 2 m may be encountered; for these situations, the flat tails of the histogram indicate insufficient performance. Usually, in these cases, the speed limit is set as low as 60 km/h; there should be a special routine available for handling these conditions, which is definitely in range with the methods developed.

Figure 9.25 shows for comparison a typical lateral deviation curve over run length while a human was driving on a normal stretch of state road [Behringer 1996]. Plus/minus 40 cm lateral deviation is not uncommon in relaxed driving; autonomous lateral guidance by machine vision compares favorably with these results.

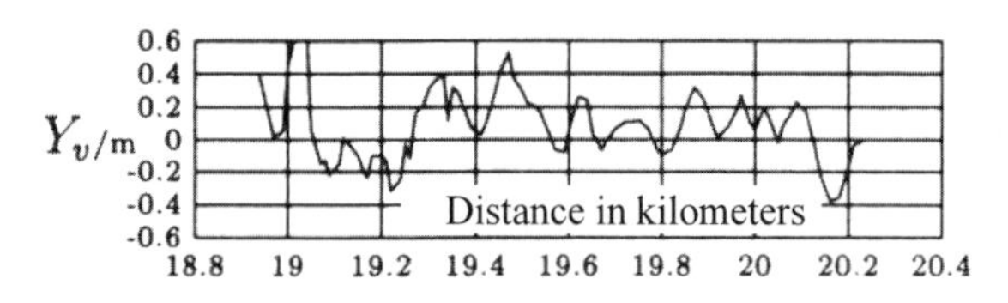

Figure 9.25. Typical lateral offsets for manual human steering control over distance driven

The last two figures in this section show results from sections of high-speed state roads driven autonomously in Denmark on this trip. Lane width varies more frequently than on the Autobahn; widths from 2.7 to 3.5 m have been observed over a distance of about 3 km (Figure 9.26). The variance in width estimation is around 5 cm on sections with constant width.

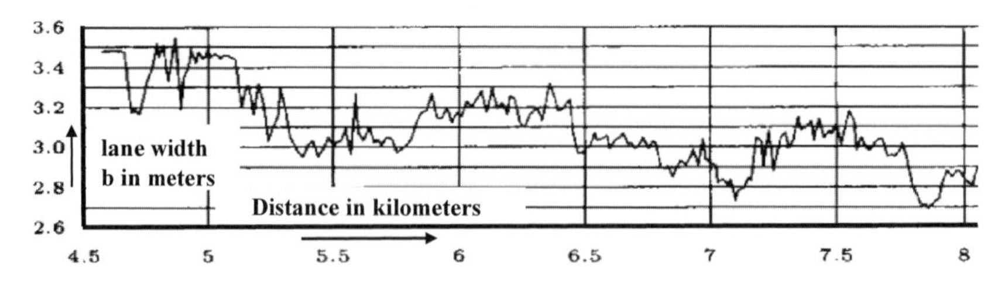

Figure 9.26. Varying width of a state road can be distinguished from the variance of width estimation by spatial frequency; standard variation of lane width estimation is about 5 cm

The left part of Figure 9.27 gives horizontal curvatures estimated for the same stretch of road as the previous figure. Radii of curvature vary from about 1 km to 250 m. Straight sections with curvature oscillating around zero follow sections with larger (constant?) curvature values that are typically perceived as oscillating (as in Figure 9.17 on our test track). The system interprets the transitions as clothoid arcs with linear curvature change. It may well be that the road was pieced together from circular arcs and straight sections with step-like transitions in curvature; the perception process with the clothoid model may insist on seeing clothoids due to the effect of low-pass filtering with smoothing over the look-ahead range (compare upper part of Figure 9.17).

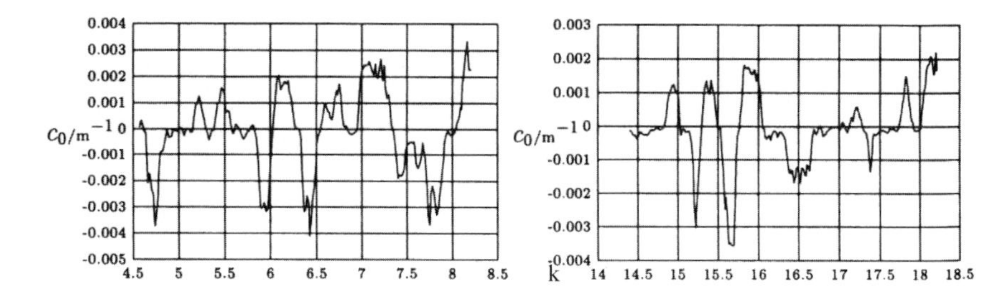

Figure 9.27. Perceived horizontal curvature profile on two sections of a high-speed state road in Denmark while driving autonomously: Radius of curvature comes down to a minimum of ~ 250 m (at km 6.4). Most radii are between 300 and 1000 m ($R = 1/\,c_0$).

The results in accuracy of road following are as good as if a human were driving (deviations of 20 to 40 cm, see Figure 9.28). The fact that lateral offsets occur to the 'inner' side of the curve (compare curvature in Figure 9.27 left with lateral offset in Figure 9.28 for same run length) may be an indication that the underlying road model used here for perception may be wrong (no clothoids); curves seem to

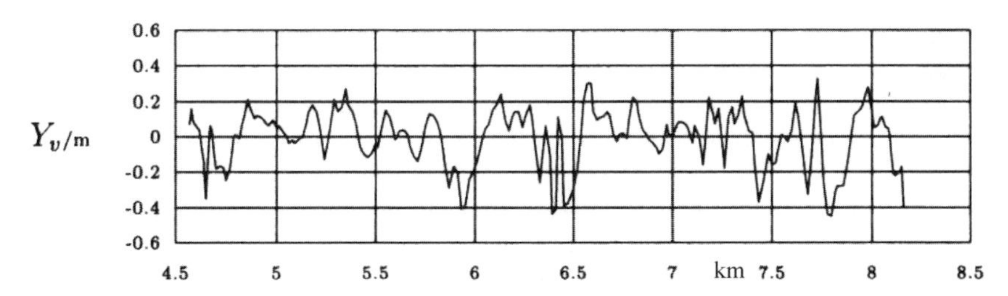

Figure 9.28. Lateral offset on state road driven autonomously; compare to manual driving results in Figure 9.25 and curvature perceived in 9.27.

be 'cut,' as is usual for finite steering rates on roads pieced together from arcs with stepwise changes in curvature. This is the price one has to pay for the stabilizing effect of filtering over space (range) and time simultaneously. Roads with real clothoid elements yield better results in precise road following.

[As a historic remark, it may be of interest that in the time period of horse carts, roads used to be made from exactly these two elements. When high-speed cars driven with a finite steering rate came along, these systematic 'cuts' of turns by the trajectories actually driven have been noticed by civil engineers who – as a progressive step in road engineering – introduced the clothoid model (linear curvature change over arc length).]

9.5 High-precision Visual Perception

With the capability of perceiving both horizontal and vertical curvatures of roads and lanes together with their widths and the ego- state including pitch angle, it is important to exploit precision achievable to the utmost to obtain good results. Subpixel accuracy in edge feature localization on the search path has been used as standard for a long time (see Section 5.2.2). However, with good models for vehicle pitching and yawing, systematic changes extended edge features in image sequences can be perceived more precisely by exploiting knowledge represented in the elements of Jacobian matrices. This is no longer just visual feature extraction as treated in Section 5.2.2 but involves higher level knowledge linked to state variables and shape parameters of objects for handling the aperture problem of edge features; therefore, it is treated here in a special section.

9.5.1 Edge Feature Extraction to Subpixel Accuracy for Tracking

In real-time tracking involving moving objects, predictions are made for efficient adjustment of internal representations of the motion process with both models for shape and for motion of objects or subjects. These predictions are made to subpixel accuracy; edge locations can also be determined easily to subpixel accuracy by the methods described in Chapter 5. However, on one hand, these methods are geared to full pixel size; in CRONOS, the center of the search path always lies at the center of a pixel (0.5 in pixel units). On the other hand, there is the aperture problem on an edge. The edge position in the search path can be located to sub-pixel accuracy, but in general, the feature extraction mask will slide along a body-fixed edge in an unknown manner. Without reference to an overall shape and motion model, there is no solution to this problem. The 4-D approach discussed in Chapter 6 provides this information as an integral part of the method. The core of the solution is the linear approximation of feature positions in the image relative to state changes of 3-D objects with visual features on their surfaces in the real world. This relationship is given by concatenated HCTs represented in a scene tree (see Section 2.1.1.6) and by the Jacobian matrices for each object–sensor pair.

For precise handling of subpixel accuracy in combination with the aperture problem on edges, one first has to note that perspective mapping of a point on an edge does not yield the complete measurement model. Due to the odd mask sizes of $2^n + 1$ pixels normal to the search direction in the method CRONOS, mask locations for edge extraction are always centered at 0.5 pixel. (For efficiency reasons, that is, changing of only a single index, search directions are either horizontal or vertical in most real-time methods). This means that the row or column for feature search is given by the integer part of the pixel address computed (designated as 'entier(y or z)' here). Precise predictions of feature locations according to some model have to be projected onto this search line. In Figures 9.29 and 9.30, the two predicted points, P^*_{1N} (upper left) and P^*_{2N} (lower right), define the predicted edge line drawn solid.

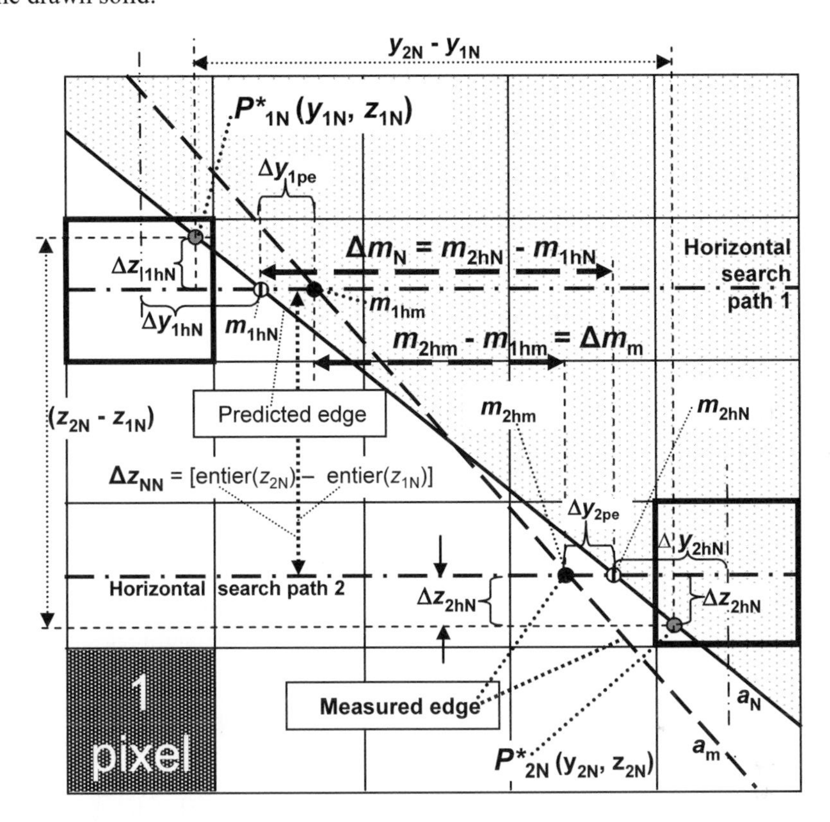

Figure 9.29. Application-specific handling of aperture problem in connection with edge feature extractor in rows (like UBM1; nominal search path location at center of pixel): Basic grid corresponds to 1 pixel. Both predictions of feature locations and measurements are performed to subpixel accuracy; Jacobian elements are used for problem specific interpretation (see text). Horizontal search direction: Offsets in vertical direction are transformed into horizontal shifts exploiting the slopes of both the predicted and the measured edges; slopes are determined from results in two neighboring horizontal search paths.

Depending on the search direction chosen for feature extraction (horizontal h, Figure 9.29 or vertical v, Figure 9.30), the nominal edge positions (index N) taking the measurement process into account are m_{1hN} and m_{2hN} (9.29) respectively m_{1vN} and m_{2vN} (9.30, textured circles on solid line).

The slope of the predicted edge is

$$a_N = (z_{2N} - z_{1N})/(y_{2N} - y_{1N}) \,. \tag{9.46}$$

For *horizontal search directions*, the vertical differences Δz_{1hN}, Δz_{2hN} to the center of the pixel z_{iN} defining the search path are

$$\Delta z_{1\text{hN}} = z_{1\text{N}} - entier(z_{1\text{N}}) - 0.5,$$
$$\Delta z_{2\text{hN}} = z_{2\text{N}} - entier(z_{2\text{N}}) - 0.5; \quad (9.47)$$

in conjunction with the slope a_N they yield the predicted edge positions on the search paths as the predicted measurement values

$$m_{ihN} = y_{iN} - \Delta z_{ihN} / a_N, \quad i = 1,\ 2. \quad (9.48)$$

In Figure 9.29 (upper left), it is seen that the feature location of the predicted edge *on the search path* (defined by the integer part of the predicted pixel) actually is in the neighboring pixel. Note that this procedure eliminates the z-component of the image feature from further consideration in horizontal search and replaces it by a corrective y-term for the edge measured. For vertical search directions, the opposite is true.

For *vertical search directions*, the horizontal differences to the center of the pixel defining the search path ($\Delta y_{1\text{vN}}$, $\Delta y_{2\text{vN}}$) are

$$\Delta y_{\text{ivN}} = y_{\text{iN}} - entier(y_{\text{iN}}) + 0.5\ ; \quad (9.49)$$

together with the slope a_N, this yields the predicted edge positions *on the search path* as the predicted measurement values to subpixel accuracy:

$$m_{\text{ivN}} = z_{\text{iN}} - \Delta y_{\text{ivN}} \cdot a_\text{N}\ , \quad i = 1, 2. \quad (9.50)$$

For point 1 this yields the predicted position $m_{1\text{vN}}$ in the previous vertical pixel (above), while in the case of point 2 the value $m_{2\text{vN}}$ lies below the nominal pixel (lower right of Figure 9.30). Again this procedure eliminates the y-component of the image feature from further considerations in a vertical search and replaces it by a corrective z term for the edge measured.

9.5.2 Handling the Aperture Problem in Edge Perception

Applying horizontal search for precise edge feature localization yields the measurement points $m_{1\text{hm}}$ for point 1 and $m_{2\text{hm}}$ for point 2 (dots filled in black in Figure 9.29). Taking knowledge about the 4-D model and the aperture effect into account, the sum of the squared prediction errors shall be minimized by changing the unknown state variable x_S. However, the sliding effect of the feature extraction masks along the edges has to be given credit. To do this, the linear approximation of perspective projection by the Jacobian matrix is exploited. This requires that deviations from the real situation are not too large.

The Jacobian matrix (abbreviated here as J), as given in Section 2.1.2, approximates the effects of perspective mapping. It has $2{\cdot}m$ rows for m features (y and z components) and n columns for n unknown state variables $x_{\text{S}\alpha}$, $\alpha = 1, \ldots$ n. Each image point has two variables y and z for describing the feature position. Let us adopt the convention that all odd indices of the $2{\cdot}m$ rows ($i_\text{y} = 2{\cdot}i - 1$, $i = 1$ to m) of J refer to the y-component (horizontal) of the feature position, and all following even indices ($i_\text{z} = 2{\cdot}i$) refer to the corresponding z-component (vertical). All these couples of rows multiplied by a change vector for the n state variables to be adjusted, $\delta x_{\text{S}\alpha}$, $\alpha = 1, \ldots\ n$ yield the changes δy and δz of the image points due to δx_S:

$$J_{i\alpha} \cdot \delta x_{S\alpha} = (\delta y,\ \delta z)_i^T . \quad (9.51)$$

Let us consider adjusted image points (y_{iA}, z_{iA}) after recursive estimation for locations 1 and 2 which have been generated by the vector products

$$\begin{aligned} y_{1A} &= y_{1N} + J_{1y\alpha} \cdot \delta x_{S\alpha}; \qquad z_{1A} = z_{1N} + J_{1z\alpha} \cdot \delta x_{S\alpha}; \\ y_{2A} &= y_{2N} + J_{2y\alpha} \cdot \delta x_{S\alpha}; \qquad z_{2A} = z_{2N} + J_{2z\alpha} \cdot \delta x_{S\alpha}. \end{aligned} \tag{9.52}$$

These two points yield a new edge direction a_A for the yet unknown adjustments $\delta x_{S\alpha}$. However, this slope is measured on the same search paths given by the integer values of the search row (or column) through measurement values m_{ihm} (or m_{ivm}). The precise location of the image point for a minimum of the sum of the squared prediction errors depends on the $\delta x_{S\alpha}$, $\alpha = 1, \ldots n$, to be found, and it has thus to be kept adaptable.

Analogous to Equation 9.46, one can write for the new slope, taking Equation 9.52 into account,

$$\begin{aligned} a_A &= \frac{z_{2A} - z_{1A}}{y_{2A} - y_{1A}} = \frac{z_{2N} - z_{1N} + (J_{2z\alpha} - J_{1z\alpha}) \cdot \delta x_{S\alpha}}{y_{2N} - y_{1N} + (J_{2y\alpha} - J_{1y\alpha}) \cdot \delta x_{S\alpha}} \\ &= a_N \cdot \frac{1 + \Delta J_{21z\alpha} / \Delta z_N \cdot \delta x_{S\alpha}}{1 + \Delta J_{21y\alpha} / \Delta y_N \cdot \delta x_{S\alpha}}. \end{aligned} \tag{9.53}$$

For $|\varepsilon_i| << 1$, the following linear approximation is valid for the ratio:

$$(1+\varepsilon_1)/(1+\varepsilon_2) \approx (1+\varepsilon_1)\cdot(1-\varepsilon_2) = 1+\varepsilon_1 - \varepsilon_2 - \varepsilon_1 \cdot \varepsilon_2 \approx 1 + \varepsilon_1 - \varepsilon_2 .$$

Applying this to Equation 9.53 yields a linear approximation in $\delta x_{S\alpha}$:

$$a_A = a_N \cdot \left[1 + \left(\frac{\Delta J_{21z\alpha}}{\Delta z_N} - \frac{\Delta J_{21y\alpha}}{\Delta y_N} \right) \cdot \delta x_{S\alpha} - \text{h.o.t.} \right] \approx a_N + a_N \cdot C_{\text{mod}} \cdot \delta x_{S\alpha}, \tag{9.54}$$

with $C_{\text{mod}} = \Delta J_{21z\alpha} / \Delta z_N - \Delta J_{21y\alpha} / \Delta y_N$.

Horizontal and vertical search paths will be discussed in separate subsections.

9.5.2.1 Horizontal Search Paths

The slope of the edge given by Equation 9.46 can also be expressed by the predicted measurement values m_{1hN} and m_{2hN} on the nominal search paths 1 and 2 (*dash–dotted lines* in Figure 9.29 at a distance $\Delta z_{NN} = entier(z_{2N}) - entier(z_{1N})$ from each other); this yields the solid line passing through all four points P^*_{1N}, P^*_{2N}, m_{1hN} and m_{2hN}. The new term for the predicted slope then is

$$a_N = \Delta z_{NN} / (m_{2hN} - m_{1hN}) = \Delta z_{NN} / \Delta m_{hN} . \tag{9.55}$$

Similarly one obtains for the measured slope a_m from the two measured feature locations on the same search paths

$$a_m = \Delta z_{NN} / (m_{2hm} - m_{1hm}) = \Delta z_{NN} / \Delta m_{hm} . \tag{9.56}$$

Dividing Equation 9.56 by Equation 9.55 and multiplying the ratio by a_N yields

$$\begin{aligned} a_m &= a_N \cdot \Delta m_N / \Delta m_{hm} = \\ &a_N \cdot (m_{2hN} - m_{1hN}) / (m_{2hm} - m_{1hm}). \end{aligned} \tag{9.57}$$

Setting this equal to Equation 9.54 yields the relation

$$a_m = a_A = a_N \cdot \frac{m_{2hN} - m_{1hN}}{\Delta mh_m} = a_N + a_N \cdot C_{\text{mod}} \cdot \delta x_{S\alpha}. \tag{9.58}$$

Dividing by a_N and bringing the resulting 1 in the form $(m_{2hm} - m_{1hm})/\Delta m_m$ onto the left side yields after sorting terms,

$$\frac{m_{2hN} - m_{1hN} - m_{2hm} + m_{1hm}}{\Delta m_{hm}} = C_{\text{mod}} \cdot \delta x_{S\alpha}. \tag{9.59}$$

With the prediction errors Δy_{ipe} on the nominal search paths

$$\Delta y_{ipe} = m_{ihm} - m_{ihN}, \tag{9.60}$$

Equation 9.59 can be written

$$(\Delta y_{1pe} - \Delta y_{2pe}) / \Delta m_{hm} = C_{\text{mod}} \cdot \delta x_{S\alpha}. \tag{9.61}$$

This is the iteration equation for a state update taking the aperture problem and knowledge about the object motion into account. The position of the feature in the image corresponding to this innovated state would be (n × n vector product of the corresponding row in the Jacobian matrix and the change in the state vector):

$$\delta y_i = \Delta J_{iy\alpha} \cdot \delta x_{S\alpha}; \quad \delta z_i = \Delta J_{iz\alpha} \cdot \delta x_{S\alpha}. \tag{9.62}$$

Note, however, that this image point is not needed (except for checking progress in convergence) since the next feature to be extracted depends on the predicted state resulting from the updated state 'now' and on single step extrapolation in time. This modified measurement model solves the aperture problem for edge features in horizontal search. This result can be interpreted in view of Figure 9.29: The term on the left-hand side in Equation 9.61 is the difference in the predicted and the measured position along the (forced) nominal horizontal search paths 1 and 2 at the center of the pixel. If both prediction errors are equal, the slope does not change and there is no aperture problem; the Jacobian elements in the y-direction at the z-position can be taken directly for computing $\delta x_{S\alpha}(\delta y_i)$. If the edge is close to vertical ($\Delta m_m \approx 0$), Equation 9.61 will blow up; however, in this case, Δy_N is also close to zero, and the aperture problem disappears since the search path is orthogonal to the edge. These two cases have to be checked in the code for special treatment by the standard procedure without taking aperture effects into account. The term on the right-hand side is the modified Jacobian matrix (Equation 9.54). The terms in the denominator of this equation indicate that for almost vertical edges in horizontal search and for almost horizontal edges in vertical search, this formulation should be avoided; this is no disadvantage, however, since in these cases the aperture problem is of no concern.

9.5.2.2 *Vertical Search Paths*

The predicted image points P^*_{1N} and P^*_{2N} in Figure 9.30 define both the expected slope of the edge and the position of the search paths (*vertical dash-dotted lines*); the distance of the search paths from each other is $\Delta y_{NN} = entier(y_{2N})$ - $entier(y_{1N})$, four pixels in the case shown. The intersections of the straight line through points P^*_{1N} and P^*_{2N} with the search paths define the predicted measurement values (m_{1vN} and m_{2vN}); in the case given, with the predicted image points in the upper right corner of the pixel labeled with index (1, upper left in the figure) and the lower left corner of the pixel labeled (2, lower right in the figure), m_{1vN} lies in the previous pixel of the search direction, and m_{2vN} in the following pixel of the corresponding search path (top-down search).

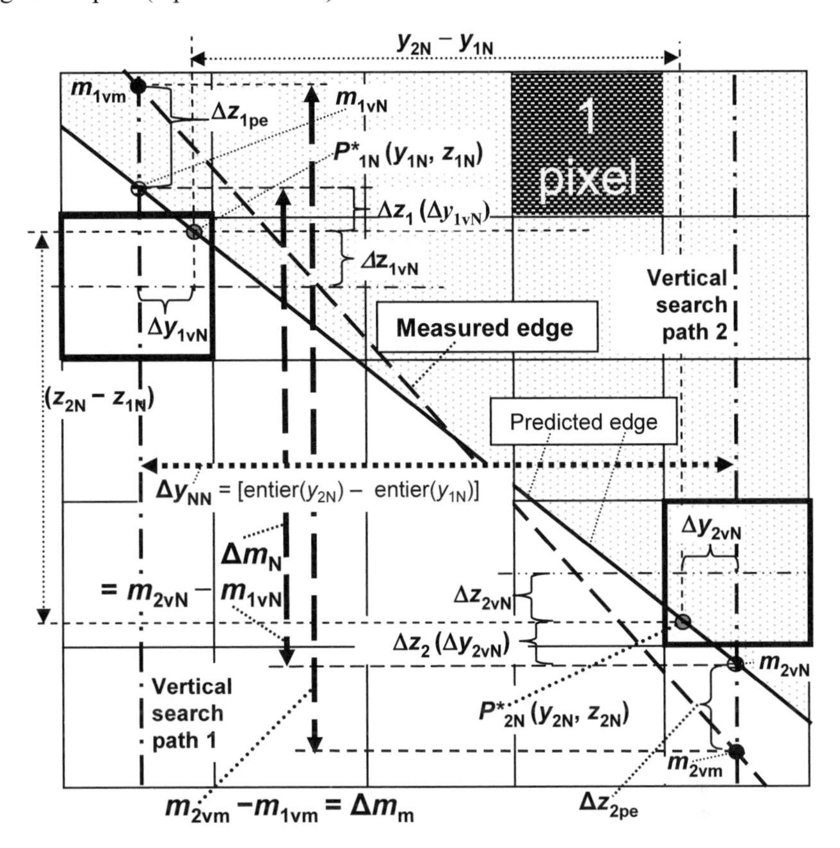

Figure 9.30. Application-specific handling of the aperture problem in connection with edge feature extractor in columns: Both predictions of feature locations and measurements are performed to subpixel accuracy; Jacobian elements are used for problem specific interpretation (see text). Vertical search direction: Offsets in horizontal direction are transformed into vertical shifts exploiting the slopes of both the predicted and the measured edges; slopes are determined from results in two neighboring horizontal search paths.

The exact position is given by Equation 9.50. The predicted slope of the edge a_N (see Equation 9.46) can thus also be written

$$a_N = (m_{2vN} - m_{1vN}) / \Delta y_{NN} = \Delta m_N / \Delta y_{NN}. \tag{9.63}$$

The edge locations actually found on the search paths are the points m_{1vm} and m_{2vm}; the prediction errors on the nominal search paths thus are

$$\Delta z_{1pe} = m_{1vm} - m_{1vN} \tag{9.64}$$

and

$$\Delta z_{2pe} = m_{2vm} - m_{2vN}.$$

The difference in the measured z-location on the search paths is

$$\Delta m_{vm} = m_{2vm} - m_{1vm}; \tag{9.65}$$

in conjunction with Δy_{NN} this yields the slope of the measured edge (*dashed line* in Figure 9.30) as

$$a_m = \Delta m_{vm} / \Delta y_{NN}. \tag{9.66}$$

Dividing Equation 9.66 by Equation 9.63 and multiplying the ratio by a_N yields

$$\begin{aligned} a_m &= a_N \cdot \Delta m_{vm} / \Delta m_N \\ &= a_N \cdot (m_{2vm} - m_{1vm}) / (m_{2vN} - m_{1vN}). \end{aligned} \tag{9.67}$$

Setting this equal to Equation 9.54 yields the relation (similar to Equation 9.58)

$$\frac{m_{2hm} - m_{1hm}}{\Delta m_{vm}} - 1 = C_{\text{mod}} \cdot \delta x_{S\alpha}. \tag{9.68}$$

Replacing the 1 by Equation 9.65 divided by Δm_{m} and observing Equation 9.64, the equation for vertical search analogous to Equation 9.61 (horizontal) is

$$(\Delta z_{1pe} - \Delta z_{2pe}) / \Delta m_{vm} = C_{\text{mod}} \cdot \delta x_{S\alpha}. \tag{9.69}$$

T compute the exact position of the projected, updated point in the image, Equation 9.62 is equally valid. With respect to singularities, the comments made for the horizontal case are valid with proper substitution of terms.

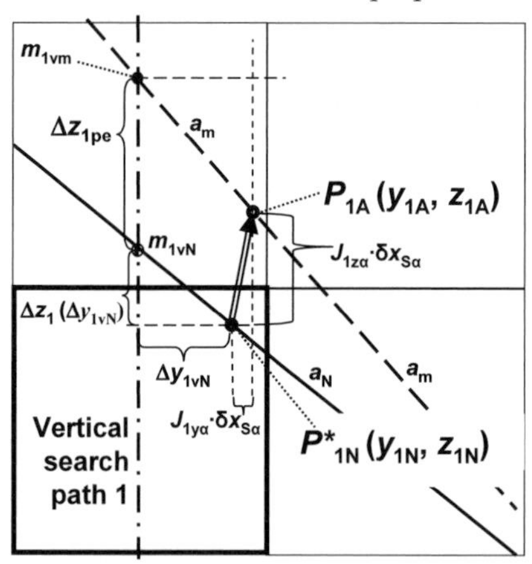

Figure 9.31. Subpixel edge iteration taking the aperture effects of edge feature extraction with method CRONOS locally into account according to a 3-D shape model of the object with points 1 and 2 on straight edges

Figure 9.31 shows the basic idea underlying the modified iteration process. The state update is computed from the prediction errors on the fixed search paths for two points on a straight edge (Equations 9.61 for horizontal and Equation 9.69 for vertical search). The corresponding position of the updated (innovated) feature point P_{iA} on the edge in the image is given by Equation 9.62. Again, computing the position of this pointy in the image is not needed for the recursive estimation process; only if a check of iteration results by visual inspection is wanted, should the point be determined and inserted into the overlay of the original video image for monitoring.

10 Perception of Crossroads

For navigation in networks of roads, the capability of recognizing different types of crossroads and road forks as well as the capability of negotiating them in the direction desired is a key element. On unidirectional highways with multiple parallel lanes, selecting the proper lane – supported by navigation signs – is the key to find the connection to crossroads. On roads of lower order with the same-level connections between the crossing roads, new performance elements are necessary containing components of both perception and motion control.

10.1 General Introduction

Making a turn onto a crossroad on the side of standard driving (to the right in continental Europe and the Americas, to the left in the United Kingdom, *etc.*) is the easier of the two possibilities; crossing oncoming traffic lanes, usually, requires checking the traffic situation in these lanes too, which on high-speed roads means perception up to a large distance. The maneuvering capability developed for turnoffs is currently confined to the case where there is no interference with any other vehicles or obstacles, either on one's own or on the crossroad. This field has been pioneered by K. Müller; the reader interested in more details is referred to this dissertation [Müller 1996].

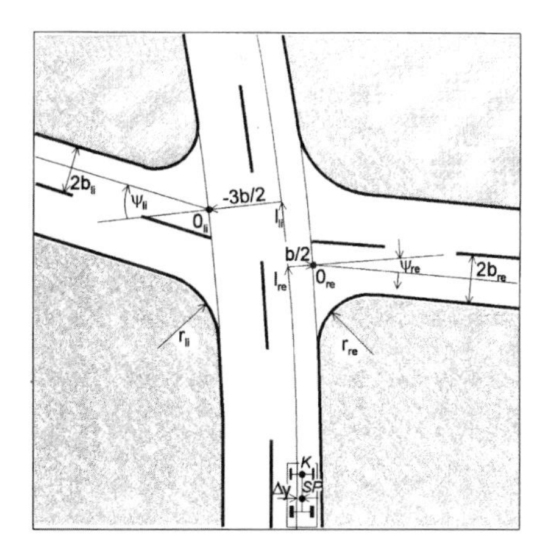

Figure 10.1. General geometry of an intersection: The precise location along the road driven, the width, and the intersection angle of the crossroad as well as the radii of curvature at the corners are not known in general; these parameters have to be determined by vision during approach.

It is assumed here that the higher levels of mission control in the overall system have been able to determine from odometry (or GPS) and map reading that the next upcoming crossroad (with certain visual features) will be the one to turn onto; the precise location, width and relative orientation, however, are unknown (see Figure 10.1).

These have to be determined while approaching the crossroad; therefore, speed will be reduced to make more processing time available per distance traveled and for slowing down to the speed allowed for curve driving without exceeding lateral acceleration limits (usually ≈ 2 m/s^2).

10.1.1 Geometry of Crossings and Types of Vision Systems Required

To estimate the distance to the intersection of two characteristic lines of crossing roads precisely, both the point of intersection and a sufficiently long part of the crossroad for recognizing its width and direction relative to the subject's road have to be viewed simultaneously. As a line characteristic of the road driven, the right or left boundary line is selected; at crossings, there may be no visible boundary line, so a "virtual" one has to be interpolated by a smooth curve connecting those before and after the intersection. Two points and two tangents allow deriving the parameters of a clothoid in the image (see Section 5.2.2.3); however, the real-world road parameters should be available from road-running (see below). As a line characteristic of the crossroad, the centerline of the lane intended to be turned onto is chosen, yielding the intersection points O_{re} for turning to the right and O_{li} for turning to the left (Figure 10.1).

The angle of intersection is measured relative to a right-angle intersection as the standard case; it is dubbed ψ_{re} on the right and ψ_{li} on the left side. For simplicity, the crossroad is modeled as a straight section in the region viewed. The road driven is characterized by its curvature parameters (known from methods of Chapter 9) and the road width (here designated as $2 \cdot b$); the desired driving state in the subject's lane is tangential at the center ($b/2$) with speed V. This leads to the space to be crossed on the subject's road of $b/2$ for turnoffs to the right and $-3b/2$ for those to the left. The actual offset Δy (and $\Delta\psi$) at the location of the cg is assumed to be known as well from the separate estimation process running for the road driven.

It has to be kept in mind that at larger look-ahead distances (of, say, 100, 60, and 20 m), a crossroad of limited width (say, 7 m) appears as a small, almost horizontal stripe in the image. Even intersection angles of ψ_{re} up to $\pm 50°$ lead to deviations from the image horizontal of $< 10°$, usually. A brief computation like that underlying Figure 5.4 yields the number of pixels on the crossroad (vertically) as a function of camera elevation H above the ground and focal length f (or, equivalently, resolution = number of pixel per degree). Table 10.1 shows results for the test vehicles VaMoRs and VaMP based on Table 7.1; evaluating only video fields, the number of rows available is about 240.

Table 10.1. Number of pixels vertically covering a crossroad of width 7 m at different look-ahead distances on planar ground for two focal lengths f (typical for video fields evaluated with ~ 240 rows) and two camera elevations above the ground

Vehicle	VaMoRs (H = 1.8 m)		VaMP (H = 1.3 m)	
Resolution in pixel/degree (f)	8	30	8	30
Look-ahead distance / m ↓	Number of pixels on crossroad of 7 m width (vert.)			
100	0.6	2.2	0.4	1.6
60	2.9	11	1.2	4.4
20	14	54	10	39

The number of pixels on the crossroad in the table increases by about a factor of 2 for evaluation of full video frames. Results for slightly different mapping conditions are shown in Figure 10.2 as a continuous function of range.

It can be seen that for the conditions of Table 10.1 at 100 m distance the crossroad covers just one and a half pixel for the car VaMP and about two for the van in the teleimage; in a noisy image from standard focal length (40° horizontal field of view) the crossroad is mapped onto a half pixel and cannot be detected. At 60 m distance, this lens yields ~1 pixel on the crossroad for the car and ~3 for the van. In the tele-image, with 4.4 pixels on the crossroad for the car VaMP, it may just become robustly distinguishable, while 11 pixels in the van allow easy recognition and tracking. At 20 m distance, the standard lens allows a similar appearance in the image (10, respectively, 14 pixels on the crossroad). However, at this distance, the teleimage covers just 3.7 m laterally on the ground, and a camera with standard lens mounted with its optical axis parallel to the longitudinal axis of the body will have a lateral range of about 12 m into the crossroad.

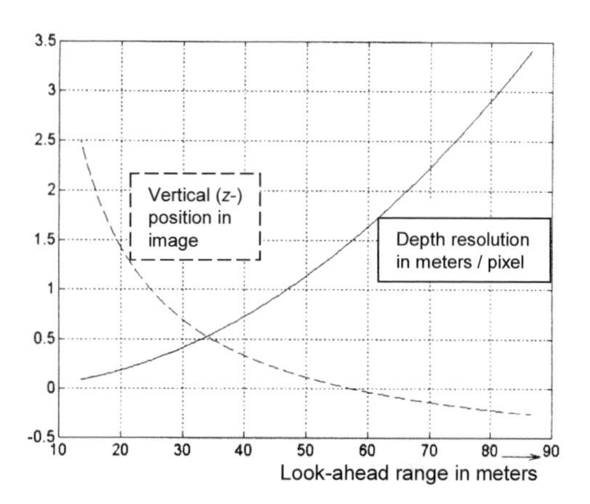

Figure 10.2. Coverage of horizontal distance by a single pixel as a function of look-ahead distance for a video image with improved mapping parameters compared to Table 10.1: f = 2200 pixel (= 25 mm), H = 2 m, and the camera looking 2° downward in pitch [Müller 1996]

These few numbers should make clear that a single camera or a pair of one standard and one telecamera mounted directly onto the body of the vehicle will not be sufficient for tight maneuvering at a road crossing. When approaching the crossing, the cameras have to be turned in the direction of the crossroad so that a sufficiently large part of it can be seen, allowing precise determination of its relative direction and width. The resulting "vehicle eye" will be discussed in Chapter 12.

10.1.2 Phases of Crossroad Perception and Turnoff

These considerations had led to active gaze control for a bifocal camera system for road vehicles from the beginning of these activities at UniBwM [Mysliwetz, Dickmanns 1986; Mysliwetz 1990; Schiehlen 1995]. Beside "looking into the curve" on curved roads and "fixation of obstacles on the road" while driving, developing the "general capability of perceiving crossroads and turning off onto them" was the major application area in the early 1990s.

K. Müller arrived at the following sequence of activities for this purpose; it was intended to be so flexible that most situations could be handled by minor adaptations. Seven phases can be distinguished; the first four are predominated by perception while the last three are strongly intertwined activities of perception, gaze control, and vehicle control.

The first phase is *prepare for crossroad perception according to the mission plan*, which is a slight turn of gaze direction toward the side of the expected crossroad. In the second phase, *visual search for candidate features* is performed with the favorite methods for feature extraction bottom-up. After sets of promising features have been found in a few consecutive images, the third phase is *hypothesis generation*, in which tracking the crossroad is initiated; now the speed for turning off is set which, usually, means slowing down so that a lateral acceleration of $a_{y,max}$ is not exceeded. In the fourth phase of *increasing precision in perception of parameters while approaching*, more complex gaze control strategies are started and proper adjustment of methods and tuning parameters is performed.

With the intersection parameters determined, (coarse) planning the behavioral modes and corresponding parameters is performed, yielding the distance from the crossing when to start the sequence of maneuver elements for turning off. The fifth phase then is *Start of motion control (steering for turnoff maneuver) with continuing (separate) gaze control and perception*; this part is rather involved, and timing has to be carefully tuned for the feed-forward components used. Well into this complex maneuver (60 to 70 %), the sixth phase is started by *switching to the crossroad as a (new) reference system*; feedback is superimposed both for gaze and for vehicle control to counteract unpredictable errors and perturbations encountered. The final phase of the turnoff maneuver is the transition to roadrunning on

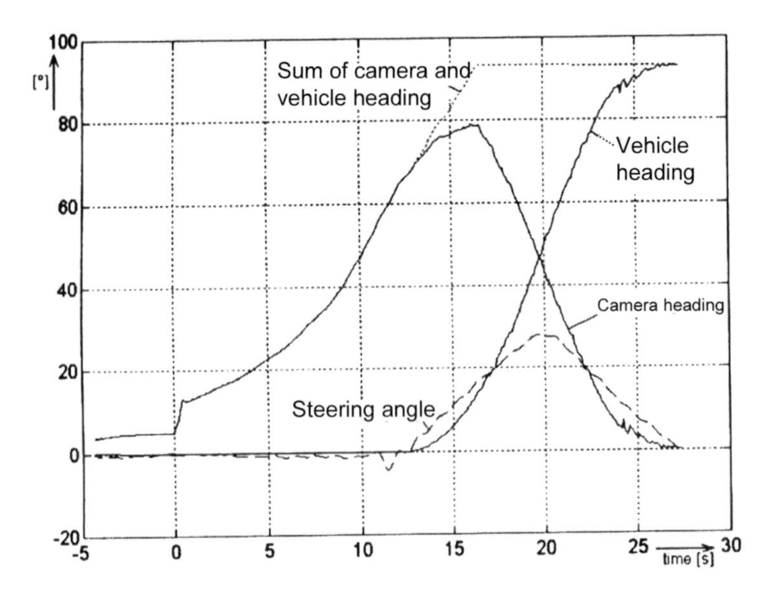

Figure 10.3. Camera heading relative to vehicle, steering angle, and absolute vehicle heading during a turnoff maneuver with VaMoRs [Müller, Baten 1995]

the new road element: *Finish feed-forward components* and store results observed during the turnoff maneuver for future evaluation.

Figure 10.3 shows a typical maneuver from real-world test results by displaying time histories of the camera heading angle, the vehicle heading, and steering angle as well as the summed angle of camera and vehicle heading. The maneuver lasts for 27 seconds (s); the last 16 include both gaze and lateral motion control. It starts with a fast shift of camera yaw angle (~ 8° at $t = 0$). Over the next 16 s, this angle continues to increase, initially fully controlled by feed-forward terms. When the steering angle starts turning the vehicle (noticeable at ~13 s), the sum of both angles determines the gaze direction in absolute space; at about 17 s, this sum reaches the orientation angle of the crossroad, and gaze has to be turned back at the rate of vehicle turn (17 to 27 s). Visual fixation of the lane center at a larger look-ahead range achieves this automatically; the vehicle turns underneath the crossroad-oriented cameras.

Before discussing these phases in detail, a survey of the hardware base (Section 10.1.3) and the theoretical background is given (Section 10.2).

10.1.3 Hardware Bases and Real-world Effects

Since on one hand the experimental vehicle VaMoRs was very busy in the EUREKA project *Prometheus* in the early 1990s, being used by other groups, and since, on the other hand, it is essential to have fully repeatable conditions available for testing complex maneuvers and their elements, initial development of *curve steering* (CS), as it was called, was done in a vision laboratory.

Hardware-In-the-Loop (HIL) simulation: This at that time unique installation for the development of autonomous vehicles with the sense of vision, derived from "hardware-in-the-loop" (HIL) testing for guided missiles with infrared sensors. Real vision hardware and gaze control was to be part of this advanced simulation intended to allow easy transfer of integrated hardware to test vehicles afterward (shaded area in lower center and right corner of Figure 10.4). It shows the HIL-simulation system developed and used for *autonomous visual curve steering* at UniBwM. Before this application, it had been in use for about a decade for developing the capability of visual autonomous landing approaches of unmanned air vehicles.

A simulation computer (top right) determines the coordinate time histories of all moving components of the dynamic scene under the differential equation constraints valid for them. The angular orientation of the subject vehicle is transferred to the DBS controller (center left) which translates this information via a three-axis angular motion simulator (DBS, hydraulic/electric, lower center left) into real physical angles of a platform onto which the vehicle eye is mounted. On top of the orientation angles computed for the subject vehicle, stochastic perturbations may be superimposed representing the effects of roughness of the road surface.

The coordinates describing position and orientation in the environment are sent to a real-time system for *computer-generated images* (CGI, center top) which generates the sequences of images at video rate to be projected onto a curved screen in front of the cameras (left). Machine vision with the original cluster of *Transputers* now closes the loop by sending the control inputs derived both to gaze control (center bottom, real feedback) and for control of the subject vehicle to the simulation computer via a special subsystem (center right).

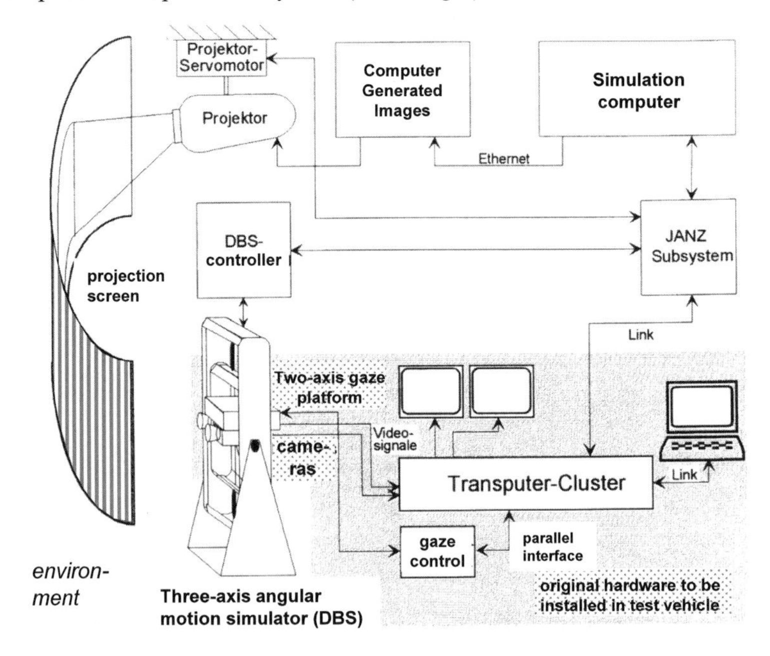

Figure 10.4. Hardware-in-the-loop simulation facility of UniBwM for the development of dynamic vision in autonomous vehicles: The shaded area (*lower right*) shows the original hardware intended for the test vehicle in a later development stage. The 'vehicle eye' with active gaze control and all computer systems for visual perception are tested with realistic time delays, disturbances, and nonlinearities. The visual environment (left) is generated by the simulation computer with CGI and projection.

A lot of the detailed developments of the CS-module have been done with this simulation loop, very similar to training pilots in simulation loops with visual displays of the dynamically changing environment. Nonlinear control limits of the real vehicle like limited steering rate and minimum turn radius possible can easily be imposed. Also the effects of time delays in measurement, interpretation, and control (up to several hundred ms) can easily be varied and studied. Surface roughness effects (uneven surface) can be simulated effectively by proper control of the mechanical unit DBS (capable of realizing fast disturbances with corner frequencies of up to 10 Hz).

There may be some artifacts from image generation and projection which have to be taken special care of; this was not too difficult. In summary, this simulation loop has saved quite a bit of development time and cost during the transition from pure computer simulation of everything to real autonomous driving.

Test vehicle VaMoRs: When the basic challenges of autonomous visual guidance of road vehicles were solved in the early 1990s, driving in public traffic on high-speed roads was emphasized in the second half of the "Prometheus" project; for this purpose, together with our partner Daimler-Benz AG, two S-class cars were equipped with new *Transputer*-based vision systems (vehicles VaMP and ViTA_2). The European development of the communication-oriented processors ("trans-" instead of com-puter) with easy scalability in larger networks allowed a big step forward, independent of the massively parallel computer architectures studied elsewhere.

For this reason, also the second-generation vision system for the older vehicle VaMoRs was to be based on transputers; this vehicle now had the assignment of continuing pioneering steps in vision in the framework of driving on networks of minor roads with and without lane markings and on unsealed surfaces or dirt roads.

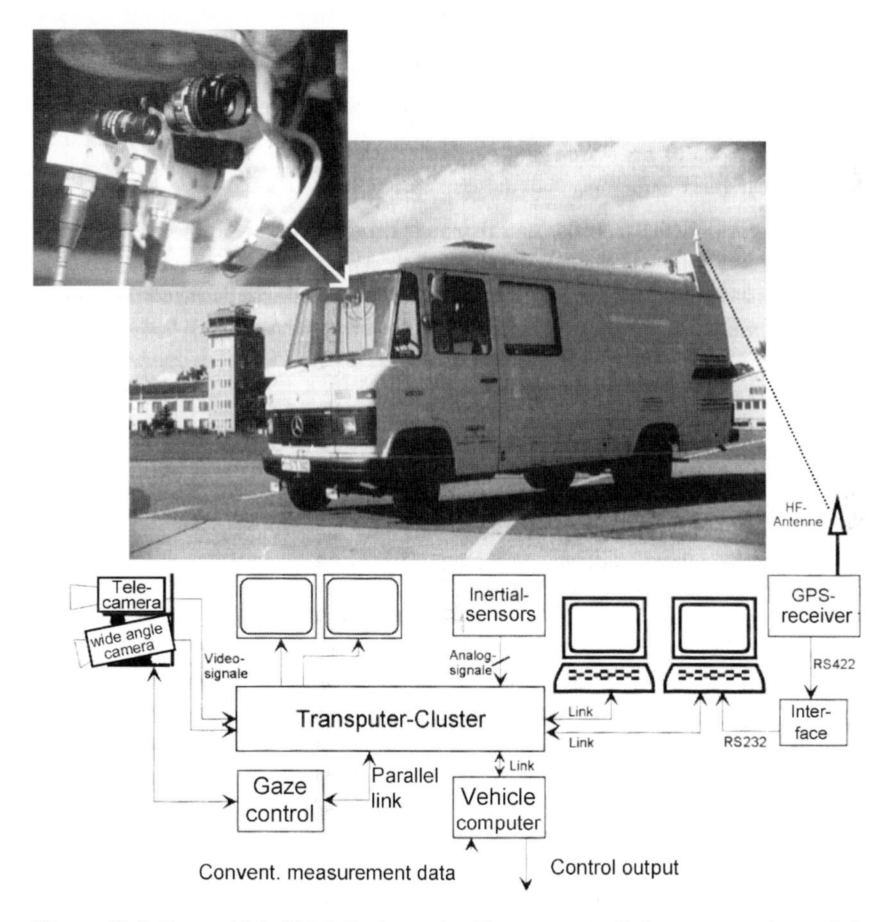

Figure 10.5. Test vehicle VaMoRs (center) with gaze-controlled camera set (upper left corner, exploded view of center top behind windshield, follow white arrow). The lower part of the collage is a block diagram of equipment carried onboard in the mid 1990s for the turnoff experiments on the closed-down airport serving as a test track for autonomous driving.

Tight maneuvering naturally comes into play on these types of roads. Here, it has to be kept in mind that the vehicle now generates different tracks for each wheel (refer to Section 3.4.2.2). The outer front wheel shows the largest radius of curvature while the inner rear wheel has the smallest. Both should not leave the intended lane (danger from curbstones during a left or right turn depending on left- or right-hand traffic!).

The precise shape of the tracks depends on the axle distance and the width of the wheel base as well as on the steering angle input and the speed driven. Figure 10.5 shows the test vehicle VaMoRs with its gaze controlled camera system (vehicle eye) as an exploded view (top left) and a coarse block diagram of the equipment onboard as the collage below (compare to Figure 10.4).

The upper camera has a telelens with focal length of $f = 25$ mm; of the two standard cameras below, just one with $f = 7.5$ mm has been used in these experiments. Gaze control can be very fast; 20° shifts can be performed in a fraction of a second. Camera stabilization in pitch is achieved by feedback from an inertial rate sensor on the gaze platform [Schiehlen 1995]. The steering rate of the vehicle is limited to $\lambda\text{-dot}_{max} = 15°/s$ around the pivot axle by power available; the minimal turn radius of the vehicle is $R_{min} \approx 6$ m. Time delays of different magnitude occur in most data paths. Conventional measurements (odometry, inertial data) are so fast that they can be considered instantaneous without making sensible errors, usually. Video has 40 ms cycle time, and since these data are shifted through the system after specific processing steps at this rhythm, three to five video cycles may elapse until final interpretation is available for control decision. Control output via a sequence of special processors in VaMoRs requires a few tenths of a second; these time delays cannot be neglected for precise steering.

10.2 Theoretical Background

Before the integrated performance of the maneuver can be discussed, performance elements for motion control of the vehicle (Section 10.2.1), for gaze control (Section 10.2.2), and for recursive estimation (Section 10.2.3) have to be discussed. Basic material has been covered in Chapters 2 – 6; here, specific items of interest for curve steering are detailed.

10.2.1 Motion Control and Trajectories

To gain basic insight into dynamic curve steering, some simplifications like driving at constant speed V and with piecewise constant steering rates $\lambda\text{-dot} = A$ are applied throughout, here. For passenger comfort, maximal lateral acceleration of $a_{y\text{-}max} = 2$ m/s^2 is set as an upper limit; the basic relation $a_y = V^2/R$ then fixes the maximal speed V allowed as a function of radius R. With the minimum turn radius R_{min} fixed by vehicle design, the maximal speed for this tight turn is given by

$$V = \sqrt{R \cdot a_{y\max}}; \quad \text{yielding } V_{\max}(R_{\min}). \tag{10.1}$$

For VaMoRs with the minimal turn radius of 6 m mentioned, this leads to $V_{max}(R_{min}) = \sqrt{12} = 3.46$ m/s, corresponding to ≈ 12.5 km/h. The inverse relation $R = V^2 / a_{y\max}$ fixes the minimal radius of turn R_{min} allowed as a function of speed V driven. The distinction between cg and wheel position will be discussed below; at these tight turns, outer front wheel, cg position, and inner rear wheel have quite different radii. Table 10.2 shows a few speeds of interest for tight curve steering based on cg coordinates. At the minimal turn radius of VaMoRs, the speed maximally allowed (~ 3.5 m/s) for $a_{ymax} = 2$ m/s^2 results in a steady turn rate of 33°/s (from simple relations for a body concentrated at the cg); slower speeds are never required to satisfy the lateral acceleration limit.

Table 10.2. Minimal turn radius allowed at speed V in order not to exceed the lateral acceleration limit $a_{ymax} = 2$ m/s^2 in tight curve steering (from simple kinematic relations); corresponding turn (yaw) rates. Last but one row gives the time needed at the steering rate of 15°/s to achieve the turn radius in row 3; the distance traveled can be seen in the last row.

Speed V (m/s)	1.5	2.5	3.5	5.0	7.5
" (km/h)	5.4	9.0	12.6	18.0	27.0
R_{min} ($a_{ymax} = 2$)	(1.125)	(3.125)	6.125	12.5	28.0
yaw rate (rad/s)	1.333	0.8	0.57	0.4	0.27
$d\psi/dt$ (°/s)	76.0	46.0	33.0	23.0	15.0
$t_{clothoid}$ at A_{max}	(time in	seconds)	2.0	1.04	0.49
Length (m)	(arc length	of clothoid)	7.0	5.2	3.67

Higher turn rates can never be achieved for this a_{ymax}. Increasing speed has the minimal turn radius growing drastically (third row, last two columns); this results in $R_{min} = 35$ m for $V = 30$ km/h. With a limit of the steering rate of $|A| = 15$°/s, the time needed for achieving the turn radius allowed ($\lambda_{max} = 30$ °) is given in the second row from the bottom. The corresponding distance traveled on the clothoid arc is shown in the last row.

These results have been achieved with the simple dynamic model discussed in Chapter 3 for the cg. In tight turns of larger vehicles, however, local conditions change considerably for different parts of the vehicle. The turn radius usually quoted is the one of the outer wheel track when driving a tight circle. For example, the van VaMoRs has an axle distance of $a = 3.5$ m, a track width of $b_{Tr} \sim 1.8$ m, and the minimal radius of turn is around 6 m, measured for the outer front wheel. From Figure 3.10, it can immediately be seen that if the turn radius of the outer front wheel is fixed as $R_{fout} = 6$ m, the turn radius of the inner rear wheel is

$$R_{ri} = \sqrt{(R_{fout}^2 - a^2)} - b_{Tr}, \tag{10.2}$$

which – with the numbers just given – yields $R_{ri} = 3.07$ m, almost half the value of R_{min}. With the center of gravity at the center between the axles, the turn radius of the cg is half the sum of both, about 4.5 m for VaMoRs. Since, for a constant turn rate, circumferential speed goes with R from the turn center, centrifugal acceleration (V^2/R) increases linearly with R; this is to say that the outer front wheel (about 30% further away than the cg) experiences a centrifugal acceleration correspondingly higher, while that of the inner rear wheel is as much lower. Limiting lateral acceleration at the cg thus leads to lower speed bounds; $R_{cg} \approx 4.5$ m thus requires V_{max} lowering V_{max} to $\sqrt{4.5 \cdot 2} = 3$ m/s. The turn rate goes up to 2/3 rad/s ~ 38°/s, which has the outer front wheel experience 2.67 m/s^2 and the inner rear wheel only 1.33 m/s^2 centrifugal acceleration.

The inner rear wheel, of course, is not allowed to leave the road and hit the curbstones; therefore, its track is critical in curve steering. From these considerations, for lateral accelerations limited to less than ~1/4 g (~ 2.5 m/s^2), the speed range up to about $V = 5$ m/s (18 km/h) is of most practical interest in tight curve steering.

10.2.1.1 Influence of Speed V Driven

As seen in the previous section, turn rates ψ-dot of up to 2/3 rad/s around the cg occur which means there are opposing lateral velocity components at the front and rear axles. For cg location at l_h from the rear and l_v from the front axle, $l_h + l_v = a$, (see Figure 3.23) the magnitudes are

$$\Delta\dot{y}_{\text{rearax}} = -\dot{\psi} \cdot l_h; \quad \Delta\dot{y}_{\text{frontax}} = +\dot{\psi} \cdot l_v \,. \tag{10.3}$$

These components, superimposed on the shifted velocity vector V with slip angle β at the cg, yield the actual velocity components V_f and V_r at the wheels, which – through angle of attack and tire stiffness – condition the forces between the ground and the tires. The point on the vehicle moving in direction of the longitudinal axis is P; it is the one closest to the pivot point M and is located at distance l_p in front of the rear axle. At this point, the following balance holds

$$V \cdot \sin\beta = (l_h - l_p)\dot{\psi} \,. \tag{10.4}$$

Assuming that the cg is at $l_h = l_v = a/2$ and β remains small so that sin β ≈ β, in connection with the dynamic model of Section 3.4.5.2, this results in the relation

$$l_p = V \cdot T_\beta = V^2 / k_{ltf} \tag{10.5}$$

for the location of point P in front of the rear axle. The outer point of the body on the line $\underline{PM}$ is the one coming closest to the vertical line above the curbstone in tight curves.

For the van VaMoRs the characteristic numbers needed are (Section 3.4.5.2) $m = 4000$ kg; $I_z = 9000$ kg m^2 (yielding $i_{zB}^2 = [i_z/(a/2)]^2 = 0.735$); $T_\beta = V/60$ and $T_\psi = V/81.7$, with V in m/s. At speeds up to $V = 5$ m/s, the two time constants will be < 0.1 seconds (eigenvalues on the negative real axis in the Laplace-transform domain with magnitude > 12). This means that within 0.3 seconds (the typical human time delay in visual perception), the stationary value is approached to better than 3%. For this reason, the dynamic equations for yaw rate $\dot{\psi}$ and slip angle β in Chapter 3.38 can be well approximated by algebraic ones:

$$\dot{\psi} = (V/a) \cdot \lambda; \tag{10.6}$$

$$\beta = (1/2 - T_\beta \cdot V/a) \cdot \lambda \,. \tag{10.7}$$

Since these equations contain speed, different trajectories will result for constant maximal steering rate A.

10.2.1.2 Maneuver Elements

These are supposed to consist of piecewise constant arcs at constant speed with steering rate λ-dot = ± *A*, respectively, zero for turnoff. Starting from driving straight ahead ($\lambda_0 = 0$) or on a circular arc ($\lambda_0 \neq 0$), the steering angle will increase or decrease linearly with time

$$\lambda = \lambda_0 \pm A \cdot t \,. \tag{10.8}$$

Inserting this into Equation 10.6 and integrating the resulting differential equation with respect to time yields

$$\begin{aligned} \psi(t) &= V / a \cdot (\lambda_0 \cdot t + A \cdot t^2 / 2), \\ \beta(t) &= (1/2 - T_\beta \cdot V / a) \cdot A \cdot t. \end{aligned} \tag{10.9}$$

Since the slip angle β is defined as the angle between the trajectory of the cg (tangent) and the heading of the vehicle body, for the clothoid arc defined by Equation 10.9 (with *A* and *V* = const.), the trajectory heading χ of the cg can now be written for curve initiation ($\lambda_0 = 0$) with Equations 10.7 and 10.9:

$$\begin{aligned} \chi = \psi + \beta &= \frac{VA}{2a} t^2 + \left(\frac{1}{2} - \frac{VT_\beta}{a}\right) A \cdot t \\ &= \frac{VA}{2a} t \cdot \left[t + \left(\frac{a}{V} - 2 \cdot T_\beta\right)\right]. \end{aligned} \tag{10.10}$$

With $V\,T_\beta / a = V^2/(60 \cdot 3.5) < 0.119$ for VaMoRs and $V < 5$ m/s, the second term in the last bracket is always smaller than the first one; this means that vehicle heading lags the trajectory heading during curve steering. The slip angle is positive here (turn right), as can be seen from Figure 10.6.

The minimum turn radius occurs for maximum steering angle λ_{max}. Dividing λ_{max} by the maximally possible steering rate A_{max} yields the minimal maneuver time to the tightest turning. For example, the data for VaMoRs: A_{max} = 15°/s and λ_{max} = 30° yield the time to the tightest turn T_{Rmin} = 2 seconds irrespective of the

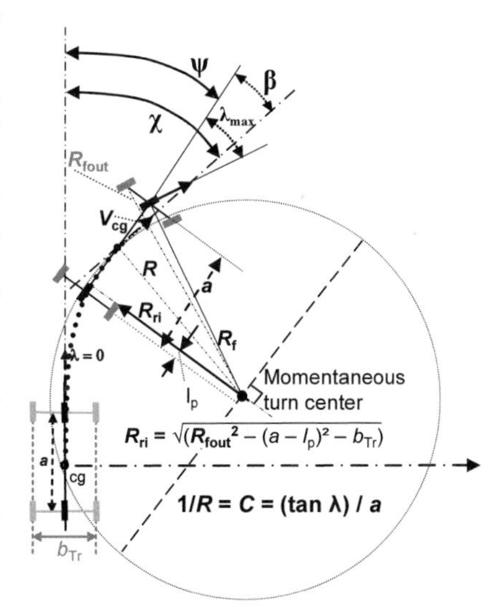

Figure 10.6. Curve initiation at constant speed *V* with constant steering rate *A* till λ_{max} (minimum turn radius). The critical parts are the outer front wheel (max. radius R_{fout}) and the inner rear wheel (min. radius R_{ri}). At slow speeds for tight curving, the slip angle β is positive, and trajectory heading χ is larger than vehicle body heading ψ. All wheels run at different speeds because of the different radii. Note that the vehicle does not turn around a center on the extended rear axle (as assumed from simple considerations, usually).

speed driven. This is to say that tighter turns are possible at lower speeds; however, in the limiting case of $V = 0$, the steered wheels will be turned on the spot and no distance is covered so that no turn of the vehicle occurs. If the vehicle starts driving with λ_{max}, the tightest possible turn is obtained: A circular arc with R_{min} as turn radius (higher order dynamic effects are neglected).

Note from Equation 10.5 and Figure 10.7 that for very small speeds, the actual center for translation and rotation on a circle lies on the extended rear axle. Equation 10.7 indicates that the speed vector at the cg is not aligned with the body axis but occurs at an angle proportional to the steering angle.

Driving a curve with $V_{cg} = 3.5$ m/s means that a distance of $V_{cg} \cdot T_{Rmin}$ (= 7 m in the example) is traveled until the minimal turn radius is reached. This situation is sketched in Figure 10.6. Equation 10.6 then yields the turn rate ψ-dot = 30°/s and Equation 10.9 the yaw angle ψ = 30° and the slip angle β = 13.25°. The heading angle χ_{Rmin} of the trajectory thus is 43.25°, almost half a right-angle turn. Since the second part of a 90°-curve-steering maneuver can be viewed as mirrored at π/2 (45°), the vehicle would have to drive on a circle with minimal turn radius and constant slip angle for an angle

$$\Delta\chi_{circle} = 2 \cdot (\pi/4 - \chi_{Rmin})$$

$$\text{or a time span} \qquad \Delta t_{circle} = \Delta\chi_{circle} / \dot{\psi} \tag{10.11}$$

before the inverse control input has to be started. In the example, $\Delta\chi = 3.5°$ and $\Delta =$ 0.117 s, during which the vehicle travels about 0.41 m along the circular arc.

The lateral offset of the cg from the initial straight line driven is obtained from the differential equation for y (last row of Equation 3.38):

$$\dot{y} = V \sin(\psi + \beta),$$

which with Equation 10.10 can be written as

$$\dot{y} = V \sin\left(\frac{VA}{2a}\left[t^2 + (\frac{a}{V} - 2T_\beta) \cdot t\right]\right) = V \sin(k_1 \cdot [t^2 + k_2 \cdot t]). \tag{10.12}$$

The approximation $\sin\chi \approx \chi - \chi^3/6$ yields (at π/4 the error is < 1%)

$$\chi^3 = k_1^3 \cdot [t^6 + 3k_2 t^5 + 3k_2^2 t^4 + k_2^3 t^3]$$
$$\dot{y}/V \approx k_1 \cdot [t^2 + k_2 \cdot t] - k_1^3 \cdot \{t^6/6 + k_2 t^5/2 + k_2^2 t^4/2 + k_2^3 t^3/6\}, \tag{10.13}$$

which can be integrated analytically to give (for V and A = const.)

$$y(t) = V \cdot k_1 \left(-k_1^2 \cdot \left[\frac{t^7}{42} + k_2 \frac{t^6}{12} + k_2^2 \frac{t^5}{10} + k_2^3 \frac{t^4}{24}\right] + \frac{t^3}{3} + k_2 \frac{t^2}{2}\right). \tag{10.14}$$

For the example of VaMoRs with $V = 3.5$ m/s and $A = 15°$/s, that is, $k_1 = 0.131$ rad/s^2 and $k_2 = 0.883$ s, the lateral offset when reaching R_{min} is $y_{fin} = 1.95$ m (at $t = T_{Rmin} = 2$ s); here, A (λ-dot) is set back to zero. The distance traveled along the arc is $T_{Rmin} \cdot V = 7$ m, of course. At other speeds below 3.5 m/s, A and T_{Rmin} remain the same; for speeds above 3.5 m/s, T_{Rmin} is reduced (see Table 10.2). Table 10.3 shows some results for curve initiation, for the central circular arc to be inserted, and for the total maneuver time at various constant (slow) speeds. It can be seen that the shortest maneuver time is achieved when the lateral acceleration limit is exploited (last row), but then the space needed is largest.

Table 10.3. Tight 90° curves at various speeds and maximal steering rate till R_{min} (test vehicle VaMoRs): Rows 2 to 9: Values when reaching R_{min}; row 8: Circular arc to be driven with R_{min}; row 9: Time on circular arc; row 10: Total time for 90° turn

1	V in m/s	0.5	1	1.5	2.5	3	3.5
2	ψ-dot in °/s	4.29	8.57	12.9	21.4	25.7	30
3	k_1 in rad/s^2	0.0187	0.374	0.561	0.095	0.112	0.131
4	k_2 in 1/s	6.98	3.47	2.28	1.32	1.067	0.883
5	ψ in °	4.29	8.57	12.9	21.4	25.7	30
6	β in °	15.0	14.9	14.7	14.1	13.7	13.25
7	$\chi = \psi + \beta$	19.25	23.5	27.6	35.5	39.4	43.25
8	$\Delta\chi_{circle}$	51.5	43	34.8	19	11.2	3.5
9	Δt_{circle}	12	5.0	2.7	0.89	0.44	0.117
10	$T_{90°}$ in s	16	9	6.7	4.89	4.44	4.12

The other extreme is driving at very slow speed so that distance traveled during steering is minimal. The theoretical limit is turning the wheel while the vehicle stands still; but of course then the vehicle has to be accelerated to move at all. It should then stop again when the body axis has reached the direction of the crossroad. Here, the steering angle has to be turned back to zero, and roadrunning in the crossroad can start with these new conditions. Figure 10.7 shows this limit case.

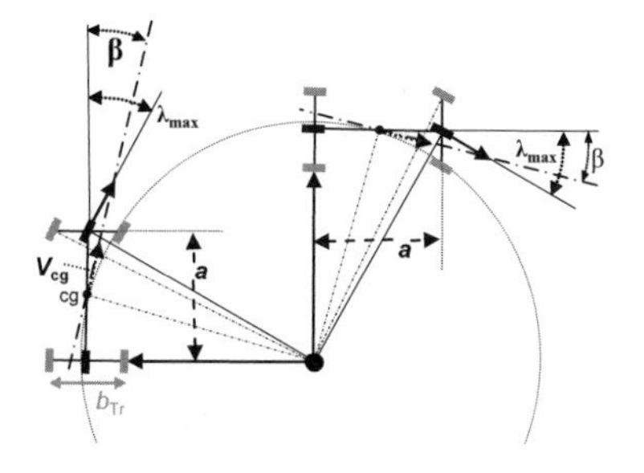

Figure 10.7. Theoretical limit for tightest turn possible: (1) Turn steering angle to maximum at standstill; (2) Drive circular arc until the direction of the crossroad is reached (here 90°); stop again. (3) Turn steering angle back to zero and start road running. Note that changing the steering angle λ at slow speed immediately turns the velocity vector; therefore, the circular arc driven starts at β and has to be driven till $(90 + \beta)°$

Note that the velocity vector turns directly with the steering angle; this implies that the cg moves on the circular arc from angle β to $(90 + \psi_{CR} + \beta)°$. This is due to the fact that the axle distance is a large percentage of the minimal turn radius. In the real world, accelerating and decelerating the vehicle with maximal steering angle will lead to different force distributions at the tires and much more complex behavior including shifts of the actual pivot point for circular motion. However, Figure 10.7 may give us an idea of what is really going to happen in this case. From Table 10.3, it can be seen that making the turn at speed $V = 2.5$ to 3 m/s (about 10 km/h) does not require too much time for tight turns at constant speed. The same is true for space needed (not shown in table), so that this speed range is a good compromise for driving tight curves.

A different approach when space is tight in the crossroad but not in the road driven is first to make a partial turn with lateral offset to the opposite direction of the crossroad and then turn into the crossroad with proper spacing; Figure 10.8 gives a sketch of the maneuver.

Müller (1996) has studied this case extensively; it will not be covered here.

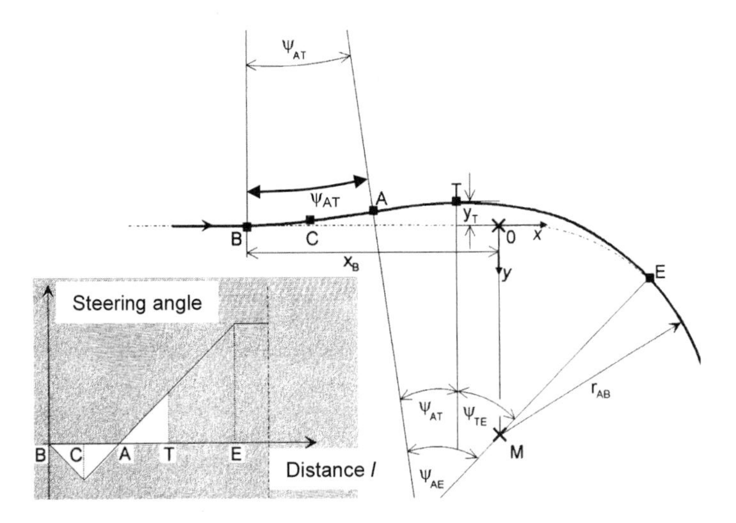

Figure 10.8. Curving into a tight crossroad with initial steering in opposite direction; the circular arc with minimum radius starts at point E

10.2.1.3 Behavioral Capabilities Required

The maneuver elements discussed in the previous section are special cases of feed-forward control which have to be complemented by feedback components whenever perturbations have some influence. The essential point of feed-forward control is the correct trigger point for initiation. Accuracy for entering the crossroad should be in the decimeter range. The example quoted from VaMoRs indicates that for a speed of $V = 3.5$ m/s, the actual pivot point for turning lies about 2 decimeters in front of the rear axle. This, as well as the delay time from control decision to actual movement of the actuator (over 200 ms corresponding to a 7-decimeter distance traveled at that speed), have to be taken into account in triggering the steering rate.

During the maneuver elements, the actual trajectories have to be observed and compared to the intended ones; if deviations are too large, either corrective feedback components have to be superimposed, or a new feed-forward control law has to be selected. If, for some reason, a dangerous situation occurs, *e.g.*, the vehicle starts leaving the intended lane, it should be stopped. (VaMoRs experienced this case when there was a failure in the power supply for the steering motor, reducing maximum steering rate achievable to ~ 50% of the nominal value.) As corrective feedback in standard cases, the usual one with pole assignment and gain scheduling as a function of speed [Zapp 1988; Brüdigam 1994] is sufficient for curve steering also [Müller 1996].

Toward the end of the turnoff maneuver, when the vehicle eye is locked stably onto the crossroad and when the lateral displacement of the vehicle from the center of the lane is not too large, corrective (additive) control output should be superimposed in any case to drive the deviations to zero. For the final part of the maneuver, the feed-forward control input may be dropped altogether.

10.2.2 Gaze Control for Efficient Perception

As mentioned in Section 10.1.2, there are at least four phases in gaze control for crossroad detection and turnoff in the scheme developed; they are discussed in detail here.

10.2.2.1 Attention-controlled Search for Crossroads Detection

In phase 1, additional windows for detection of the crossroad are positioned to that side of the road to be turned to, and feature search will be started at a trigger point given from navigation. Both edge- and area-based features are sought; the latter are more robust under the aspect conditions initially given. With little computing power available, edge features and adjacent average gray values are efficiently extracted with ternary masks (templates) as shown in Figure 5.5, since edge direction can be rather well predicted. The mask parameters have to be intelligently controlled by the interpretation process; they may vary with time of year and day and with the distance of interpretation (range).

With the computing power usually available nowadays, full sets of area-based features should not be omitted since they tend to be more robust and easier to interpret. Crossroads may have different gray values or texture; methods like UBM (Section 5.3) have also been developed with this task in mind. To get a larger part of the crossroad into the image, gaze direction is biased to the side of the turnoff by a few degrees, and search in vertical (column) stripes is started.

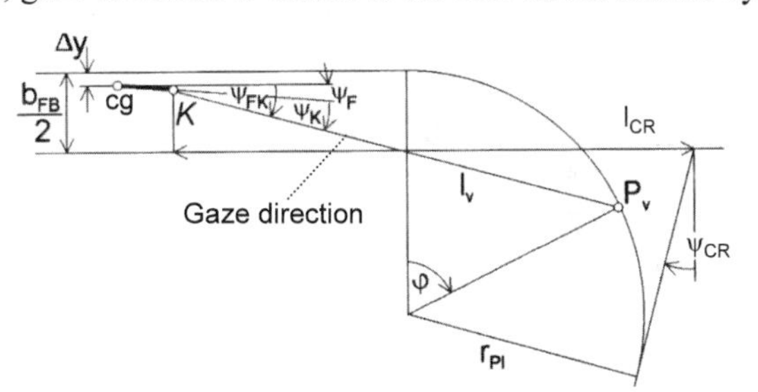

Figure 10.9. After detection of crossroad candidate connect the straight lines assumed to be driven on both roads (lanes) by a circular arc; find point P_v where gaze direction intersects this arc. Distance to P_v is fixed, which for a given speed V determines the yaw rate for gaze control.

In the case that a candidate for a crossroad is detected by a series of consistent feature sets over time, transition to phase 2 starts. Figure 10.9 shows this situation when the distance to the intersection of the center line of the crossroad with the boundary line of the road driven (l_{CR}) as well as the intersection angle ψ_{CR} are known approximately. If ψ_{CR} is corrupted by noise, starting with an intersection angle of 90° is sufficient for fast convergence while approaching the intersection, usually.

10.2.2.2 Phase 2: Fixation Point P_v on Circular Arc

In this phase, the viewing direction starts turning into the crossroad while approaching it at an approximately right angle. Speed V determines the yaw rate for gaze control if point P_v moves on the circular arc given (constant radius r_{Pl}) [Mueller 1996]:

$$\dot{\psi}_{K2} = V\cos\psi_F / (l_v \cdot (\sin\psi_{FK} + \cos\psi_{FK} \cdot \cot\varphi) - \dot{\psi}_F . \qquad (10.15)$$

Figure 10.10 visualizes the idea for computing the feed-forward yaw rate for gaze control approaching the junction.

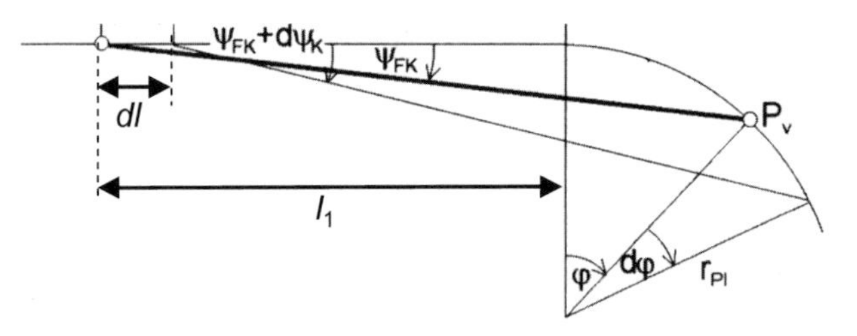

Figure 10.10. Determining yaw rate of gaze direction in phase 2 from fixation of point P_v at constant distance from the turning camera but moving on a circular arc connecting the straight center lines of the roads crossing.

10.2.2.3 Phase 3: View Fixation on Centerline of the Crossroad

Phase 3 of gaze control for turningoff into a crossroad starts where the circular arc and the following straight section in the crossroad meet. At this point, the fixation point starts moving at a proper speed V_{Pv} to avoid excessive gaze turn rates. The resulting feed-forward gaze turn rate for phase 3 then becomes

$$\begin{aligned}\dot{\psi}_{K3} &= V / l_v \cdot \sin(\psi_K - \beta) + V_{Pv} / l_v \cdot \cos(\psi_F - \psi_{CR} + \psi_K) \\ &\quad - (1 + l_K / l_v \cdot \cos\psi_K)\, \dot{\psi}_F .\end{aligned} \qquad (10.16)$$

Figure 10.11 shows the geometry; point P_v moves independently at speed V_{Pv} that has to be selected in a certain range for an appropriate turn rate of the gaze direction. The increasing yaw angles steadily improve the aspect conditions for precisely determining the two unknown parameters width and angle of intersection of the crossroad that determine the turnoff trajectory.

From the side constraint that look-ahead distance l_v should not grow during this phase, an upper limit for the speed V_{Pv} can be given;

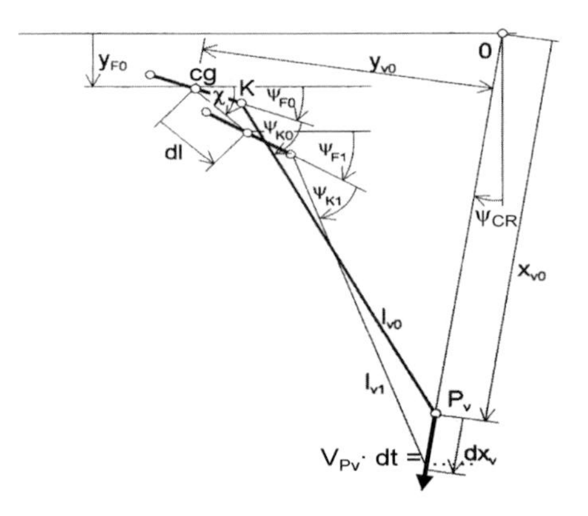

Figure 10.11. Gaze control in phase 3 for precisely determining angle ψ_{CR} and width b of the crossroad at an intersection.

$$V_{Pv,\max} = \left(\frac{\cos(\psi_K - \beta)}{\sin(\psi_F - \psi_{CR} + \psi_K)} \right) \cdot V + \left(\frac{l_k \cdot \sin \psi_K}{\sin(\psi_F - \psi_{CR} + \psi_K)} \right) \cdot \dot{\psi}_F \,. \tag{10.17}$$

Point P_v moves with constant speed $V_{Pv} < V_{Pv,max}$ on the crossroad until a predefined distance x_{fix} from the origin (O) is reached; it then fixates this point until the look-ahead distance is shortened to a predefined minimal value l_{vmin}. This limit is given by a certain image row under the condition of nominal pitch angle and planar ground using perspective projection by a pinhole model; it thus depends on the focal length of the camera used.

10.2.2.4 Phase 4: Evaluate Constant Image Row into the Crossroad

For this look-ahead distance l_{vmin} = const, a yaw rate for gaze control is performed leading to normal roadrunning on the crossroad. The essential component now is counteracting the vehicle turn rate [Müller 1996] (see second term):

$$\begin{aligned} \dot{\psi}_{K4} = & \left(\frac{\cos(\psi_F - \psi_{CR} + \beta)}{l_v \cdot \sin(\psi_F - \psi_{CR} + \psi_K)} \right) \cdot V \\ & - \left(1 + \frac{l_k \cdot \sin(\psi_F - \psi_{CR})}{l_v \cdot \sin(\psi_F - \psi_{CR} + \psi_K)} \right) \cdot \dot{\psi}_F \,. \end{aligned} \tag{10.18}$$

The vehicle turns underneath the cameras which are not yet locked onto the crossroad by visual feedback (next phase) but receive a feed-forward signal corresponding to this situation from theory. In Figure 10.3 this corresponds to $t >$ about 17 s.

10.2.2.5 Superimposed Feedback Control

While traveling into the crossroad, the feed-forward component of phase 4 fades out and a feedback component from centering the gaze on the lane center at l_{vmin} takes over; this also takes care of perturbations and the effects of imprecise perception and modeling accumulated up to now. Since the viewing direction is fixed into the crossroad, toward the end, it hardly turns relative to the spine of this road; however, the vehicle turns underneath the camera.

Here, only the basic considerations for gaze control have been discussed. Feed-forward control for pitch and feedback components in yaw also in phases 2 to 4 have been studied for making gaze behavior more robust; details may be found in [Müller 1996, Section 5.4].

10.2.3 Models for Recursive Estimation

Two sets of data have to be determined while approaching the crossroad: (1) the shape parameters of the crossing and the crossroad and (2) the vehicle position relative to the crossroad.

10.2.3.1 Shape Parameters of Crossing

The parameters, crossing angle (deviation from 90°) ψ_{CR} and lane width in the crossroad, selected as $2{\cdot}b_{CS}$ here (see Figure 10.12 below), have to be determined from visual observation. The radius r_{CS} of the borderline (curbstones) connecting the two roads in the curve is assumed to be standard and known. The distance to the crossroad l_{CS} has to be estimated precisely to find the right trigger point for the initiation of the steering angle feed-forward maneuver. All parameters of, and the relative egostate to the road actually driven are assumed to be known from application of the methods discussed in Chapter 9.

System model: Figure 10.1 has shown the basic geometry of a general crossing including the rounded corners assumed to be circular arcs of radius r. The coordinate x points in the driving direction, y to the right, and z downward (in the direction of gravity, international standard from air vehicle analysis). Initially, when the yaw angle for gaze relative to the road driven is small, its cosine is assumed to be 1, and measurement of features on the crossroad only in the vertical direction of the images is sufficient. These features in the real world as well as the intersection point of the two roads then move only in the x-direction. Therefore, for each of the two intersection parameters, a second-order dynamic model driven by noise is used here; for sampled data, this is written as

$$\mathrm{x}_{k+1} = \Phi_k \cdot \mathrm{x}_k + \mathrm{q}_k = \begin{pmatrix} 1 & T \\ 0 & 1 \end{pmatrix} \cdot \begin{pmatrix} x_k \\ \dot{x}_k \end{pmatrix} + \begin{pmatrix} T \\ 1 \end{pmatrix} q_1(t_k). \tag{10.19}$$

T is the cycle time of sampling (for CCIR-video = 40 ms), and $q_1(\mathrm{t_k})$ is the discrete noise term driving estimation.

The distance l_{CR} to the intersection is affected directly by longitudinal control inputs to the vehicle (throttle and brakes); accelerations and decelerations should thus be included in the model, increasing the order of the system by one (to three). Because of the strong dependence on the point of operation and on parameters, there is no simply computable relationship between actuator input and effective acceleration/deceleration. Therefore, "colored noise" has been chosen as a model, described in the following equation by the term $-\alpha$:

$$\underline{\dot{x}}(t) = \begin{pmatrix} 0 & 1 & 0 \\ 0 & 0 & 1 \\ 0 & 0 & -\alpha \end{pmatrix} \cdot \begin{pmatrix} l_{CR} \\ \dot{l}_{CR} \\ \ddot{l}_{CR} \end{pmatrix} + \begin{pmatrix} 0 \\ 0 \\ q_l \end{pmatrix}. \tag{10.20}$$

α represents the reciprocal correlation time constant of colored noise; based on experimental results for VaMoRs, the value $\alpha = 1/2\ \mathrm{s}^{-1}$ has been adopted. Assuming that accelerations/decelerations are constant over one cycle, the following discrete model results:

$$x_{k+1} = \begin{pmatrix} 1 & T & \varphi_{13} \\ 0 & 1 & \varphi_{23} \\ 0 & 0 & \varphi_{33} \end{pmatrix} \cdot x_k + \begin{pmatrix} \varphi_{13} \\ \varphi_{23} \\ \varphi_{33} \end{pmatrix} \cdot q_l(t_k), \tag{10.21}$$

with $\varphi_{13} = (\alpha T - 1 + e^{-\alpha T})/\alpha^2$; $\varphi_{23} = (1 - e^{-\alpha T})/\alpha$; $\varphi_{33} = e^{-\alpha T}$ for colored noise; for white noise, these terms are

$$\varphi_{13} = T^2/2; \qquad \varphi_{23} = T; \qquad \varphi_{33} = 1.$$

In tests, it has been found advantageous not to estimate the shape parameters and distance separately, but in a single seventh-order system; the symmetry of the crossroad boundaries measured by edge features tends to stabilize the estimation process:

$$x_{k+1} = \begin{pmatrix} 1 & T & \varphi_{13} & 0 & 0 & 0 & 0 \\ 0 & 1 & \varphi_{23} & 0 & 0 & 0 & 0 \\ 0 & 0 & \varphi_{33} & 0 & 0 & 0 & 0 \\ 0 & 0 & 0 & 1 & T & 0 & 0 \\ 0 & 0 & 0 & 0 & 1 & 0 & 0 \\ 0 & 0 & 0 & 0 & 0 & 1 & T \\ 0 & 0 & 0 & 0 & 0 & 0 & 1 \end{pmatrix} \cdot \begin{pmatrix} l_{CR} \\ \dot{l}_{CR} \\ \ddot{l}_{CR} \\ b_{CS} \\ \dot{b}_{CS} \\ \psi_{CR} \\ \dot{\psi}_{CR} \end{pmatrix}_k + \begin{pmatrix} \varphi_{13} \cdot q_l \\ \varphi_{23} \cdot q_l \\ \varphi_{33} \cdot q_l \\ T \cdot q_b \\ q_b \\ T \cdot q_\psi \\ q_\psi \end{pmatrix}_k . \tag{10.22}$$

The last vector (noise term) allows determining the covariance matrix Q of the system. The variance of the noise signal is

$$\sigma = \mathrm{E}\{(q - \bar{q})\} = \mathrm{E}\{(q)\} \quad \text{for } \bar{q} = 0 . \tag{10.23}$$

Assuming that the noise processes q_l, q_b, and $q_{l\psi}$ are uncorrelated, the covariance matrix needed for recursive estimation is given by Equation 10.24.

The standard deviations σ_l, σ_b, and σ_ψ have been determined by experiments with the real vehicle; the values finally adopted for VaMoRs were $\sigma_l = 0.5$, $\sigma_b = 0.05$ and $\sigma_\psi = 0.02$.

$$Q = E\{q \cdot q^T\} =$$

$$\begin{pmatrix} \varphi_{13}^2 \cdot \sigma_l^2 & \varphi_{13} \cdot \varphi_{23} \cdot \sigma_l^2 & \varphi_{13} \cdot \varphi_{33} \cdot \sigma_l^2 & 0 & 0 & 0 & \sigma_\psi \\ \varphi_{23} \cdot \varphi_{13} \cdot \sigma_l^2 & \varphi_{23}^2 \cdot \sigma_l^2 & \varphi_{23} \cdot \varphi_{33} \cdot \sigma_l^2 & 0 & 0 & 0 & 0 \\ \varphi_{33} \cdot \varphi_{13} \cdot \sigma_l^2 & \varphi_{33} \cdot \varphi_{23} \cdot \sigma_l^2 & \varphi_{33}^2 \cdot \sigma_l^2 & 0 & 0 & 0 & 0 \\ 0 & 0 & 0 & T^2 \cdot \sigma_b^2 & T \cdot \sigma_b^2 & 0 & 0 \\ 0 & 0 & 0 & T \cdot \sigma_b^2 & \sigma_b^2 & 0 & 0 \\ 0 & 0 & 0 & 0 & 0 & T^2 \cdot \sigma_\psi^2 & T \cdot \sigma_\psi^2 \\ 0 & 0 & 0 & 0 & 0 & T \cdot \sigma_\psi^2 & \sigma_\psi^2 \end{pmatrix} . \tag{10.24}$$

Measurement model: Velocity measured conventionally is used for the vision process since it determines the shift in vehicle position from frame to frame with little uncertainty. The vertical edge feature position in the image is measured; the one-dimensional measurement result thus is the coordinate z_B. The vector of measurement data therefore is

$$\mathrm{y}^T = (V, z_{B0}, z_{B1}, z_{B2}, \ldots\ldots, z_{Bi}, \ldots\ldots, z_{Bm}) . \tag{10.25}$$

The predicted image coordinates follow from forward perspective projection based on the actual best estimates of the state components and parameters. The partial derivatives with respect to the unknown variables yield the elements of the Jacobian matrix C (see Section 2.1.2):

$$\underline{C} = \partial y / \partial x \,. \tag{10.26}$$

The detailed derivation is given in Section 5.3 of [Müller 1996]; here only the result is quoted. The variable *dir* has been introduced for the direction of turn-off:

$$\begin{aligned} dir &= +1 \text{ for turns to the right} \\ dir &= -1 \text{ for turns to the left ;} \end{aligned} \tag{10.27}$$

the meaning of the other variables can be seen from Figure 10.12. The reference point for perceiving the intersection and handling the turnoff is the point 0 at which the right borderline of the (curved) road driven and the centerline of the crossroad (assumed straight) intersect. The orthogonal line in point 0 and the centerline of the crossroad define the relative heading angle ψ_{CR} of the intersection (relative to 90°).

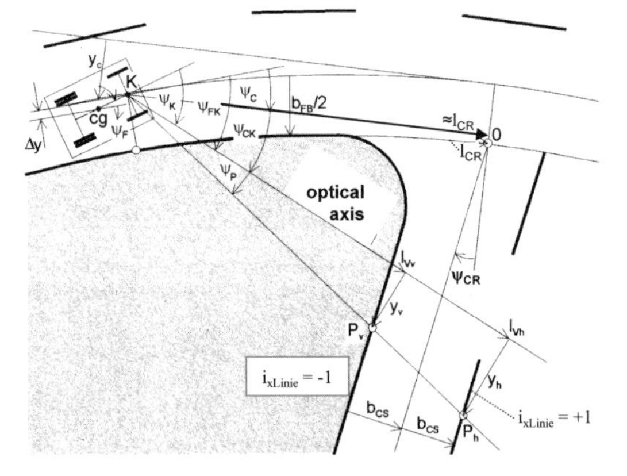

Figure 10.12. Visual measurement model for crossroad parameters during the approach to the crossing: Definition of terms used for the analysis

Since for vertical search paths in the image, the horizontal coordinate y_B is fixed, the aspect angle of a feature in the real world is given by the angle ψ_{Pi}

$$\tan \psi_{Pi} = y_{Bi} / f \cdot k_y \,. \tag{10.28}$$

In addition, an index for characterizing the road or lane border of the crossroad is needed. The "line index" ix_{Linie} is defined as follows:

$$ix_{Linie} = \begin{cases} -1 \ldots \text{ right border of right lane of crossroad} \\ +1 \ldots \text{ left border of right lane of crossroad} \\ +3 \ldots \text{ left border of left neighboring lane} \end{cases} . \tag{10.29}$$

With the approximation that l_{CR} is equal to the straight-line distance from the camera (index K) to the origin of the intersection coordinates (see Figure 10.12), the following relations hold:

$$\begin{aligned} l_{CR} &= \frac{l_{vi}}{\cos \psi_{Pi}} \cos(\psi_{CK} + \psi_{Pi}) + ix_{Linie} \frac{b_{CS}}{\cos \psi_{CR}} + y_{Pi} \tan \psi_{CR}, \\ y_{Pi} &= \frac{l_{vi}}{\cos \psi_{Pi}} \sin(\psi_{CK} + \psi_{Pi}) + dir \cdot y_K, \\ y_K &= l_K \sin \psi_F + \Delta y \cos \psi_c + y_c - (n_{spur} + 1/2) \cdot b_{FB}, \end{aligned} \tag{10.30}$$

with n_{spur} = number of lanes to be crossed on the subject's road when turning off.

Setting $\cos\psi_c \approx 1$ allows resolving these equations for the look-ahead range l_{vi}:

$$l_{vi} = \frac{l_{CR} \cdot \cos \psi_{CR} - ix_{Linie} \cdot b_{CS} - dir \cdot y_K \sin \psi_{CR}}{\cos(\psi_{CK} - \psi_{CR}) - [y_{Bi} / (f \cdot k_y)] \cdot \sin(\psi_{CK} - \psi_{CR})} . \tag{10.31}$$

Each Jacobian element gives the sensitivity of a measurement value with respect to a state variable. Since the vertical coordinate in the image depends solely on the look-ahead range l_{vi} directly, the elements of the Jacobian matrix are determined by applying the chain rule,

$$\partial z_B / \partial x_j = (\partial z_B / \partial l_{vi}) \cdot (\partial l_{vi} / \partial x_j). \tag{10.32}$$

The first multiplicand can be determined from Section 9.2.2; the second one is obtained with

$$k_1 = \{cos(\psi_{CK} - \psi_{CR}) - [y_{Bi} / (f \cdot k_y)] \cdot \sin(\psi_{CK} - \psi_{CR})\}^{-1} \tag{10.33}$$

as

$$\begin{aligned}
\frac{\partial l_{vi}}{\partial l_{CR}} &= k_1 \cdot (cos\psi_{CR} - dir \cdot \psi_c \cdot \sin\psi_{CR}),\\
\frac{\partial l_{vi}}{\partial b_{Cs}} &= k_1 \cdot (-ix_{Linie}),\\
\frac{\partial l_{vi}}{\partial \psi_{CR}} &= k_1 \cdot \{-l_{CR} \cdot \sin\psi_{CR} - dir \cdot y_{Bi} \cdot cos\psi_{CR}\\
&\quad - l_{vi}[\sin(\psi_{CK} - \psi_{CR}) + [y_{Bi} / (f \cdot k_y)] \cdot \cos(\psi_{CK} - \psi_{CR})]\}.
\end{aligned} \tag{10.34}$$

With Equation 10.32, the C-matrix then has a repetitive shape for the measurement data derived from images, starting in row 2; only the proper indices for the features have to be selected. The full Jacobian matrix is given in Equation 10.35. The number of image features may vary from frame to frame due to changing environmental and aspect conditions; the length of the matrix is adjusted corresponding to the number of features accepted.

$$\underline{C} = \begin{pmatrix}
0 & -1/\cos\psi_{CR} & 0 & 0 & 0 & 0 & 0\\
\frac{\partial z_{B0}}{\partial l_{CR}} & 0 & 0 & \frac{\partial z_{B0}}{\partial b_{CR}} & 0 & \frac{\partial z_{B0}}{\partial \psi_{CR}} & 0\\
\vdots & \vdots & \vdots & \vdots & \vdots & \vdots & \vdots\\
\frac{\partial z_{Bi}}{\partial l_{CR}} 0 & 0 & 0 & \frac{\partial z_{Bi}}{\partial b_{CR}} & 0 & \frac{\partial z_{Bi}}{\partial \psi_{CR}} & 0\\
\vdots & \vdots & \vdots & \vdots & \vdots & \vdots & \vdots
\end{pmatrix}. \tag{10.35}$$

Statistical properties of measurement data: They may be derived from theoretical considerations or from actual statistical data. Resolution of speed measurement is about 0.23 m/s; maximal deviation thus is half that value: 0.115 m/s. Using this value as the standard deviation is pessimistic; however, there are some other effects not modeled (original measurement data are wheel angles, slip between tires and ground, *etc.*). The choice made showed good convergence behavior and has been kept unchanged.

Edge feature extraction while the vehicle was standing still showed an average deviation of about 0.6 pixels. While driving, perturbations from uneven ground, from motion blur, and from minor inaccuracies in gaze control including time lags increase this value. Since fields have been used in image evaluation (every second row only), the standard deviation of 2 pixels was adopted.

Assuming that all these measurement disturbances are uncorrelated, the following diagonal measurement covariance matrix for recursive estimation results

$$\underline{R} = \mathrm{Diag}(\sigma_V^2, \sigma_{z_B}^2, ...\sigma_{z_B}^2, ...) . \qquad (10.36)$$

The relations described are valid for features on the straight section of the crossroad; if the radius of the rounded corner is found, more complex relations have to be taken into account.

Feature correlation between real world and images: Image interpretation in general has to solve the big challenge of how features in image space correspond to features in the real world. This difficulty arises especially when distances have to be recovered from perspective mapping (see Figure 5.4 and Section 7.3.4). Therefore, in [Müller 1996] appreciable care was taken in selecting features for the object of interest in real space.

Each extracted edge feature is evaluated according to several criteria. From image "windows" tracking the same road boundary, an extended straight line is fit to the edge elements yielding the minimal sum of errors squared. This line is accepted only if several other side constraints hold. It is then used for positioning of measurement windows and prediction of expected feature locations in the windows for the next cycle. Evaluation criteria are prediction errors in edge element location, the magnitude of the correlation maximum in CRONOS, and average gray value on one side of the edge. For proper scaling of the maximal magnitude of the correlation results in all windows, $korr_{max}$ as well as maximal and minimal intensities of all edge elements found in the image are determined. For each predicted edge element of the crossroad boundaries, a "window of acceptance" is defined (dubbed "*basis*") in which the features found have to lie to be accepted. The size of this window changes with the number of rows covered by the image of the crossroad (function of range). There is a maximal value $basis_{\mathrm{max}}$ that has an essential influence on feature selection.

In preliminary tests, it has turned out to be favorable to prefer such edges that lie below the predicted value, *i.e.*, which are closer in real space. This results in an oblique triangle as a weighting function, whose top value lies at the predicted edge position (see Figure 10.13).

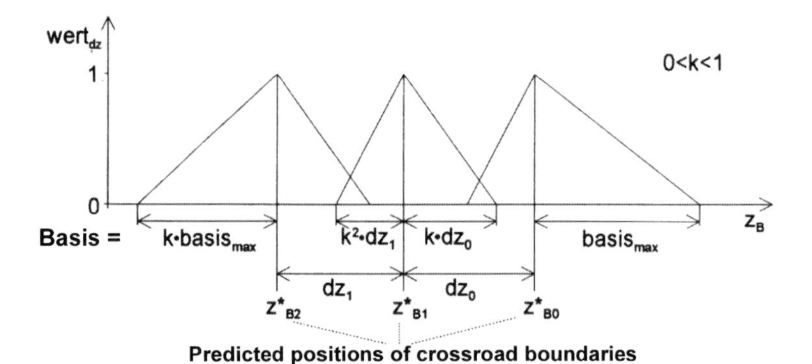

Figure 10.13. Scheme for weighting features as a function of prediction errors

The weight in window i for a prediction error thus is

$$wert_{dz,i} = \begin{cases} 1 - \dfrac{(z_{B,mess} - z^*_B)}{basis_{pos}} \ldots \text{for} \ldots z^*_B \le z_{B,mess} \le z^*_B + basis_{pos} \\ 1 + \dfrac{(z_{B,mess} - z^*_B)}{basis_{neg}} \ldots \text{for} \ldots z^*_B > z_{B,mess} \ge z^*_B - basis_{neg} \\ 0 \qquad \text{else.} \end{cases} \tag{10.37}$$

Here, $basis_{pos}$ designates the baseline of the triangle in positive z_B-direction (downward) and $basis_{neg}$ in the negative z_B direction. The contribution of the mask response $korr_i$ and the average intensity I_i on one side of the CRONOS-mask to the overall evaluation is done in the following way: Subtraction of the minimal value of average intensity increases the dynamic range of the intensity signal in non-dimensional form: $(I_i - I_{min})/(I_{max} - I_{min})$. The total weight $wert_i$ is formed as the weighted sum of these three components:

$$wert_i = \begin{cases} k_{dz} \cdot wert_{dz,i} + k_{korr} \dfrac{korr_i}{korr_{max}} + k_{grau} \dfrac{I_i - I_{min}}{I_{max} - I_{min}} & \text{for} \quad wert_{dz,i} > 0, \\ 0 & \text{for} \quad wert_{dz,i} = 0. \end{cases} \tag{10.38}$$

The factors k_{dz}, k_{korr}, and k_{grau} have been chosen as functions of the average distance between the boundaries of the crossroad in the image: $\overline{dz_B}$ (see Figure 10.14).

The following considerations have led to the type of function for the factors k_i: Seen from a large distance, the lines of the crossroad boundaries are very close together in the image. The most important condition to be satisfied for grouping edge features is their proximity to the predicted coherent value according to the model ($k_{dz} > k_{korr}$ and k_{grau}). The model thus supports itself; it remains almost rigid.

Approaching the crossroad, the distance between the boundaries of the crossroad in the image starts growing. The increasingly easier separation of the two boundary lines alleviates grouping features to the two lines by emphasizing continuity conditions in the image; this means putting more trust in the image data relative to the model (increasing k_{korr}). In this way, the model parameters are adjusted to the actual situation encountered.

Beside the correlation results, the average intensity in one-half of the CRONOS mask is a good indicator when the distance to the crossing is small and several pixels fall on bright lines for lane or boundary marking. A small distance means a large value of $\overline{dz_B}$; in Figure 10.14 this intensity criterion k_{grau} is used only when $\overline{dz_B} > basis_{max}$. Values of $basis_{max}$ in the range of 20 to 30 pixels are satisfactory. Beyond this point, the boundary lines are treated completely separately.

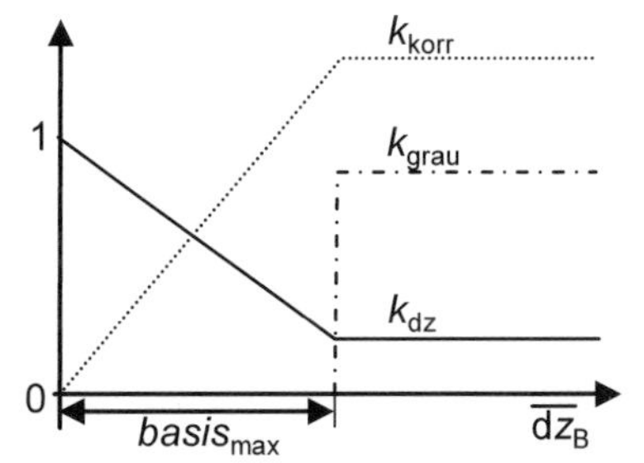

Figure 10.14. Parameters of weighting scheme for edge selection as function of width of crossroad dz_B in the image (increases with approach to the crossing)

The edge features of all windows with the highest evaluation results around the predicted boundary lines are taken for a new least-squares line fit, which in turn serves for making new predictions for localization of the image regions to be evaluated in the next cycle. The fitted lines have to satisfy the following constraints to be accepted:

1. Due to temporal continuity, the parameters of the line have to be close to the previous ones.
2. The distance between both boundaries of the crossroad is allowed to grow only during approach.
3. The slopes of both boundary lines in the image are approximately equal; the more distant line has to be less inclined relative to the horizontal than the closer one.

With decreasing distance to the crossroad, bifocal vision shows its merits. In addition to the teleimage, wide-angle images have the following advantages:

- Because of the reduced resolution motion blur is also reduced, and the images are more easily interpreted for lateral position control in the near range.
- Because of the wider field of view, the crossroad remains visible as a single object down to a very short distance with proper gaze control.

Therefore, as soon as only one boundary in the teleimage can be tracked, image evaluation in the wide-angle image is started. Now the new challenge is how to merge feature interpretation from images of both focal lengths. Since the internal representation is in (real-world) 3-D space and time, the interpretation process need not be changed. With few adaptations, the methods discussed above are applied to both data streams. The only changes are the different parameters for forward perspective projection of the predicted feature positions and the resulting changes in the Jacobian matrix for the wide-angle camera (a second measurement model); there is a specific Jacobian matrix for each object–sensor pair.

The selection algorithm picks the best suited candidates for innovation of the parameters of the crossroad model. This automatically leads to the fade out of features from the telecamera; when this occurs, further evaluation of tele images is discarded for feedback control. Features in the far range are continued because of their special value for curvature estimation in roadrunning.

10.2.3.2 Vehicle Position Relative to Crossroad

During the first part of the approach to an intersection, the vehicle is automatically visually guided relative to the road driven. At a proper distance for initiation of the vehicle turn maneuver, the feed-forward control time history is started, and the vehicle starts turning; a trajectory depending on environmental factors will result. The crossroad continues to be tracked by proper gaze control.

When the new side of the road or the lane to be driven into can be recognized in the wide-angle image, it makes sense immediately to check the trajectory achieved relative to these goal data. During the turn into the crossroad, its boundary lines tend to move away from the horizontal and become more and more diagonal or even closer to vertical (depending on the width of the crossroad). This means that in the edge extractor CRONOS, there has to be a switch from vertical to horizontal search paths (performed automatically) for optimal results. Feature interpretation (especially the precise one discussed in Section 9.5) has to adapt to this procedure. The state of the vehicle relative to the new road has to be available to correct errors accumulated during the maneuver by feedback to steering control. For this purpose, a system model for lateral vehicle guidance has to be chosen.

System model: Since speed is small, the third-order model may be used. Slightly better results have been achieved when the slip angle also has been estimated; the resulting fourth-order system model has been given in Figure 7.3b and Equation 7.4 for small trajectory heading angles χ ($\cos\chi \approx 1$ is sufficient for roadrunning with χ measured relative to the road). When turning off onto a crossroad, of course, larger angles χ have to be considered. In the equation for lateral offsets y_V, now the term $V{\cdot}\cos\chi$ occurs twice (instead of just V).

After transition to the discrete form for digital processing (cycle time T) with the state vector $\underline{\Delta x_q}^T = [y_q, \psi_q, \beta_q, \lambda_q]$ (here in reverse order of the components compared to Equation 7.4 and with *index q* for the turnoff maneuver, see Equation 10.40), the dynamic model directly applicable for recursive estimation is, with the following abbreviations entering the transition matrix Φ_k and the vector b_k multiplying the discrete control input

$$p_1 = V\cdot\cos\chi; \qquad p_2 = V/a;$$

$$p_3 = [1/(2T_\beta) - p_2]; \qquad p_4 = [1 - \exp(-T/T_\beta)];$$

$$\Delta x_{k+1} = \Phi_k(T)\cdot\Delta x_k + b_k(T)\cdot u_k;$$

where

$$\Phi_k(T) = \begin{bmatrix} 1 & p_1 T & p_1 p_4 T_\beta & p_1 p_2\cdot T^2/2 + p_1 p_3\cdot T_\beta^2 (T/T_\beta - p_4) \\ 0 & 1 & 0 & p_2 T \\ 0 & 0 & 1 - p_4 & p_3 p_4 T_\beta \\ 0 & 0 & 0 & 1 \end{bmatrix},$$

$$b_k(T) = \begin{bmatrix} p_1 p_2\cdot T^3/6 + p_1 p_3\cdot T_\beta^3 [T^2/(2T_\beta^2) - T/T_\beta - p_4] \\ p_2\cdot T^2/2 \\ p_3\cdot T_\beta^2 (T/T_\beta - p_4) \\ T \end{bmatrix}. \qquad (10.39)$$

Since the transition matrix Φ is time variable (V, χ), it is newly computed each time. Prediction of the state is not done with the linearized model but numerically with the full nonlinear model. The covariance matrix Q has been assumed to be diagonal; the following numerical values have been found empirically for VaMoRs: $q_{yy} = (0.2\text{m})^2$, $q_{\psi\psi} = (2.0°)^2$, $q_{\beta\beta} = (0.5°)^2$, and $q_{\lambda\lambda} = (0.2°)^2$.

Initialization for this recursive estimation process is done with results from relative ego-state estimation on the road driven and from the estimation process described in Section 10.2.3.1 for the intersection parameters:

$$\begin{aligned} y_{q0} &= dir\cdot[l_{CR}\cos\psi_{CR} + y_K\sin\psi_{CR} + l_K\cos(\psi_F - \psi_{CR})], \\ \psi_{q0} &= \psi_F - \psi_{CR} - dir\cdot\pi/2, \\ \beta_{q0} &= (1/2 - T_\beta V/a), \\ \lambda_{q0} &= \lambda_F. \end{aligned} \qquad (10.40)$$

Measurement model: Variables measured are the steering angle λ_F (mechanical sensor), the yaw rate of the vehicle ψ_V-dot (inertial sensor), vehicle speed V (as a

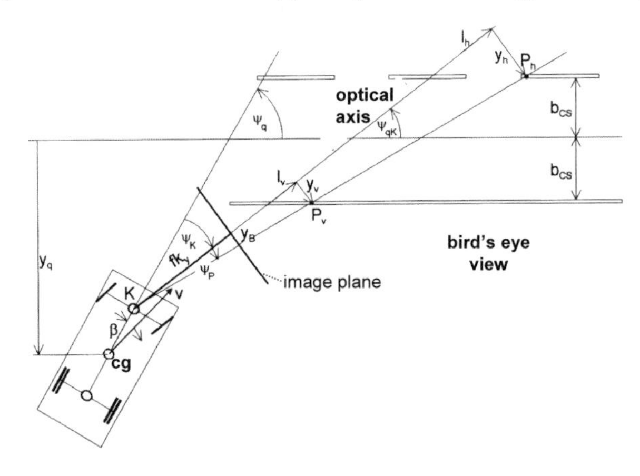

Figure 10.15. Measurement model for relative egostate with active gaze control for curve steering: Beside visual data from lane (road) boundaries, the gaze direction angles relative to the vehicle body (ψ_K ,θ_K), the steer angle λ, the inertial yaw rate from a gyro $\dot{\psi}_F$, and vehicle speed V are available from conventional measurements.

parameter in the system model, derived from measured rotational angles of the left front wheel of VaMoRs), and the feature coordinates in the images.

Depending on the search direction used, the image coordinates are either y_B or z_B. With k_B as a generalized image coordinate, the measurement vector $\underline{y}$ has the following transposed form

$$\underline{y}^T = [\lambda_F, \dot{\psi}_F, k_{B0}, \dots k_{Bi}, \dots k_{Bm}]. \tag{10.41}$$

From Figure 10.15, the geometric relations in the measurement model for turning off onto a crossroad (in a bird's-eye view) can be recognized:

$$y_i \cdot \cos\psi_{qK} = l_{vi} \cdot \sin\psi_{qK} - l_K \cdot \sin\psi_q - y_q + ix_{Linie} \cdot b_{CS}. \tag{10.42}$$

Perspective transformation with a pinhole model (Equations 2.4 and 7.20) yields

$$\frac{y_{Bi}}{f \cdot k_y} = \frac{y_i}{l_{vi}}; \qquad \frac{z_{Bi}}{f \cdot k_z} = \frac{H_K + l_{vi} \cdot \tan\theta_K}{l_{vi} - H_K \cdot \tan\theta_K}. \tag{10.43}$$

For planar ground, a search in a given image row z_{Bi} fixes the look-ahead range l_{vi}. The measurement value then is the column coordinate

$$y_{Bi} = \frac{-f \cdot k_y}{\cos\psi_{qK}} \left(\sin\psi_{qK} + \frac{l_K \sin\psi_q + y_q - ix_{Linie} \cdot b_{CS}}{l_{vi}(z_{Bi})} \right). \tag{10.44}$$

The elements of the Jacobian matrix are

$$\begin{aligned} \frac{\partial y_{Bi}}{\partial y_q} &= \frac{-f \cdot k_y}{l_{vi}(z_{Bi}) \cos\psi_{qK}} \\ \frac{\partial y_{Bi}}{\partial \psi_q} &= \frac{-f \cdot k_y}{l_{vi}(z_{Bi}) \cos\psi_{qK}} [l_{vi} + l_K \cdot \cos\psi_q + (y_q + ix_{Linie} \cdot b_{CS}) \sin\psi_{qK}]. \end{aligned} \tag{10.45}$$

For measurements in a vertical search path y_{Bi} (columns), the index in the column z_{Bi} is the measurement result. Its dependence on the look-ahead range l_{vi} has been given in Equation 10.31. For application of the chain rule, the partial derivatives of l_{vi} with respect to the variables of this estimation process here have to be determined: With

$$k_2 = [\sin\psi_{qK} + (y_{Bi} / f \cdot k_y) \cdot \cos\psi_{qK}]^{-1},$$

$$\frac{\partial l_{vi}}{\partial y_q} = -k_2; \quad \frac{\partial l_{vi}}{\partial \psi_q} = -k_2[l_K \cdot \cos\psi_q + l_{vi}(\cos\psi_{qK} - \frac{y_{Bi}}{f \cdot k_y}\sin\psi_{qK})]. \tag{10.46}$$

In summary, the following Jacobian matrix results (repetitive in the image part)

$$\underline{C}^T = \begin{pmatrix} 0 & 0 & \frac{\partial k_{B0}}{\partial y_q} & \cdots & \frac{\partial k_{Bi}}{\partial y_q} & \cdots & \frac{\partial k_{Bm}}{\partial y_q} \\ 0 & 0 & \frac{\partial k_{B0}}{\partial \psi_q} & \cdots & \frac{\partial k_{Bi}}{\partial \psi_q} & \cdots & \frac{\partial k_{Bm}}{\partial \psi_q} \\ 0 & 0 & 0 & \cdots & 0 & \cdots & 0 \\ 1 & V/a & 0 & \cdots & 0 & \cdots & 0 \end{pmatrix}. \tag{10.47}$$

The measurement covariance matrix R is assumed to be diagonal:

$$R = diag[\sigma^2_{\lambda_F}, \sigma^2_{\psi_F}, \sigma^2_{k_B}, \ldots \sigma^2_{k_B}, \ldots]. \tag{10.48}$$

From practical experience with the test vehicle VaMoRs, the following standard deviations showed good convergence behavior

$$\sigma_{\lambda_F} = 0.05°; \quad \sigma_{\psi_F} = 0.125°/s; \quad \sigma_{k_B} = 2 \text{ pixels}.$$

The elements of the Jacobian matrix may also be determined by numerical differencing. Feature selection is done according to the same scheme as discussed above. From a straight-line fit to the selected edge candidates, the predictions for window placement and for computing the prediction errors in the next cycle are done.

10.3 System Integration and Realization

The components discussed up to now have to be integrated into an overall (distributed) system, since implementation requires several processors, *e.g.*, for gaze control, for reading conventional measurement data, for frame grabbing from parallel video streams, for feature extraction, for recursive estimation (several parallel processes), for combining these results for decision-making, and finally for implementing the control schemes or signals computed through actuators.

For data communication between these processors, various delay times occur; some may be small and negligible, others may lump together to yield a few tenths of a second in total, as in visual interpretation. To structure this communication process, all actually valid best estimates are collected – stamped with the time of origination – in the dynamic data base (DDB [or DOB in more recent publications, an acronym from dynamic object database]). A fast routing network realizes communication between all processors.

10.3.1 System Structure

Figure 10.16 shows the (sub-) system for *curve steering* (CS) as part of the overall system for autonomous perception and vehicle guidance. It interfaces with visual data input on the one side (bottom) and with other processes and for visual perception (road tracking RT), for symbolic information exchange (dynamic data base), for vehicle control (VC), and for gaze control by a two-axis platform (PL) on the other side (shown at the top). The latter path is shown symbolically in duplicate form by the dotted double-arrow at the center bottom.

The central part of the figure is a coarse block diagram showing the information flow with the spatiotemporal knowledge base at the center. It is used for hypothesis generation and -checking, for recursive estimation as well as for state prediction, used for forward perspective projection ("imagination") and for intelligent control of attention; feature extraction also profits from these predictions.

Features may be selected from images of both the tele- and the wide-angle camera depending on the situation, as previously discussed. Watching delay times and compensating for them by more than one prediction step, if necessary, is required for some critical paths. Trigger points for initiation of feed-forward control in steering are but one class of examples.

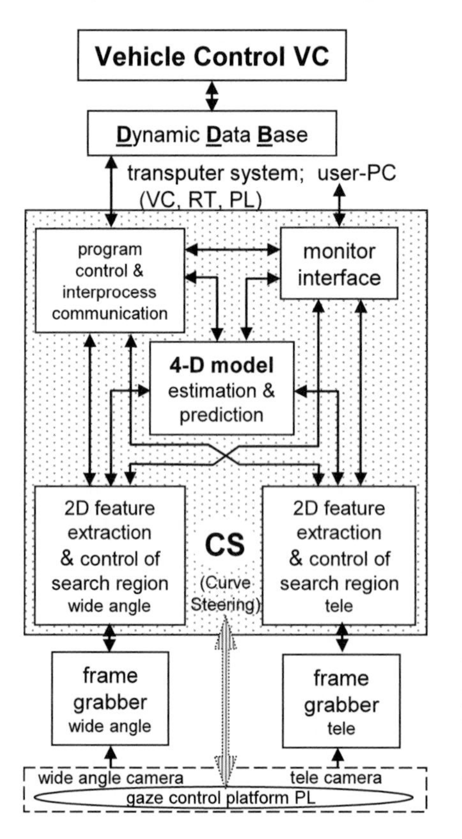

Figure 10.16. System integration for *curve steering* (CS): The textured area contains all modules specific for this task. Image data from a tele- and a wide-angle camera on a gaze control platform PL (bottom of figure) can be directed to specific processors for (edge) feature extraction. These features feed the visual recognition process based on recursive estimation and prediction (computation of expectations in 4D, center). These results are used internally for control of gaze and attention (top left), and communicated to other processes via the dynamic database (DDB, second from top). Through the same channel, the module CS receives results from other measurement and perception processes (*e.g.*, from RT for road tracking during the approach to the intersection). The resulting internal "imagination" of the scene as "understood" is displayed on a video monitor for control by the user. Gaze and vehicle control (VC) are implemented by special subsystems with minimal delay times.

10.3.2 Modes of Operation

The CS module realizes three capabilities: (1) Detection of a crossroad, (2) estimation of crossroad parameters, and (3) perception of egostate relative to the crossroad. These capabilities may be used for the following behavioral modes:

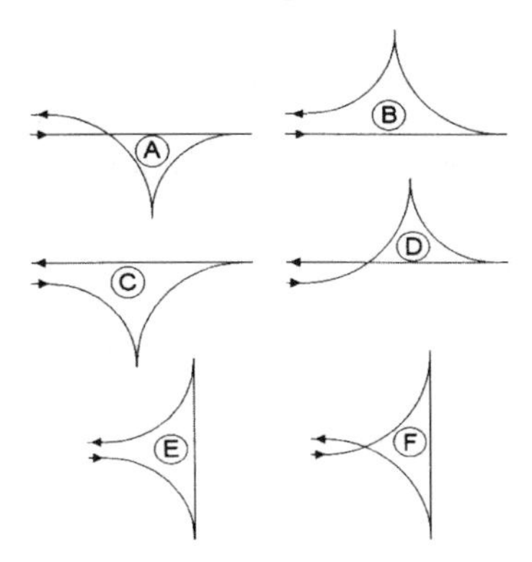

Figure 10.17. U-Turn maneuvers requiring capabilities for turning off; the maneuvers A to D require turning into a crossroad in reverse gear, while maneuvers E and F just require backing up in a straight line. [Backing up has not been realized with our vehicles because of lacking equipment.]

- The mode "ZWEIG-AB" ("turnoff") uses all three capabilities in consecution; this maneuver ends when the vehicle drives in the direction of the crossroad with small errors in relative heading (for example $|\Delta\psi| < 6° \approx 0.1$ rad) and a small lateral offset ($|\Delta y| < 0.5$ m).
- The mode "FIND-ABZW" ("find-crossroad") serves for detecting a crossroad as a landmark. With the distance to the intersection, the navigational program can verify the exact position on a map (for example, when GPS data are poor). Here only the first two capabilities are used.
- The remaining two modes "WENDE-G-EIN" and "WENDE-G-AUS" may be used for U-turns at a crossroad. In Figure 10.17 six different realizable maneuvers are sketched.

10.4 Experimental Results

Initially, this capability of turning off has been developed as a stand-alone capability, first in (HIL) simulation then with the test vehicle VaMoRs in a dissertation [Mueller 1996], based on the second-generation ("Transputer") hardware. Because of the modularity planned from the beginning, it could be transferred to the third-generation "EMS vision" system with only minor changes and adaptations [Lützeler, Dickmanns 2000; Pellkofer, Dickmanns 2000; Siedersberger 2000].

Here, only some results with the second-generation system will be discussed; results with the integrated third-generation system EMS vision will be deferred to Chapter 14.

10.4.1 Turnoff to the Right

Figure 10.18 gives an impression of the complexity of the turnoff maneuver with active gaze control for bifocal vision. The upper part shows the distant approach with the telecamera trying to pick up features of the crossroad [after "mission control (navigation)" has indicated that a crossroad to the right will show up shortly). The second to fourth snapshot (b to d, marked in the central column), which are several frames apart, show how gaze is increasingly turned into the direction of the crossroad after concatenation of edge features to extended line elements has been achieved for several lines. For easier monitoring, the lengths of the markings superimposed on the image are proportional to measured gradient intensity.

The upper four images stem from a camera with a telelens during the initial approach. *Top left*: Search in window (white rectangle) for horizontal edge features in the direction of the expected crossroad. *Top right*: The fixation point starts moving into the crossroad. *Second left*: Three white lane markings are tracked at a preset distance. *Second right*: With better separation of lines given, only the markings of the lane to be turned into are continued. The lower four images are from the wide-angle camera, whose images start being evaluated in parallel; the vehicle is now close to the intersection, looking to the right and has started turning off. The dark bar is a column of the vehicle structure separating the front- from the side windshield). *Third row left*: Three lane markings are found in vertical search regions at different distances. *Right and bottom left*: Only the future lane to be driven is continued; *bottom right*: Vehicle has turned in the direction of the new road (the bar has disappeared), feature search has been switched to horizontal, and the system performs the transition to standard roadrunning [slight errors in yaw angle and lane width can be recognized from the prediction errors (white dots left)].

Figure 10.18. Series of snapshots from a curve-steering maneuver to the right on test track Neubiberg (see text)

The building in the background allows the human observer to establish temporal continuity when proceeding to the wide-angle snapshots in the lower part of the figure. The vehicle looks "over the shoulder" to the right. In the lower part of the figure, the borderlines have become close to vertical, and the feature search with CRONOS is done in rows. Due to a small lateral offset from the center of the lane in the crossroad, the predicted points for the lane markings show a slight error (white dots at center of dark lines and *vice versa*). Especially in the transition phase, the 4-D model-based approach shows its strengths.

Figure 10.19 shows the time histories of the corresponding estimated crossroad parameters and vehicle states.

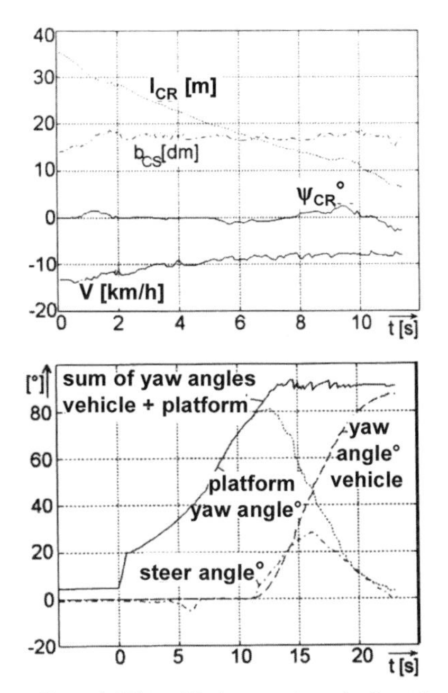

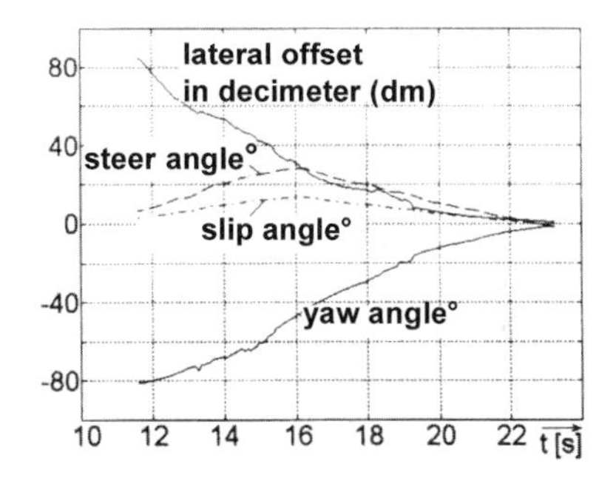

Figure 10.19. Parameter and state estimation for turning off to the right with VaMoRs (Figure 10.18): Top left: Crossroad parameters and speed driven; top right: Time history of vehicle lateral offset and yaw angle relative to crossroad. Left: State and control variables are given in vehicle coordinates relative to initial state.

Speed V is still decreasing during the approach (top left, lower curve); the turn is performed at $V \approx 2.3$ m/s. The yaw angle of the gaze platform is turned to 20° when the crossroad is picked up (at t = 0 in lower left figure); it then increases up to 80° at around 12 seconds. This is when the estimation process of vehicle state relative to the crossroad starts (upper right subfigure). The vehicle is still 8 m (= 80 decimeters) away from the center of the right lane of the crossroad, and vehicle heading relative to the crossroad (yaw angle) is − 80°. It can be nicely seen that the slip angle β is about half the steering angle λ; they all tend toward zero at the end of the maneuver at about 23 seconds. The lower left figure shows that the sum of vehicle and platform yaw angle accumulates to 90° finally; between 13 and 20 seconds, the yaw rates of vehicle (*dashed curve*) and platform (*dotted*) have about the same magnitude but opposite sign. This means that gaze is fixated to the crossroad, and the vehicle turns underneath the platform.

At the end of parameter estimation for the crossroad (top left) the best estimate for half the lane width b_{CS} is ≈ 1.75 m (correct: 1.88 m) and for the heading angle ψ_{CR} is ≈ -2.6° (correct: +0.9°). Since lateral feedback is added toward the end of the maneuver, this is of no concern.

10.4.2 Turn-off to the Left

Figure 10.20 shows how the general procedure developed for the maneuver "turn off" works for a turn across lanes for oncoming traffic onto a crossroad at an angle of about −115° (negative yaw angle is defined as left). There is much more space for turning to the left rather than to the right (in right-hand traffic), therefore, the maximal platform yaw angle is only about 50° (lower right sub-figure, from ~ 6 to 14 s), despite the larger total turn angle. The initial hypothesis developed for a gaze angle of ~ −30° at a distance l_{CR} of 17 m was an intersection angle of ~ −98° (*top left, lowest solid curve*) and a half-lane width of $b_{CS} \approx 1.2$ m (*dash-dotted curve*).

During the approach to an estimated 12 m, the initial crossroad parameters change only slightly: ψ_{CR} decreases from −8 to −11°, and b_{CS} increases to ~1.4 m. However, gaze direction is turned steadily to over 40°. Under these aspect conditions with increasingly more features becoming visible further into the crossroad, at around 4.5 to 5 s the sharp turn angle is recognized (top left subfigure), and in a

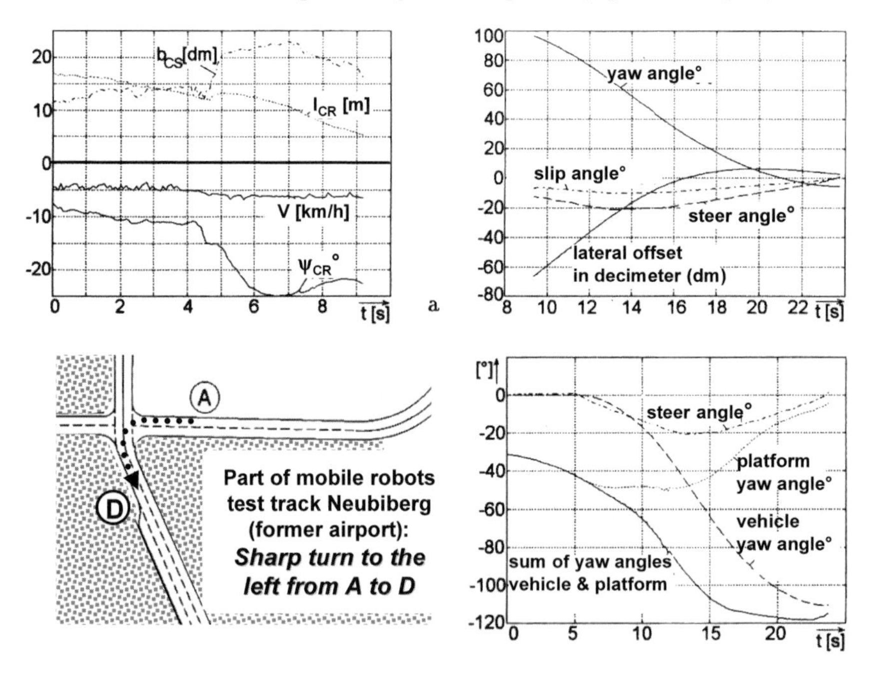

Figure 10.20. Turnoff to the left onto a road branching at an angle deviating considerably from 90° (lowest curve, top left, and bird's-eye view in subfigure bottom left); the best estimate shows some overshoot at around 7 s. At about 4.3 s, when the platform angle is about 40° (lower right), the higher turnoff angle is discovered (lowest curve, top left), and the estimated width of the crossroad jumps to over 2 m; at 5 s, control output is started (*dash-dotted line, lower right subfigure*). Total turn angle is about 115°; around 20 s, an overshoot in lateral displacement in the new lane of about a half meter occurs (*top right, solid curve*). Since the new lane is far more than 3 m wide ($2b_{CS}$), this looks quite normal to a human observer.

transient mode, estimated lane width, intersection angle, and distance to the intersection point of the two lanes show transient dynamics (note speed changes also!).

At around 9 s (that is 4 s into the steering rate feed-forward maneuver), the vehicle has turned around −15°, and the estimation process relative to the crossroad as a new reference is started based on wide-angle image data (top right subfigure). At around 24 s, all variables tend toward zero again, which is the nominal state for roadrunning in the new reference frame.

In more general terms, this maneuver should be labeled "Turning off with crossing the lanes of oncoming traffic". To do this at a crossing without traffic regulation, it is necessary that the oncoming traffic is evaluated up to greater distances. This has not been possible by vision up to now; only the basic perception and control steps for handling the geometric part of the turnoff maneuver have been discussed here.

10.5 Outlook

It has been shown that the mission element "turn off at the next crossroad (right or left)" is rather involved; it requires activity sequences both in viewing direction and in feature-extraction control as well as in control outputs for vehicle steering. These activities have to be coordinated relative to each other, including some feedback loops for fixing the viewing direction; all these activities may be symbolized on a higher level of representation by the symbol "make turn (right/left)."

Table 10.4 shows a summary in coarse granularity for the maneuver "Turn-off to the left." The bulk of the work for implementation lies in making the system robust to perturbations in component performance, including varying delay times and nonlinearities not modeled. This maneuver element has been ported to the third-generation vision and autonomous guidance system in which the general capability network for both visual perception and locomotion control has been implemented [Lützeler 2002, Pellkofer 2003, Maurer 2000, Gregor 2002, Siedersberger 2004].

A similar local maneuver element (behavioral capability) has to be available for handling road forks (see Figure 5.3).

Table 10.4. Perceptual and behavioral capabilities to be activated with proper timing after the command from central decision (CD): "Look for crossroad to the left and turn onto it" (no other traffic to be observed).

Perception	Monitoring	Gaze Control	Vehicle Control
1. Edge- and area-based features of crossroad (CR); start in teleimage: **1a)** directly adjacent to left road boundary; **1b)** some distance into CR-direction for precisely determining • distance to CR center- line: l_{CR} • angle $\Delta\psi_{CR}$ and width of CR: $2 \cdot b_{CS}$	Time of command, saccades, insertion of accepted hypothesis in DOB; convergence parameters, variances over time for l_{CR}, $\Delta\psi_{CR}$, b_{CS}	Saccade to lateral position for CR detection; after hypothesis acceptance: Fixate point on circular arc, then on CR at growing distance.	Lane keeping (roadrunning) till l_{CR} reaches trigger point for initiation of steering angle feed-forward control program at proper speed; start curve steering at constant steering rate.
2a) Continue perceiving CR in teleimage; estimate distance and angle to CR boundary on right-hand side, width of CR. **2b)** Track left boundary of road driven (in wide-angle image), own relative position in road.	Store curve initiation event, maneuver time history; compute $\underline{x} - \underline{x}_{exp}$ (from knowledge base for the maneuver)	Compensate for effect of vehicle motion: a) inertial angular rate b) position change x, y; (feed-forward phases), c) fixate on CR at l_{vi}	At λ_{max}, for transition to constant steering angle till start of negative steering rate. At 60 to 70 % of maneuver time, start superposition of feedback from right-hand CR boundary .
3. Set up new road model for (former) CR: **3a)** In near range fit of straight-line model from wide-angle cameras; determine range of validity from near to far. **3b)** Clothoid model of arbitrary curvature later on.	CR parameters, relative own position, road segment limit; statistical data on recursive estimation process.	Stop motion compensation for gaze control when angular yaw rate of vehicle falls below threshold value; resume fixation strategy for roadrunning	Finish feed-forward maneuver; switch to standard control for roadrunning with new parameters of (former) CR: Select: driving speed and lateral position desired in road.

11 Perception of Obstacles and Vehicles

Parallel to road recognition, obstacles on the road have to be detected sufficiently early for proper reaction. The general problem of object recognition has found broad attention in computer (machine) vision literature (see, *e.g.*, http://iris.usc.edu/Vision-Notes/bibliography/contents.html); this whole subject is so diverse and has such a volume that a systematic review cannot be given here.

In the present context, the main emphasis in object recognition is on detecting and tracking stationary and moving objects of rather narrow classes from a moving platform. This type of dynamic vision has very different side constraints from so-called "pictorial" vision where the image is constant (one static "snapshot"), and there are no time constraints with respect to image processing and interpretation. In our case, in addition to the usually rather slow changes in aspect conditions due to translation, there are also relatively fast changes due to rotational motion components. In automotive applications, uneven ground excites the pitch (tilt) and roll (bank) degrees of freedom with eigendynamics in the 1-Hz range. Angular rates up to a few degrees per video cycle time are not unusual.

11.1 Introduction to Detecting and Tracking Obstacles

Under automotive conditions, short evaluation cycle times are mandatory since from the time of image taking in the sensor till control output taking this information into account, no more than about one-third of a second should have passed, if human-like performance is to be achieved. On the other hand, these interpretations in a distributed processor system will take several cycles for feature extraction and object evaluation, broadcasting of results, and computation as well as implementation of control output. Therefore, the basic image interpretation cycle should not take more than about 100 ms. This very much reduces the number of operations allowable for object detection, tracking, and relative state estimation as a function of the limited computing power available.

With the less powerful microprocessor systems of the early 1990s, this has led to a pipeline concept with special processors devoted to frame-grabbing, edge feature extraction, hypothesis generation/state estimation, and coordination; the processors of the mid-1990s allowed some of these stages to run on the same processor. Because of the superlinear expansion of search space required with an increase in cycle time due to uncertainties in prediction from possible model errors and to unknown control inputs for observed vehicles or unknown perturbations, it pays off to keep cycle time small. In the European video standard, preferably 40 ms (video frame time) have been chosen. Only when this goal has been met already, additional computing power becoming available should be used to increase the complexity of image evaluation.

Experience in real road traffic has shown that crude but fast methods allow recognizing the most essential aspects of motion of other traffic participants. There are special classes of cases left for which it is necessary to resort to other methods to achieve full robust coverage of all situations possible; these have to rely on region-based features like color and texture in addition. The processing power to do this in the desired time frame is becoming available now.

Nevertheless, it is advantageous to keep the crude but fast methods in the loop and to be able to complement them with area-based methods whenever this is required. In the context of multifocal saccadic vision, the crude methods will use low-resolution data in set the stage for high-resolution image interpretation with sufficiently good initial hypotheses. This coarse-to-fine staging is done both in image data evaluation and in modeling: The most simple shape model used for another object is the encasing box which for aspect conditions along one of the axes of symmetry reduces to a rectangle (see Figure 2.13a/2.14). This, for example, is the standard model for any type of car, truck, or bus in the same lane nearby where no road curvature effects yield an oblique view of the object.

11.1.1 What Kinds of Objects Are Obstacles for Road Vehicles?

Before proceeding to the methods for visual obstacle detection, the question posed in the heading should be answered. Wheels are the essential means for locomotion of ground vehicles. Depending on the type of vehicle, wheel diameter may vary from about 20 cm (as on go-carts) to over 2 m (special vehicles used in mining). The most common diameters on cars and trucks are between 0.5 and 1 m. Figure 11.1 shows an obstacle of height H_{Obst}. The circles represent wheels when the edge of the rectangular obstacle (*e.g.*, a curbstone) is touched. With the tire taking about one-third of the wheel radius $D/2$, obstacles of a height H_{Obst} corresponding to $D/H > 6$ may be run over at slow speed so that tire softness and wheel dynamics can work without doing any harm to the vehicle. At higher speeds, a maximal obstacle height to $D/H > 12$ or even higher may be required to avoid other dynamic effects.

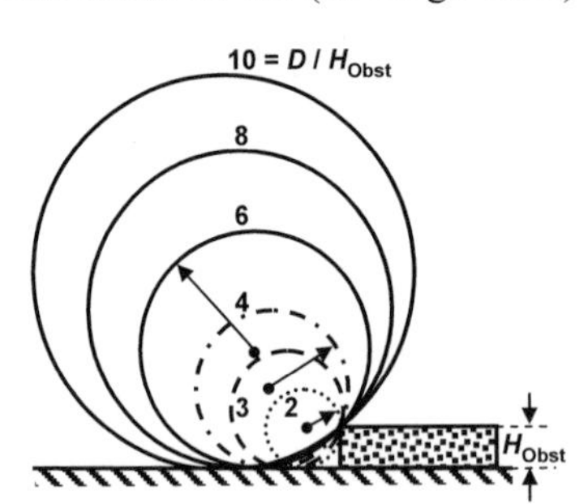

Figure 11.1. Wheel diameter D relative to obstacle height H_{Obst}

However, an "obstacle" is not just a question only of size. A hard, sharp object in or on an otherwise smooth surface may puncture the tire and must thus be avoided, at least within the tracks of the tires. All obstacles above the surface on which the vehicle drives are classified as "positive" obstacles, but there are also "negative" obstacles. These are holes in the planar surface into which the wheel may (partially) fall. Figure 11.2 shows the width of a ditch or pothole relative to the wheel diameter; in this case, $W > D/2$ may be a serious obstacle, especially at low speeds. At higher speeds, the inertia of the wheel will keep it from falling into the hole if this is not too large; otherwise there is support of the ground again underneath the tire before the wheel travels a significant distance in the vertical direction. Holes or ditches of width larger than about 60 % of tire diameter and correspondingly deep should be avoided anyway.

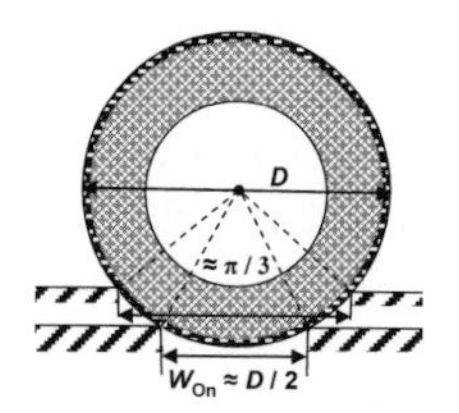

Figure 11.2. Wheel diameter D relative to width W of a negative obstacle

11.1.2 At Which Range Do Obstacles Have To Be Detected?

There are two basic ways of dealing with obstacles: (1) Bypassing them, if there is sufficient free space and (2) stopping in front of them or keeping a safe distance if the obstacles are moving. In the first case, lateral acceleration should stay within bounds and safety margins have to be observed on both sides of its own body. The second case, usually, is the more critical one at higher speeds since the kinetic energy ($\sim m \cdot V^2$) has to be dissipated and the friction coefficient to the ground may be low. Table 11.1 gives some characteristic numbers for typical speeds driven (a) in urban areas (15 to 50 km/h), (b) on cross-country highways (80 to 130 km/h), and (c) for high-speed freeways. [Note that for most premium cars top speed is electronically limited to 250 km/h; at this speed on a freeway with 1 km radius of curvature, lateral acceleration $a_{y,250/1}$ will be close to 0.5 g! To stop in front of an obstacle with a constant deceleration of 0.6 g, the obstacle has to be detected at a range of ~ 440 m; initially at this high speed in a curve with $C = 0.001\ \text{m}^{-1}$, the total horizontal acceleration will be 0.78 g (vector sum: $0.5^2 + 0.6^2 = 0.78^2$).]

Table 11.1. Braking distances with a half second reaction time and three deceleration levels of −3 , −6, and −9 m/s^2

Speed km/h	0.5 s $\Delta L_{react.}$	ΔL_{br} with $a_x = 0.3\ g$	L_{brake} in m	ΔL_{br} with $a_x = 0.6\ g$	L_{brake} in m	ΔL_{br} with $a_x = 0.9\ g$	L_{brake} in m
15	2.1	2.9	5	1.5	3.6	1	3.1
30	4.2	11.6	15.8	5.8	10	3.9	8.1
50	6.9	32.2	39.1	16.1	23	10.7	17.6
80	11.1	82.3	93.4	41.2	52.3	27.4	38.5
100	13.9	128.6	142.5	64.3	78.2	42.9	56.8
130	18	217.3	235.3	108.7	126.7	72.4	90.4
180	25	416.7	442.7	208.3	233.3	138.9	163.9
250	34.7	803.8	838.5	401.9	436.6	268	302.6

Even for driving at 180 km/h (50 m/s or 2 m per video frame), the detection range has to be about 165 m for harsh deceleration (0.9 g) and about 235 m for medium-harsh deceleration (0.6 g); with pleasant braking at 0.3 g, the look-ahead range necessary goes up to ~ 450 m. For countries with maximum speed limits around 120 to 130 km/h, look-ahead ranges of 100 to 200 m are sufficient for normal braking conditions (dry ground, not too harsh).

A completely different situation is given for negative obstacles. From Figure 5.4, it can be seen that a camera elevation above the ground of $H = 1.3$ m (typical for a car) at a distance of $L = 20 \cdot H = 26$ m (sixth column) leads to coverage of a hole in the ground of size H in the gaze direction by just 1.9 pixels. This means that the distance of one typical wheel diameter ($\approx H/2 = 65$ cm) is covered by just one pixel; of course, under these conditions, no negative obstacle detrimental to the vehicle can be discovered reliably. Requiring this range of 65 cm to be covered with a minimum of four pixels for detection leads to an L/H-ratio of 10 (fifth column, Figure 5.4); this means that the ditch or the pothole can be discovered at about 13 m distance. To stop in front of it, Table 11.1 indicates that the maximal speed allowed is around 30 km/h.

Taking local nonplanarity effects or partial coverage with grass in off-road driving into account (and recalling that half the wheel diameter may be the critical dimension to watch for; see Figure 11.2), speeds in these situations should not be above 20 km/h. This is pretty much in agreement with human cross-country driving behavior. When the friction coefficient must be expected to be very low (slippery surface), speed has to be reduced correspondingly.

11.1.3 How Can Obstacles Be Detected?

The basic assumption in vehicle guidance is that there is a smooth surface in front of the vehicle for driving on. Smoothness again is a question of scale. The major yardsticks for vehicles are their wheel diameter and their axle distance on a local scale and riding comfort (spectrum of accelerations) for high-speed driving. The former criteria are of special interest for off-road driving and are not of interest here. Also, negative obstacles will not be discussed (the interested reader is referred to [Siedersberger *et al.* 2001; Pellkofer *et al.* 2003] for ditch avoidance).

For the rest of the chapter, it is assumed that the radii of vertical curvature R_{Cv} of the surface to be driven on are at least one order of magnitude larger than the axle distance of the vehicles (typically $R_{Cv} > 25$ m). Under these conditions, the perception methods discussed in Section 9.2 yield sufficiently good internal representations of the regular surface for driving; larger local deviations from this surface are defined as obstacles. The mapping conditions for cameras in cars have the favorable property that features on the inner side of the silhouette of obstacles hardly (or only slowly) change their appearance, while on the adjacent outer side, features from the background move by and change appearance continuously, in general. For stationary objects, due to egomotion, texture in the background is covered and uncovered steadily, so that looking at temporal continuity helps detecting the obstacle; this may be one of the benefits of "optical flow". For moving objects, several features on the surface of the object move in conjunction. Again, local temporal changes or smooth feature motion give hints on objects standing out of the surface on which the subject vehicle drives. On the other hand, if there are inhomogeneous patches in the road surface, lacking feature flow at the outer side of their boundaries is an indication that there is no 3-D object causing the appearance.

Stereovision with two or more cameras exploits the same phenomenon, but due to the stereo baseline, the different mapping conditions appear at one time. In the near range, this is known to work well in most humans; but people missing the capability of stereo vision are hardly hampered in road vehicle guidance. This is an indication in favor of the fact that motion stereo is a powerful tool. In stereovision using horopter techniques with image warping, those features above the ground appear at two locations coding the distance between camera and object [Mandelbaum *et al.* 1998].

In laser range finding and radar ranging, electromagnetic pulses are sent out and reflected from surfaces with certain properties. Travel time (or signal phase) codes the distance to the reflecting surface. While radar has relatively poor angular resolution, laser ranging is superior from this point of view. Obstacles sticking out of the road surface will show a shorter range than the regular surface. Above the horizon, there will be no signals from the regular surface but only those of obstacles. Mostly up to now, laser range finding is done in "slices" originating from a rotating mirror that shifts the laser beam over time in different directions. In modern imaging laser range finding devices, beside the "distance image" also an "intensity image" for the reflecting points can be evaluated giving even richer information for perception of obstacles. Various devices with fixed multiple laser beams (up to 160 × 120 image points) are on the market.

However, if laser range finding (LRF) is compared to vision, the angular resolution is still at least one order of magnitude less than in video imaging, but there is no direct indication of depth in a single video image. This fact and the enormous amount of video data in a standard video stream have led the application-oriented community to prefer LRF over vision. Some references are [Rasmussen 2002; Bostelman *et al.* 2005; PMDTech 2006]. There are recently developed systems on the market that cover the full circular environment of 360° in 64 layers with 4000 measurement points ten times a second. This yields a data rate of 2.56 million data points per second and a beam separation of 0.09° or 1.6 mrad in the horizontal direction; in practical terms, this means that at a distance of 63 m two consecutive beams are 10 cm apart in the circumferential direction. In contrast, the resolution of telecameras range to values of ~ 0.2 mrad/pixel; the field of view covered is of course much lower. The question, which way to go for technical perception systems in road traffic (LRF or video or a combination of both), is wide open today.

On the other hand, humans have no difficulty understanding the 3-D scene from 2-D image sequences (over time). There are many temporal aspects that allow this understanding despite the fact that direct depth information has been lost in each single video image. In front of this background, in this book, all devices using direct depth measurements are left aside and interpretation concentrates on spatio-temporal aspects for visual dynamic scene understanding in the road traffic domain. Groups of visual features and their evolution (motion and changes) over time in conjunction with background knowledge on perspective mapping of moving 3-D objects are the medium for fully understanding what is happening in "the world". Because of the effects of pinhole mapping, several cameras with different focal lengths are needed to obtain a set of images with sufficient resolution in the near, medium, and far ranges.

Before we proceed to this aspect in Chapter 12, the basic methods for detecting and tracking of stationary and moving objects are treated first. Honoring the initial developments in the late 1980s and early 1990s with very little computing power onboard, the historical developments will be discussed as points of entry before the methods now possible are treated. This evolution uncovers the background of the system architecture adopted.

11.2 Detecting and Tracking Stationary Obstacles

Depending on the distance viewed, the mask size for feature extraction has to be adapted correspondingly; to detect stationary objects with the characteristic dimensions of a human standing upright and still, a mask size in CRONOS of one-half to one-tenth of the lane width (of ~ 2 to 4 m in the real world) at the distance L observed seems to be reasonable. In road traffic, objects are usually predominantly convex and close to rectangular in shape (encasing boxes); gravity determines the preferred directions horizontally and vertically. Therefore, for obstacle detection, two sets of edge operators are run above the road region in the image: detectors for vertical edges at different elevations are swept horizontally, and extractors for horizontal edges are swept vertically over the candidate region. A candidate for an object is detected by a collection of horizontal or vertical edges with similar average intensities between them.

For an object, observed from a moving vehicle, to be stationary, the features from the region where the object touches the ground have to move from one frame to the next according to the egomotion of the vehicle carrying the camera. Since translational motion of the vehicle can be measured easily and reliably by conventional means, no attempt is made to determine egomotion from image evaluation.

11.2.1 Odometry as an Essential Component of Dynamic Vision

The translational part of egomotion can be determined rather well from two mechanically implemented measurements at the wheels (or for simplicity at just one wheel) of the vehicle. Pulses from dents on a wheel for measuring angular displacements directly linked to one of the front wheels deliver information on distance traveled; the steer angle, also measured mechanically, gives the direction of motion. From the known geometry of the vehicle and camera suspension, translational motion of the camera can be determined rather precisely. The shift in camera position is the basis for motion stereointerpretation over time.

Assuming no rotational motion in pitch and roll of the vehicle (nominally), the known angular orientation of the cameras relative to the vehicle body (also mechanically measured) allows predicting the shift of features in the next image. Small perturbations in pitch and bank angle will average out over time. The predicted feature locations are checked against measurements in the image sequence.

In connection with the Jacobian elements, the resulting residues yield information for systematically improving the estimates of distance and angular orientation of the subject vehicle relative to the obstacle.

Assuming that the object has a vertical extension above the ground, this body also will have features on its surface. For a given estimated range, the relative positions of these local features on the body, geared to the body shape of the obstacle, can be predicted in the image; prediction-errors from these locations allow adapting the shape hypothesis for the obstacle and its range.

11.2.2 Attention Focusing on Sets of Features

For stationary obstacles, the first region to be checked is the location where the object touches the ground. The object is stationary only when there is no inhomogeneous feature flow on the object and in a region directly outside its boundaries at the ground. (Of course, this disregards obstacles hanging down from above, like branches of trees or some part from a bridge; these rare cases are not treated here.)

To find the overall dimension of an obstacle, a vertical search directly above the region where the obstacle touches the ground in the image is performed looking for homogeneous regions or characteristic sets of edge or corner features. If some likely upper boundary of the object (its height H_O in the image) can be detected, the next step is to search in an orthogonal direction (horizontally) for the lateral boundaries of the object at different elevations (maybe 25, 50, and 75% H_O). This allows a first rough hypothesis on object shape normal to the optical axis. For the features determining this shape, the expected shift due to egomotion can also be computed. Prediction-errors after the next measurement either confirm the hypothesis or give hints how to modify the assumptions underlying the hypothesis in order to improve scene understanding.

For simple shapes like beams or poles of any shape in cross section, the resulting representation will be a cylindrical body of certain width (diameter d) and height (H_O) appearing as a rectangle in the image, sufficiently characterized by these two numbers. While these two numbers in the image will change over time, in general, the corresponding values in the real world will stay constant, at least if the cross section is axially symmetrical. If it is rectangular or elliptical, the diameter d will depend also on the angular aspect conditions. This is to say that if the shape of the cross section is unknown, its change in the image is not a direct indication of range changes. The position changes of features on the object near the ground are better indicators of range. For simplicity, the obstacle discussed here is assumed to be a rectangular plate standing upright normal to the road direction (see Figure 11.3). The detection and recognition procedure is valid for many different types of objects standing upright.

11.2.3 Monocular Range Estimation (Motion Stereo)

Even though the obstacle is stationary, a dynamic model is needed for egomotion; this motion leads to changing aspect conditions of the obstacle; it is the base for

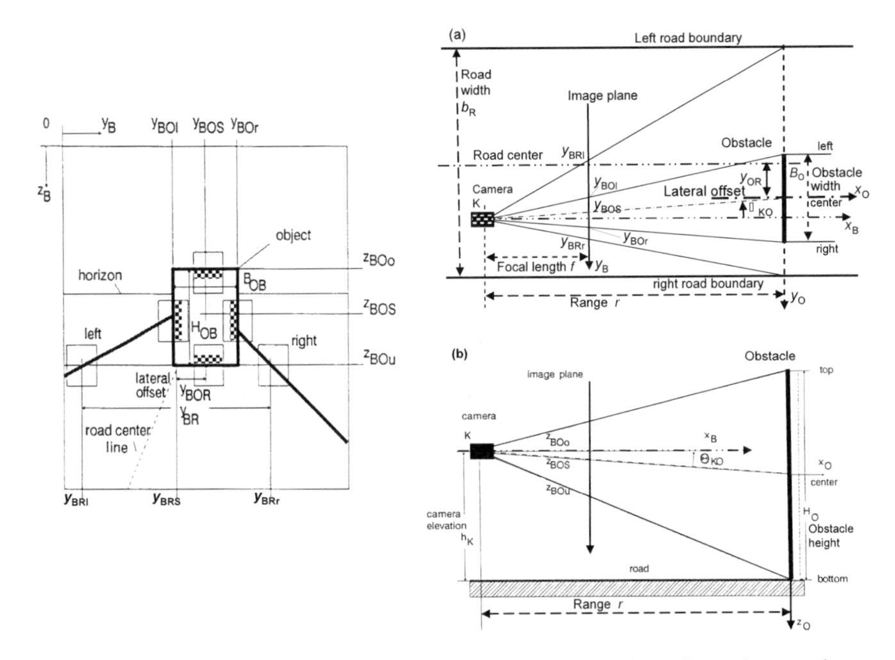

Figure 11.3. Nomenclature adopted for detection and recognition of a stationary obstacle on the road: *Left:* Perspective camera view on obstacle with rectangular cross section (B_{OB}, H_{OB}). *Top right:* Top down ("bird's-eye") view with the obstacle as a flat vertical plane, width B_O. *Lower right:* View from right-hand side, height H_O.

motion stereointerpretation over time. Especially the constant elevation of the camera above the ground and the fact that the obstacle is sitting on the ground are the basis for range detection; the pitch angle for tracking the features where the obstacle touches the ground changes in a systematic way during approach.

11.2.3.1 Geometry (Measurement) Model

In Figure 11.3, the nomenclature used is given. Besides the object dimensions, the left and right road boundaries at the position of the lower end of the obstacle are also determined (y_{BRl}, y_{BRr}); their difference $y_{BR} = (y_{BRr} - y_{BRl})$ yields the width of the road b_R at the look-ahead range r_o.
From Equation 9.9 and 9.10, there follows

$$r_o = h_K / \tan\{\theta_K + \arctan[z_{BOu}/(f \cdot k_z)]\}$$
$$\text{and for} \quad \theta_K = 0: \; r_o = h_K \cdot f \cdot k_z / z_{BOu}. \tag{11.1}$$

In Figure 11.3 (bottom) the camera looks horizontally ($\theta_K = 0$) from an elevation h_K above the ground (the image plane is mirrored at the projection center); for small azimuthal angles ψ to all features, the road width then is approximately

$$b \approx r \cdot (y - y)/(f \cdot k). \tag{11.2}$$

A first guess on obstacle width then is

$$b_O \approx r_o \cdot (y_{BOr} - y_{BOl})/(f \cdot k_y). \tag{11.3}$$

Without perspective inversion already performed in the two equations above, this immediately yields the obstacle size in units of road width b_O / b_R.

Half the sum of both feature pairs yields the road center and the horizontal position of the obstacle center in the image

$$y_{BRS} = (y_{BRl} + y_{BRr})/2; \quad y_{BOS} = (y_{BOl} + y_{BOr})/2. \tag{11.4}$$

y_{BOS} directly determines the azimuthal angle ψ_{KO} of the obstacle relative to the camera. The difference $y_{BOR} = (y_{BOS} - y_{BRS})$ yields the initial estimate for the position of the obstacle relative to the road center:

$$y_{OR} \approx r_o \cdot (y_{BOS} - y_{BRS})/(f \cdot k_y). \tag{11.5}$$

This information is needed for deciding on the reaction of the vehicle for obstacle avoidance: whether it can pass at the left or the right side, or whether it has to stop. Note that lateral position on the road (or in a lane) cannot be derived from simple LRF if road and shoulders are planar, since the road (lane) boundaries remain unknown in LRF. In the approach using dynamic vision, in addition to lateral position, the range and range rate can be determined from prediction-error feedback exploiting the dynamic model over time and only a sequence of monocular intensity images.

The bottom part of Figure 11.3 shows perspective mapping of horizontal edge feature positions found in vertical search [side view (b)]. Only the backplane of the object, which is assumed to have a shape close to a parallelepiped (rectangular box), is depicted. Assuming a planar road surface and small angles (as above: cos ≈ 1, sine ≈ argument), all mapping conditions are simple and need not be detailed here. Half the sum yields the vertical center cg_v of the obstacle. The tilt angle between cg_v and the horizontal gaze direction of the camera is θ_{KO}; the difference of feature positions between top and bottom yields obstacle height.

The elements of the Jacobian matrix needed for recursive estimation are easily obtained from these relations.

11.2.3.2 The Dynamic Model for Relative State Estimation

Of prime interest are the range r and the range rate $\dot{r}$ to the obstacle; range is the integral of the latter. The lateral motion of the object relative to the road v_{OR} is zero for a stationary object. Since iteration of the position is necessary, in general, the model is driven by a stochastic disturbance variable s_i. This yields the dynamic model (V = speed along the road: index O = object (here $V_O = 0$); index R = road)

$$\begin{aligned} \dot{r} &= V_O - V + s_r, & \dot{V}_O &= s_{VO}, \\ \dot{y}_{OR} &= v_{OR}, & \dot{v}_{OR} &= s_{yOR}. \end{aligned} \tag{11.6}$$

In addition, to determine the obstacle size and the viewing direction relative to its center, the following four state variables are added (index K = camera):

$$\begin{aligned} \dot{H}_O &= s_{HO}, & \dot{B}_O &= s_{BO}, \\ \dot{\psi}_{KO} &= s_{KO}, & \dot{\theta}_{KO} &= s_{\theta KO}, \end{aligned} \tag{11.7}$$

where again the s_i are assumed to be unknown Gaussian random noise terms. In shorthand vector notation, these equations are written in the form

$$\dot{x}(t) = f[x(t), u(t), s(t)], \tag{11.8}$$

with the state variables

$$x^T = (r_o, V_O, \psi_{KO}, \theta_{KO}, B_O, H_O, y_{OR}, v_{OR}), \quad [V_O \; and \; v_{OR} = 0]. \tag{11.9}$$

After transformation into the discrete state transition form for the cycle time T used in image processing, standard methods for state estimation, as discussed in Chapter 6, are applied. Note that nothing special has to be done to achieve motion stereointerpretation; it is an integral part of the approach.

11.2.3.3 The Estimation Process

Figure 11.3 top shows the window arrangement set up for determining the relative state of an obstacle. Initially, to detect a relatively large, uniformly gray obstacle, a horizontal search for close-to-vertical edge features is done in the region above the road known from the road tracker running separately; some edge features with similar average intensity values on one side can then be grouped with the center lying at column positions y_{BOl} and y_{BOr}. Their position is shown in the figure by the textured area indicating similar average gray values [Dickmanns, Christians 1991].

Simultaneously, in a rectangular window above the road and around a nominal look-ahead distance for safe stopping, a vertical search for horizontal edge features is performed with the same strategy yielding the row positions z_{BOo} and z_{BOu}. Again, the textured areas show similar average image intensity. The centers of these two groups of features (y_{BOS} and z_{BOS}) are supposed to mark the center of an obstacle. After this initial hypothesis has been set up, further search for feature collections belonging to the same hypothesized object is done in crosswise orthogonal search paths centered on y_{BOS} and z_{BOS}, as indicated in the figure; after each measurement update in the row direction, the search in the column direction is shifted according to y_{BOS} and vice versa, thus keeping attention focused on the obstacle.

At the same time, the position of the road boundary is determined in the row given by the lower horizontal edge feature of the hypothesized obstacle z_{BOu}. This information is essential: (1) for scaling obstacle size ($y_{BOr} - y_{BOl}$) in terms of road size ($y_{BRr} - y_{BRl}$) (see Equations 11.2 and 11.3) and (2) for initially determining range to the object when the vertical curvature of the road is known (usually assumed to be zero, that is, the road is planar). The initial guess for distance r to the obstacle is obtained by inverting the mapping model for a pinhole camera and perspective projection with the data for z_{BOu} (see bottom Subfigure 11.3 and Equation 11.1). Similarly, the initial values for the lateral position of the obstacle on the road y_{OR}, obstacle width B_O, and height H_O are obtained, all scaled by road width B_R, for which a best estimate exists from the parallel tracking process for the road (Chapter 9 and Figure 11.4 right). The bearing angles to the center of the obstacle (ψ_{BO} in azimuth and θ_{BO} in elevation) are given by the offset of this center from the image center (optical axis). Note that these variables contain redundant information if the state of the subject vehicle relative to the road (y_V, ψ_V and θ_V) is estimated separately (as done in Chapter 9). Of course, the variables ψ_{BO} and θ_{BO} contain all the perturbation effects of the road on the vehicle carrying the cameras. Measuring the position of the obstacle relative to the local road (y_{OR}) thus is more stable than relying on bearing information relative to the suject vehicle on roads with a nonsmooth (noisy) surface.

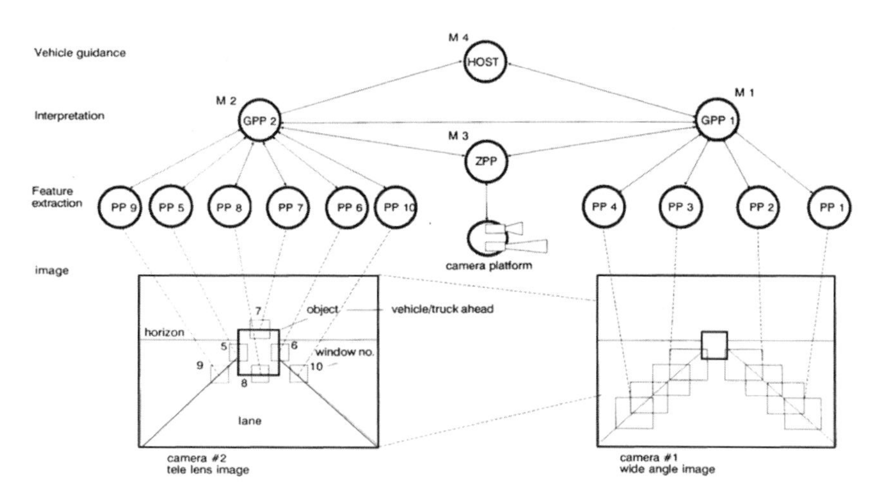

Figure 11.4. First-generation system architecture for dynamic vision in road scenes of the late 1980s [Dickmanns, Christians 1991]; the parallel processors for edge feature extraction (PPi) were 16-bit Intel 8086 microprocessors capable of extracting just a few edge features per observation cycle of 80 ms

The estimation cycle on processor GPP2, initially in VaMoRs a microprocessor Intel 80386®, ran at 25 Hz (40 ms) while feature extraction and crosswise tracking, as mentioned, ran at full video rate (50 H on Intel 8086®) for better performance under high-frequency perturbations in pitch. This was a clear indication of the fact that high-frequency image evaluation may be more important than image understanding at the same rate; the temporal models are able to bridge the gap if the quality of the features found is sufficiently good. The initial transient in image interpretation took 10 to 20 cycles until a constant error level had been achieved.

In a prediction step, the expected position of features for the next measurement is computed by applying forward perspective projection to the object as “imagined” by the interpretation process. Only those feature positions delivered by the PPs that are sufficiently close to these values (within the 3σ range) have been accepted as candidates; others are rejected as outliers. This contributed considerably to stabilizing the interpretation in noisy natural environments.

11.2.3.4 Modular Processing Structure

In Figure 11.4, the modular processing structure resulting naturally in the 4D-approach, oriented toward physical objects, is emphasized. There are four processing layers and two object-oriented groups of processors shown: The pixel-level (bottom), where 2-D spatial data structures (intensity images and subimages) have to be handled. Then, at the PP level, edge elements and adjacent intensity features are extracted with respect to the 2D-position and orientation; any relation to 3-D space or time is still missing. Only in the third layer, implementing object interpretations on the GPP, spatial and temporal constraints are introduced for associating objects with groupings of features and their relative change over time. In our case, objects are the road (with the egovehicle) and the obstacle on the road; it is easily seen how this approach can be extended to multiple objects by adding more groups of processors.

It may be favorable for the initialization phase to insert an additional 2-D object layer (for feature aggregations) between layers 2 and 3, as given here [Graefe 1989]. All layers have grown in conjunction with more computing power and experience gained over the years to deal with more complex tasks and situations (see Chapters 13 and 14). The growth rate in performance of general purpose microprocessors (GPP) of about one order of magnitude every 4 to 5 years has allowed quick development of powerful real-time systems; though the distribution of architectural elements on processor hardware has changed quite a bit over time, the general structure shown in Figure 11.4 has remained surprisingly stable.

11.2.4 Experimental Results

The three-stage process of obstacle detection, hypothesis generation (recognition), and relative spatial state estimation has been tested with VaMoRs on an unmarked two-lane campus road at speeds up to 40 km/h with an obstacle of about 0.5 m^2 cross section (a dark trash can ≈ 0.5 m wide and ≈ 1 m high). The detection range was expected to be about 30 to 40 m. The vehicle had to stop autonomously about 15 m in front of the obstacle. In the example shown, driving speed was about 4 m/s (15 km/h). As Figure 11.5 shows, range estimation started at $r = 33$ m (upper right graph) derived from the bottom feature of the obstacle under the assumption of a flat surface (known camera elevation above the ground). The transient in relative speed estimation to the negative value of vehicle speed (-4) took about 1 second (lower right in Figure 11.5) with two oscillations before stabilizing at the correct value. In the first ten to fifteen video cycles, *i.e.*, ≈ 0.3 seconds from detection during hypothesis generation and -testing (activation of the obstacle-processor group) the results in spatial interpretation are noise-corrupted and not useful for decision–making. Allowing about 1 second for stabilization of the interpretation process seems reasonable (similar to human performance).

During the test sequence shown, the range decreased from 33 to 16 m, and the speed diminishing toward the end due to braking. Note that vehicle speed for prediction was available from conventional mechanical measurements; the speed shown in the lower right part is the visually estimated speed based on image data. The result is rather noisy; this has led to the decision that relative speed was set at the negative speed of the vehicle (measured by the tachometer) in future tests.

Obstacle height was estimated as very stable (110 cm). The pitch angle increased in magnitude during the approach since the elevation of the camera above the ground (≈ 2 m) was higher than the object center. Apparently, a slight curve was steered since the azimuth angle ψ_{KO} shows a dip (around 3 seconds). With these experiments, the 4-D approach to real-time machine vision was shown to be well suited for monocular depth estimation. Since image sequence evaluation is done with time explicitly represented in the model underlying the recognition process, motion stereo is an inherent property of the approach, including odometry.

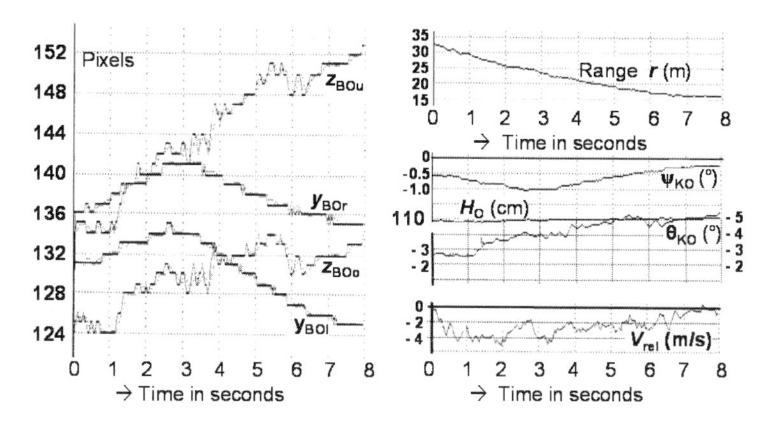

Figure 11.5 Historic first experiment on monocular depth perception and collision avoidance using 'dynamic vision' (4-D approach) with test vehicle **VaMoRs**: A trash can of approximate size 0.5 × 1 m was placed about 35 m in front of the vehicle. Tracking the upper and lower as well as the left and right edge of the can (left part of figure), the vehicle continuously estimated the distance to the obstacle (top right) as well as the horizontal and the vertical bearing angles to the center of the object (center right). Bottom right shows the speed profile driven autonomously; the vehicle stopped 16 m from the obstacle after 8 seconds.

Accuracy in the percent range has been demonstrated; it becomes better the closer the obstacle is approached, a desirable property in obstacle avoidance. The approach based on edge features is computationally very economical and has led to a processor architecture oriented toward physical objects in a modular way. This has allowed refraining from additional direct measurements of distance by laser range finders or radar, even for moving obstacles.

11.3 Detecting and Tracking Moving Obstacles on Roads

Since it is not known, in general, where new objects will appear, all relevant image regions have to be covered by the search for relevant features. In some application domains such as highway traffic, a new object can enter the image only from well-defined regions: Overtaking vehicles will at first appear in the leftmost corner of the image above the road (in right-hand traffic). Vehicles being approached from the back will at first be seen in the teleimage far ahead. Depending on the lane they drive in, the feature flow goes either to one side (vehicle in a neighboring lane) or a looming effect appears, indicating that there may be a collision if no proper action is taken. For hypothesis generation, this information allows starting with proper aspect conditions for tracking; these aspect conditions have a strong influence on feature distribution. The aspect graphs are part of generic object shape representation (maybe even specialized for multiple scales to be used at different distances).

Figure 11.6 shows six of the eight clearly separable standard aspect conditions for cars with a typical set of features for each case. The resulting feature distributions will, of course, depend on the type of vehicle (van, bus, truck, recreation vehicle, *etc.*), usually recognizable by their size and the kind of features exposed in certain relations to each other. For cars, very often the upper part of the body with glass and shiny surfaces is not well suited for tracking since trees and buildings in the environment of the road are mirrored by the (usually two-dimensionally curved) surface yielding very noisy feature distributions.

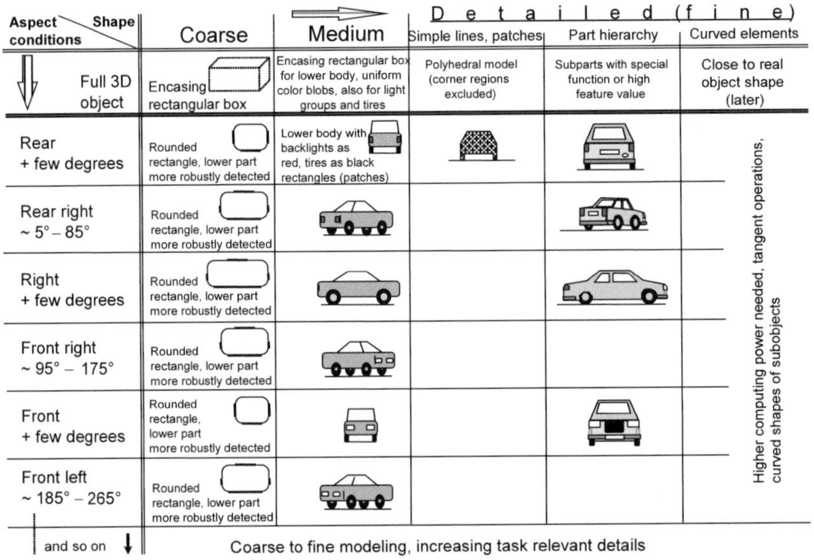

Aspect conditions \ Shape	Coarse	Medium	Detailed (fine): Simple lines, patches	Part hierarchy	Curved elements
Full 3D object	Encasing rectangular box	Encasing rectangular box for lower body, uniform color blobs, also for light groups and tires	Polyhedral model (corner regions excluded)	Subparts with special function or high feature value	Close to real object shape (later)
Rear + few degrees	Rounded rectangle, lower part more robustly detected	Lower body with backlights as red, tires as black rectangles (patches)			Higher computing power needed, tangent operations, curved shapes of subobjects
Rear right ~ 5° – 85°	Rounded rectangle, lower part more robustly detected				
Right + few degrees	Rounded rectangle, lower part more robustly detected				
Front right ~ 95° – 175°	Rounded rectangle, lower part more robustly detected				
Front + few degrees	Rounded rectangle, lower part more robustly detected				
Front left ~ 185° – 265°	Rounded rectangle, lower part more robustly detected				
and so on	Coarse to fine modeling, increasing task relevant details				

Figure 11.6. Cars modeled on different scales for various aspect conditions: Coarse: Encasing box, with rounded corners omitted; medium: Only lower part of body with some simple details; fine: Reasonably good silhouette, groups of very visible features from subparts, => close to real rendering.

For this reason, concentration only on the lower part of the body has been more stable for visual perception of cars (upper limit of about 1 m above the ground); this is shown in column two of Figure 11.4. However, for big trucks and tanker vehicles, the situation is exactly the opposite. The lower part is rather complex due to irregular structural elements, while the upper part is rather simple, usually, like rectangular boxes or cylinders, often with a homogeneous appearance.

So the first step after feature detection and maybe some feature aggregation for elongated line elements should always be to look for the set of features moving in conjunction; this gives an indication of the size of the object and of the vehicle class it may belong to. Three to five images of a sequence are sufficient to arrive at a stable decision, usually. The second step then is to stabilize the interpretation as a member of a certain vehicle class by tapping more and more knowledge from the internal knowledge base and proving its validity in the actual case. Tracking certain groups of features may already give a good indication of the relative motion of the object without it being fully "re-cognized". Partial occlusion by other vehicles or due to a curved road surface (both horizontally and vertically) will complicate the general case. Therefore, it is always recommended to take into account the best estimates for the road state and for the relative state of other vehicles.

The last three columns in Figure 11.6 will be of interest for the more advanced vision systems of the future exploiting the full potential of the sense of vision with high resolution when sufficient computing power will be available.

It is the big advantage of vision over radar and laser range finding that vision allows recognizing the traffic situation with good resolution and up to greater ranges if multifocal vision with active gaze control is used. This is not yet the general state of the art since the data rates to be handled are rather high (many gigabytes/second) and their interpretation requires sophisticated software.

In the case of *expectation-based, multi focal, saccadic vision* (EMS vision) it has been demonstrated that from a functional point of view, visual perception as in humans is possible; until the human performance level is achieved, however, quite a bit of development has still to be done. We will come back to this point in the final outlook.

Due to this situation, industry has decided to pick radar for obstacle detection in systems already on the market for traffic applications; LRF has also been studied intensively and is being prepared for market introduction in the near future. Radar-based systems for driver assistance in cruise control have been available for a few years by now. Complementing them by vision for road and lane recognition as well as for reduction of false alarms has been investigated for about the same time. These combined systems will not be looked at here; the basic goal of this section is to develop and demonstrate the potential of vertebrate-type vision for use in the long run. It exploits exactly the same features as human vision does, and should thus be sufficient for safe driving. Multisensor adaptive cruise control will be discussed in Section 14.6.3.

11.3.1 Feature Sets for Visual Vehicle Detection

Many different approaches have been tried for solving this problem since the late 1980s. Regensburger (1993) presents a good survey on the task "visual obstacle recognition in road traffic". In [Carlson, Eklundh 1990], an object detection method using prediction and motion parallax is investigated. In [Kuehnle 1991], the use of symmetries of contours, gray levels and horizontal lines for obstacle detection and tracking is discussed. [Zielke *et al.* 1993] investigates a similar approach. Other approaches are the evaluation of optical flow fields [Enkelmann 1990] and model-based techniques like the one described below [Koller *et al.* 1993]. Solder and Graefe (1990) find road vehicles by extracting the left, right and lower object boundary using controlled correlation. An up-to-date survey on the topic may be found in [Masaki 1992++] or in the vision bibliography [http://iris.usc.edu/Vision-Notes/bibliography/contents.html]. Some more recent papers are [Graefe, Efenberger 1996; Kalinke *et al.* 1998; Fleischer *et al.* 2002; Labayarde *et al.*2002; Broggi *et al.* 2004].

The main goal of the 4-D approach to dynamic machine vision from the beginning has been to take advantage of the full spatiotemporal framework for internal representation and to do as little reasoning as possible in the image plane and between frames. Instead, temporal continuity in physical space according to some model for the motion of objects is being exploited in conjunction with spatial shape rigidity in this "analysis-by-synthesis" approach.

Since high image evaluation rate had proven more beneficial in this approach than using a wide variety of features, only edge features with adjacent average intensity values in mask regions were used when computing power was very low (see Section 5.2). With increasing computing power, homogeneously shaded blobs, corner features, and in the long run, color and texture are being added. In any case, perturbations both from the motion process and from measurements as well as from data interpretation tend to change rapidly over time so that a single image in a sequence should not be given too much weight; instead, filtering likely (maybe not very precise) results at a high rate using motion models with low eigenfrequencies has proven to be a good way to go. So, concentration on feature extraction was on fast available ones with selection of those used guided by expectations and statistical data of the recursive estimation process running.

For this reason, image evaluation rates of less than about ten per second were not considered acceptable from the beginning in the early 1980s; the number of processors in the system and workload sharing had to be adjusted such that the high evaluation rate was achievable. This was in sharp contrast to the approaches to machine vision studied by most other groups around the globe at that time. Accumulated delay times could be handled by exploiting the spatiotemporal models for compensation by prediction. These short cycle times, of course, left no great choice of features to be used. On the contrary, even simple edge detection could not be used all over the image but had to be concentrated (attention controlled!) in those regions where objects of interest for the task at hand could be expected.

Once the road has been known from the specific perception loop for it, "obstacles" could be only those objects in a certain volume above the road region, strictly speaking only those within and somewhat to the side of the width of the wheel tracks.

11.3.1.1 Edge Features and Adjacent Average Gray Values

Edge features are robust to changes in lighting conditions; maybe this is the reason why their extraction is widespread in biological vision systems (striate cortex). Edge features on their own have three parameters for specifying them completely: position, orientation, and the value of the extreme intensity gradient. By associating the average intensity on one side of the edge as a fourth parameter with each edge, average intensities on both sides are known since the gradient is the difference between both sides; this allows coarse area-based information to be included in the feature.

Mori and Charkari (1993) have shown that the shadow underneath a vehicle is a significant pattern for detecting vehicles; it usually is the darkest region in the environment. Combining this feature with knowledge of 3-D geometric models and 4-D dynamic scene understanding leads to a robust method for obstacle detection and tracking. [Thomanek *et al.* 1994; Thomanek 1996] developed the first vision system capable of tracking a half dozen vehicles on highways in each hemisphere with bifocal vision based on these facts in closed-loop autonomous driving. This approach will be taken as a starting point for discussing more modern approaches exploiting the increase in computing power by at least two orders of magnitude (factor > 100) since.

Figure 11.7 shows a highway scene from a wide-angle camera with one car ahead in the subject's lane. A search for horizontal edge features is performed in vertical search stripes with KRONOS masks of size 5 × 7 as indicated on the right-hand side (see Section 5.2). Due to missing computer performance in the early 1990s, the search stripes did not cover the whole image below the horizon; evaluation cycle time was 80 ms (every second video field with the same index). Stripe width and spacing as well as mask parameters had to be adjusted according to the detection range desired. For improved resolution, there was a second camera with a telelens on the gaze controlled platform (see Figure 1.3) with a viewing range about three times as far (and a correspondingly narrower field of view) compared to the wide-angle camera. This allowed using exactly the same feature extraction algorithms for vehicles nearby and further away (see Figure 11.22 further below).

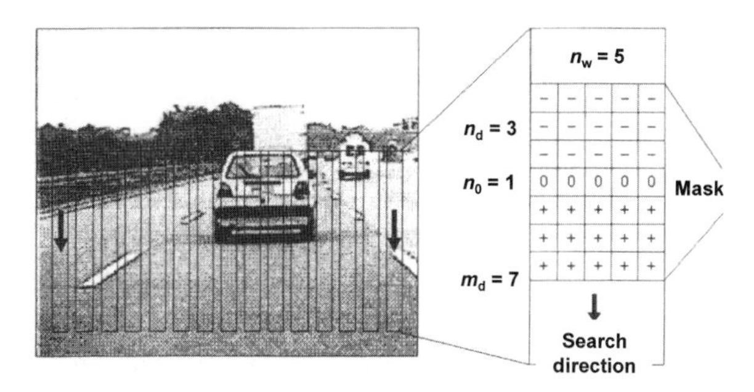

Figure 11.7. Detection of vehicle candidates by search of horizontal edges in vertical search stripes below the horizon: Mask parameters selected such that several stripes cover a small vehicle [Thomanek 1996]

Find lower edge of a vehicle: About 30 search stripes of 100 pixels length have been analyzed by shifting the correlation mask top-down to find close-to-horizontal edge features at extreme correlation values. Potential candidates for the dark area underneath the vehicle have to satisfy the criteria:

- The value of the mask response (correlation magnitude) at the edge has to be above a threshold value ($corr_{min,uv}$).
- The average gray value of the trailing mask region (upper part) has to be below a threshold value ($dark_{min,uv}$).

The first bullet requires a pronounced dark-to-bright transition, and the second one eliminates areas that are too bright to stem from the shaded region underneath the vehicle; adapting these threshold values to the situation actually given is the challenge for good performance. For tanker vehicles and low standing sun, the approach very likely does not work. In this case, the big volume above the wheels may require area-based features for robust recognition (homogeneously shaded, for example).

Generate horizontal contours: Edge elements satisfying certain gestalt conditions are aggregated applying a known algorithm for chaining. The following steps are performed, starting from the left window and ending with the right one:

1. For each edge element, search the nearest one in the neighboring stripe and store the corresponding index if the distance to it is below a threshold value.

2. Tag each edge element with the number count of previous corresponding elements (*e.g.*, six, if the contour contains six edge elements up to this point).

Read starting point $P_s(y_s, z_s)$ and end point $P_e(y_e, z_e)$ of each extracted contour and check the slope, whose magnitude $|(z_e - z_s)/(y_e - y_s)|$ must be below a threshold for being accepted (close to horizontal, see Figure 11.8).

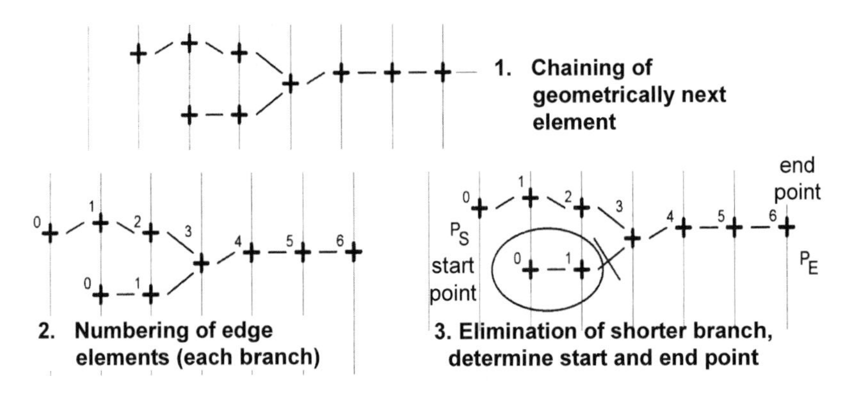

Figure 11.8. Contour generation from edge elements observing gestalt ideas of nearness and colinearity; below an upper limit for total contour length, only the longer one is kept

If lines grow too long, they very likely stem from the shadow of a bridge or from other buildings in the vicinity; they may be tracked as the hypothesis for a new stationary object (shadow or discontinuity in surface appearance), but eliminating them altogether will do no harm to tracking moving vehicles with speed already recognized. Within a few cycles, these elongated lines will have moved out of the actual image. With knowledge of 3-D geometry (projection equations link row number to range), the extracted contours are examined to see whether they allow association with certain object classes: Side constraints concerning width must be satisfied; likely height is thereby hypothesized.

Contours starting from inhomogeneous areas inside the objects (*i.e.*, bumper bar or rear window) are discarded; they lie above the lower shadow region (see Figure 11.9).

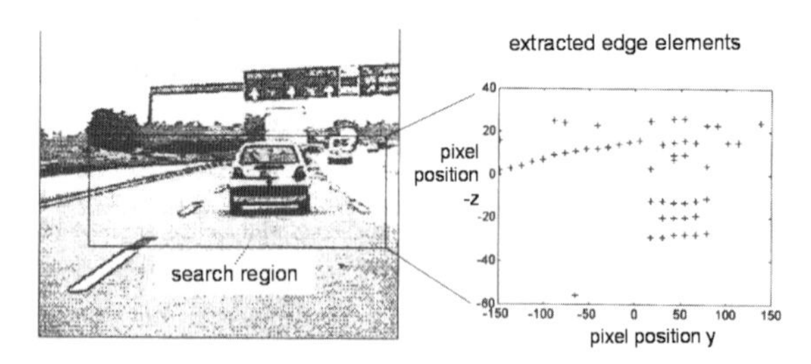

Figure 11.9. Extracted horizontal edge elements: The rectangular group of features is an indication of a vehicle candidate; the lower elements (aggregated shadow region under the car) allow estimation of the range to the vehicle

Determine lateral boundaries: Depending on the lateral position relative to the lane driven in, the vertical object boundaries are extracted additionally. This is done with an edge detector which exploits the fact that the difference in brightness on the object and from the background is not constant and can even change sign; in

Figure 11.10, the wheels and fender are darker than the light gray of the road while the white body is brighter than the road.

For this purpose, the gradient of brightness is calculated at each position in each image row, and its absolute values are summed up over the lines of interest. The calculated distribution of correlation values has significantly large maxima at the object boundaries (lower part of figure). The maxima of the accumulated values yield the width of the obstacle in the image; knowing range and mapping parameters, obstacle size in the real world is initialized for recursive estimation and updated until it is stable. With clearly visible extremes as in the lower part of Figure 11.10 the object width of the real vehicle is fixed, and changes in the image are from now on used to support range estimation.

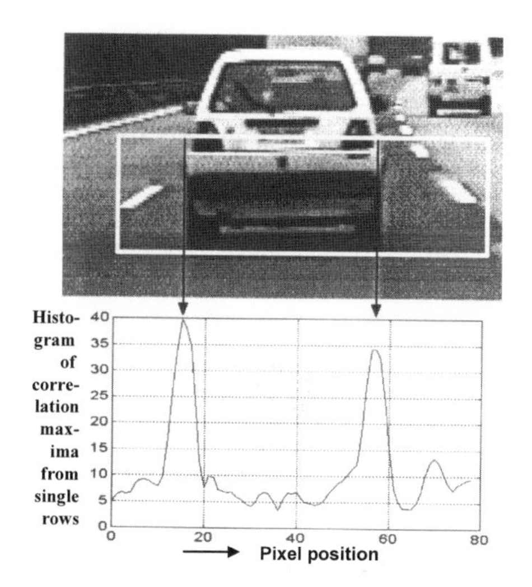

Figure 11.10. Determination of lateral boundaries of a vehicle by accumulation of correlation values at each position in each single row of the lower part of the body with a KRONOS-mask ($n_w = 1$; n_d = large). The maxima of the accumulated values yield the width of the obstacle in the image.

For vehicles driving in their own lanes, the left and right object boundary must be present to accept the extracted horizontal contour as representing an object. In neighboring lanes, it suffices to find a vertical boundary on the side of the vehicle adjacent to their own lane to prove the hypothesis of an object in connection with the lower contour. This means that in the left lane, a vertical line to the right of the lower contour has to be found, while in the right lane, one to the left has to be found for acceptance of the hypothesis of a vehicle. This allows recognition of partially occluded objects, too. The algorithm was able to detect and track up to five objects in parallel with four INMOS Transputer® 222 (16 bit) for feature extraction and one T805 (32 bit) for recursive estimation at a cycle time of 80 ms.

Applying these methods is a powerful tool for extracting vehicle boundaries in monochrome images also for modern high-performance microprocessors. Adding more features, however, can make the system more versatile with respect to type of vehicle and more robust under strong perturbations in lighting conditions.

11.3.1.2 Homogeneous Intensity Blobs

Region-based methods, extracting homogeneously shaded or textured areas are of importance especially for robust recognition of large vehicles. Color recognition very much alleviates object separation in complex scenes with many objects of different colors. But just regions of homogeneous intensity shading alleviate object separation considerably (especially in connection with other features).

In Figure 11.11 the homogeneously shaded areas of the road yield the background for detecting vehicles with different intensity blobs above a dark region on the ground, stemming from vehicle shade underneath the body. Though resolution is poor (32 pixels per mel and 128 per mask) and some artifacts normal to the search direction can be seen, relatively good hypotheses for objects are derivable from this coarse scale. Five vehicle candidates can be recognized, three of which are partially occluded. The car ahead in the same lane and the bus in the right neighboring lane are clearly visible. The truck further ahead in the subject's lane can clearly be recognized by its dark upper body. For the two cars in the left neighboring lane, resolution is too poor to recognize details; however, from the shape of the road area, the presence of two cars can be hypothesized. Low resolution allows higher evaluation frequency for limited computing power.

Figure 11.11. Highway scene with many vehicles, analyzed with UBM method (see Section 5.3.2.4) in vertical stripes with coarse resolution (22.42C) and aggregation of homogeneous intensity blobs (see text).

Performing the search on the coarse scale for homogeneously shaded regions in both vertical and horizontal stripes yields sharp edges in the search direction; thus, close to vertical blob boundaries should be taken from horizontal search results while close to horizontal boundaries should be taken from vertical search results.

Figure 11.12 shows results from a row search with different parameters (11.44R) for another image out of the same sequence (see bus in right neighboring lane and the

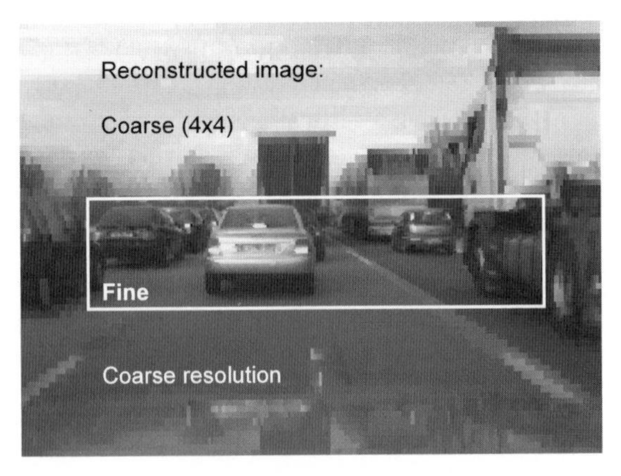

Figure 11.12. Highway scene similar to Figure 11.11 with more vehicles analyzed with UBM method in horizontal stripes; the outer regions are treated with coarse resolution (11.44R), while the central region (within the white box) covering a larger look-ahead range above the road, is analyzed on a fine scale (11.11R) (reconstructed images, see text)

dark truck in the subject's lane). Here, however, the central part of the image, into which objects further away on the road are mapped, is analyzed at fine resolution giving full details (11.11R). This yields many more details and homogeneous intensity blobs; the reconstructed image shown can hardly be distinguished from the original image. A total of eight vehicle candidates can be recognized, six of which are partially occluded. It can be easily understood from this image that large vehicles like trucks and buses should be hypothesized from the presence of larger homogeneous areas well above an elevation of one wheel diameter from the ground. For humans, it is immediately clear that in neighboring lanes, vehicles are recognized by three wheels if no occlusion is present; the far outer front wheel will be self-occluded by the vehicle body. All wheels will be only partially visible. This fact has led to the development of parameterized wheel detectors based on features defined by regional intensity elements [Hofmann 2004].

Figure 11.13 shows the basic idea and the derivation of templates that can be adapted to wheel diameter (including range) and aspect angle in pan (small tilt angles are neglected because they enter with a cosine effect (≈ 1)); since the car body occludes a large part of the wheels, the lower part of the dark tire contrasting the road to its sides is especially emphasized. For orthogonal and oblique views of the near side of the vehicle, usually, the inner part of the wheel contrasts to the tire around it; ellipticity is continuously adapted according to the best estimate for the relative yaw (pan) angle.

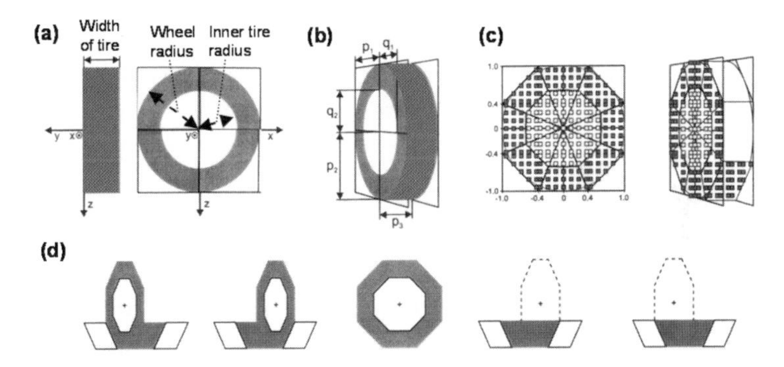

Figure 11.13. Derivation of templates for wheel recognition from coarse shape representations (octagon): (a) Basic geometric parameters: width, outer and inner visible radius of tire; (b) oblique view transforms circle into ellipses as a function of aspect angle; (c) shape approximation for templates, radii, and aspect angle are parameters; (d) template masks for typically visible parts of wheels [seen from left, right, ≈ orthogonal, far side (underneath body)]. Intelligently controlled 2-D search is done based on the existing hypothesis for a vehicle body (after [Hofmann 2004]).

The wheels on the near side appear in pairs, usually, separated by the axle distance in the longitudinal direction which lets the front wheel appear higher up in the image due to camera elevation above the wheel axle. There is good default knowledge available on the geometric parameters involved so that initialization poses no challenge. Again, being overly accurate in a single image does not make sense, since averaging over time will lead to a stable (maybe a little bit noisier) result with the noise doing no harm. To support estimation of the aspect conditions, taking into account other characteristic subobjects like light groups in relation to the license plate as regional features will help.

11.3.1.3 Corner Features

This class of features is especially helpful before a good interpretation of the scene or an object has been achieved. If corner localization can be achieved precisely and consistently from frame to frame, it allows determining feature flow in both image dimensions and is thus optimally suited for tracking without image understanding. However, the challenge is that checking consistency requires some kind of understanding of the feature arrangement. Recognition of complex motion patterns of articulated bodies is very much alleviated using these features. For this reason, their extraction has received quite a bit of attention in the literature (see Section 5.3.3). Even special hardware has been developed for this purpose.

With the computing power nowadays available in general-purpose microprocessors, corner detection can be afforded as a standard component in image analysis. The unified blob-edge-corner method (UBM) treated in Section 5.3 first separates candidate regions for corners in a very simple way from those for homogeneously shaded regions and edges. Only a very small percentage of usual road images qualify as corner candidates depending on the planarity threshold specified (see Figures 5.23 and 5.26); this allows efficient corner detection in real time together with blobs and edges. The combination then alleviates detection of joint feature flow and object candidates: Jointly moving blobs, edges, and corners in the image plane are the best indicators of a moving object.

11.3.2 Hypothesis Generation and Initialization

The center of gravity of a jointly moving group of features tells us something about the translational motion of the object normal to the optical axis; expanding or shrinking similar feature distributions contains information on radial motion. Changing relative positions of features other than expansion or shrinking carries information on rotational motion of the object. The crucial point is the jump from 2-D feature distributions observed over a short amount of time to an object hypothesis in 3-D space and time.

11.3.2.1 Influence of Domain and Actual Situation

If one had to start from scratch without any knowledge about the domain of the actual task, the problem would be hardly solvable. Even within a known domain (like "road traffic") the challenge is still large since there are so many types of roads, lighting-, and weather conditions; the vehicle may be stationary or moving on a smooth or on a rough surface.

It is assumed here that the human operator has checked the lighting and weather conditions and has found them acceptable for autonomous perception and operation. When observation of other vehicles is started, it is also assumed that road recognition has been initiated successfully and is working properly; this provides the system (via DOB, see Chapters 4 and 13) with the number and widths of lanes actually available. With GPS and digital maps onboard and working, the type of road being driven is known: unidirectional or two-way traffic, motorway or general cross-country/urban road.

The type of road determines the classes of obstacles that might be expected with certain likelihood; the levels of likelihood may be taken into account in hypothesis generation. Pedestrians are less likely on high-speed than on urban roads. Speed actually being driven and traffic density also have an influence on this choice; for example, in a traffic jam on a freeway with very low average speed, pedestrians are more likely than in normal freeway traffic.

11.3.2.2 Three Components Required for Instantiation

In the 4-D approach, there are always three components necessary for starting perception based on recursive estimation: (1) the generic object type (class and subclass with reasonable parameter settings), (2) the aspect conditions (initial values for state components, and (3) the dynamic model as knowledge (or side constraint) of evolution over time; for subjects, this includes knowledge of (stereotypical) motion capabilities and their temporal sequence. This latter component means an individual capability for animation based on onsets of maneuvers visually observed; this component will be needed mainly in tracking (see Section 11.3.3). However, a passing car cutting into the vehicle's lane immediately ahead will be perceived much faster and more robustly if this motion behavior (normally not allowed) is available also during the initialization phase, which takes about one half to one second, usually.

Instantiation of a generic object (3-D shape): The first step always is to establish good range estimation to the object. If stereovision or direct range measurements are available, this information should be taken from these sources. For monocular vision, this step is done with the row index z_{Bu} of the lowest features that belong most likely to the object. Then, the first part of the following procedure is, as it is for static obstacles, to obtain initial values of range and bearing.

With range information and the known camera parameters, the object in the image can be scaled for comparison with models in the knowledge base of 3-D objects. Homogeneously shaded regions with edges and corners moving in conjunction give an indication of the vehicle type. For example, in Figure 11.11, the car upfront, the truck ahead of it (obscured in the lower part), and the bus upfront to the right are easily classified correctly; the two cars in the lane to the left allow only uncertain classification due to occlusion of large parts of them. Humans may feel certain in classifying the car upfront left, since they interpret the intensity blobs vertically located at the top and the center of the hypothesized car: The somewhat brighter rectangle at the top may originate from the light of the sky reflected from the curved roof of the car. The bright rectangular patch between two more quadratic ones a little bit darker halfway from the roof to the ground is interpreted as a license plate between light groups at each rear side of the car.

Figure 11.12 (taken a few frames apart from Figure 11.11) shows in the inner high-resolution part that this interpretation is correct. It also can be seen by the three bright blobs reasonably distributed over the rear surface that the car immediately ahead is now braking (in color vision, these blobs would be bright red). The two cars in the neighboring lane beside the dark truck are also braking. (Note the different locations and partial obscuration of the braking lights on the three cars depending on make and traffic situation). Confining image interpretation for obstacle detection to the region marked by the white rectangle (as done in the early days) would make vehicle classification much more difficult. Therefore, both peripheral low-resolution and foveal high-resolution images in conjunction allow efficient and sufficiently precise image interpretation.

Aspect conditions: The vertical aspect angle is determined by the range and elevation of the camera in the subject vehicle above the ground. It will differ for cars, vans, and trucks/buses. Therefore, only the aspect angle in yaw has to be derived from image evaluation. In normal traffic situations with vehicles driving in the direction of the lanes, lane recognition yields the essential input for initializing the aspect angle in yaw.

On straight roads, lane width and range to the vehicle determine the yaw aspect angle. It is large for vehicles nearby and decreases with distance. Therefore, in the right neighboring lane, only the left-hand and the rear side can be seen; in the left neighboring lane, it is the right-hand and rear side. Tires of vehicles on the left have their dark contact area to the ground on the left side of the elliptically mapped vertical wheel surface (and *vice versa* for the other side; see Figure 11.13d). Aspects conditions and 3-D shape are closely linked together, of course, since both in conjunction determine the feature distribution in the image after perspective projection, which is the only source available for dynamic scene understanding.

Dynamic model: The third essential component for starting recursive estimation is the process model for motion which implements continuity conditions and knowledge about the evolution of motion over time. This temporal component was the one that allowed achieving superior performance in image sequence interpretation and autonomous driving. As mentioned before, there are two big advantages in temporal embedding:

1. Known state variables in a motion process decouple future evolution from the past (by definition); so there is no need to store previous images if all objects of relevance are represented by an individual dynamic process model. Future evolution depends only on (a) the actual state, (b) the control output applied, and (c) on external perturbations. Items (b) and (c) principally are the unknowns while best estimates for (a) are derived by visual observation exploiting a knowledge base of vehicle classes (see Chapter 3).
2. Disturbance statistics can be compiled for both process and measurement noise; knowing these characteristics allows setting up a temporal filter process that (under certain constraints) yields optimal estimates for open parameters and for the state variables in the generic process model.
3. These components together are the means by which "the outside world is transduced into an internal representation in the computer". (The corresponding question often asked in biological systems is, *how does the world get into your*

head?) Quite a bit of background knowledge has to be available for this purpose in the computer process analyzing the data stream and recognizing "the world"; features extracted from the image sequence activate the application of proper parts of this knowledge. In this closed-loop process resulting in control output for a real vehicle (carrying the sensors), feedback of prediction-errors shows the validity of the models used and allows adaptation for improved performance.

The dynamic models used in the early days in [Thomanek 1996] were the following (separate, decoupled models for longitudinal and lateral translation, no rotational dynamics):

Simplified longitudinal dynamics: The goal was to estimate the range and range rate sufficiently well for automatic transition into and for convoy driving. Since the control and perturbation inputs to the vehicle observed are unknown, a third-order model with colored noise for acceleration as given in Equations 2.34 and 2.35 has been chosen and proven to be sufficient [Bar-Shalom, Fortmann 1988]. The noise term n(t) is fed into a first-order system with time constant $T_c = 1/\alpha$. The discrete model then is (here $\Phi = A$)

$$\begin{aligned} x_{k+1} &= \Phi \cdot x_k + D_k \cdot n_k \\ \textit{with} \quad D_k^T &= [T^2/2,\ T,\ 1]. \end{aligned} \tag{11.10}$$

The discrete noise term n_k is assumed to be a white, bias-free stochastic process with normal distribution and expectation zero; its variance is

$$E[n_k^2] = s_q^2. \tag{11.11}$$

After [Loffeld 1990] σ_q should be chosen as the maximally expected acceleration of the process observed.

Simplified lateral dynamics: Since the lateral positions of the vehicles observed have only very minor effects on the subject's control behavior, a standard second-order dynamical model with the state variables, lateral position y_o relative to the subject's (its own) lane center and lateral speed v_{yo} are sufficient. The discrete model then is

$$\begin{pmatrix} y_o \\ v_{yo} \end{pmatrix}_{k+1} = \begin{pmatrix} 1 & T \\ 0 & 1 \end{pmatrix} \cdot \begin{pmatrix} y_o \\ v_{yo} \end{pmatrix}_k + \begin{pmatrix} 0 \\ 1 \end{pmatrix} \cdot n_{v,k}. \tag{11.12}$$

Again, the discrete noise term $n_{v,k}$ is assumed to be a white, bias-free stochastic process with normal distribution, expectation equal to zero, and variance σ_{qy}^2.

11.3.2.3 Initial State Variables for Starting Recursive Estimation

Figure 11.14 visualizes the transformation of the feature set marking the lower bound of a potential vehicle into estimated positions for the vehicles in Cartesian coordinates. In the left part of the figure, dark-to-bright edges of the dark area underneath the vehicle in a top-down search are shown. For the near range, assuming a flat surface is sufficient, usually. The tangents to the local lane markings are extrapolated to a common vanishing point of the near range if the road is curved.Since convergence behavior is good, usually, special care is not necessary in the general case. From the right part of the figure, the bearing angles ψ_i to the vehicle candidates can easily be determined. Initialization and feature selection for tracking have to take the different aspect conditions in the three lanes into account.

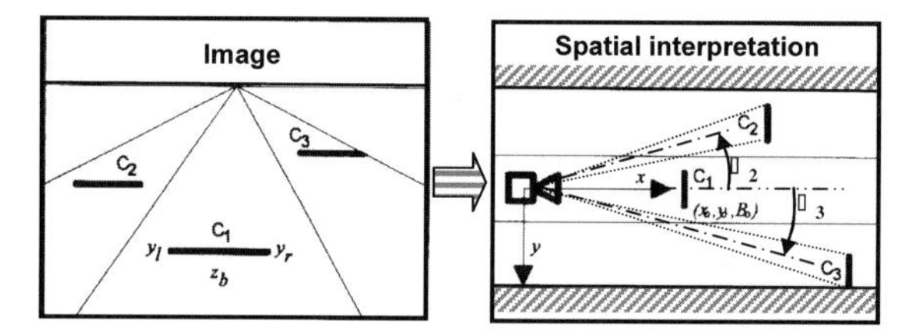

Figure 11.14. Transformation of image row, in which the lower edge of the dark region underneath the vehicle appears, and of lateral position into Cartesian coordinates, based on camera elevation above the ground assumed to be flat

The aspect graph in Figure 11.15 shows the distribution of characteristic features as seen from the **r**ear **l**eft (vehicle in front in neighboring lane to the right). The features detected, for which correspondence can be established most easily at low computational cost, are the dark area underneath the vehicle and edge features at the vehicle corners, front left and rear right (marked by bold letters in the figure). The configuration of rear groups of lights and license plate (dotted rectangle) and the characteristic set of wheel parts are the features detectable and most easily recognizable with additional area-based methods.

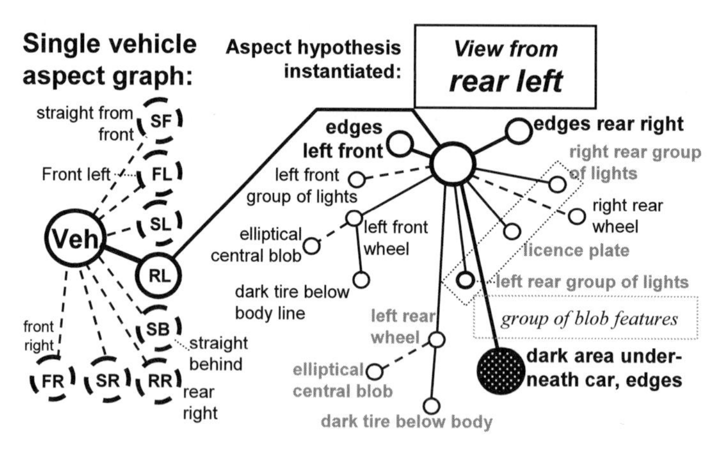

Figure 11.15. Aspect conditions determine feature sets to be extracted for tracking. On the same road in normal traffic, road curvature, distance, and the lane position relative to the subject's own lane are the most essential parameters; traffic moving in the same or in the opposite direction exhibits rear/front parts of vehicles. On crossroads, views from the side predominate. The situation shown is typical for passing a vehicle in right-hand traffic.

Getting good estimates for the velocity components needed for each second-order dynamic model is much harder. Again, trusting in good convergence behavior leads to the easy solution, in which all velocity components are initialized with zeros. Faster convergence may be achieved if an approximate estimation of the speed components can be achieved in the initial observation period of a few cycles; this is especially true if the corresponding elements of measurement covariance are set large and system covariance is set low (high confidence in the correctness of the model).

11.3.2.4 Measurement Model and Jacobian Elements

There are two essentially independent motion processes in the models given: "Longitudinal" (x_o, V_o) and "lateral" states (y_o, v_{yo}) of the vehicle observed. Pitching motion of the vehicle has not yet been taken into account. However, if the sensors (cameras) have no degree of freedom for counteracting vehicle motion in pitch, this motion will affect visual measurement results appreciably (see Section 7.3.4). Depending on acceleration and deceleration, pitch angles of several degrees (≈ 0.05 mrad) are not uncommon; at 70 m distance, this value corresponds to a height change in the real world of ≈ 3.5 m for a point in the same row of the image.

Rough ground may easily introduce pitch vibrations with amplitudes around 0.25° (≈ 0.005 mrad); at the same distance of 70 m, this corresponds to a 35 cm height change or changes in look-ahead distances on flat ground in the range of 10 to 20 m (around 25 %). At shorter look-ahead distances, this sensitivity is much reduced; for example, at 20 m, the same vibration amplitude of 0.25° leads to look-ahead changes of only about 1.3 m (6.5 %) for the test vehicle VaMP with camera elevation $H_K = 1.3$ m above the ground; larger elevations reduce this sensitivity.

Of course, this sensitivity enters range estimation directly according to Figure 11.14. A sensitivity analysis of Equation 7.19 (with distance ρ instead of L_f) shows that range changes ($\partial\rho/\partial\theta$) as a function of pitch angle θ, and ($\partial\rho/\partial z_B$) as a function of image row z_B go essentially with the square of the range:

$$\Delta\rho \sim (\rho^2 / H_K) \cdot \Delta\theta + [\rho^2 / (H_K \cdot f \cdot k_z)] \cdot \Delta z_B \,. \tag{11.13}$$

This emphasizes analytically the numbers quoted above for the test vehicle VaMP. Therefore, for large look-ahead ranges, one should not rely on average values for the pitch angle. If no gaze stabilization in pitch is available, it is recommended to measure lane width at the position of the lower dark-to-bright feature (Equation 11.2) and to evaluate an estimate for range assuming that lane width is the same as determined nearby at distance L_n and to compute the initial value x_o using the pinhole camera model.

The measurement model for the width of the vehicle is given by Equation 11.3; for each single vertical edge feature, the elements of the Jacobian matrix are

$$\begin{aligned} \frac{\partial y_{Bo}}{\partial y_o} &= \frac{\partial}{\partial y_o}\left(\frac{y_o}{x_o}\right) \cdot f \cdot k_y = \frac{f \cdot k_y}{x_o} \\ \frac{\partial y_{Bo}}{\partial x_o} &= \frac{\partial}{\partial x_o}\left(\frac{y_o}{x_o}\right) \cdot f \cdot k_y = -\frac{y_o}{x_o^2} \cdot f \cdot k_y . \end{aligned} \tag{11.14}$$

It can be seen that changes in lateral feature position in the image depend on changes of state variables in the real world by

$$(\partial y_{Bo} / \partial y_o) \cdot \Delta y_o = f \cdot k_y \cdot \Delta y_o / x_o;$$
$$(\partial y_{Bo} / \partial x_o) \cdot \Delta x_o = -f \cdot k_y \cdot y_o \cdot \Delta x_o / x_o^2. \tag{11.15}$$

The second equation indicates that changes in range, Δx_o, can be approximately neglected for predicting lateral feature position since y_o and $\Delta x_o << x_o$, and range is not updated from the prediction-error Δy_{Bo} (no direct cross-coupling between longitudinal and lateral model necessary). However, since lateral position y_o in the first equation is updated by inverting the Jacobian element $(\partial y_{Bo}/\partial y_o)$, small prediction-errors Δy_{Bo} in the feature position in the image will lead to large increments Δy_o. Note that this sensitivity results from taking the camera coordinates only as reference (x_o). Determining y_{oL} relative to the local road or lane has range x_o cancel out, and the lateral position of the vehicle in its lane can be estimated as much less noise-corrupted.

11.3.2.5 Statistical Parameters for Recursive Estimation

The covariance matrices Q of the system models are required as knowledge about the process observed to achieve good convergence in recursive estimation. The covariance matrix of the longitudinal model has been given as Equation 2.38. For the lateral model, one similarly obtains

$$Q = \begin{pmatrix} T & T \\ T & 1 \end{pmatrix} \cdot \sigma. \tag{11.16}$$

Optimal values for σ_q^2 have been determined in numerous tests with the real vehicle in closed-loop performance driving autonomously. Stable driving with good passenger comfort has been achieved for values $\sigma_q^2 \approx 0.1$ (m/s^2)^2. A detailed discussion of this filter design may be found in [Thomanek 1996].

The statistical parameters for image evaluation determine the measurement covariance matrix R. Errors in row and column evaluation are assumed to be uncorrelated. Since lateral speed is not measured directly but only reconstructed from the model, the matrix R can be reduced to a scalar r with

$$r^2 = \sigma_r^2 \tag{11.17}$$

as the variance of the feature extraction process. For the test vehicle VaMP with transputers performing feature localization to full pixel resolution (no subpixel interpolation) and with only 80 ms cycle time, best estimation results were achieved with $\sigma_r^2 = 2$ pixel^2 [Thomanek 1996].

11.3.2.6 Falsification Strategies for Hypothesis Pruning

Computing power available in the early 1990s allowed putting up just one object hypothesis for each set of features found as a candidate. The increase in computational resources by two to three orders of magnitude in the meantime (and even more in the future) allows putting up several likely object hypotheses in parallel. This reduces delay time until stable interpretation and a corresponding internal representation has been achieved.

The early jump to full spatiotemporal object hypotheses in connection with more detailed models for object classes has the advantage that it taps into the knowledge bases with characteristic image features and motion models without running the risk of combinatorial feature explosion as in a pure bottom-up approach putting much emphasis on generating "the most likely single hypothesis". Each hypothesis allows predicting new characteristic features which can then be tested in the next image taking into account temporal changes already predictable. Those hypotheses with a high rate and good quality of feature matches are preferred over the others, which will be deleted only after a few cycles. Of course, it is possible that two (or even more) hypotheses continue to exist in parallel. Increasingly more features considered in parallel will eventually allow a final decision; otherwise, the object will be published in the DOB with certain parameters recognized but others still open.

An example is a trailer observed from the rear driving at low speed; whether the vehicle towing is a truck capable of driving at speeds up to say 80 km/h or an agricultural tractor with a maximal speed of say 40 km/h cannot be decided until an oblique view in a tighter curve or performing a lane change is possible; the length of the total vehicle is also unknown. This information is essential for planning a passing maneuver in the future. The oblique view uncovers a host of new features which easily allow answering the open questions. Moving laterally in one's own lane is a maneuver often used for uncovering new features of vehicles ahead or for discovering the reason for unexpectedly slow moving traffic.

Once the tracking process is running for an object instantiated, the bottom-up detection process will rediscover its features independent of possible predictions. Therefore, the main task is to establish correspondence between features predicted and those newly extracted. A Mahalanobis distance with the matrix Λ for proper weighting of the contribution of different features of a contour is one way to go.

Let the predicted contour of a vehicle be c^*, the measured one c. Its position in the image depends on (at least) three physical parameters: distance x, lateral position y and vehicle width B. From the best estimates for these parameters, c^* is computed for the predicted states at the time of next measurement taking. The prediction-errors $c - c^*$ are taken to evaluate the set of features of a contour minimizing the distance

$$d = (c - c^*)^T \cdot \Lambda^{-1} \cdot (c - c^*) \,. \tag{11.18}$$

From the features satisfying threshold values for $(c - c^*)$, those minimizing d are selected as the corresponding ones. Proper entries for Λ have to be found in a heuristic manner by experiments.

It is the combination of robust simple feature extraction and high-level spatio-temporal models with frequent bottom-up and top-down traversal of the representation hierarchy that provides the basis for efficient dynamic vision. In this context, *time and motion* in conjunction with knowledge about spatiotemporal processes *constitute an efficient hypothesis pruning device*.

If both shape and motion state have to be determined simultaneously [Schick 1992] an interference problem may occur trading shape variations versus aspect conditions; these problems have just been tackled and it is too early to make general statements on favorable ways to proceed. But again, observing both spatial rigidity and (dynamic) time constraints yields the best prospects for solving this difficult task efficiently.

11.3.2.7 Handling Occlusions

The most promising way to understand the actual traffic situation is to start with road recognition nearby and track lane markings or road boundaries from near to far. Since parts of these "objects" may be occluded by other vehicles or self-occluded by curvatures of the road (both horizontally and vertically), obstacles and geometric features on the side of the road have to be recognized in parallel for proper understanding of "the scene". The different types of occlusions require different procedures as proper behavior. A good general indicator for occlusion is changing texture to one side of an edge and constant texture on the other side [Gibson 1979]; to detect this, high-resolution images are required, in general. This item will be an important step for developing machine vision in the future with more computing power available.

In curved and hilly terrain, driving safely until new viewing conditions allow larger visual ranges is the only way to go. In flat terrain with horizontally curved roads, there are situations where low vegetation obscures the direct view of the road, but other vehicles can be recognized by their upper parts; their trajectory over time marks the road. For example, if driving behind a slow vehicle, behind an oncoming curve to the left, a vehicle in opposite traffic direction has been observed over a straight stretch, and no indication of further vehicles is in sight, the vehicle intending to pass should verify the empty oncoming lane at the corner and start passing right away if the lane is free over a sufficiently long distance. (Note that knowledge about one's own acceleration capabilities as a function of speed and visual perception capabilities for long ranges have to be available to decide this maneuver.) In all cases of partial occlusion it is important to know characteristic subparts and their relative arrangement to hypothesize the right object from the limited feature set. Humans are very good in this respect, usually; the capability of reliably recognizing a whole object from only parts of it visible is one important ingredient of intelligence. Branches of trees or sticking snow partially occluding traffic signs hardly hamper correct recognition of them. The presence of a car is correctly assumed if only a small part of it and of one wheel are visible in the right setting. Driving side by side in neighboring lanes, a proper continuous temporal change of the gap between the front wheel (only partially visible) and fender of the car passing will indicate an upcoming sideways motion of the car, probably needing special attention for a while. Knowing the causal chain of events allows reasonable predictions and may save time for proper reactions.

The capability of drawing the right conclusions from a limited set of information visually accessible probably is one of the essential points separating good drivers from normal or poor ones. This may be one of the areas in the long run where really intelligent assistance systems for car driving will be beneficial for overall traffic; early counteraction can prevent accidents and damage. (Our present technical vision systems are far from the level of perception and understanding required for this purpose; in addition, difficult legal challenges have to be solved before application is feasible.)

11.3.3 Recursive Estimation of Open Parameters and Relative State

The basic method has been discussed in Chapter 6 and applied to road recognition while driving in Chapters 7 to 10. In these applications, the scene observed was static, but appeared to be dynamic due to relative egomotion, for which the state has been estimated by prediction-error feedback. The new, additional challenge here is to detect and track other objects moving to a large extent (but not completely) independent of egomotion of the vehicle carrying the cameras. For safe driving in road traffic, at least half a dozen nearest vehicles in each hemisphere of the environment have to be tracked, with visual range depending on the speed driven. For each vehicle (or "subject"), a recursive estimation process has to be instantiated for tracking. This means that about a dozen of these have to run in parallel (level 2 of Figure 5.1).

Each of these recursive estimation processes requires the integration of visual perception. The individual subtasks of feature extraction, feature selection, and grouping as well as hypothesis generation have been discussed separately up to here. The integration to spatiotemporal percepts has to be achieved in this step now. In the 4-D approach, this is done in one demanding large step by directly jumping to internal representations of 3-D objects moving in 3-D space over time. Since these objects observed in traffic are themselves capable of perception and motion control (and are thus "subjects" as introduced in Chapters 2 and 3), the relevant parts of their internal decision processes leading to actual behavior also have to be recognized. This corresponds to *an animation process for several subjects in parallel*, based on observed motion of their bodies. The lower part of Figure 11.16 symbolizes the individual steps for establishing a perceived subject in the internal representation exploiting features and feature flow over time (upper part). The percept comes into existence by fusing image data measured with generic models from object classes stored in a background knowledge base; using prediction-error feedback, this is a temporally extended process with adaptation of open parameters and of state variables in the generic models describing the geometric relations in the scene.

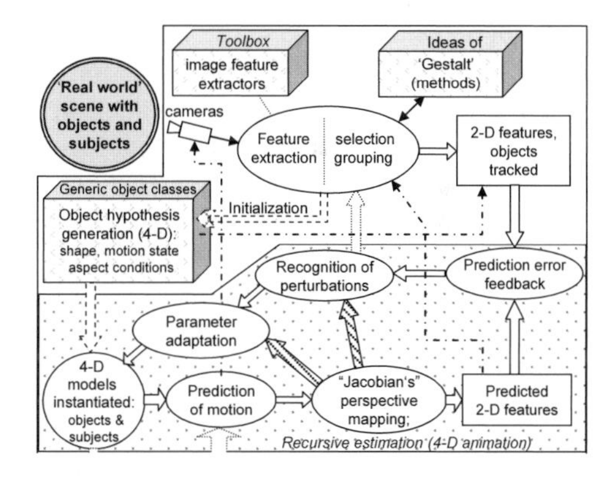

Figure 11.16 Integration of visual perception for a single object: In the upper part the coarsely grained block diagram shows conventional feature extraction, feature selection, and grouping as well as hypothesis generation. The lower part implements the 4-D approach to dynamic vision, in which background knowledge about 3-D shape and motion is exploited for animating the spatiotemporal scene observed in the interpretation process.

11.3.3.1 Which Parameters/States Can Be Observed?

Visual information comes from a rather limited set of aspect conditions over a relatively short period of time. Based on these poor conditions for recognition, either a large knowledge base about the subject class has to be available or uncertainty about the "object" perceived will be large. After hypothesizing an object/subject under certain aspect conditions, the features available from image evaluation may be complemented by additional features which should be visible if the hypothesis holds. From all possible aspect conditions in road scenes, the dark area underneath the vehicle and a left as well as a right vertical boundary on the lower part of the body should always be visible; whether these vertical edges are detectable depends on the image intensity difference between vehicle body and background. In general, this contrast may be too small to be noticed at most over a limited amount of time, due to changing background when the vehicle is moving. This is the reason why these features (written bold in Figure 11.15) have been selected as a starting point for object detection.

When another vehicle is seen from straight behind (SB), the vertical edges seen are those from the left and right sides of the body; its length is not observable from this aspect condition and is thus not allowed to be one of the parameters to be iterated. When the vehicle is seen straight from the side (SR or SL), its width is unobservable but its length can be estimated (see Figure 11.6). Viewing the vehicle under a sufficiently oblique angle, both length and width should be determinable; however, the requirement is that the "inner" corner in the image of the vehicle is very recognizable. This has not turned out to be the case frequently, and oscillations in the separately estimated values for length L and width B resulted [Schmid 1995]. Much more stable results have been achieved when the length of the diagonal ($D = (L^2 + B^2)^{1/2}$) has been chosen as a parameter for these aspect conditions; if B has been determined before from viewing conditions straight behind, the length parameter can be separated assuming a rectangular shape and constant width. For rounded car bodies, the real length will be a little longer than the result obtained from the diagonal.

The yaw angle of a vehicle observed relative to the road is hard to determine precisely. The best approach may be to detect the wheels and their individual distances to the lane marking; dividing the difference by the axle distance yields the (tangent of the) yaw angle. Of the state variables, only the position components can be determined directly from image evaluation; the speed components have to be reconstructed from the underlying dynamic models.

It makes a difference in lateral state estimation whether the correct model with Ackermann steering is used or whether independent second-order motion models in both translational degrees of freedom and in yaw are implemented (independent Newtonian motion components). In the latter case, the vehicle can drift sidewise and rotate even when it is at rest in the longitudinal direction. With the Ackermann model, these motion components are linked through the steering angle, which can thus be reconstructed from temporal image sequences (see Section 14.6.1) [Schick 1992]. If phases of standing still with perturbations on the vehicle body are encountered (as in stop-and-go traffic), the more specific Ackermann model is recommended.

As mentioned before, motion of other vehicles can be described relative to the subjet vehicle or relative to the local road. The latter formulation is less dependent on perturbations of the vehicle body and thus is preferable.

11.3.3.2 Internal Representations

Each object hypothesis is numbered consecutively and stored with time of installation marking the observation period in local memory first for checking the stability of the hypothesis; all parameters and valid state variables actually are stored in specific slots. These are the same as available for publication to the system after the hypothesis has become more stable within a few evaluation cycles; in addition, information on the initial convergence/divergence process is stored such as: percentage of predicted features actually found, time history of convergence criteria including variance, *etc.* When certain criteria are met, the hypothesis is published to the overall system in the *dynamic database* (DDB, renamed later in dynamic object database, DOB); all valid parameters and state variables are stored in specific slots. Special state variables may be designated for entry into a ring buffer of size n_{RB}; this allows access to the n_{RB} latest values of the variable for recognizing time histories. The higher interpretation levels of advanced systems may be able to recognize the onset of maneuvers looking at the time history of characteristic variables.

Other subsystems can specify information on objects they want to receive at the earliest date possible; these lists are formed by the DOB manager and sent to the receiver cyclically. (To avoid time delays too large for stable reaction, some preferred clients receive the information wanted directly from the observation specialist in parallel.)

11.3.3.3 Controlling and Monitoring the Estimation Process

Each estimation process is monitored steadily through several goodness criteria. The first is the count of predicted features for which no corresponding measured feature could be established. Usually, obtaining a few times (factor of 2 to 4) the amount of features minimally required for good observation is tried. If less than a certain percentage (say, 70%) of these features has no corresponding feature measured, the measurement data are rejected altogether and estimation proceeds with pure prediction according to the dynamic model. Confidence in the estimation results is decreased; Figure 11.17 shows a time history of the confidence measure from [Thomanek 1996]. After initialization, confidence level V_E is low for several cycles until convergence data allow lifting it to full confidence V_V; if not sufficiently many corresponding features can be found, confidence is decreased to level V_P. If this occurs in several consecutive

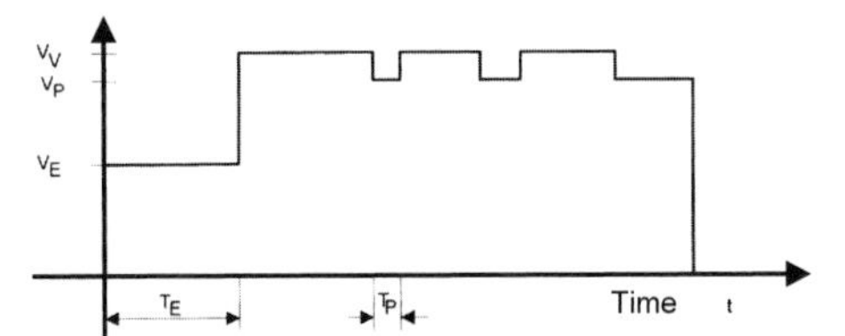

Figure 11.17. Time history of confidence measure in object tracking: V_V full confidence, V_P confidence from prediction only; V_E confidence for initialization.

cycles, confidence is completely lost and goes to zero (termination of tracking).

Monitoring is done continuously for all tracking processes running. Figure 11.18 shows a schematic diagram implemented in the transputer system of the mid-1990s. The vision system has two blocks: One for the front hemisphere with two cameras using different focal lengths (lower left in figure with paths 1 and 2, the latter not shown in detail) and one for the rear hemisphere (lower right, paths 3 and 4, only the latter is detailed). Objects are sorted and analyzed from near to far for handling occlusions in a natural way. For each image stream, a local object administration is in place. This is the unit communicating estimation results to the rest of the system, especially to the global object administration process (center top). Feature occlusion in the teleimage by objects tracked in the wide-angle images has to be resolved by this agent. It also has to check the consistency of the individual results; for example, very often the vehicle ahead in the subject's lane is the same in the tele- and the wide-angle image. In a more advanced system, this fact should be detected automatically, and a single estimation process for this vehicle should be fed with features from the two image streams; lacking time has not allowed realizing this in the "Prometheus" project.

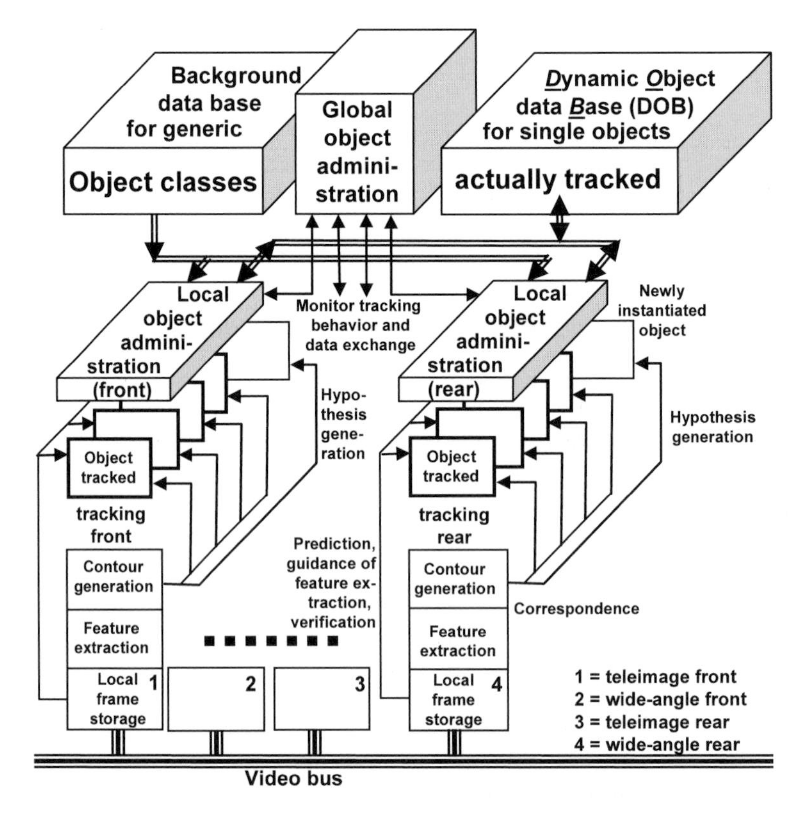

Figure 11.18. Structure and data flow of second-generation dynamic vision system for ground vehicle perception realized with 14 transputers (see Figure 11.19) in test vehicles VaMP and VITA_2 [Thomanek 1996]

All processes work with models from the background data base for generic object classes (top left in Figure 11.18) and feed results into the DOB (top right).

This detailed visualization of the *obstacle recognition* subsystem has to be abstracted on a higher level to show its embedding in the overall automatic visual control system developed and demonstrated in the *Prometheus*-project; beside *obstacle recognition*, other vision components for *road recognition* [Behringer 1996], for *3-D shape during motion* [Schick 1992], for *moving humans* [Kinzel 1995], and for *signal lights* [Tsinas 1997] were developed. This was the visual perception part of the system that formed the base for the other components for *viewing direction control* [Schiehlen 1995], for *situation assessment* [Hock 1994], for *behavior decision* [Maurer 2000], and for vehicle control [Brüdigam 1994].

Figure 11.19 shows the overall system, largely based on transputers, in coarse resolution; this was the system architecture of the second-generation vision systems of UniBwM. Here, we just discuss the subsystem for visual perception with the cameras shown in the upper left, and the road and obstacle perception system shown in the lower right corners. The upper right part for system integration/locomotion control will be treated in Chapters 13 and 14.

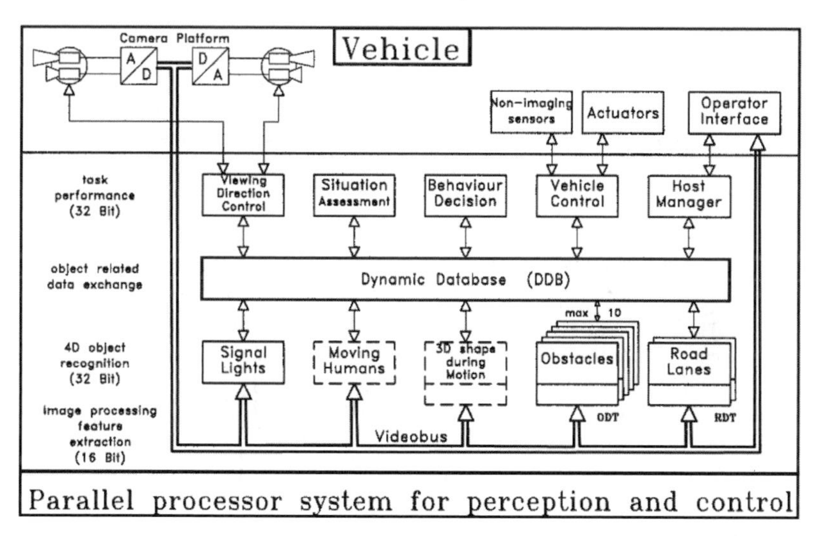

Figure 11.19. Overall system architecture of second-generation vision system based on transputers (about 60 in total) in VaMP: Road with lane markings and up to 12 vehicles, 6 each in front and rear hemisphere, could be tracked at 12.5 Hz (80 ms cycle time) with four cameras on two yaw platforms (top left). Blinking lights on vehicles tracked could be detected with the subsystem in the lower left of the figure.

11.3.4 Experimental Results

From a system architecture point of view, the dynamic database (DDB) was introduced as a flexible link and distributor allowing data input from many different devices. These data were routed to all other subsystems; any type of processor or external computer could communicate with it.

Images used were 320 × 240 pixels from every fourth field (every second frame, only odd or even fields); for intensity images coded with 1 byte/pixel, this yields a video data input rate into the computer system of about 1MB/s for each video stream. A total of 4 MB/s had to be handled on the two video buses for four cameras shown in the bottom of the figure. The fields of view of these cameras looking to the front and rear hemispheres can be seen for the front hemisphere in Figure 11.20; the ratio of focal lengths of the cameras was about 3.2 which seemed optimal for observing the first two rows of vehicles in front. The system was designed for perceiving the environment in three neighboring lanes on high-speed roads with unidirectional traffic.

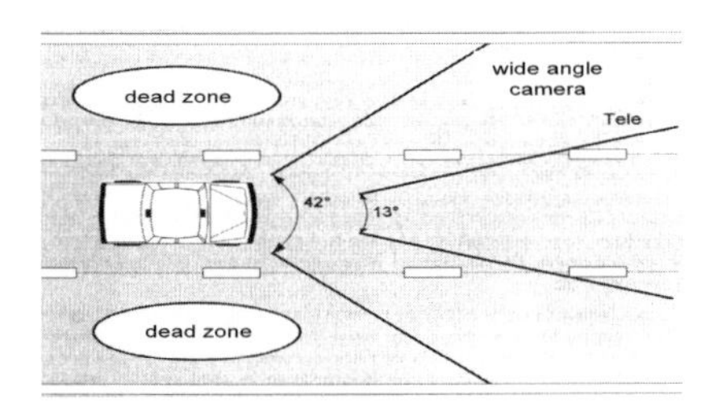

Figure 11.20. Bifocal vision of test vehicle VaMP for high-speed driving and larger look-ahead ranges (see Figure 1.3 for a picture of the 'vehicle eye' with two finger cameras on a single axis platform for gaze control in yaw)

A more detailed view of the vision part of the system with the video busses at the bottom, realized exclusively on about four dozen transputers, can be seen in Figure 11.21. In addition to the intensity images, one color image stream from one of the front cameras could be sent via a separate third bus to the unit for signal light analysis of the vehicle directly ahead (lower left corner in Figure 11.19 [Tsinas 1997]).

The arrangement of subsystems is slightly different in Figures 11.19 and 11.21. The latter shows more organizational details of the distributed transputer system of three different types (T2, T4, T8); the central role of the communication link (CL) developed as a separate unit [von Holt 1994] is emphasized here. *O*bject *d*etection and *t*racking (ODT) has been realized with 14 transputers: Two T4 (VPU) for data input from the two video buses, for data distribution into the four parallel paths analyzing one image stream each and for graphic output of results as overlay in the images (lower right); eight T2 (16-bit processors) performed edge feature extraction with the software package KRONOS written in the transputer programming language "Occam" (see Section 5.2 and [Thomanek, DDickmanns 1992]). Four T8 served as "object processors" (OP) on which recursive estimation for each object tracked and the administrative functions for multiple objects were realized.

The data flow rate from ODT and RDT into the DDB compared to the video rate is reduced by about two to three orders of magnitude (factor of ~ 300); in absolute terms it is in the range *10 to 20 KB/s for ten vehicles tracked.* With roughly 1 KB/s data rate per vehicle tracked at 12.5 Hz, time histories of a few state variables can easily be stored over tens of seconds for maneuver recognition at higher system levels. (This function was not performed with the transputer system in the mid-1990s.)

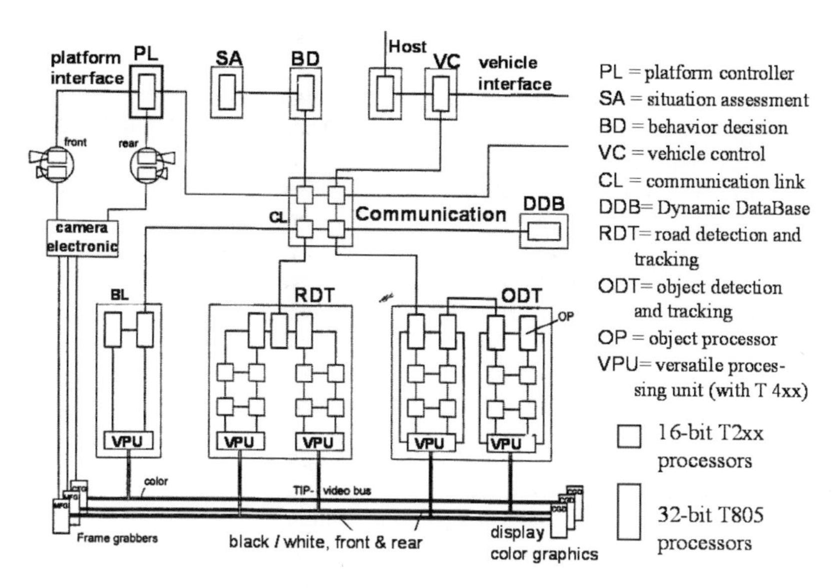

Figure 11.21. Realization of second-generation vision system with distributed transputers of various types: 16-bit processors (designated by squares) were used for feature extraction and communication; 32-bit processors (rectangles) were used for number crunching. Each transputer has four direct communication links for data exchange, making it well suited for this type of distributed system.

Local object administration had to detect inconsistencies caused by two processes working on different images but tracking the same vehicle. It also had to handle communication with other modules for exchanging relevant data. In addition, as long as a situation assessment module was not in operation on a higher level, it determined the relevant object for "vehicle control" to react to.

11.3.4.1 Distance Keeping (Convoy Driving)

Figure 11.22 shows a situation with the test vehicle VaMP driving in the center lane of a three-lane Autobahn; the vision system has recognized three vehicles in each image (wide-angle at left, teleimage on the right). However, the vehicle in the lane driven is the same in both images, while in the neighboring lanes, the vehicles picked are different ones. Of course, in the general case, some vehicles in the neighboring lanes may also be picked in both the wide angle and the teleimage; these are the cases in which local supervision has to intervene and to direct the recursive estimation loops properly. In the situation shown (left image), if the subject vehicle passes the truck seen to the right nearby, but the black car in front remains visible in the teleimage (say by a slight curve to the left), the case mentioned would occur. The analysis process of the teleimage would have to be directed to track the white truck (partially occluded here) in front of the black car; this car in turn would have to be tracked in the wide-angle image.

Figure 11.22. Five objects tracked in front hemisphere; in the own lane only the vehicle directly ahead can be seen (by both cameras). In neighboring lanes, the tele camera sees the second vehicle in front.

It can be seen that the vehicles in front occlude a large part of the road boundaries and of lane markings further away. It does not make sense that the road recognition process tries to find relevant features in the image regions covered by other vehicles; therefore, these regions are communicated from ODT to RDT and are excluded in RDT from feature search.

Range estimation by monocular vision has encountered quite a number of skeptical remarks from colleagues since direct range information is completely lost in perspective mapping. However, humans with only one eye functional have no difficulty in driving a road vehicle correctly and safely.

To quantify precisely the accuracy achievable with monocular technical vision after the 4-D approach, comparisons with results from laser range finders (LRF) in real traffic have been performed. An operator pointing a single beam LRF steadily onto a vehicle also tracked by the vision system was the simplest valid method for obtaining relevant data.

Figure 11.23 shows a comparison between range estimation results with vision (ODT as discussed above) and with a *single-beam laser range finder* pointed at the object of relevance. It is seen that, except for a short period after the sign change of relative speed (lower part), the agreement in range estimation is quite good (error around 2% for ranges of 30 to 40 m). For ranges up to about 80 m, the error increased to about 3%. The lower part in the figure shows that in the initial transient phase, reconstruction (observation) of relative speed exhibits some large deviations from reference values measured by a LRF. Speed from a LRF is a derived variable also, but it is obtained at a higher rate and smoothed by filtering; it is certainly more accurate than vision. The right subfigure for reconstructed relative speed (range rate) shows strong deviations during the initial transient phase, starting with the initial guess "0", till about 2 seconds after hypothesis generation.

The heavy kinks at the beginning are due to changing feature correspondences over time until estimation of vehicle width has stabilized. After the transients have settled, visual estimation of relative speed is close to the LRF-based one. These results are certainly good enough for guiding vehicles only by machine vision; due to increasing angles for decreasing range between features separated on the body surface by a constant distance, accuracy of monocular vision becomes better when it

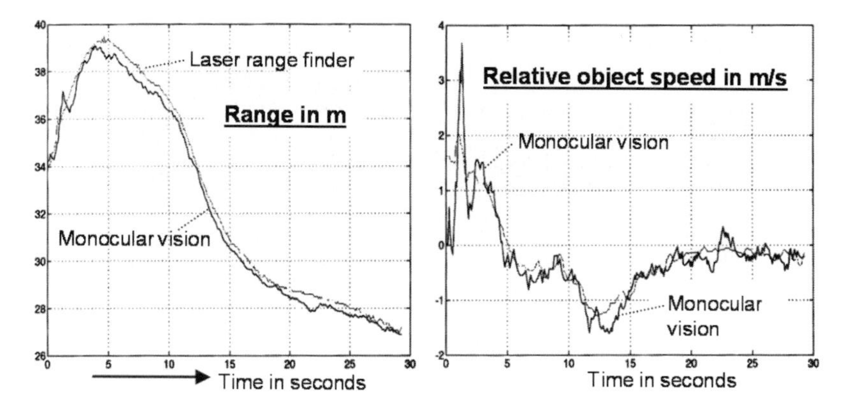

Figure 11.23. Comparison of results in range estimation between a single-beam, accurately pointed laser range finder and dynamic monocular vision after the 4-D approach in a real driving situation on a highway

is most needed (before a possible touch or crash). The results shown were achieved in 1994 with the transputer system described.

Migration to new computer hardware: In 1995, replacing transputers within the ODT block by PowerPC processors (Motorola 601) with more than ten times the processing power per unit, evaluation frequency could be doubled (25 Hz) and the number of processors needed was reduced by a factor of 6. With this system, the long distance trip to the city of Odense, Denmark, mentioned in Section 9.4.2.5, was performed in the same year.

Computing power per general-purpose microprocessor kept increasing at a fast pace in these years. Since high-performance networks for fast data exchange between these processors also became available and the more powerful transputers did not materialize, the third-generation vision system was started in 1997 on the basis of “commercial-off-the-shelf “ (COTS) hardware, Intel Pentium®. One rack-mounted PC system sufficed for doing the entire road recognition; a second was used for bifocal obstacle recognition. Taking all pixels digitized in only odd or even fields (25 Hz) allowed increasing the resolution to about 770 pixels per row with 40 ms cycle time; vertical resolution was unchanged.

Object detection and tracking with COTS-PC systems: Figure 11.24 shows tracking of two cars in a high-resolution image with edge extraction in intelligently controlled search windows based on predictions from spatiotemporal models running at 40 ms cycle time (25 Hz).

The higher tracking frequency in connection with high image resolution allowed robust tracking performance at moderate cost. For the same performance level, the COTS system has been purchased for ~ 20% of the cost of a custom-designed system used before.

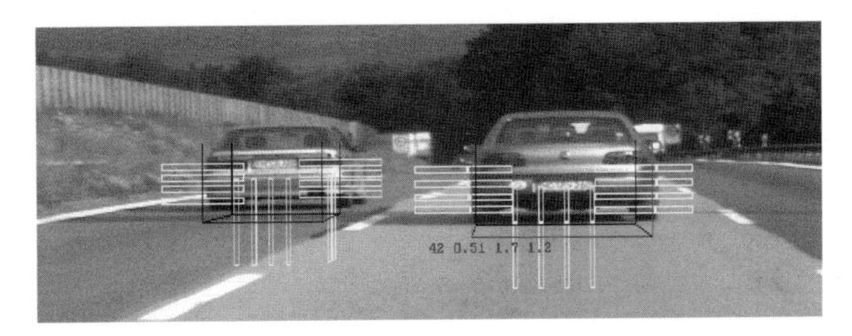

Figure 11.24. Object tracking with a COTS-PC system at the turn of the century. The software system applied for intelligently controlled edge extraction was CRONOS.

At about the same time, the first systems for *Automatic Cruise Control* (ACC) came on the market based on radar for distance measurement. The radar principle allows good measurement results for range (and range rate by smoothed differencing), but poor lateral localization and relatively many false alarms; road and lanes cannot be recognized by radar, in general. So it was quite natural that a combination of radar and vision has been investigated to improve results [Hofmann *et al.* 2003].

Figure 11.25 shows a snapshot of a scene on the Autobahn where the subject vehicle is laterally controlled by a human driver in the third lane (leftmost); only longitudinal control is done fully automatically based on the fused evaluation results of radar and vision. Radar is used for hypothesis generation and range estimation; vision checks all of these hypotheses for objects (vehicles) and eliminates those that cannot be substantiated by corresponding sets of visual features. For those confirmed, their precise lateral extension of the lower part and their positions relative to the lanes are estimated. The reference vehicle for distance keeping is marked by a red rectangle (at the left in the wide-angle image, left part of figure, at the center in the teleimage, right part of figure).

Other vehicles recognized are marked by a light blue rectangle (center of left image). Lane markings recognized are also painted into the image as overlays. The

Figure 11.25. Bifocal lane and object tracking with a COTS-PC system after object candidates have been detected by radar, including false alarms which have to be rejected based on (comparatively high-resolution) vision [Hofmann 2004]

bright (yellow) line at the center of the image displays the vertical curvature estimated by vision; the outer parts mark the reference "zero", and the central part moves up and down for upward or downward curvature. At the moment shown, a slight negative vertical curvature (downward) has been estimated. This vertical curvature component is necessary for obtaining consistent range results by vision and radar; recall that look-ahead range depends strongly on vertical curvature (see Section 9.2).

When the driver changes lane, the vision system automatically switches to the new lane and picks it as reference. Figure 11.26 schematically sketches the advantage in accuracy of joint radar and vision evaluation. This development is being continued in industry towards products for the automotive market in the near future.

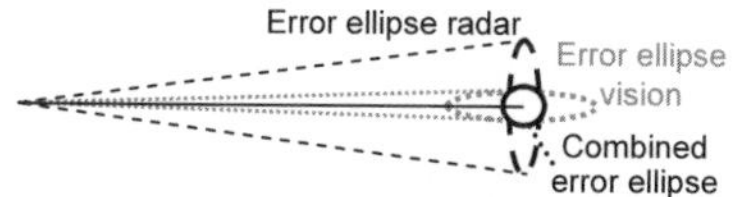

Figure 11.26. Schematic visualization of advantages in hybrid radar and vision sensing: Improved accuracy beside reduction in false alarm rate

Road scene recognition with region-based methods added: The computing power of PCs is sufficient nowadays for analyzing entire road scenes including lane and vehicle recognition in real time. In Section 5.3, the corresponding methods developed by Hofmann at UniBwM and the author have been described. Merged results for adjacent edge elements have been given in Figure 5.44. Merging shaded regions across search stripes yields homogeneous 2-D blobs. Figure 11.27 shows a typical scene with the sets of blobs and edges extracted with the Hofmann operator.

Figure 11.27. Example of joint detection of edges (left) and homogeneous regions (right) in vertical search with 'Hofmann operator'. The vertical centers of homogeneous regions are marked with different colors [Hofmann 2004].

Corners becoming available with the UBM-method (Section 5.3.3) at little additional cost allow making object tracking with the unified feature set much more robust than achievable up to now in real-time vision.

Vehicles under oblique viewing conditions: Vehicles in neighboring lanes or further away on curved roads may appear as rather complex arrangements of simple features, if all types of road vehicles such as trucks, tractors with trailers, tanker, or recreation vehicles beside cars are taken into account. The most characteristic set of subobjects or groups of features for all of these vehicles are the pairs of wheels they need for stable driving (except the rare tricycles hardly seen in normal traffic). Recognizing these wheels under partially occluded conditions therefore is essential for telling vehicles apart from other objects (obstacles like boxes) on the road.

For this purpose, [Hofmann 2004] has developed a special wheel detection algorithm described in Section 11.3.1.2, (see Figure 11.13). Figure 11.28 shows results of its application in a highway scene. Since several pixels are lost for template-based evaluation at the horizontal image boundaries, wheels in these regions are not detected.

Figure 11.28. Confirming vehicle recognition under oblique viewing conditions based on characteristic wheel patterns. Area-based feature correlation in a 2-D search helps finding wheels by characteristic bright-dark patterns.

Even in image sequences with relatively poor resolution, the results look promising; uncertainties in the wide-angle image may be resolved by directing the tele-camera toward this region.

11.3.4.2 Lane Change and Passing

Figure 11.29 shows a situation after passing. (The poor quality is to a large part due to the state of technology available for miniaturized cameras and to the side constraints on image evaluation in the test vehicle one and a half decade ago.) Information about vehicles in neighboring lanes is essential when running up to a

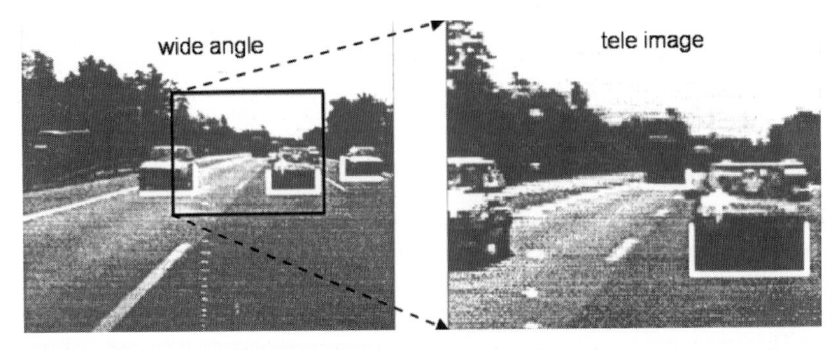

Figure 11.29. Rear-looking view of bifocal camera set (VaMP 1994) while changing back into the old lane after passing. It has to be checked whether the vehicle passed has accelerated in the meantime and whether the range gained is safe for this maneuver.

slower vehicle in front and when one does not want to make a transition to convoy driving at this slow speed. After passing another vehicle, a similar situation exists for changing back to the right lane (in right-hand traffic). In these cases, not only other vehicles to the side or ahead in the neighboring lane have to be detected, but also vehicles behind (approaching or having been passed).

The methods applied for evaluation of image sequences from the rear cameras are very similar to those from the front for the simple features used. When groups of light or the license plate are to be detected, specific generic shape models should be used; the aspect conditions (see Figure 11.15) for hypothesis generation and wheel patterns have to be adapted correspondingly, of course.

To detect vehicles efficiently in the neighboring lane nearby, Rieder (2000) has studied several approaches. One of the more promising ones is shown in Figures 11.30 and 11.31. The areas to be monitored are defined in 3-D space relative to the subject's body such that without another vehicle only lane surface can be seen (Figure 11.30); this lane surface is assumed to be approximately homogeneous, usually. The threshold for homogeneity has to be chosen correspondingly. Outside the lane, the background will be noisy and image content would change at high frequency, in general. The idea is that vehicles with their dark tires and body brightness will differ from the average gray values of the road surface most of the time. This discrepancy is an indication of their presence.

Figure 11.30. Definition of regions for detecting passing vehicles: In the road direction, extended vertical rectangles of predefined height are positioned (virtually) side by side in the near range along the neighboring lane just above the ground

The height of the rectangles defined determines *lateral resolution in position for the feature detected*; however, it should allow sufficiently many pixels within the rectangle for reliable results. The features checked over time are the sums of pixel values in the columns of each rectangle sampled. This compromise has to be found depending on the specific task. In all of the rectangular regions (distorted in the image by perspective projection), the sum of local pixel intensity is formed. If these values are approximately constant, this means that the subject vehicle passes through a region with a homogeneous surface in the neighboring lane. If these values change over an extended region with consistent motion to the right, a vehicle can be assumed to pass through this region. In Figure 11.31, these columns are marked by dashed dark lines.

The number of columns selected for evaluation (ten here) determines total computing load and spatial resolution for detection (lower part). For a homogeneously looking lane, these individual sums are nearly constant (shown brighter, to the right). This detection procedure is not claimed to be especially accurate; it is meant for quick and efficient detection of candidates. Changing gaze direction toward this region will allow more precise vehicle recognition with the methods discussed previously, especially wheel detection, which will allow precise observation of the lateral motion of the vehicle relative to its lane. This situation is one of the few where stereointerpretation of binocular images is a big advantage if no range finders (radar or laser) are available.

Figure 11.31. A vehicle moving through the columns in the image most likely will produce changing values of the sums; these columns are marked as *dashed dark lines*. Their vertical extension symbolized distance at the lane boundary. To determine range to a changing feature, its horizontal distance from the lane marking has to be taken into account.

11.3.4.3 Stop-and-Go Traffic

Stop-and-go-behavior is a basic maneuvering capability when driving behind a single vehicle in front. The first fully vision-based demonstrations of this capability were done in 1991 in the framework of an intermediate demo for the project Prometheus in Turino, Italy, on a separate test track. Performing this maneuver safely in an urban environment with a multitude of different moving subjects (cars, vans, trucks, bicycles, humans, and animals of all kinds) is a different challenge not yet solved. The U.S.-Defense Advanced Research Project Agency (DARPA) has set a price of $ 2 million for the autonomous vehicle capable of driving a 60-mile distance in an urban environment in the least amount of time (below 6 hours as additional constraint); this shall include reacting to other vehicles, also in *stop-and-go* traffic. The final test of this "Urban Grand Challenge" with a maximal allowed speed of 20 mph is planned for the 3rd of November 2007 in a mock-up town in the United States.

Car manufacturers and suppliers to the automotive industry have been considering the extension of adaptive cruise control (ACC) to *stop-and-go*-driving for some time. Most of these activities are based on two different types of radar (long- and short range, different frequencies) and on various types of laser range finders (LRF). Multiple planes in LRF with both scanning and multiple beam designs are under consideration. Typical angular resolutions for modern LRF designs go down to about 0.1° (≈ 2 mrad). This means that at 50 m distance, the resolution is about 10 cm, a reasonable value for slow speeds driven. If interference problems with active sensing can be excluded, these modern LRF sensors just being developed and tested may be sufficient to solve the problem of obstacle recognition.

However, human vision and multifocal, active technical vision can easily exploit ten times this resolution with systems available today. It will be interesting to observe, which type of technical vision system will win the race for industrial implementation in the long run. In the past and still now, computing power and knowledge bases needed for reliable visual perception of complex scenes have been marginal.

11.3.5 Outlook on Object Recognition

With several orders of magnitude in computing power per processor becoming available in the next one or two decades (las in the past according to "Moore's law"), the prospects are bright for high-resolution vision as developed by vertebrates. Multifocal eyes and special "glasses", under favorable atmospheric conditions, will allow passive viewing ranges up to several kilometers. High optical resolution in connection with "passive" perception of colors and textures will allow understanding of complex scenes much more easily than with devices relying on reflected electromagnetic radiation sent out and reflected at far distances.

Generations of researchers and students will compile and structure the knowledge base needed for passive vision based on spatiotemporal models of motion processes in the world. Probably other physical properties of light like direction of polarization or other spectral ranges may become available to technical vision systems as for some animal species. This would favor passive vision in the sense of no active emission of rays by the sensor. Active gaze control is considered a "must" for certain (if not most) application areas; Near (NI) or far infrared radiation are such fields of practical importance for night vision and night driving.

In the approach developed, bifocal vision has become the standard for low to medium speeds; differences in focal length from three to about ten have been investigated. It seems that trifocal vision with focal lengths separated by a factor of 3 to 5 is a good way to go for fast driving on highways. If an object has been detected in a wide-angle image and is too small for reliable recognition, attention focusing by turning the camera with a larger focal length onto the object will yield the improved resolution required. Special knowledge based algorithms (rules and inference schemes) are required for recognizing the type of object discovered. These object recognition specialists may work at lower cycle times and analyze shape details while relative motion estimation may continue to be done in parallel at high frequency with low spatial resolution exploiting the "encasing box" model. This corresponds to two separate paths to the solution of the "where" problem and of the "what" problem.

Systematic simultaneous *interpretation* of image sequences on different pyramid levels of images has not been achieved in our group up to now, though data processing for correlation uses this approach successfully, *e.g.*, [Burt 1981; Mandelbaum *et al.* 1998]. This approach may be promising for robust blob- and corner tracking and for spatiotemporal interpretation in complex scenes.

For object detection in the wide-angle camera, characteristic features of minimum size are required. Ten to twenty pixels on an object seem to be a good compromise between efficiency and accuracy to start with. Control of gaze and attention can then turn the high-resolution camera to this region yielding one to two orders of magnitude more pixels on this object depending on the ratio of focal lengths in use. This is especially important for objects with large relative speed such as vehicles in opposite traffic direction on bidirectional high-speed roads.

Another point needing special attention is discovery of perturbations: Sudden disappearance of features predicted to be very visible, usually, is an indication of occlusion by another object. If this occurs for several neighboring features at the same time, this is a good hint to start looking for another object which has newly appeared at a shorter range. It has to be moving in the opposite direction relative to the side where the features started disappearing. If just one feature has not been measured once, this may be due to noise effects. If measurements fail to be successful at one location over several cycles, there may be some systematic discrepancy between model and reality and, therefore, this region has to be scrutinized by allocating more attention to it (more and different feature extractors for discovering the reason). This will be done with a new estimation process (new object hypothesis) so that tracking and state estimation of the known object is not hampered. First results of systematic investigations for situations with occlusions were obtained in the late 1980s by M. Schmid and are documented in [Schmid 1992]. This area needs further attention for the general case.

12 Sensor Requirements for Road Scenes

In the previous chapters, it has been shown that vision systems have to satisfy certain lower bounds on requirements to cover all aspects of interest for safe driving if they shall come close to human visual capabilities on all types of roadways in a complex network of roads, existing in civilized countries.

Based on experience in equipping seven autonomous road vehicles with dynamic machine vision systems, an arrangement of miniature TV cameras on a pointing platform was proposed in the mid-1990s which will satisfy all major requirements for driving on all types of roads. It has been dubbed *multifocal, active, reflex-like reacting vehicle eye* (MarvEye). It encompasses the following properties:

1. large binocular horizontal field of view (*e.g.*, ≥ 110°),
2. bifocal or multifocal design for region analysis with different resolution in parallel,
3. ability of view fixation on moving objects while the platform base also is moving; this includes high-frequency, inertial stabilization ($f > 200$ Hz),
4. saccadic control of region of interest with stabilization of spatial perception in the interpretation algorithms;
5. capability of binocular (trinocular) stereovision in near range (stereo base similar to the human one, which is 6 – 7 cm);
6. large potential field of view in horizontal range (*e.g.*, 200° with sufficient resolution) such that the two eyes for the front and the rear hemisphere can cover the full azimuth range (360°); stereovision to the side with a large stereo base becomes an option (longitudinal distance between the "vehicle eye" looking forward and backward, both panned by ~ 90° to the same side).
7. high dynamic performance (*e.g.*, a saccade of ≈ 20° in a tenth of a second).

In cars, the typical dimension of this "vehicle eye" should not be larger than about 10 cm; two of these units are proposed for road vehicles, one looking forward, located in front of the inner rearview mirror (similar to Figure 1.3), the other one backward; they shall feed a 4-D perception system capable of assessing the situation around the vehicle by attention control up to several hundred meters in range.

This specification is based on experience from over 5000 km of fully autonomous driving of both partners (Daimler-Benz and UniBwM) in normal traffic on German and French freeways as well as state and country roads since 1992. A human safety pilot – attentively watching and registering vehicle behavior but otherwise passive – was always in the driver's seat, and at least one of the developing engineers (Ph.D. students with experience) checked the interpretations of the vision system on computer displays.

Based on this rich experience in combination with results from aeronautical applications (onboard autonomous visual landing approaches till touch down with the same underlying 4-D approach), the design of MarVEye resulted. This chapter first discusses the requirements underlying the solution proposed; then the basic design is presented and sensible design parameters are discussed. Finally, steps towards first realizations are reviewed. Most experimental results are given in Chapter 14.

12.1 Structural Decomposition of the Vision Task

The performance level of the human eye has to be the reference, since most of the competing vision systems in road vehicle guidance will be human ones. The design of cars and other vehicles is oriented toward pleasing human users, but also exploiting their capabilities, for example, look-ahead range, reaction times, and fast dynamic scene understanding.

12.1.1 Hardware Base

The first design decision may answer the following: Is the human eye with its characteristics also a good guide line for designing technical imaging sensors, or are the material substrates and the data processing techniques so different that completely new ways for solving the vision task should be sought? The human eye contains about 120 million light-sensitive elements, but two orders of magnitude fewer fibers run from one eye to the brain, separated for the left and the right halves of the field of view. The sensitive elements are not homogeneously distributed in the eye; the fovea is much more densely packed with sensor elements than the rest of the retina. The fibers running via "lateral geniculate" (an older cerebral structure) to the neo-cortex in the back of the head obtain their signals from "receptive fields" of different types and sizes depending on their location in the retina; so preprocessing for feature extraction is already performed in retinal layers [Handbook of Physiology: Darian-Smith 1984].

Technical imaging sensors with some of the properties observed in biological vision have been tried [Debusschere *et al.* 1990; Koch 1995], but have not gained ground. Homogeneous matrix arrangements over a very wide range of sizes are state of the art in microelectronic technology; the video standard for a long time has been about $640 \times 480 \approx 307\,000$ pixels; with 1 byte/pixel resolution and 25 Hz frame rate, this results in a data rate of ≈ 7.7 MB/s. (Old analogue technology could be digitized to about 770×510 pixels, corresponding to a data rate of about 10MB/s.) Future high-definition TV intends to move up to 1920×1200 pixels with more than 8-bit intensity coding and a 75 Hz image rate; data rates in the gigabit/second-range will be possible. In the beginning of real-time machine vision (mid-1980s) there was much discussion whether there should be preprocessing steps near the imaging sensors as in biological vision systems; "massively parallel processors" with hundreds of thousands of simple computing elements have been proposed (DARPA: "On Strategic Computing" [Klass 1985; Roland, Shiman 2002].

With the fast advancement of general-purpose microprocessors (clock rates moving from MHz to GHz) and communication bandwidths (from MB/s to hundreds of MB/s) the need for mimicking carbon-based data processing structures (as in biology) disappeared for silicon-based technical systems.

With the advent of high-bandwidth communication networks between multiple general-purpose processors in the 1990s, high-performance, real-time vision systems became possible without special developments for vision except frame grabbers. The move to digital cameras simplified this step considerably. To develop methods and software for real-time vision, relatively inexpensive systems are sufficient. The lower end video cameras cost a few dollars nowadays, but reasonably good cameras for automotive applications with increased dynamic intensity range have also come down in price and do have advantages over the cheap devices. For later applications with much more emphasis on reliability in harsh environments, special "vision hardware" on different levels may be advantageous.

12.1.2 Functional Structure

Contrary to the hardware base, the functional processing steps selected in biological evolution have shown big advantages: (1) Gaze control with small units having low inertia is superior to turning the whole body. (2) Peripheral-foveal differentiation allows reducing maximal data rates by orders of magnitude without sacrificing much of the basic transducer-based perception capabilities if time delays due to saccadic gaze control are small. (Eigenfrequencies of eyes are at least one order of magnitude higher than those for control of body movements.) (3) Inertial gaze stabilization by negative feedback of angular rates, independent of image evaluation, reduces motion blur and extends the usability of vision from quasi-static applications for observation to really dynamic performance during perturbed egomotion. (4) The construction of internal representations of 3-D space over time based on previous experience (models of motion processes for object classes) triggered by visual features and their flow over time allows stabilizing perception of "the world" despite the very complex data input resulting from saccadic gaze control: Several frames may be completely noninterpretable during saccades.

Note that controllable focal length on one camera is not equivalent to two or more cameras with different focal lengths: In the latter case, the images with different resolution are available in parallel at the same time, so that interpretation can rely on features observed simultaneously on different levels of resolution. On the contrary, changing focal length with a single camera takes time, during which the gaze direction in dynamic vision may have changed. For easy recognition of the same groups of features in images with different resolution, a focal length ratio of three to four experimentally yields the best results; for larger factors, the effort of searching in a high-resolution image becomes excessive.

The basic functional structure developed for dynamic real-time vision has been shown in Figure 5.1. On level 1 (bottom), there are feature extraction algorithms working fully bottom-up without any reference to spatiotemporal models. Features may be associated over time (for feature flow) or between cameras (for stereointerpretation).

On level 2, *single objects* are hypothesized and tracked by prediction-error feedback; there are parallel data paths for different objects at different ranges, looked at with cameras and lenses of different focal lengths. But the same object may also be observed by two cameras with different focal lengths. Staging focal lengths by a factor of exactly 4 allows easy transformation of image data by pyramid methods. On all of these levels, *physical objects* are tracked "here and now"; the results on the object level (with data volume reduced by several orders of magnitude compared to image pixels and features) are stored in the DOB. Using ring buffers for several variables of special interest, their recent time history can be stored for analysis on the third level which does not need access to image data any longer, but looks at objects on larger spatial and temporal scales for recognition of maneuvers and possibly cues for hypothesizing intentions of subjects. Knowledge about a subject's behavioral capabilities and mission performance need be available only here.

The physical state of the subject body and the environmental conditions are also monitored here on the third level. Together they provide the background for judging the quality and trustworthiness of sensor data and interpretations on the lower levels. Therefore, the lower levels may receive inputs for adapting parameters or for controlling gaze and attention. (In the long run, maybe this is the starting point for developing some kind of self-awareness or even consciousness.)

12.2 Vision under Conditions of Perturbations

It is not sufficient to design a vision system for clean conditions and later on take care of steps for dealing with perturbations. In vision, the perturbation levels tolerable have to be taken into account in designing the basic structure of the vision system from the beginning. One essential point is that due to the large data rates and the hierarchical processing steps, the interpretation result for complex scenes becomes available only after a few hundred milliseconds delay time. For high-frequency perturbations, this means that reasonable *visual* feedback for counteraction is nearly impossible.

12.2.1 Delay Time and High-frequency Perturbation

For a time delay of 300 ms (typical of inattentive humans), the resulting phase shift for an oscillatory 2-Hz motion (typical for arms, legs) is more than 200°; that means that in a simple feedback loop, there is a sign change in the signal (cos (180°) = −1). Only through compensation from higher levels with corresponding methods is this type of motion controllable. In closed-loop technical vision systems onboard a vehicle with several consecutive processing stages, 3 to ≈ 10 video cycles (of 40 or 33 ms duration) may elapse until the control output derived from visual features hits the physical device effecting the command. This is especially true if a perturbation induces motion blur in some images.

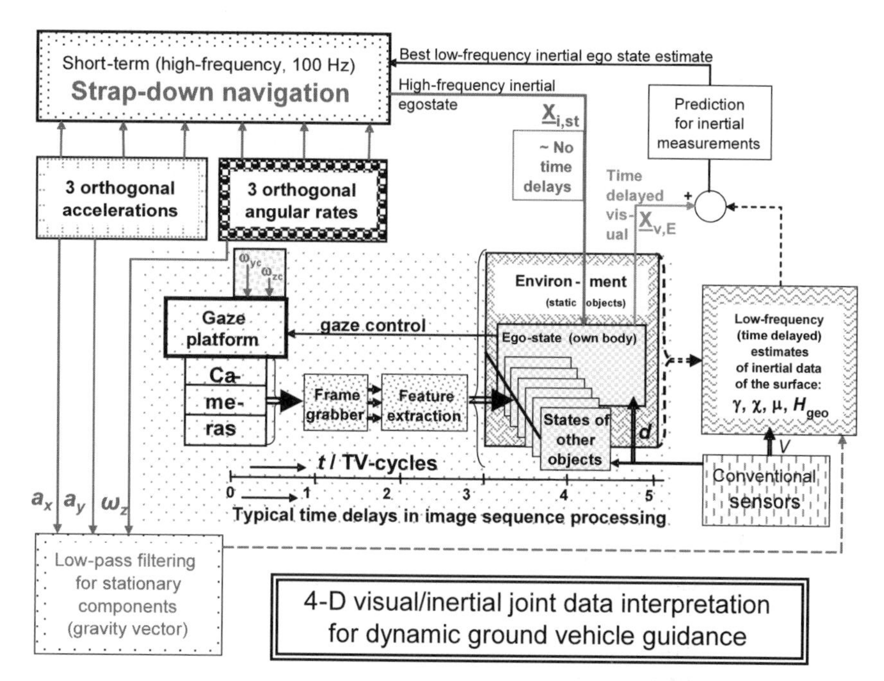

Figure 12.1. Block diagram for joint visual/inertial data collection (stabilized gaze, center left) and interpretation; the high-frequency component of a rotational ego-state is determined from integration of angular rates (upper center), while long-term stability is derived from visual information (with time delay, see lower center) from objects further away (*e.g.*, the horizon). Gravity direction and ground slope are derived from x- and y-accelerations together with speed measured conventionally.

This is the reason that direct angular rate feedback in pitch and yaw from sensors on the same platform as the cameras is used to command the opposite rate for the corresponding platform component. Reductions of perturbation amplitudes by more than a factor of 10 have been achieved with a 2 ms cycle time for this inner loop (500 Hz). Figure 12.1 shows the block diagram containing this loop: Rotational rates around the *y*- and *z*-axes of the gaze platform (center left) are directly fed back to the corresponding torque motors of the platform at a rate of 500 Hz if no external commands from active gaze control are received. The other data paths for determining the inertial egostate of the vehicle body in connection with vision will be discussed below. The direct inertial feedback loop of the platform guarantees that the signals from the cameras are freed from motion blur due to perturbations. Without this inertial stabilization loop, visual perception capability would be deteriorated or even lost on rough ground.

If gaze commands are received from the vision system, of course, counteraction by the stabilization loop has to be suppressed. There have to be specific modes available for different types of gaze commands (smooth pursuit of saccades); this will not be treated here. The beneficial effect of gaze stabilization for a braking maneuver with 3° of perturbation amplitude (min to max) in vehicle pitch angle is shown in Figure 12.2. The corresponding reduction in amplitude on the stabilized platform experienced by the cameras is more than a factor of 10. The strong deviation of the platform base from level, which is identical with vehicle body motion and can be seen as the lower curve, is hardly reflected in the motion of the camera sitting on the platform head (*upper, almost constant curve*).

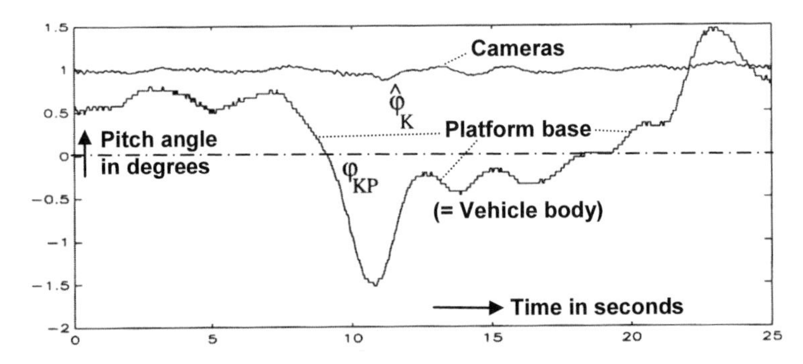

Figure 12.2. Gaze stabilization in pitch by negative feedback of angular rate for test vehicle **VaMoRs** (4-ton van) during a braking maneuver

The most essential state components of the body to be determined by integration of angular rate signals with almost no delay time are the angular orientations of the vehicle. For this purpose, the signals from the inertial rate sensors mounted on the vehicle body are integrated, shown in the upper left of Figure 12.1; the higher frequency components yield especially good estimates of the angular pose of the body. Due to low-frequency drift errors of inertial signals, longer term stability in orientation has to be derived from visual interpretation of (low-pass-filtered) features of objects further away; in this data path, the time delay of vision does no harm. [It is interesting to note that some physiologists claim that sea-sickness of humans (nausea) occurs when the data from both paths are strongly contradicting.]

Joint inertial/visual interpretation also allows disambiguating relative motion when only parts of the subject body and a second moving object are in the fields of view; there have to be accelerations above a certain threshold to be reliable, however.

12.2.2 Visual Complexity and the Idea of Gestalt

When objects in the scene have to be recognized in environments with strong visual perturbations like driving through an alley with many shadow boundaries from branches and twigs, picking “the right” features for detection and tracking is essential. On large objects such as trucks, coarse-scale features averaging away the fine details may serve the purpose of tracking better than fine-grained ones. On cars with polished surfaces, disregarding the upper part and mildly inclined surface elements of the body altogether may be the best way to go; sometimes single highlights or bright spots are good for tracking over some period of time with given aspect conditions. When the aspect or the lighting conditions change drastically, other combinations of features may be well suited for tracking.

This is to say that image evaluation should be quickly adaptable to situations, both with respect to single features extracted and to the knowledge base establishing correspondence between groups of features in the images and the internal representation of 3-D objects moving over time through an environment affecting the lighting conditions. This challenge has hardly been tackled in the past but has to be solved in the future to obtain reliable technical vision systems approaching the performance level of trained humans. The scale of visual features has to be expanded considerably including color and texture as well as transparency; partial mirroring mixed with transparency will pose demanding challenges.

12.3 Visual Range and Resolution Required for Road Traffic Applications

The human eyes have a simultaneous field of view of more than 180°, with coarse resolution toward the periphery and very high resolution in the foveal central part of about 1 to 2° aperture; in this region, the grating resolution is about 40 to 60 arc-seconds or about 2.5 mrad. The latter metric is a nice measure for practical applications since it can be interpreted as the length dimension normal to the optical axis per pixel at 1000 times its distance (width in meters at 1 km, in decimeters at 100 m, or in millimeters at 1 m, depending on the problem at hand). Without going into details about the capability of subpixel resolution with sets of properly arranged sensor elements and corresponding data processing, let us take 1 mrad as the human reference value for comparisons.

Both of the human eye and head can be turned rapidly to direct the foveal region of the eye onto the object of interest (attention control). Despite the fast and frequent viewing direction changes (saccades) which allocate the valuable high-resolution region of the eye to several objects of interest in a time slicing multiplex procedure, the world perceived looks stable in a large viewing range. This biological system evolved over millennia under real-world environmental conditions: the technical counterpart to be developed has to face these standards.

It is assumed that the *functional* design of the biological system is a good starting point for a technical system, too: however, technical realizations have to start from a hardware base (silicon) quite different from biological wetware. Therefore, with the excellent experience from the conventional engineering approach to dynamic machine vision, our development of a technical eye continued on the well proven base underlying conventional video sensor arrays and dynamic systems theory of the engineering community.

The seven properties mentioned in the introduction to this chapter are detailed here to precise specifications.

12.3.1 Large Simultaneous Field of View

There are several situations when this is important. First, when starting from stop, any object or subject within or moving into the area directly ahead of the vehicle should be detectable; this is also a requirement for *stop-and-go* traffic or for very slow motion in urban areas. A horizontal slice of a complete hemisphere should be covered with gaze changes in azimuth (yaw) of about $\pm$ 35°. Second, when looking tangentially to the road (straight ahead) at high speeds, passing vehicles should be detected sufficiently early for prompt reaction when they start moving into the subject lane directly in front. Third, when a lane change or a turnoff is intended, simultaneous observation and tracking of objects straight ahead and about 90° to the side are advantageous; with nominal gaze at 45° and a field of view (f.o.v.) > 100°, this is achievable.

In the nearby range, a resolution of about 5 mm per pixel at 2.5 m or 2 cm at 10 m distance is sufficient for recognizing and tracking larger subobjects on vehicles or persons (about 2 mrad/pixel); however, this does not allow reading license plates at 10 m range. With 640 pixels per row, a single standard camera can cover about a 70° horizontal f.o.v. at this resolution ($\approx$ 55° vertically). Mounting two of these (wide-angle) cameras on a platform with optical axes in the same plane but turned in yaw to each side by $\psi_{obl} \sim 20°$ (oblique views), a total f.o.v. for both cameras of 110° results; the difference between half the f.o.v. of a single camera and the yaw angle ψ_{obl} provides an angular region of central overlap ($\pm$ 15° in the example). Separating the two cameras laterally generates a base for binocular stereo evaluation (Section 12.3.4).

The resolution of these cameras is so low that pitch perturbations of about 3° (accelerations/decelerations) shift features by about 5% of the image vertically. This means that these cameras need not be vertically stabilized and do not induce excessively large search ranges; this reduces platform design considerably. The numerical values given are just examples; they may be adapted to the focal lengths available for the cameras used. Smaller yaw angles ψ_{obl} yield a larger stereo f.o.v. and lower distortions from lens design in the central region.

12.3.2 Multifocal Design

The region of interest does not grow with range beyond a certain limit value; for example, in road traffic with lane widths of 2.5 to 4 m, a region of simultaneous interest larger than about 30 to 40 m brings no advantage if good gaze control is available. With 640 pixels per row in standard cameras, this means that a resolution of 4 to 6 cm per pixel can be achieved in this region with proper focal lengths. Considering objects of 10 to 15 cm characteristic length as serious obstacles to be avoided, this resolution is just sufficient for detection under favorable conditions (2 to 3 pixel on this object with sufficient contrast). But what is the range that has to be covered? Table 11.1 contains braking distances as a function of speed driven for three values of deceleration.

About a 240-m look-ahead range should be available for stopping in front of an obstacle from V = 180 km/h (50 m/s) with an average deceleration of 0.6 Earth gravity (g) or from 130 km/h (36 m/s) with an average deceleration of 0.3 g. To be on the safe side, a 250 to 300 m look-ahead range is assumed desirable for high-speed driving. For the region of interest mentioned above, this requires a f.o.v. of 5 to 7° or about 0.2 mrad resolution per pixel. This is one order of magnitude higher than that for the near range. With the side constraint mentioned for easy feature correspondence in images of different resolution (ratio of focal lengths no larger than 4), this means that a trifocal camera arrangement should be chosen. Figure 12.3 visualizes the geometric relations.

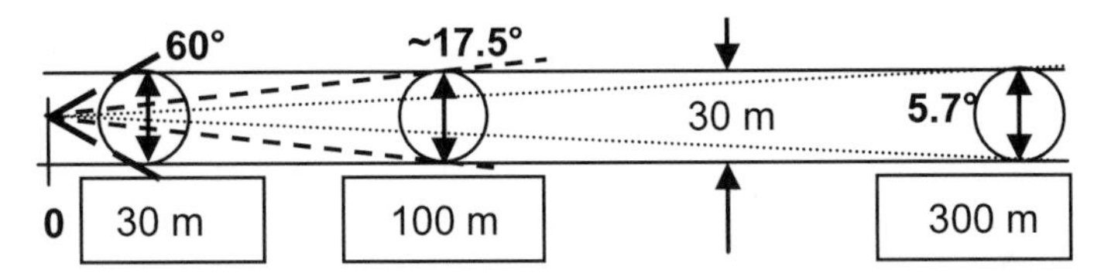

Figure 12.3. Fields of view and viewing ranges for observing a lateral range of 30 m normal to the road

If lane markings of 12 cm width shall be recognizable at 200 m distance, this requires about 2.5 pixel on the line, corresponding to an angular resolution of 0.25 mrad per pixel. For landmark recognition at far distances, this resolution is also desirable. For maximal speeds not exceeding 120 km/h, a bifocal camera system may be sufficient.

12.3.3 View Fixation

Once gaze control is available in a vision system, it may be used for purposes other than the design goal. Designed initially for increasing the potential f.o.v. or for counteracting perturbations on the vehicle (inertial stabilization), it can in addition be used for gaze fixation onto moving objects to reduce motion blur and to keep one object centered in an image sequence. This is achieved by negative visual feedback of the deviation of the center of characteristic features from the center of the image; horizontal and vertical feature search may be done every second image if computing resources are low. Commanding the next orthogonal search around the column or row containing the last directional center position has shown good tracking properties, even without an object model installed. A second-order tracking model in the image plane may improve performance for smooth motion.

However, if harsh directional changes occur in the motion pattern of the object, this approach may deteriorate the level of perturbation tolerable. For example, a ball or another object being reflected at a surface may be lost if delay times for visual interpretation are large and/or filter tuning is set to too strong low-pass filtering. Decreasing cycle time may help considerably: In conventional video with two consecutive fields (half frames), using the fields separately but doubling interpretation frequency from 25 to 50 Hz has brought about a surprising increase in tracking performance when preparing the grasping experiment in orbit onboard the Space Shuttle Columbia [Fagerer *et al.* 1994].

View fixation need not always be done in both image dimensions; for motion along a surface as in ground traffic, just fixation in yaw may be sufficient for improved tracking. It has to be taken into account, however, that reducing motion blur by tracking one object may deteriorate observability for another object. This is the case, for example, when driving through a gap between two stationary obstacles (trees or posts of a gate). Fixation of the object on one side doubles motion blur on the other side; the solution is reducing speed and alternating gaze fixation to each side for a few cycles.

12.3.4 Saccadic Control

This alternating attention control with periods of smooth pursuit and fast gaze changes at high angular rates is called "saccadic" vision. During the periods of fast gaze changes, the entire images of all cameras are blurred; therefore, a logic bit is set indicating the periods when image evaluation does not make sense. These gaps in receiving new image data are bridged by extrapolation based on the spatiotemporal models for the motion processes observed. In this way, the 4-D approach quite naturally lends itself to algorithmic stabilization of space perception, despite the fast changing images on the sensor chip. Building internal representations in 3-D space and time allows easy fusion of inertial and other conventional data such as odometry and gaze angles relative to the vehicle body, measured mechanically.

After a saccadic gaze change, the vision process has to be restarted with initial values derived from the spatiotemporal models installed and from the steps in gaze angles. Since gaze changes, usually, take 1 to 3 video cycles, uncertainty has increased and is reflected in the corresponding parameters of the recursive estimation process. If the goal of the saccade was to bring a certain region of the outside world into the f.o.v. of a camera with a different focal length (*e.g.*, a telecamera), the measurement model and computation of the Jacobian elements have to be exchanged correspondingly. Since the region of special interest also remains in the f.o.v. of the wide-angle camera, tracking may be continued here, too, for redundancy until high-resolution interpretation has become stable.

According to the literature, human eyes can achieve turn rates up to several hundred degrees per second; up to five saccades per second have been observed. For technical systems in road traffic applications, maximum turn rates of a few hundred degrees per second and about two saccades per second may be sufficient. A thorough study of controller design for these types of systems has been done in [Schiehlen 1995]. The interested reader is referred to this dissertation for all details in theoretical and practical results including delay time observers. Figure 12.4 shows test results in saccadic gaze control based on this work; note the minimal overshoot at the goal position. A gaze change of 40° is finished within 350 ms (including 67 ms delay time from command till motion onset). Special controller design minimizes transient time and overshoot.

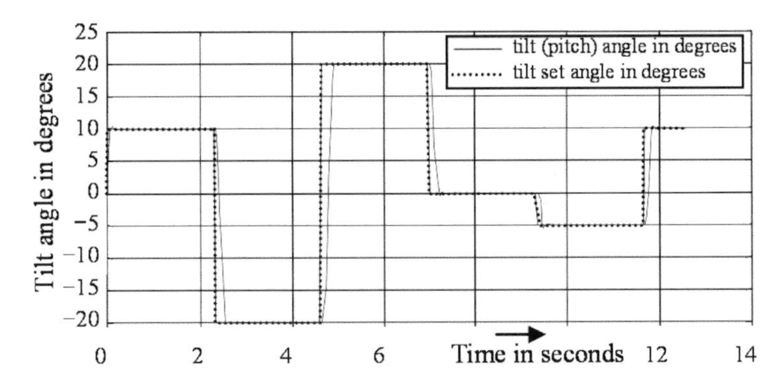

Figure 12.4. Saccadic gaze control in tilt (pitch) for the inner axis of a two-axis platform in the test vehicle **VaMoRs** (see Figure 14.16)

12.3.5 Stereovision

At medium distances when the surface can be seen where the vehicle or object touches the ground, spatial interpretation may be achieved relatively easily by taking into account background knowledge about the scene and integration over time; in the very near range behind another vehicle, where the region in which the vehicle touches the ground is obscured by the subject's own motor hood, monocular range estimation is impossible. Critical situations in traffic may occur when a passing vehicle cuts into the observer's lane right in front; in the test vehicle VaMP, the ground ahead was visible at a range larger than about 6 m (long motor hood, camera behind center top of windshield).

Range estimation from a single, well-recognizable set of features is desirable in this case. A stereo base like the human one (of about 6 to 7 cm) seems sufficient; the camera arrangement as shown in Figure 1.4 (one to each side of the telecamera(s) satisfies this requirement; multiocular stereo with improved performance may be achievable also by exploiting the tele-images for stereointerpretation. Using a stereo camera pair with non-parallel optical axes increases the computing load somewhat but poses no essential difficulties; epipolar lines have to be adjusted for efficient evaluation [Rieder 1996].

By its principle, stereovision deteriorates with range (inverse quadratic); so binocular stereo for the near range and intelligent scene interpretation for larger ranges are nicely complementary. Figure 12.6 shows images from a trinocular camera set (see Figure 14.16); the stereo viewing ranges, which might be used for understanding vertical structure on unpaved hilly roads without markings, are shown by dashed white lines. The stereo base was selected as 30 cm here. Slight misalignments of multiple cameras are typical; their effects have to be compensated by careful calibration which is of special importance in stereointerpretation. The telecamera allows detecting a crossroad from the left, here, which cannot be discovered in the wide-angle images; trinocular stereo is not possible with the pitch angle of the telecamera given here.

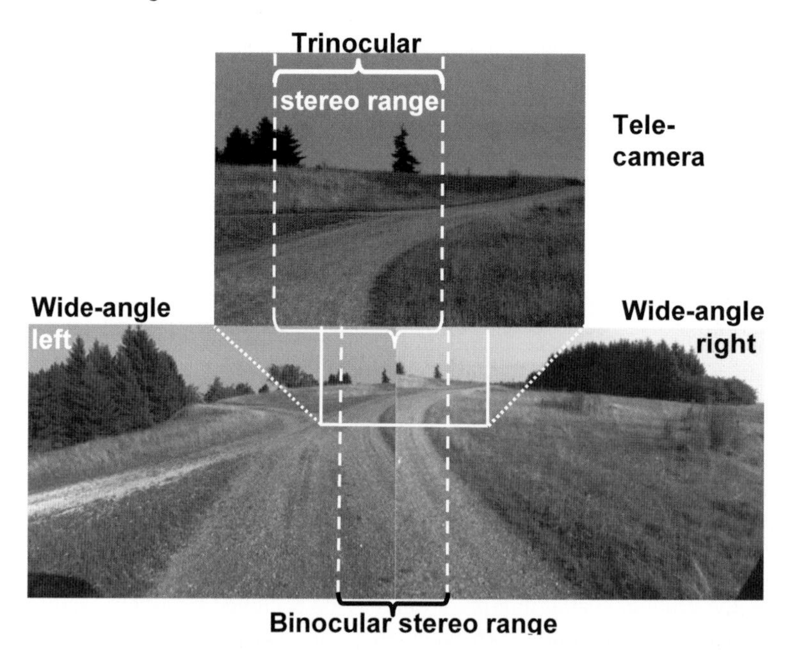

Figure 12.5. Sample images from MarVEye in VaMoRs with three cameras: Two wide-angle cameras with relatively little central overlap for binocular stereo evaluation (bottom); image from mild telecamera (top). Trinocular stereointerpretation is possible in the region marked by the white rectangle with solid horizontal and dashed vertical lines.

12.3.6 Total Range of Fields of View

Large potential f.o.v. in azimuth are needed in traffic; for low-resolution imaging, about 200° are a must at a T-junction or a road crossing without a traffic light or round-about. If the crossroad is of higher order and allows high-speed driving, the potential f.o.v. for high-resolution imaging should also be about 200° for checking oncoming traffic at larger ranges from both sides. Precise landmark recognition over time, under partial obscuration from objects being passed in the near vicinity, will also benefit from large potential f.o.v. However, the simultaneous f.o.v. with high resolution need not be large, since objects to be viewed with this resolution are far away, usually; if the bandwidth of gaze control is sufficiently high, the high-resolution viewing capability can always be turned to the object of interest before the object comes near, and thus good reactions are required.

The critical variable for gaze control, obstacle detection, and behavior decision is the "reaction time available" T_{rea} for dealing with the newly detected object. The critical variable is the radial speed component when the components normal to it are small; in this case, the object is moving almost on a direct collision path. If the object were a point and its environment would show no features, motion would be visually undetectable. It is the features off the optical line of sight that indicate range rate. An extended body, whose boundary features move away from each other with their center remaining constant, thus indicates decreasing range on a collision trajectory ("looming" effect). Shrinking feature distributions indicate increasing range (moving away). If the feature flow on the entire circumference is not radial and outward but all features have a flow component to one side of the line of sight, the object will pass the camera on this side. With looming feature sets, the momentary time to collision (TTC) is (with r = range and the dot on top for the time derivative)

$$TTC = r / \dot{r} . \tag{12.1}$$

Assuming a pinhole camera model and constant object width B, this value can be determined by measuring the object width in the image (b_{B1} and b_{B2}) at two times t_1 and t_2, $\Delta t = t_2 - t_1$ apart. A simple derivation (see Figure 2.4) with derivatives approximated by differences and b_{Bi} = measured object width in the image at time t_i yields

$$TTC = r_2 / \dot{r} = \Delta t \cdot b_{B1} /(b_{B1} - b_{B2}) . \tag{12.2}$$

The astonishing result is that this physically very meaningful term can be obtained without knowing either object size or actual range. Biological vision systems (have discovered and) use this phenomenon extensively (*e.g.*, gannets stretching their wings at a proper time before hitting the water surface in a steep dash to catch fish).

To achieve accurate results with technical systems, the resolution of the camera has to be very high. If the objects approaching may come from any direction in the front hemisphere (like at road junctions or forks), this high resolution should be available in all directions. If one wants to cover a total viewing cone of 200° by 30° with telecameras having a simultaneous f.o.v. of about 5 to 7° horizontally, each with a side ratio of 4:3 (see Section 12.3.2), the total number of cameras required would be 150 to 200 on the vehicle periphery. Of course, this does not make sense.

Putting a single telecamera on a pan and tilt platform, the only requirement for achieving the same high resolution (with minor time delays, usually) is to allow a ±97° gaze change from straight ahead in the vehicle. To keep the inertial momentum of the platform small, tilt (pitch) changes can be effected by a mirror which is rotated around the horizontal axis in front of the telelens.

The other additional (mechanical) requirement is, of course, that the simultaneous f.o.v. can be directed to the region of actual interest in a time frame leaving sufficient time for proper behavior of the vehicle; as in humans, a fraction of a second for completing saccades is a reasonable compromise between mechanical and perceptual requirements. Figure 12.6 shows two of the first realizations of the "MarVEye"-idea for the two test vehicles.

To the left is an arrangement with three cameras for the van VaMoRs; its maximal speed does not require very large look-ahead ranges. The stereo base is rather large (~ 30 cm); a color camera with medium telelens sits at the center. To the right is the pan platform for VaMP with the camera set according to Figure 1.4. Since a sedan Mercedes 500 SEL is a comfortable vehicle with smooth riding qualities, gaze control in pitch has initially been left off for simplicity.

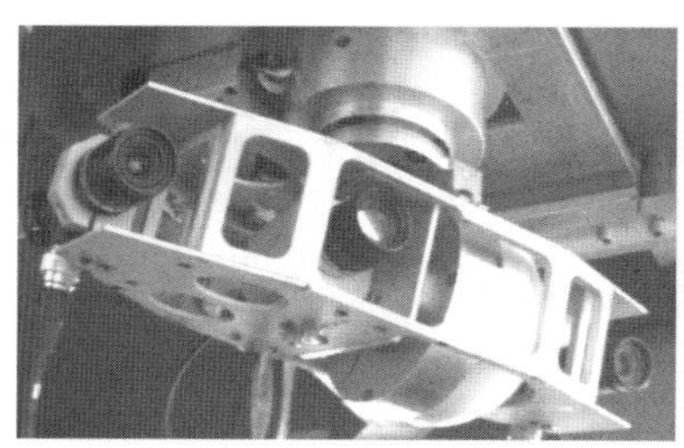

Figure 12.6. Two of the first 'MarVEye' realizations: Left: Two-axis platform for VaMoRs with three cameras and a relatively large stereo baseline (for images see Figure 12.4); right: Single axis (pan/yaw) platform with four cameras and three different focal lengths for VaMP.

12.3.7 High Dynamic Performance

With TTC as described in the previous section (Equations 12.1 and 12.2) and with knowledge about actual reaction times of subsystems for visual perception, for behavior decision, as well as for control derivation and implementation, a rather precise specification for saccadic vision in technical systems can be given.

Assuming one second for stabilizing perceptual interpretation, another two for situation assessment and behavior decision as well as a few seconds as an additional safety margin, a time horizon of 5 to 10 seconds, as assumed standard for human behavior in ground traffic, also seems reasonable for these technical systems. (In the human driver population, there is a wide variety of reaction times to be observed, depending on alertness and attention.)

Standing still at a T-junction and checking whether it is safe to move onto the road, oncoming vehicles with a speed of 108 km/h (30 m/s) have to be detected at about 150 to 300 m distance to satisfy the specifications; for speeds around 20 m/s (70 km/h), this range is reduced to 100 to 200 m. This is one of the reasons that in areas with road junctions, maximal speed allowed on the through -road is reduced.

Relative to these times for reaction, times for gaze control of at most a larger fraction of a second do not deteriorate performance of the overall perception system. Under these conditions, saving over two orders of magnitude in the data flow rate by selecting saccadic vision instead of arrays of (maybe inexpensive) cameras mounted (peripherally) on the vehicle body is a very economical and safe choice.

Figure 12.4 showed some test results in performing saccades with one of the technical eyes studied for road vehicle applications. Due to the relatively old technology used and heavy load on the platform (see Figure 14.16), reaction times can be considered as upper limits for systems based on modern technology; in any case, they are more than sufficient for normal road vehicle guidance. When stopping at T-junctions, time for gaze orientation plays no critical role; horizontal turns from 90° on one side to 90° on the opposite site are necessary for checking traffic on the road encountered. With proper design, this can easily be done in less than a second. (The example platform given had a gaze range of ± 70° in azimuth; optical distortions by the windscreen of the vehicle have to be checked for larger angles.)

12.4 *MarVEye* as One of Many Possible Solutions

From the requirements discussed above, the following design parameters for the "multifocal active, reflex-like reacting vehicle eye" (MarVEye) have been derived. The principal goals of this arrangement of three to four cameras are sketched in Figure 12.7 (compare Figure 1.4). Two divergently looking wide-angle cameras should have their optical axes in one plane for stereo evaluation; the angle δ between these axes is a design parameter determining the simultaneous field of view (σ_{total}) and the overlapping region for binocular stereo (σ_{ster}). Note that the conditions for stereo interpretation at the boundaries of the images worsen with the aperture angle of the wide-angle cameras (increasingly higher order correction terms are necessary for good results).

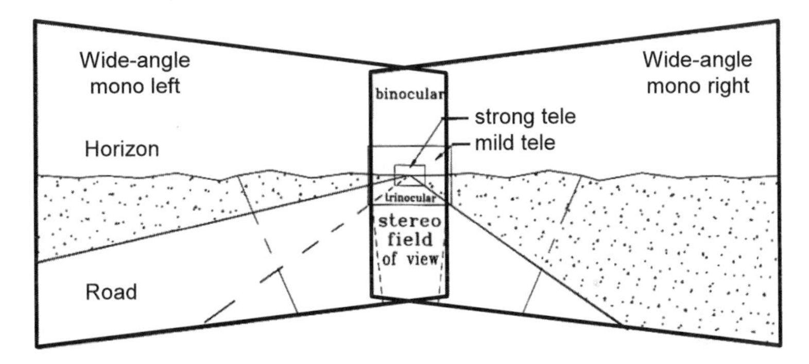

Figure 12.7. Sketch of fields of view of each camera and of the total simultaneous field of view (in azimuth and depth) of the arrangement 'MarVEye'; all cameras may have the same pixel count, but three of them have different focal lengths

When trinocular stereo is intended for using the image of the mild telecamera in addition, it makes sense to specify the angle of central overlap of the wide-angle cameras (σ_{ster}) the same as the horizontal f.o.v. of this camera σ_{tm} (center of figure). If the second pyramid level of the mild teleimage shall have the same resolution as the wide angle images, the teleimage should be one fourth of the wide-angle image in size.

The actual choice of parameters should depend on the special case at hand and on the parameters of the dominating task to be performed (maximum speed, maneuvering in tight space, *etc.*). Experience with the test vehicles VaMoRs (all cameras on a *two-axis* platform, Figure 12.6 left) and VaMP (yaw platform only, Figure 12.6 right) has shown that vertical gaze stabilization for reducing the search range in feature extraction is not necessary for wide-angle images; however, it is of advantage also for cars with smooth riding qualities if larger focal lengths are required for high-speed driving. Therefore, the design shown in Figure 12.8 has resulted:

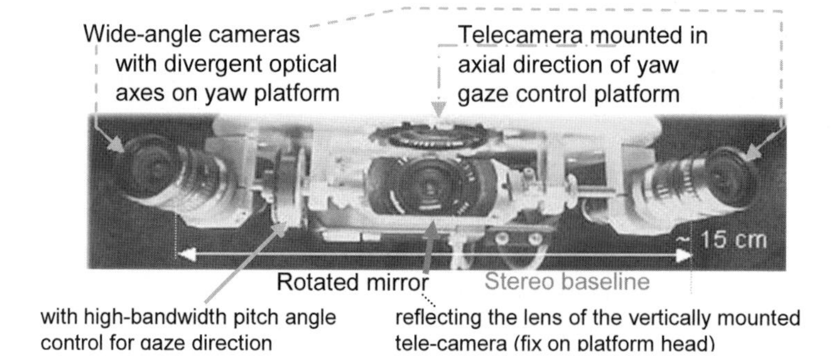

Figure 12.8. Experimental gaze control platform designed as a compromise between different mechanical and optical mapping requirements for a large horizontal viewing range in azimuth (pan). Gaze stabilization and control in tilt (pitch) is done only by a mirror in front of the telelens. This yields a relatively simple and compact design.

Only the image of the tele-camera is stabilized by a mirror with a horizontal axis of rotation; the laws of optical reflection result in cutting the amplitude of mirror angles required in half for the same stabilization effect. Mounting the telecamera with its main axis vertical minimizes its inertia acting around the yaw axis. Since the mirror has very little inertia, a small motor suffices for turning it, allowing high-bandwidth control with low power consumption. Once the basic design parameters have been validated, more compact designs with smaller stereo base will be attempted.

Using mirrors also for gaze control in pan is known from pointing devices in weapon systems. The disadvantage associated with these types of double reflections is the rotation of the image around the optical axis. For human interpretation, this is unacceptable and has to be corrected by expensive devices; computers could be programmed to master this challenge. It will be interesting to see whether these systems can gain ground in technical vision systems.

12.5 Experimental Result in Saccadic Sign Recognition

Figure 12.9 shows the geometry of an experiment with test the vehicle VaMoRs for saccadic bifocal detection, tracking, and recognition of a traffic sign while passing at a speed of 50 km/h. The tele-camera tracks the road at a larger look-ahead distance; it does not have the task of detecting the traffic sign in this experiment. The sign is to be detected and initially tracked by the standard near range camera (focal length $f = 8$ mm) with the camera platform continuing to track the road far ahead with the telelens; the road is assumed to be straight, here.

While approaching the traffic sign, its projected image travels to the side in the wide-angle image due to the increasing bearing angle given as $\psi(t) = \arctan(d/s)$

(see Figure 12.9). In the experiment, d was 6 m and the vehicle moved with constant speed V = 50 km/h (~ 14 m/s). The corresponding graphs showing the nominal aspect conditions of the traffic sign are given in Figure 12.10; it shows the bearing angle to the sign in degrees (left), the pixel position of the center of the sign (right), the distance between camera and sign in meters (top), and time till passing the sign in seconds (bottom).

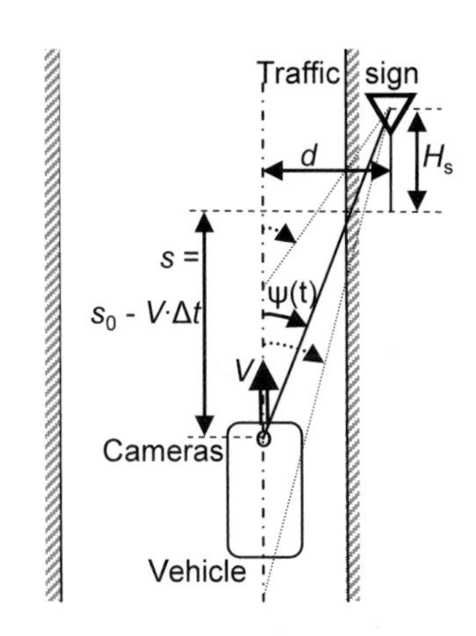

Figure 12.9. Geometry for experimental validation of saccadic bifocal sign recognition while passing (H_s is normal to plane of road)

The boundary marking of the triangle in red is 8 cm wide; it is mapped onto two pixels at a distance of about 28 m (1.95 s before passing). The triangle is searched for in phase 1 (see arrow) and detected at time 1.6 s before passing (first vertical bar), at an angle of ~ 15°. During phase 2, it is tracked in five frames 40 ms apart to learn its trajectory in the image (curve 1 in Figure 12.11, left).

This figure shows measurement results deviating from the nominal trajectory expected. After the fifth frame, a saccade is commanded to about 20°; this angle is reached in two video cycles of 40 ms (Figure 12.11, left side).

Now the traffic sign has to be found again and tracked, designated as phase 3. After about a half second from first tracking (start of phase 2) the sign of 0.9 m edge length is picked up in an almost centered position (curve 1 at lower center of Figure 12.11). It is now mapped in the teleimage also, where it covers more than 130 pixels. This is sufficient for detailed analysis. The image is stored and sent to a specialist process for interpretation.

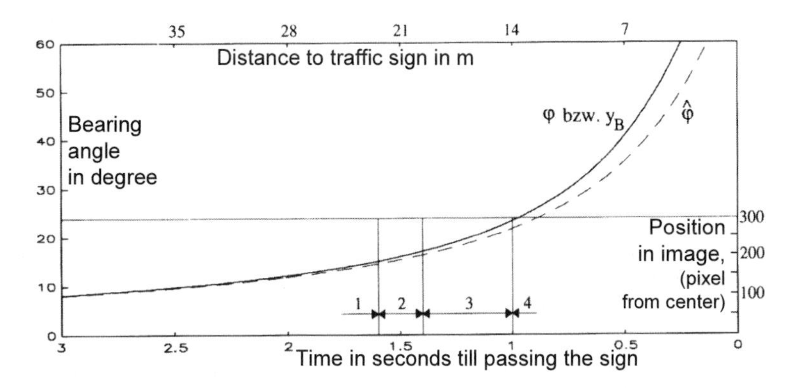

Figure 12.10. Nominal evolution of state variables (distance, bearing) and image values (lateral position) for the approach of a traffic sign and gaze control with saccades

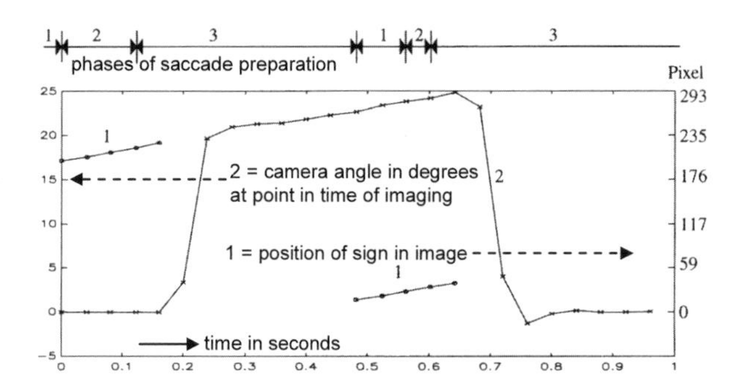

Figure 12.11. Position of traffic sign in standard image (curves 1) and gaze direction in yaw of camera platform for detecting, tracking, and high-resolution imaging of the sign

A saccade for returning to the standard viewing direction is commanded which is started a half second after the first saccade (branch 2 in Figure 12.10, right); about 0.6 s after initiating the first saccade, gaze direction is back to the initial conditions. This shows that the design requirements for the eye have been met.

The video film documenting this experiment nicely shows the quickness of the gaze maneuver with object acquisition, stable mapping during fixation, and complete motion blur during the saccades.

13 Integrated Knowledge Representations for Dynamic Vision

In the previous chapters, the individual components for visual perception of complex scenes in the context of performing some mission elements have been presented. In this chapter, the intention is to bring together all these components to achieve goal-directed and effective mission performance. This requires decomposition of the overall mission into a sequence of mission elements which can be performed satisfactorily exploiting the perceptual and behavioral capabilities of the subject's "body and mind".

Body clearly means sensor and actuator hardware attached to the structural frame being moved as a whole (the physical body). But what does mind mean? Is it allowed to talk about "mind" in the context of a technical system? It is claimed here that when a technical system has a certain degree of complexity with the following components in place, this diction is justified:

1. Measurement devices provide data about the resulting motion after control inputs to individual actuators; this motion will have certain "behavioral" characteristics linked to the control time history $u(t)$.
2. A system for perception and temporal extrapolation is in place which allows correct judgment whether or not a goal state is approached and whether or not performance is safe.
3. A storage device is available that stores the actual percepts and results and can make them available later on for comparison with results from other trials with different parameters.
4. To evaluate and compare results, payoff functions (quality criteria, cost functions) have to be available to find out the more favorable control inputs in specific situations. Environmental conditions like lighting (for visual perception) and weather (for wheel-ground contact) can affect appropriate behavior decisions significantly.
5. Knowledge about situational aspects that have to be taken into account for planning successful actions realizes the feedback of previous experience to actual decision-making.
6. Knowledge about behavioral capabilities and limitations both with respect to perception and to locomotion has to be available to handle the challenge given.
7. Capabilities similar to those of the subject vehicle are assumed to be available to other vehicles observed; this is an "animation capability" that brings deeper understanding into a scene with several objects/subjects observed.

A useful mind is assumed to exist when, based on the sensor data input, an internal representation of the situation given is built up that leads to decision-making to handle the actual real-world task successfully in a dynamically changing scene.

Just decision-making and control output without reference to "reality" would be called "idiotic" in common sense terminology.

In this context, it is easily understood why closed-loop perception–action cycles (pac) in the real world with all its dirty perturbation effects are so important. All the "mental" input that is necessary to make successful or even good decisions in connection with a given set of measurement data is termed "knowledge", either derived from previous experience or transferred from other subjects.

In Figure 13.1, the different components are shown and the main routes through which they cooperate are indicated by arrows. Only the lower bar represents the physical body; everything above this level is "mind", running of course on computer hardware as part of the body, but coming into existence only when the closed-loop pac is functioning properly. [Note that this system can function without sensory input from the real world if, instead, stored data are fed into the corresponding data paths. Control output to the real world has to be cut; its usual effects have to be computed and correspondingly substituted (as "real-world feedback") by temporal extrapolations using the dynamic models available. In analogy to biological systems, this could be called "dreaming". Physical realization in neuronal networks is, of course, completely different from that in electronic computers.]

The figure tries to visualize the interaction between measurement data and background knowledge. Measurement data stream upward from the body, processed on level 1 (number in dark circles at the left side). Background knowledge is lumped together in the dark area labeled 0 at the top of the figure. Three blocks of components are shown on this level: (1) Generic classes of objects and subjects (left), (2) environmental conditions (center), and (3) behavioral capabilities for both visual perception and locomotion (right).

In the left part of the figure, data and results flow upward; on the evaluation and decision level 4, the flow goes to the right and then downward on the right-hand side of the figure. Basic image feature extraction (level 1) is number crunching without feedback of specific knowledge or temporal coherence on a larger scale.

When it comes to object/subject hypotheses on the recursive estimation level for single units (level 2), quite a bit of knowledge is needed with respect to 3-D shape, appearance under certain aspect conditions, continuity in motion and – for subjects – control modes as well as typical behaviors. The basic framework has been given in Chapters 2 and 3; the next section summarizes items under the system integration aspect.

> The fusion task for perception is to *combine measurement data with background knowledge* such that the situation of the system in the actual environment (*i.e.*, the real world) relevant to the task at hand is closely represented in the "dynamic knowledge base" (level 3) as support for decision–making.

Recall that an open number of objects/subjects can be treated individually in parallel on level 2; the results for all units observed are collected in the DOB (left part on level 3 in the figure). Parallel to (high-frequency) tracking of individual objects/subjects, "actual environmental conditions" (AEC) have to be monitored (center on levels 2 and 3).

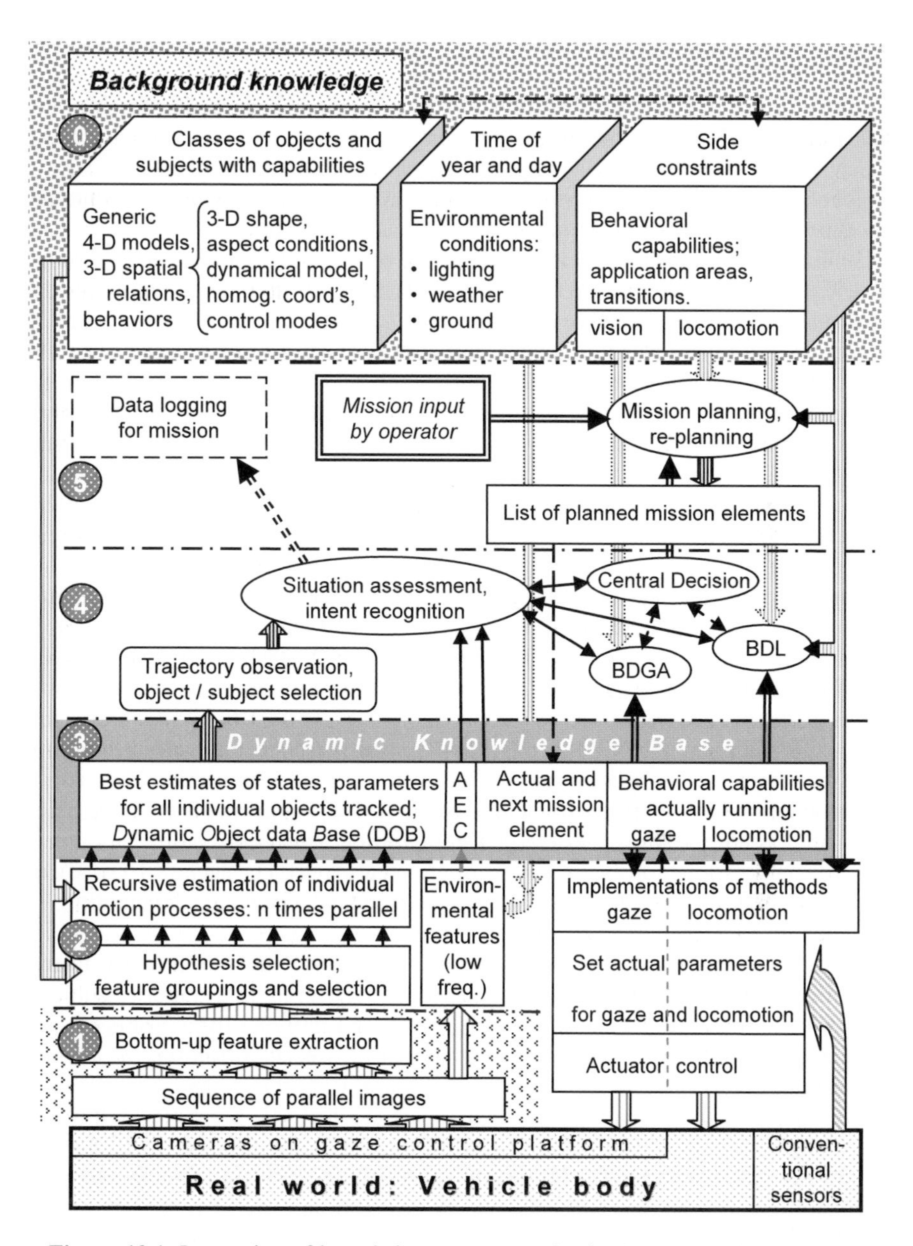

Figure 13.1. Integration of knowledge components in the 4-D approach to dynamic vision: Background knowledge on object and subject classes, on potential environmental conditions, and on behavioral capabilities (top row) is combined with measurement data from vision (lower left) and conventional sensors (lower right) to yield the base for intelligent decisions and control output for gaze and locomotion. All actual states of objects/subjects tracked (DOB), of environmental conditions (AEC), and of the subject's activities are stored in the dynamic knowledge base (DKB).

Since these conditions change only slowly, in general, their evaluation frequency may be one to two orders of magnitude slower (once every 1 to 5 seconds). Note that this is not true for lighting conditions when driving into or out of a tunnel; therefore, these objects influencing perception and action have to be recognized in a way similar to that for other objects, for example, obstacles. But for weather conditions, it may be sufficient to evaluate these within the rest of the cycle time not needed for performing the main task of the processors. This field of recognizing environmental conditions has not received sufficient attention in the past, but will have to for future more advanced assistance or autonomous systems.

For behavior decision in the mission context, the overall task is assumed to be decomposed into one global nonlinear part with nominal conditions and a superimposed linear part dealing with small perturbations only under local aspects. The basic scheme has been discussed in connection with Figure 3.7. The off-line part of nonlinear mission planning (upper right corner of level 5 in Figure 13.1, performed with special optimization methods) results in a nominal plan of sequential mission elements that can be handled by exploiting behavioral capabilities of the subject vehicle. Note that mission elements are defined in this way! This means that behavioral capabilities available determine mission elements, not the other way around, starting from abstract, arbitrarily defined mission elements. For local mission performance, only the actual mission element and maybe the next one need to be taken into account; therefore, these become part of the "dynamic knowledge base" (DKB, level 3 in Figure 13.1, center right). Actual mission performance will be discussed in Section 13.4.

The complex global nonlinear mission plan is assumed to be unaffected by minor adjustments to local perturbations (like obstacle avoidance or minor detours); if these perturbations become sufficiently large, strategic replanning of the mission may be advisable, however. These decisions run on level 4 in Figure 13.1. Local ("tactical") decisions for gaze control are grouped together in the unit BDGA (*b*ehavior *d*ecision for *g*aze and *a*ttention); those tactical decisions for locomotion are lumped into BDL (*b*ehavior *d*ecision for *l*ocomotion).

Sometimes, there may be conflicting requirements between these units. In this case and when global replanning is necessary, the *c*entral *d*ecision unit (CD) takes over which otherwise just monitors all activities and progress in realization of the plan (right part on level 4). These units are just for decision-making, not for implementing control output (upper level in Figure 3.17). Of course, they have to know not only which behavioral capabilities are available "in principle", but which are available actually "here and now". If there are minor malfunctions due to failure of a component which has not necessitated ending the mission, the actually realizable capabilities (maneuvers) have to be posted in the DKB (right part on level 3 in Figure 13.1).

Implementation of control output is done on level 2, right, with control engineering methods taking conventional measurements directly into account (minimization of delay times). This level monitors the correct functioning of all components and communicates deviations to the decision level. Representation of behavioral capabilities on levels 2 to 4 for mission performance is discussed in Section 13.3. The decisions on these levels are based on both perceptual results (left part on level 4) and background knowledge (arrows from top on the right of Figure 13.1) as well as the actual task to be performed (center level 3).

To judge the situation, it is not sufficient just to look at the actual states of objects and subjects right "now". More extended temporal maneuvers and their spatial effects on a larger scale have to be taken into account. Looking at time histories of state variables may help recognize the onset of new maneuvers (such as a lane change or turnoff); early recognition of these intentions of other subjects helps developing a safer style of "defensive driving". The representational base for recognizing these situations and maneuvers is the "scene tree" (Section 13.2) allowing fast "imagination" of the evolution of situations in conjunction with spatiotemporal models for motion. Actual decision-making on this basis will be discussed in Section 13.5.

Data logging and monitoring of behavior are touched upon in Section 13.6; in the long run, this may become the base for more extended learning in several areas.

13.1 Generic Object/Subject Classes

As mentioned in Chapters 2 and 3, humans tend to affix knowledge about the world to object and subject classes. Class properties such as gross shape and having wheels on axles to roll on are common to all members; the lower part of the car body looks like a rectangular box with rounded edges and corners, in general. The individual members may differ in size, in shape details, in color, *etc.* With respect to behavioral capabilities, maximal acceleration at different speed levels and top speed are characteristic. All in all, except for top speed of vehicle classes, variations in performance are so small that one general motion model is sufficient for all vehicles in normal traffic. Therefore, most characteristic for judgment of behavioral capabilities of vehicles with four or more wheels are size and shape, since they determine the vehicle class. The number of wheels, especially two, three, or more is characteristic for lateral control behavior; bicycles may have large bank angles in normal driving conditions.

Here, only simple objects and ground vehicles with four and more wheels are considered as simple subjects. Figure 13.2 summarizes the concept of the way the representation of these vehicles is used in the 4-D approach to dynamic vision. The lowest three rows are associated with 3-D shape and how it may appear in images. This description may become rather involved when shape details are taken into account. Simple discrimination between trucks, on the one hand, and cars/vans/buses, on the other, keeps shape complexity needed for basic understanding rather low: They all have wheels touching the ground; trucks have large ones, in general. Trucks can best be recognized and tracked by following features on the (large) upper parts of their bodies [Schmid 1992]. Cars yield more stable recognition when only their (box-like) lower part is tracked [Thomanek 1996]. A rectangular box has only three parameters: length, width, and height, the latter of which is hardly of importance for proper reactions to cars in traffic. A more refined generic model for cars has been given in Figure 2.15.

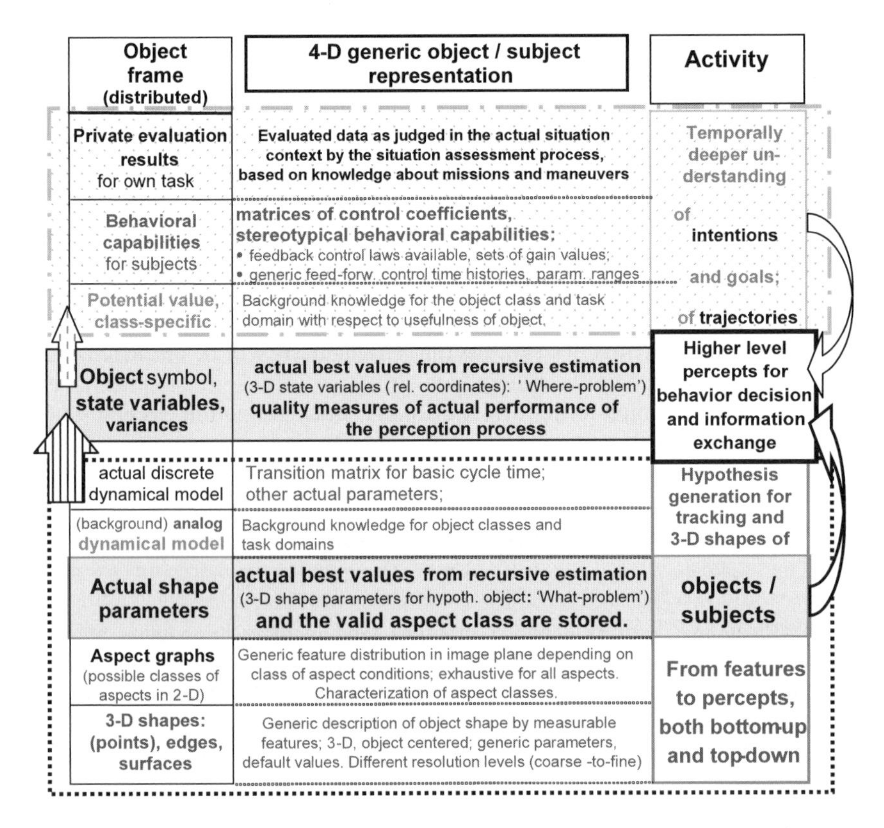

Figure 13.2. Summary of representational framework for objects/subjects in the 4-D approach to real-time vision (distributed realizations in the overall system, in general). Legend: black = actual values of dynamic object database (DOB); gray = background knowledge.

Reduction of the number of vehicle parameters makes vehicle tracking possible with relatively little computing effort. More detailed shape models, starting from the box description but taking groups of lights, license plate, and missing body regions for mounting wheels into account, can be handled with minor additional effort once the basic aspect conditions have been determined correctly. Default values for usual body size and relative positions of subobjects are part of the knowledge base (lowest row in Figure 13.2).

The appearance of body features as a function of (coarsely granular) aspect conditions is the next important knowledge item (second row from bottom). In Figure 11.15, one set of features has been shown for one of the eight aspect classes (rear left). For stabilizing image sequence interpretation, it should be respected, which ones of the features can be iterated meaningfully and which ones cannot. For example, looking straight from behind it does not make sense to try to iterate the length parameter; looking straight from the side, width estimation should be suspended. The interesting result in [Schmid 1993] was that under oblique viewing conditions, under which both length and width should be observable theoretically, estimating the length of the diagonal was much more stable than estimating both of its components; this is partly due to the rounded corners of vehicles and to the difficulty of precisely measuring the position of the vertical inner edge of the vehicle in the image.

The motion capabilities of objects/subjects (their dynamic models) – in their basic form – are given by differential equation constraints (in analogue time). Depending on the cycle time used for visual estimation, the transition and control input matrices result for sampled data (digital control). The basic frequency in the system is the field frequency: 50 Hz (20 ms cycle time) for CCIR and 60 Hz (16 2/3 ms) for the American NTSC standard. The factor of 2 for entire frames (40 ms resp., 33 1/3 ms) yields the cycle time mostly used in the past. However, in some present and in future digital cameras, cycle time may be more freely selectable; the transition matrices actually used have to be adjusted correspondingly (fifth row from bottom in Figure 13.2).

Motion state and shape parameters are determined simultaneously by prediction-error feedback of visual features. The actual best estimates for both are stored in the DOB and marked by solid black letters in the figure (left side). For objects and subjects of special interest, the time history of important variables (last m_T entries) may be stored in a frame buffer. This time history is used on level 4 in Figure 13.1 for recognition of trajectories of objects and possibly intentions of subjects. For situation assessment, there has to be a knowledge component telling the potential value or danger of the situation observed for mission performance (third row from top in Figure 13.2). For decision-making, another knowledge component has to know which behavioral capabilities to use in this situation and when to trigger them with which set of parameters [Maurer 2000; Siedersberger 2004].

Having observed another subject over some period of time, the individual (same identity tag) may be given special properties like preferring short distances in convoy driving (aggressive style of driving), or reacting rather late (being lazy or inattentive); these properties may be added in special slots for this individual (top left of figure) and are available as input for later decision–making. All of these results written in bold letters have to be taken into account as higher level percepts for decision-making (vertical center, right in Figure 13.2).

13.2 The Scene Tree

For relating spatial distances between objects and subjects (including the egovehicle) to visual measurements in the images, homogeneous coordinates have been chosen such as those used as standards in computer graphics for forward projection; there, all objects and their coordinates are given. In vision, the tricky challenge is that the unknowns of the task are the entries into the transformation matrices; due to the sine and cosine relations for rotation and due to perspective projection, the feature positions in the image depend on the entries in a nonlinear way. For this reason, the transformation relations have to be iterated, as discussed in Chapters 2 and 6, leading to the Jacobian matrices as key elements for knowledge representation in this recursive approach.

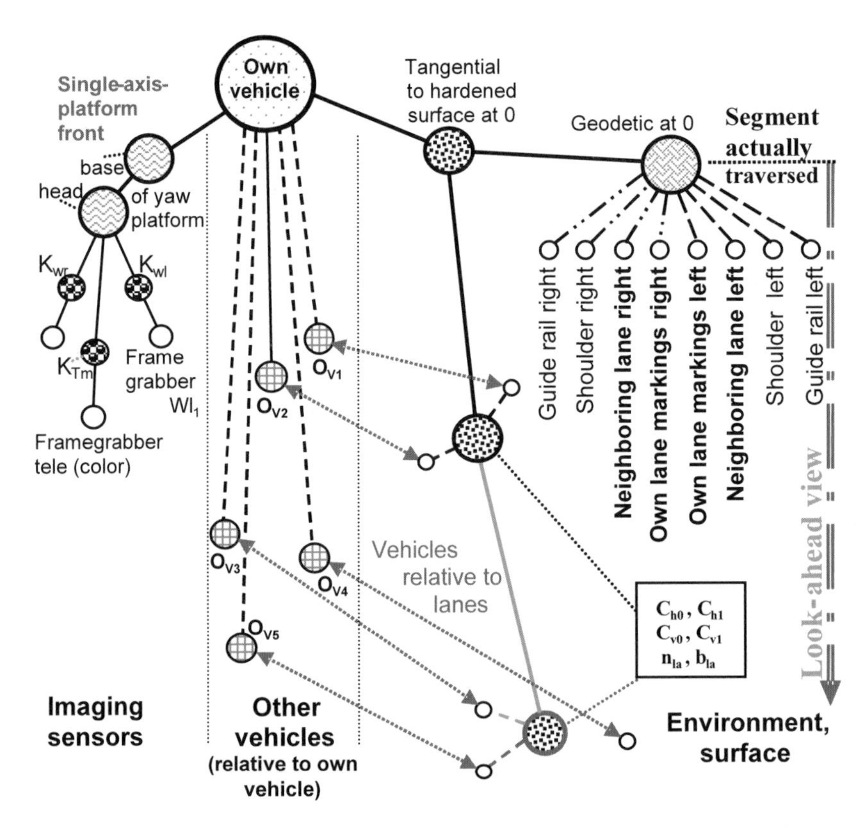

Figure 13.3. Scene tree for visual *Stop and Go* with EMS-vision on a yaw platform

A simple scene tree for representing the concatenated transformations has been given in Section 2.1.1.6 and Figure 2.7. The general scheme for proceeding with the vision task, once the scene tree has been defined, has been developed by D. Dickmanns (1997) and was given in Figure 2.10. Figure 13.3 gives a scene tree for the task of *stop-and-go* driving in a traffic jam based on *EMS vision* with three cameras on a yaw platform (see Chapter 12). The root node of the scene tree is the subject vehicle (top); the rest of the objects involved are lumped into three groups:

(1) At the left is the yaw platform with three cameras feeding three framegrabbers, which transport the images into computer storage. These images are the source of any derived information. The yaw platform is mounted behind the front windshield a certain distance away from the cg, around which rotations are defined. The platform base moves rigidly with the vehicle; the platform head is free to rotate around its vertical axis (one degree of freedom). The cameras are mounted at different locations on the head, away from the axis of rotation and with different orientations; this is represented in Figure 13.3 by the three edges linking platform, head, and cameras K_i. [For the telecamera, there may be an additional transformation necessary if the view to the outside world is gaze stabilized by a mirror (see Figure 12.7).]

(2) On the right-hand side in Figure 13.3, the road from the actual cg position (0) to the look-ahead range is represented. The geodetic representation to the right is not needed if no reference is made to these coordinates (either through maps or GPS signals); in this case, the aspects of road infrastructure given by the eight small nodes below may be shifted to the node at left. The parameters of the road observed in the look-ahead range are shown in the rectangular box: horizontal and vertical curvatures, number of lanes n_{la} and lane widths b_{la}.

(3) Other vehicles (center stripe in Figure 13.3) may be represented in two ways: Their positions may either be linked (a) directly to the subject vehicle (egocentric with azimuth (bearing), range, and range rate to each vehicle), or (b) to the local road or lane at its present position. The former approach is represented in the vertical column marked; it has the disadvantage that perturbations on one's own body directly affect the estimation process if no gaze stabilization is active; in the latter approach, for which the dotted double-arrows show the correspondence, only local information from the vicinity of the other vehicle is used, and egomotion is canceled out beforehand (see Figure 11.4, left). It is therefore more stable, especially at larger ranges. To avoid collision in the very near range, approach (a) has the advantage of immediate usability of the data [Thomanek *et al.* 1994].

13.3 Total Network of Behavioral Capabilities

Networks for representing special capabilities have been discussed in various sections of the book: Section 3.3 has dealt with perceptual capabilities, and Figure 3.6 showed the capability network for gaze control as one subtask. In Section 3.4, capabilities for locomotion have been considered for the special case of a ground vehicle; Table 3.3 showed a typical set of frequently needed "skills" in mission performance based on control engineering methods. A capability network for locomotion of a road vehicle has been shown as Figure 3.28. In this chapter, the goal is to fit these components together to achieve an overall system capable of goal-oriented action in the framework of a complex mission to be performed.

The entire Chapter 5 has been devoted to the capability of extracting visual features from image sequences, even though only a very small fraction of methods known has been treated; the selection has been made with respect to another basic perceptual capability treated in Chapter 6: Given a temporal sequence of groups of features, find a 4-D interpretation for a motion process in the real world (3-D objects moving in 3-D space over time) that has generated the feature set, taking the mapping process (lighting and aspect conditions as well as a pinhole camera model) into account. This "animation process" based on a temporal sequence of sets of visual features, maybe, is the most advanced "mental" capability of a technical system yet. Stored background knowledge on generic object/subject classes allows the association of mental representations of moving objects in the real world with the feature sets arriving from the cameras. Open parameters to be adapted for matching the measurement data allow the specification of individual members of the object/subject class. In this way, doing this *n*-fold in parallel for *n* objects/subjects of interest, the outside scene is reconstructed as an “imagined dynamically changing world” in the interpretation process.

Each interpretation is checked and adapted (stabilized or discarded) by prediction-error feedback exploiting first-order derivative matrices (so called “Jacobians” of the optical mapping process). Gaps in measurement data can be bridged by pure prediction over some cycles if the models installed (hypothesized) are adequate. In part, the visually determined interpretations can be supported by conventional measurements such as odometry or inertial sensing. The cross-feed of data and percepts allows, for example, monocular motion stereointerpretation while driving. Separating high-frequency and low-frequency signal contents helps achieve a stable and fast overall interpretation of dynamic scenes with each separate loop remaining unstable: On the one hand, due to motion blur in images and relatively large delay times in image understanding, vision cannot handle harsh perturbations on the body carrying the cameras; on the other hand, integrating inertial signals without optical (low-frequency) feedback will suffer from drift problems. Only the combined use of properly selected data can stabilize the overall percept.

The next essential capability for autonomous systems is to gain deeper understanding of the scene and the motion processes observed to arrive at “reasonable” or even optimal behavior decisions with respect to control output. Recall that control output is the only way the system can influence the future development of the scenario. However, it is not just the actual (isolated) control output that matters, but the sequence in which it is embedded, known to lead to desired changes or favorable trajectories. Therefore, there have to be representations of typical maneuvers that allow the realization of transitions from a given initial state to a desired (new intermediate or) final state. In other parts of the mission, as in roadrunning, it is not a desired transition that has to be achieved but a desired reference state has to be kept, such as “driving at a set speed in the center of the lane”, whatever the curvature of the road. There may even be an additional side constraint like: “If road curvature becomes too large so that – at the speed set – a limit value for allowed lateral acceleration (a_{ymax}) would be exceeded, reduce speed correspondingly as long as necessary”.

All these behavioral capabilities (driving skills) are coded in the conventional procedural way of control engineering. Some examples have been discussed in Section 3.4.3; Table 3.3 indicates which of these skills are realized by feed-forward (left column) or feedback control (right column). Figure 3.17 has shown the dual scheme for representation: On the top level for decision-making and triggering, rather abstract representations suffice. The detailed procedural methods and decisions depending on vehicle state actually measured conventionally reside on the lower level of the figure; they are usually implemented on different processor hardware close to the actuators to achieve minimal time delays.

The task of decision level 4 in Figure 13.1 then is to monitor whether mission performance is proceeding as planned and to trigger the transition between mission elements that can be performed with the same set of capabilities activated. The detailed organization of which component is in charge of which decision depends on the actual realization of the components and on their performance characteristics. In the structure shown, the list of mission elements determines the regular procedure for mission performance. However, since safety aspects predominate over mission performance, the *c*entral *d*ecision unit (CD) has to monitor these aspects and eventually has to make the decision to abandon the planned mission sequence to improve safety aspects; this may even result in stopping the mission and calling for inspection by an operator.

13.4 Task To Be Performed, Mission Decomposition

The knowledge needed for mission performance depends on the task domain and on the environmental conditions most likely to be encountered. The satisfaction of these requirements has to be taken care of by the human operator today; learning may be an important component for the future. Assuming that the capabilities available meet the requirements of the mission, the question then is, "How is the mission plan generated taking both sides (overall mission requirements and special behavioral capabilities) into account?"

First, it is analyzed for all consecutive parts of the mission, whose perceptual and behavioral capabilities have to be available to perform the mission safely. The side constraints regarding both perception and locomotion have to be checked; both lighting and weather conditions play a certain role. For example, the same route to be taken may be acceptable during daytime and when the surface is dry. Driving autonomously during nighttime when there are puddles of water on the road reflecting lights from different urban sources or from oncoming vehicles is not possible with today's technology. The result of this step is a first list of mission elements with all individual capabilities needed, disregarding how they are organized in the system.

As an example, let us look at a mission starting at a parking lot in a courtyard which has an exit onto a minor urban road with two-way traffic (mission element ME1). After a certain distance traveled on this road (ME2), a major urban road (three lanes) has to be entered to the right (ME3). From this road, a left turn has to be made at a traffic light with a separate turn-off lane (ME5) onto a feeder road for a unidirectional highway. Mission element ME4 contains the challenge to cross two lanes to the left to enter the turnoff lane in front of the traffic light (three lane change maneuvers). Turning off to the left when the traffic light turns green is ME6. Merging into highway traffic after a phase of roadrunning is ME7. Driving on this highway (ME8), at the next intersection with another highway, the lane leading to the desired driving direction is one of the center ones (neither the left nor the right one). The correct information, which one to take, has to be read from bridges with large navigation signs posted right above the lanes (ME9); this implies that on unidirectional highways, navigation is performed by proper lane selection and change into it. On the new highway, merging traffic from the right has to be observed (ME10); changing lane is one option for keeping traffic flowing (maneuver "lane change" again). Since most behavioral capabilities necessary for performing extended missions have been encountered up to here, let us end the example at this state of "roadrunning".

As a second step, merging of the original mission elements into larger ones, with some basic and some extra activities added if required, reduces the number of elements; for example, ME1 to ME6 may be lumped together under the label "driving in urban areas"; this part of the mission requires not just one maneuver capability but a bunch of similar ones and seems thus reasonable. In the context of these larger mission elements all coherent short activities for making a state transition will be called "maneuvers". *Stop-and-go* driving in heavy traffic is part of these extended mission elements. Maneuvers like turning onto a road from a private site, turning onto a crossroad, performing lane changes and in between: "roadrunning with convoy driving" are activated as they come, and need not (frequently cannot) be planned beforehand.

Highway driving between connection points may also be lumped together as one of the larger mission elements. At connection points (intersections), attention needed is much higher than just for roadrunning, usually; therefore, these points mark boundaries of larger mission elements. Figure 13.4 shows a graphical sketch of the decomposition of an overall mission (planned for 2.5 hours = 150 minutes) into mission elements (upper part, vertical bars mark the boundaries). On the next scale of resolution (15 minutes = 900 seconds shown here, say, for driving on a three-lane road as mentioned above), the behaviors running are the same; however, when the traffic situation yields an opportunity for lane change, this maneuver lasting about 9 seconds shall be performed. Note, that in this case (other than shown in the figure), it is not a fixed time for triggering the maneuver, but a favorable opportunity, which developed by chance, was used.

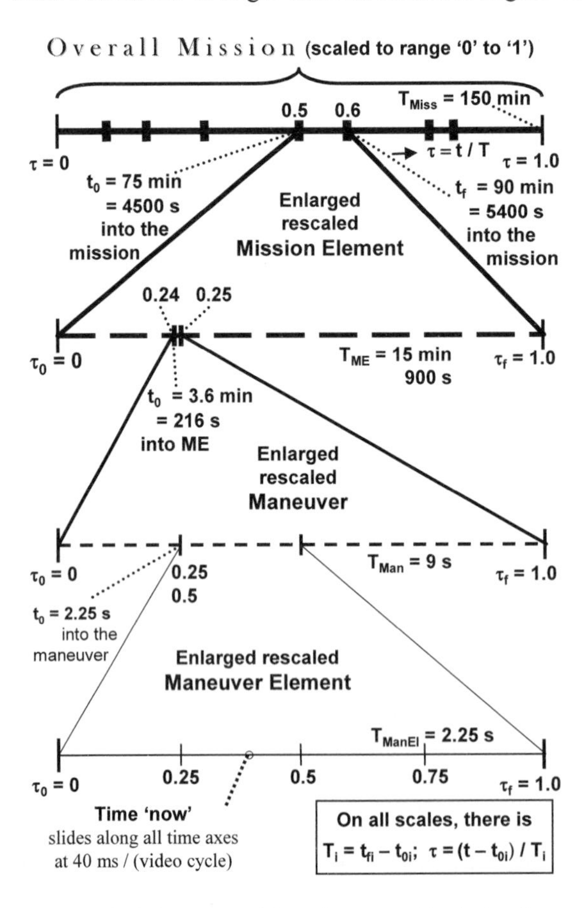

Figure 13.4. Decomposition of an overall mission into mission elements, maneuvers, and maneuver elements, all scaled to the range 0…1 for easy judgment of status in relation to the whole

For monitoring maneuvers and maneuver elements (lower part in Figure 13.4), again special scales are used. Normalizing the temporal range of each maneuver or mission element to the range from

0 to 1 allows, for each time, a quick survey, how much of the unit has been performed and what the actual state is (percentage accomplished) in the mission element or maneuver. This may be of interest for behavior decision, for example, when some failure in the system was discovered and reported during this maneuver, say, a lane change. The reaction has to be different, when the maneuver was just started, from the case when it was just before being finished or right at the center.

The upper two scales are of more interest for the navigation and mission performance level; the lower two are of special interest for the implementation level when the dynamic state of the vehicle depends on the time into the maneuver. In this case, energy off the equilibrium state may be stored in the body or the suspension system (including tires) which may couple with new control inputs and thus limit the types of safe reactions applicable; the infamous "moose test" for avoiding an obstacle suddenly appearing on the road by driving two maneuvers similar to fast lane changes right after each other in opposite directions is most critical in one of the later phases of the maneuver. Therefore, knowing the exact phase of the maneuver can be important for the lower implementation levels; keeping track of the maneuver, as shown on the lowest scale in Figure 13.4, helps avoid these situations.

The introduction of semantic terms for maneuvers in connection with corresponding knowledge representation of dynamic effects and of control activities to be avoided in certain phases (represented by rules) allows handling rather complex situations with this approach. Much of this development is still open for future activities.

13.5 Situations and Adequate Behavior Decision

The term situation is used here to designate the collection of all aspects that are important for making good behavior decisions. So it encompasses (1) The state of egomotion, absolute and relative to the road (lane), road parameters like actual curvatures of the road, and lane width (this is state of the art); (2) the surface state of the road, especially the friction coefficient at all wheels (only rough estimates are possible here); (3) other objects/vehicles of relevance in the environment: (a) Relative motion state to vehicle directly ahead in the same lane; this vehicle is the reference for convoy driving. Observation of blinking lights for turnoff and of stoplights when braking is required for safe and defensive driving; (b) vehicles nearby in neighboring lanes and their state of driving (roadrunning, relative speed, starting a lane change, *etc.*) should be tracked. Depending on the type of road (freeway, state road cross country, urban road or minor rural road), other traffic participants such as humans on bicycles, pedestrians, or a large variety of animals have to be detected and tracked; (c) to gain reaction time and deeper understanding of the evolving situation, vehicles in front of those tracked under (a) and (b) should be observed; (d) stationary obstacles on the road have to be detected early; look-ahead ranges necessary depend on speed driven (see Table 11.1); (4) extraordinary weather situations like heavy rain or snowfall, thunderstorms with hail, or dense fog have to be recognized; (5) lighting conditions have to be monitored to judge how much visual perceptions can be trusted; (6) the subject's intentions or goals are important for assessing a situation, since different emphasis on certain aspects of the payoff function may shift behavior decision. For example, if transport time is of importance and each second gained is valuable (thereby accepting a somewhat higher risk level), the decision will be different from a leisurely mission where risk minimization may be of highest importance.

Situation assessment, therefore, is a complex mixture of evaluations on different temporal scales. General environmental conditions will change most slowly, in general; they will enter short-time decision processes as quasi-static parameters without causing any harm due to some time delay (*e.g.*, weather conditions, in general). Maximal road curvatures and their onsets, on the other hand, require reaction time and thus limit maximal speed allowed. Because of the difficulty for humans in recognizing these situations sufficiently early, warning signs have been introduced, for example, in hilly terrain with large values of curvature ahead. Similar warnings are given ahead of transition points from long stretches of straight road into a tight curve, or when smoke or dust is to be expected due to activities on the side of the road (industrial installations), *etc*. These signs are posted such that at least several seconds up to minutes are available for adjusting driving behavior.

The shortest reaction times in traffic stem from other vehicles in one's own and neighboring lanes nearby. For this reason, these vehicles have to be watched with high attention all the time. In convoy driving, special rules are recommended for keeping distance from the vehicle in front depending on the speed driven: "half the tachometer reading" is an example in the "metric world". It means that the distance to be kept (measured in meters) should be at least half the speed value shown on the tachometer (in km/h). Substituting these numbers in the same dimensions [meter m and seconds s] and dividing the distance recommended by the speed driven, the result will be $T_{RG} = 1.8$ s, a constant "reaction time gap" independent of speed.

When we let our test vehicles drive autonomously according to this rule in German public traffic, the result was that other vehicles driven by humans passed the vehicle and "closed the gap" which of course had the test vehicle decelerate to adjust its behavior to the new vehicle in front. In denser traffic, this leads to an unacceptable situation. So the result was to reduce T_{RG} to the value which had only a few "aggressive" drivers still performing the gap-closing maneuver; the values yielding acceptable results (depending on the mood of the safety driver) were around $T_{RG} = 0.8$ to 1.2 s. Note that a value of 0.9 (half that recommended!) still means a distance to the vehicle in front of 25 m at a speed 100 km/h! This shows that in the realization developed, simple parameters can be used in connection with measured relative state variables to achieve rather complex-looking behavior in a simple way (in connection with underlying feedback loops from conventional control engineering): Only the parameter T_{RG} has to be specified on the upper (AI-oriented) level (4 in Figure 13.1); the implementation is done on level 2 (lower right).

Figure 3.17 has shown an example of how this cooperation is realized in extended state charts; a few "AI states" for longitudinal control are shown on the upper level. Semantic terms designate special "behavioral elements" realized on the lower level either by parameterized feed-forward time histories or feedback loops.

Possible transitions from one of these states to another are designated by arrows. These transitions are constrained by rules in which complex terms such as "reaction time gap" or "relative speed" and "distance in the same lane", *etc*., may enter. (References are given in the figure.)

The collection of all these capabilities for vehicle control yields the capability network for locomotion (Figure 3.28). It shows which complex maneuvers or driving states depend on which underlying "skills", and how these skills are linked to the actuators (lowest layer). The challenge in this concept is proper activation both with respect to temporal triggering (taking specific delay times into account) and with respect to selecting optimal parameters (on which system level?). Once the basic characteristics have been understood and parameters correspondingly adapted, the system is very flexible for modifications, which opens up an avenue for learning.

Explicitly specifying the interdependencies in the code of the overall system allows checking the actual availability of components before they are activated. This has proven valuable for failure detection before things go wrong.

13.6 Performance Criteria and Monitoring of Actual Behavior

Proper operation of complex systems depends on correct functioning of the components involved on many different levels. Since these systems never will function flawlessly over longer periods of time, failures should be detected before the component is activated or before its output is being used after a mode change. If a component is in use and the failure occurs while running, the failure should be noticed as early as possible. If measures for safe behavior can be taken immediately on the lower levels, this should be initiated while the higher levels are informed about the new situation with respect to hardware or processes available.

Correct functioning of basic hardware can be indicated by "valid"-bits (status); all essential hardware components should have these outputs for checking system readiness. Polling these bits is standard practice. More hidden failures, for example, in signal output from sensors, can often be detected by redundant sensing and cross-checking signals. In inertial sensing, this is common practice, for example, by adding a single sensor, mounted skewed to the orthogonal axes of the three other ones; this allows immediate checking of all three components by this single additional measurement.

If diverse sensors, such as inertial sensors and vision (cameras) are being used, cross-checks become possible by more involved computations. For example, integrating inertial signals (accelerations, turn rates) yields pose components; these pose angles and positional shifts are (relatively directly) measurable from features known to be stationary in the outside world (such as the horizon line or edges of buildings). These types of consistency checks should be used as much as affordable to prevent critical situations from developing.

Other inputs for system monitoring are statistical data from different processing paths. For example, if it is known that for devices functioning normally, the standard deviation of an output variable does not exceed a certain limit, sudden violations of this limit indicate that something strange has happened. It has to be checked then, whether environmental conditions have changed suddenly or whether a component failure is about to develop. In recursive estimation for dynamic vision, standard deviations for all state variables estimated are part of the regular computation. This information should be looked at for monitoring proper functioning of this perception loop. [In connection with feature extraction (Chapters 5 and 8), the use of this information for feature rejection has been mentioned.]

When the vehicle performs maneuvers in locomotion, there correspond certain characteristic time histories of state variables to each feed-forward control time history as input. For example, performing a turnoff onto a crossroad, triggering a steer rate time history at a certain distance away from the new centerline at a given forward speed of the vehicle should bring the vehicle tangential to the new road direction in the right lane. If this worked well many times, and suddenly in the next trial the maneuver has to be stopped for safety reasons before the vehicle runs off the road, some failure must have occurred. VaMoRs experienced this situation; analysis of the system turned out that the motor actuating the steering column had lost a larger part of its power (failure in an electronic circuit). The remaining power was sufficient for roadrunning toward the crossroad (nothing unusual could be observed), but insufficient for achieving the normal turn rate during curve steering. This critical self-monitoring of behaviors (comparison to expected trajectories) should be part of an intelligent system. Of course, this could have been detected by watching the time history of the steer angle on the lower level; however, this has not been done online. So this failure was detected by (in this case human) vision seeing the vehicle moving toward the edge of the road at a certain angle. In future systems, this performance monitoring should also be done by the autonomous vehicle itself.

Other performance criteria for ground vehicles of importance for behavior decision and mission performance are the following (not exhaustive):

1. Standard deviations in roadrunning: Figure 9.24b gave a histogram of lateral offsets of the test vehicle VaMP driving on the Autobahn (lane width of 3.75 m usually) over a very long distance. It can be seen that a maximal deviation of 0.6 m occurred which is secure for a vehicle having a width of less than 2 m. However, at construction sites, reduced lane widths on the leftmost lane may go down to 2.2 or even 2 m; so, with the parameter settings given, the vehicle could not use these lanes. Whether there is a different system tuning (perception and control) guaranteeing the small deviations maximally allowed under the circumstances given there (a well-marked lane) has not been tried.
2. Total travel time: If this has to be minimal for some reason, always the maximally possible safe speed has to be selected; fuel consumption is of no concern in this case.
3. Minimal fuel consumption for the mission: If this point of view is of dominant importance, quite a different behavior from point 2 results. The dependence of momentary fuel flow on the speed and gear selected determines the evolution of the mission. Actual fuel flow measured and low-pass filtered is available in most vehicles today. However, it is not the absolute value of temporal flow which is of interest, but – since a given distance has to be traveled during the

mission – the ratio of actual fuel flow (m-dot) divided by present vehicle speed V (m-spec = m-dot/V) has to be minimal. In modern cars, this value is displayed to the human driver. A problem for low-speed driving or halting is the division by speed (close to or equal to zero!). When the percentage of time spent in this state is small, these effects may be neglected; in inner city traffic with jams, they play a certain role (if the engine is kept running) and overall fuel consumption is hard to predict as a function of route selection.

4. More practical than the theoretical extremes mentioned above is a weighted mix as a payoff function: $\Phi = K_1 \cdot$(total time) + $K_2 \cdot$(fuel used). The verbal description is: "Drive with relatively low fuel consumption but also look at the time needed". When a larger gain in time can be achieved at the expense of a little more fuel, choose this driving style. This means that as input data for optimization not just the value m-spec is sufficient, but also its sensitivity to speed changes and throttle setting. Due to finite gear ratios and specific engine characteristics, these values are hard to determine; simple rules for selecting speed and throttle settings yield practically satisfying solutions.
5. Safety first: In the extreme, this may lead to overly cautious driving. When safety margins are selected too generously for the subject's style of driving in the nominal case, the reactions of less patient drivers may lead to less safe traffic in the real world, also for the cautious driver. (One example mentioned is the safety distance in convoys inducing other vehicles to pass, maybe in a daring maneuver.)

13.7 Visualization of Hardware/Software Integration

Because of the complexity of autonomous mobile robots with the sense of vision, the local distribution and functional cooperation of the parts can be visualized differently depending on the aspect angle preferred. Figures 1.6, 5.1, 6.1, 6.2, 11.19, 11.21 and 13.1 are examples of different stages of development of UniBwM vision systems.

Figure 13.5 gives an unconventional survey on the third-generation vision system and the turnoff example as one special mission element (right-hand side). The scene representation (broad column, center right) changes depending on the environment encountered and the mission element being performed. On the left side of the figure, the computers in the system and the main processes running on them are listed and linked by arrows indicating which processes run on which processor and which sensor hardware is connected to which processor. The object classes treated by the processes are also indicated by arrows into the scene tree.

The figure is not intended to give a complete introduction to or an overview of the system at one glance; it rather intends to visualize the central importance of the scene tree and how the rest of the closed-loop perception–action cycles are grouped around it. The Jacobian matrices as core devices for recursive estimation in dynamic vision are first-order relationships between couples of nodes in the scene tree: Image features on the one side (delivered in raw pixel form by the cameras C_i and the framegrabbers FG, lower center), and object states or parameters on the

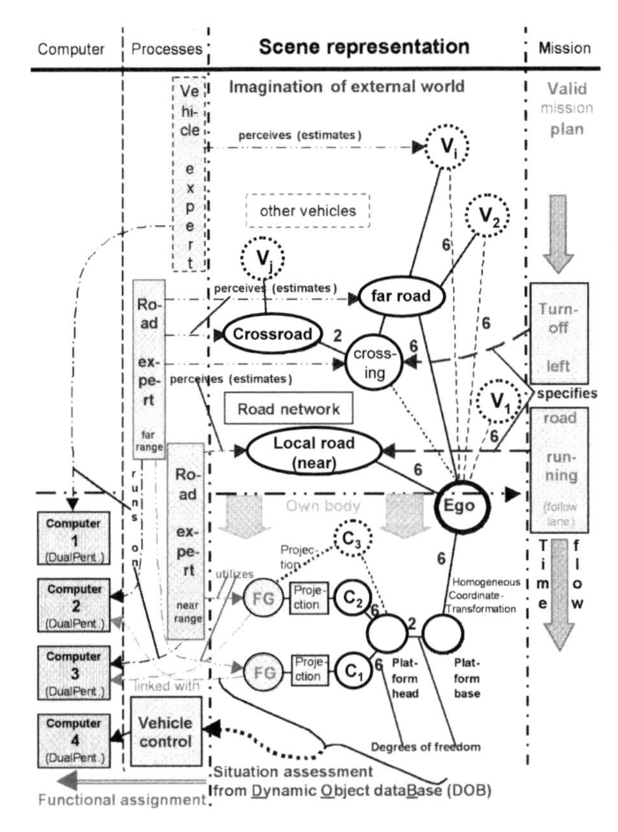

Figure 13.5. Visualization of hardware/software connectivity in EMS-vision for driving on networks of roadways

other side such as elements of the road network (shown as ellipses) or other vehicles (dotted circles, relative to near road elements). These objects can be created or deleted in the scene tree as corresponding features are detected or lost.

While driving, the variables in the homogeneous coordinate transformations HCT (edges between nodes) change continuously; they are constrained, however, by the dynamic models representing knowledge about motion processes not usually taken advantage of in "computational vision". Since these transformations are in 3-D space with spatial object models carrying visual features, a full spatio-temporal reconstruction takes place in model fitting, thereby avoiding the loss of the depth information critical in snapshot-interpretation with motion understanding as a second (separate) step derived from sequential images.

In our approach, the motion models, with prediction-error feedback for understanding, are used directly for reconstructing an "animated spatiotemporal world" in the form of symbolic descriptions based on background knowledge shown in Figure 13.1, top. However, for arriving at "imagined scenes" the conventional computer graphics approach developed in the last quarter of the last century is sufficient; no image evaluation is needed if objects and transformations in the HCT are "dreamed-up". In accordance with what some psychologists claim for humans, one might say that – also for EMS vision –

vision is (feedback-) controlled hallucination.

14 Mission Performance, Experimental Results

The goal of all previous chapters has been to lay the foundations for mission performance by autonomous vehicles. Some preliminary results, which have helped find the solution presented, have been mentioned in previous sections. After all structural elements of the system developed have been introduced, now their joint use for performing mission elements and entire missions autonomously will be discussed.

Before this is done, some words about the term "autonomous" seem in order. It originally does not mean just "no human onboard" or "without actual (including remote) human intervention". The linguistic meaning of "auto" is that the system does not have to rely on any external input during mission performance but has all resources needed onboard. Looked at from this (purist) point of view, a system that, for example, has to rely on steady input of GPS data during the mission cannot be called autonomous. At most, it is autonomous in the sense of unmanned vehicles driving along a local electric field generated by wires buried in the ground ("cable guided"); the only difference for a dense map of GPS waypoints is that buried wires give continuous analogue input to sensors onboard the vehicle, while GPS provides sampled data representations (mesh points of a polygon) into which a smooth curve for determining control output has to be interpreted by digital processing onboard. If the external GPS system fails, "autonomy" is lost, which is in contrast to the definition of the term.

Using GPS very sparsely, similar to infrequent landmark navigation with conventional technology would change the picture, since large fractions of the mission have to be driven really autonomously. In commonsense language, "autonomous" would mean that the system is able to generate all the information on its own that conventionally is provided by the human driver. This includes finding the precise path to drive and checking for obstacles of any kind (including negative ones such as ditches or large potholes) as well as finding detours for avoiding collisions. If the path to be driven and marked by many GPS waypoints is even prepared to guarantee the lack of obstacles of certain kinds, mission performance can hardly be termed autonomous. This would only be a relatively easy to achieve step in the direction of full autonomy.

On the other hand, if GPS signals could be guaranteed to be available like sunlight, the situation would change. The big challenge then would be to have the actual state of the environment correctly available for planning the positioning of the waypoints for the mission in an optimal (or at least sufficiently good) way.

This part of off line mission analysis for decomposition of a global mission into a sequence of mission elements, stored in a list for execution one after the other, is not discussed here; it is assumed to have been done correctly, satisfying the behavioral capabilities of the vehicle.

14.1 Situational Aspects for Subtasks

To perform a mission autonomously, several classes of subtasks should be distinguished. The initialization for starting an autonomous mission is different from continuously performing the mission after the start. In both subareas again, different situations and classes of other subtasks have to be distinguished. Once in operation, the subtasks are grouped into classes geared to certain subsystems for perception, situation assessment, behavior decision, and implementation of behaviors including fast first reactions to failures in the performance of components.

14.1.1 Initialization

For initialization, the major difference is whether the vehicle is at rest (fully autonomous start) or whether it has been operated till now by a human driver who has kept it in a reasonable driving state; this is the standard starting condition for an assistance system.

14.1.1.1 Starting from Normal Driving State

The manufacturer of an assistance system for normal driving will require that the vehicle is in a safe driving state, achieved by the human driver, when the system is switched on. Those assistance systems that are meant to help prevent or reduce the severity of accidents in dangerous situations are included (like the "Electronic Stability Program", ESP or similar acronym); they only become "active in control output" in a dangerous state but have been observing vehicle motion and checking the situation before.

A normal driving state alleviates the initialization process since positions of essential features with approximately known parameters can be found in relatively small search regions. However, some uncertainty will always be there so that it is advisable to collect as much information as possible from conventional sensors first: Speed from a tachometer (or derived from an odometer) is a "must"; in connection with accelerometer readings in the horizontal plane (a_x, a_y, vehicle body oriented), after a short time, both the orientation of the vehicle body relative to the gravity vector (surface slope) and vehicle accelerations (longitudinal and lateral) can be perceived. Angular rate sensors inform the system about the smoothness of the ground and, in connection with the steering angle, about the curvature of the trajectory actually driven. The variance of the vertical acceleration component also contains information on surface smoothness (vertical vibrations).

While this evaluation of conventional sensor signals is done, image processing should collect general information on lighting and weather conditions to find proper threshold values to be used in feature extraction algorithms. Then, initialization of the vision system for road detection and recognition, as discussed in Chapter 8, should be done.

14.1.1.2 Fully Autonomous Start from Rest

For a really autonomous vehicle, it is also important first to collect all information it can get from conventional sensors before higher level vision is tried; it is even more important here that the imaging sensors should first be used for collecting information on general environmental conditions.

Background information from conventional sensors: With little computational effort, a large amount of important information for visual perception can be obtained from the following sensors:

1. Odometer/tachometer: Is the vehicle stationary or moving? If moving, at what speed and with which steering angle (turn rate)?
2. Inclinometer, accelerometers in the horizontal plane: What is the inclination of the vertical axis of the vehicle relative to the gravity vector? If the vehicle is at rest, the sensor readings determine the inclination of the surface on which the vehicle stands. If the vehicle is moving, observation of acceleration in the longitudinal direction and of the speed history over time also allows determining surface slope.
3. A thermometer (outside) tells us about the state in which water on the surface or as precipitation will appear likely.
4. Inertial sensors for translation and rotation give an indication of surface roughness (vibration levels).

Background information from imaging sensors: Before visual perception of objects is started, here too, the general visual situation with respect to lighting and weather conditions should be checked. If the vehicle is at rest, the surface state of the environment (smoothness, vegetation, *etc.*) should be checked. Precise perception cannot be expected with state-of-the-art sensors and algorithms; quite a bit of effort has to go into this field to achieve close-to-human performance. For example, smooth ground with tall grass or other vegetation moving due to wind, which would allow crossing the region with no danger to the vehicle, cannot be judged correctly today neither by laser range finders (LRF) nor by vision. While LRF show a solid obstacle or a surface at the elevation of the tips of the grass, vision could have sufficient resolution for recognizing single plants, but computing power and algorithms are still missing for the spatiotemporal recognition process required for correct perception.

To recognize drivability of the terrain on which the vehicle is standing, the tangential plane defined by the vehicle body at rest, but at the elevation of the lower parts of the wheels, should be determined first. Modeling the vertical deviations from a horizontal plane (surface structure) is just about to be developed, both with LRF and with stereovision; robust real-time performance has yet to be achieved.

Recognizing the state of illumination can be done by looking at average image intensities and the distributions of contrast. Recognizing weather conditions is much harder for the general case; fog recognition has been looked at but is not solved for the general case. Diminishing contrasts in the image for areas further away are the key feature to look at. Recognizing different kinds of rain and snowfall still has to be achieved, but should be in range in the near future.

The hardest challenge both for human drivers and for autonomous systems is the question whether or not the ground is conditioned (sufficiently hard) to support the vehicle. This problem will remain to be decided by a human operator walking into these regions and checking by using his legs and corresponding pressure tests on the surface. Observing other vehicles driving through these regions is the other way to go (but may take longer).

Once the decision for a certain behavior in locomotion has been taken, the methods for prediction-error feedback are activated and the continuous perception-and-action loop is entered (see Chapter 8).

14.1.2 Classes of Capabilities

For efficient mission performance, there has to be a close interaction between the three classes of capabilities: Situation assessment, behavior decision for gaze and attention (BDGA), and behavior decision for locomotion (BDL). One might reason that because of the tight interactions these three should be performed in one single process. However, both BDGA and BDL have to rely on situation assessment for selecting objects and subjects of high relevance for the task to be performed. On one hand, BDGA has to concentrate on short-term aspects for collecting the right information on these objects and subjects most efficiently and precisely, based on image features. On the other hand, BDL has to understand the semantics of the situation to arrive at good decisions in the maneuver or mission context; this requires looking at larger scales in both space and time.

So the *situation* is the common part, but BDGA has to ensure its correct perception by providing rich selections of image features to the visual tracking processes with the least delay time, while BDL (or alternatively BDAs in the assistance mode) gives the essential input for defining and judging the situation in the mission context and then decides which of the locomotion capabilities (or of the assistance functions) is best suited for handling the actual situation and for achieving the goal of the mission most efficiently and safely. To stay in the performance range of humans, a few tenths of a second delay time are acceptable here.

This interplay is schematically shown in Figure 14.1: All evaluation processes (shown as rectangles) have access to the situational aspects derived previously and to the state time histories of visual perception (stored in the DOB by the individual tracking processes). The arrows E symbolize this basis for actual evaluation of changes in the situation. The arrows A indicate that new aspects of general relevance are added to the common description of the situation (including deletion of old aspects no longer relevant). Certain special results will be kept local to avoid communication overload.

Each tracking process and the object detection system can request a certain gaze direction for achieving their actual tasks; in addition, an intended maneuver such as a turnoff has to be taken into account by BDGA. If contradicting requests are received and cannot be reconciled by saccadic time-slicing of attention, the "central decision" unit (CD, center top) in charge of overall mission performance has to be

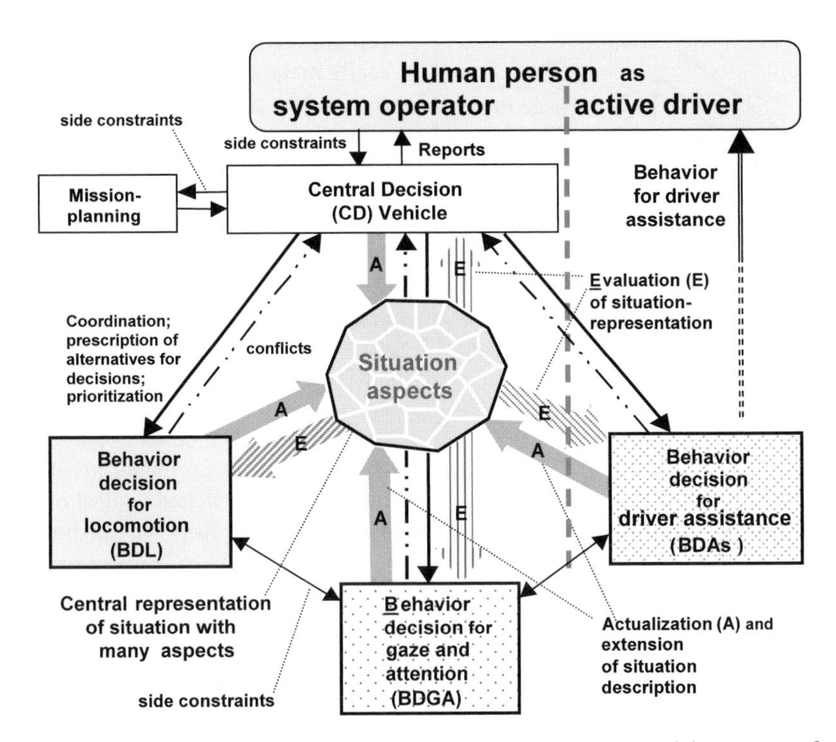

Figure 14.1. Organizational structure of interactions between special processes for situation assessment: "Behavior decision for gaze and attention" (BDGA, center bottom) and one of the two alternative processes "Behavior decision for locomotion" (BDL, left, for the autonomous driving mode) or "Behavior decision for driver assistance" (BDAs right, if used as an assistance system). (Note that only one of the two can be active at a time.) In case of conflict between BDGA and BDL, "central decision" (CD, center top) may be invoked for coordination and possibly adaptation of the mission plan or other behavioral parameters.

invoked and has to come up with a specific maneuver, if necessary; for example, slowing down or shifting the lateral offset parameter from the center of the lane for "roadrunning" by feedback control may solve the problem. These are the cases where control of gaze and locomotion are no longer independent of each other.

14.1.2.1 Visual Perception Capabilities

Basic aspects have been discussed in Section 3.3.2.3 (see Figure 3.6). It is essential that an autonomous system has a knowledge base telling it when to use which capability with which parameters in given situations. For example, when tight maneuvering is performed, it makes sense to link gaze direction to the steering angle selected for forward driving (see Figure 14.2). The correct gaze angle to be commanded is not the intended steering angle but the one actually reached because this is the angle to which changes of the body direction actually correspond in normal driving (no slip). Therefore, the command signal for gaze control in yaw should be

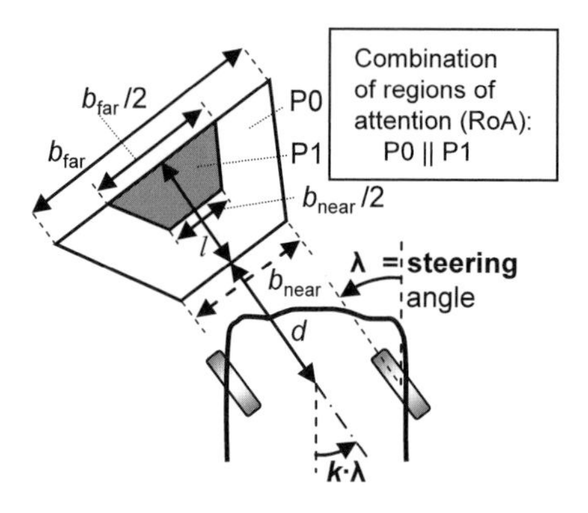

Figure 14.2. Linking gaze direction to steering angle for maneuvering

proportional to the steering angle actually measured (and maybe smoothed by a low-pass filter). Figure 14.2 shows two viewing regions of attention (RoA) by different shades which correspond to the fields of view of two cameras with different focal lengths (here a factor of 2 apart) [Pellkofer 2003]. So, intelligent gaze behavior may be simply realized by internal cross-feed of measurement data.

Triggering these behaviors is done by rules checking the preconditions for activation. Once in operation, the decision level is no longer involved in determining the actual control outputs. Other rules depending on situation parameters are then checked for finishing this type of behavior and for switching to a new one; transitions to feed-forward maneuvers or to other feedback modes are possible (such as tracking a certain object in the DOB).

Other modes in gaze control for driving cross-country and avoiding ditches (negative obstacles) are shown in Figure 14.3. For interpreting the size of the ditch, the perception system has to look alternatively to the right and the left border of the ditch (P1 and P2), while internally the encasing box for the entire ditch has to be determined, covering the region P0 containing all of the ditch (last row of the ta-

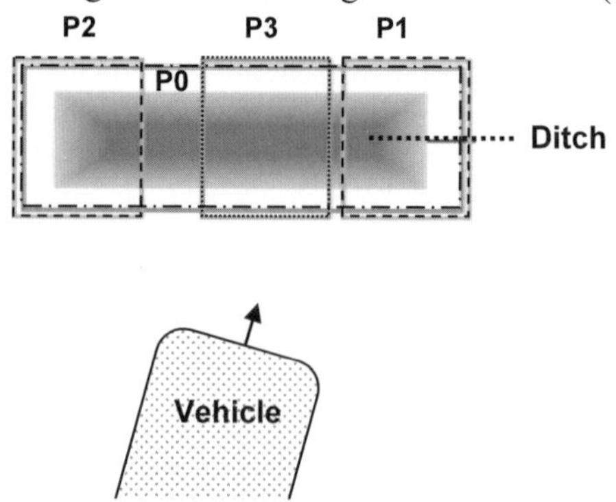

Perception modes	Combinations of RoA	Chosen during
Right border	P0 || P1	Evasion to right
Left border	P0 || P2	Evasion to left
Part in front of vehicle	P0 || P3	Stop in front
Left and right borders	P0 || (P1 & P2)	Interpretation

Figure 14.3. Regions of attention (RoA) for saccadic gaze control approaching a ditch of size larger than the field of view

ble). A complete internal representation of the ditch is constructed even though it is never seen as a whole at the same time; the changes in gaze direction together with the positions of features in the images are jointly used for reconstruction of the 3-D geometry [Pellkofer 2002, Pellkofer *et al.* 2003].

When a decision for behavior has been achieved, the vision system will continue with one of three possible modes: (a) fixation of the right corner (keeping the entire ditch "in mind") for evasion to the right (first row of table), (b) fixation of the left corner (keeping the entire ditch "in mind") for evasion to the left (second row of table), or with (c) looking straight ahead when stopping in front of the ditch (third row of table). Different feedback strategies of specific image features help realize the modes.

A third example, perceiving a crossroad and turning onto it has been discussed in Chapter 10; in Section 14.6.3, the same maneuver will be discussed, now integrated in the capability concept of saccadic vision.

14.1.2.2 Situation Assessment Capabilities

The discussion has not really been settled within our team whether it is advantageous to have situation assessment as a separate process beside BDGA for gaze and BDL for locomotion control. There are many aspects of situations specific to perception and others specific to locomotion, so these processes will do large parts of the overall situation assessment work on their own, especially when these tasks are developed by different persons. Maybe experience is not yet sufficiently large to find a stable structure of task distribution; in the long run, combining experience from the different domains in a unified approach seems advantageous (at least from an abstract architectural point of view). As shown in Figure 14.1 (center), representation of the results of situation assessment, the situation aspects, should be accessible to all processes involved in decision-making on a higher level.

For today's assistance systems, this part is rather small, usually; most effort goes into finding solutions satisfying the user from an ergonomic point of view. Realizations are mostly procedural statements. Only with an advanced sense of vision in the future will behavior decision for assistance (BDAs) become more demanding.

14.1.2.3 Locomotion Capabilities

This topic has been treated to some extent in Section 3.4 (see Figure 3.28 for the corresponding capability network summarizing results). Here, the overall task is broken down into two layers: A representational (abstract) one for decision–making in the strategic task or mission context on one hand, and a procedural one for efficient implementation with minimal time delays taking most recent measurement values into account, on the other. Figure 3.17 shows just one example of behavior decision for longitudinal control; it contains the conditions for transitions possible between the different modes on the upper level. These are coded in rules with decisions depending on parameters evaluated in situation assessment.

Table 3.3 shows a typical collection of behavioral capabilities of road vehicles and their way of realization by either feed-forward or feedback control. Section

14.6.4 will discuss the joint use of different capabilities for deciding on and performing lane change maneuvers in normal highway traffic.

The detailed codes for realization will not be discussed here. Quite a bit of effort is expended in this field by car manufacturers or suppliers for developing systems to come on the market soon. The next section gives a short survey of the general concept.

14.2 Applying Decision Rules Based on Behavioral Capabilities

A behavioral capability results from matching general algorithms for the field considered with specific hardware of a vehicle (or vehicle class) through adaptation of parameters and specification of the range of applicability. There are sets of rules specific to each of the three classes of capabilities mentioned in the previous section plus the set for central decision (CD).

CD is the first unit addressed when a new mission is to be performed ("Task assignment", arrow 'I-1' in Figure 14.4, center top). It first initiates off line "Mission

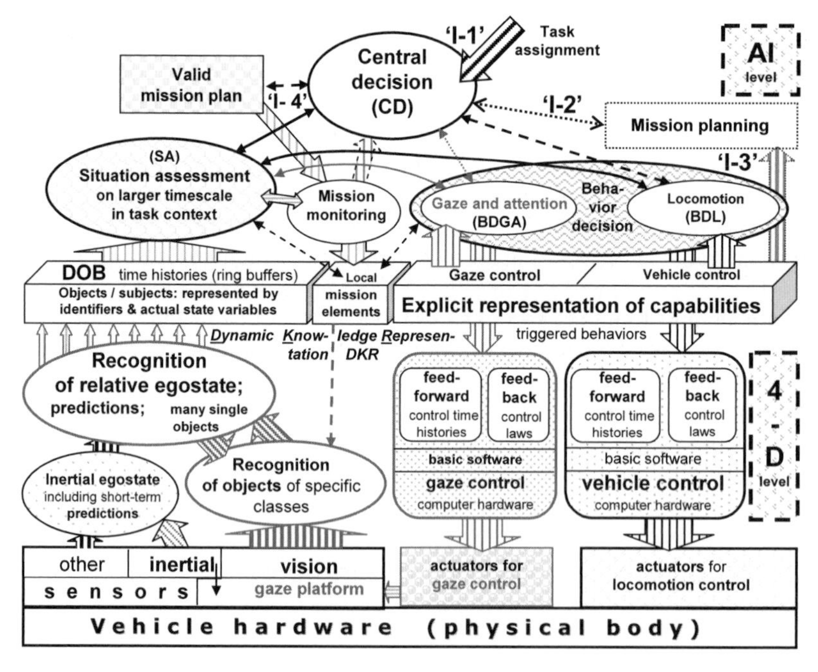

Figure 14.4. Coarse block diagram of the system architecture for performing entire missions: Arrows 'I-1' to 'I-4' point out mission initialization before the recursive loops are started. The inner light-gray loop represents expectation-based, multifocal, saccadic (EMS) vision, while the outer dark loop represents conventional automotive control with a separated decision level shown above the horizontal bar for "dynamic knowledge representation" (DKR).

planning" ('I-2') to analyze the mission and to come up with one or several alternative mission plans taking the nominal capabilities of the system into account (arrow 'I-3', upper right). For each plan consisting of: (1) a list of mission elements, (2) a list of performance criteria with expected values, and (3) a list of the subsystems needed, CD evaluates which plan is best suited for the subsystems actually available. This "valid mission plan" is stored as a guideline for actually performing and monitoring the mission (arrows 'I-4', top left) [Hock 1994; Gregor 2002]. The list of mission elements of this validated plan is then consecutively activated by copying one piece after the other into "dynamic knowledge representation" [(DKR, center of the horizontal bar shown in Figure 14.4 separating the decision level (top) from the procedural evaluation level (lower part; see also Figure 13.5 right)]. The perceptual capabilities actually needed are initiated via BDGA; each path knows its own initiation procedures and will try to represent stable interpretations in the DOB. Now, "situation assessment" (SA) can start its evaluation taking the requirements of the first mission element into account (top left corner).

The evaluation results are first used for checking the perception mode by BDGA, and second for starting locomotion activities through BDL (initiate physical mission performance) by triggering behavioral skills for locomotion on the lower (procedural) level (bottom right corner of Figure 14.4). The progress of mission performance is monitored on several levels. If situations occur that do not allow planned mission progress, as, for example, in roadrunning on a multilane highway with a slower vehicle in the same lane ahead, BDL has to come up with a decision whether a transition into convoy driving or a lane change maneuver with passing should be performed. Before the latter decision can be made, perception to the rear and to the side and corresponding evaluations of "situation assessment" have to ensure that this maneuver is possible safely; this evaluation will take some time. If the gap to the vehicle in front closes too rapidly, a safety mode on the lower level for vehicle control may start a transition to convoy driving in the same lane.

This example shows that the coordination of decisions in the different modes and on the time line have to be done carefully; this is by no means an easy task and has taken quite a bit of effort in developing this approach for practical applications with the test vehicles in standard traffic scenes. Once the parameter ranges for safe operation are known, it is very flexible, and it is easy to make improvements by new modular units for perception and control.

14.3 Decision Levels and Competencies, Coordination Challenges

As mentioned in Section 14.1.2 and shown in Figure 14.1, there are several decision levels emphasizing different aspects of the overall task. The mission and the actual state of the vehicle as well as of the environment including other objects/subjects define the situation for the system. BDGA has to ensure that perception of this situation is performed as well as possible and is sufficient for the task. If the situation deteriorates, it has to warn CD and BDL to adapt to the new environmental conditions. If the perturbations are sufficiently severe, CD has to adapt parameter settings or may even have to start "mission replanning" to adapt the future list of mission elements to the new conditions.

To a certain extent, the procedural implementation levels can react directly to perturbations noticed. It is an engineering standard to keep the system in a safe or agreeable operational state by adapting gain factors in feedback control, for example. The challenge is to keep all system components informed about the actual performance level; therefore, the "explicit representation of capabilities" has to be updated each time a capability running has been adapted for whatever reason.

For example, if it has been noted that during the last braking action, the effective friction coefficient was lower than expected (due to wet soil or slush or whatever), the updated friction coefficient should be used for planning and for realizing the next braking maneuver. Engineers will tend to keep the details of these decisions on the procedural level with direct access to the most recent measurement data (minimal delay time); they expect the decision level to trigger transitions or new behaviors sufficiently early so that implementation can be performed optimally, directly based on the latest data; conventional measurement data may be available at a higher rate than video evaluations (the maximum update rate for the higher system levels). [In our test vehicles, inertial data come at 100 Hz with almost no delay time, while video yields new results at 25 Hz (every 40 ms) with at least an 80 to 120 ms time delay.]

For this reason, in critical situations, both the fast lower level and the more intelligent higher levels will contribute to safety: First reactions with the goal to increase safety margins are immediately performed on the lower level, and the rest of the system is informed about facts leading to this decision and to the adaptations made. A more thorough analysis and assessment of the new situation is then performed on the higher level which then may trigger new behavioral modes or even abandon the mission. This is an area where much development work remains to be done for systems more complex than those up to now.

Some aspects of this integration task have been performed for several mission elements and a small mission as a final demo of the development project; this will be described in the remaining sections.

14.4 Control Flow in Object-oriented Programming

The first two generations of vision systems realized and experimentally tested have been programmed in procedural languages (FORTRAN, Occam, and C). The different subtasks have been treated as self-contained units, and the main goal of the tests has been to demonstrate the fitness of the approach to handle the well-defined task domains. In the mid-1990s with the final demo of the "Prometheus" project and the long distance drive from Munich to Odense, Denmark, (see Section 9.4.2.5) successfully performed, it was felt that the next-generation system should be designed as an integrated entity right from the beginning. As a programming style, object-oriented coding in C++ has been selected for practical reasons of availability and trained personnel. A strong tendency toward the language Ada for more reliable code for the complex system to be expected did not find majority approval of the group.

Figure 14.4 has shown the basic structure of the system designed and the general flow of activities in the perception–action cycle. Multiple feedback loops realized in the system have been given in Figure 6.1; all of this implements the "4-D" approach to dynamic vision, the basic idea of which has been graphically summarized in Figure 6.2. A more detailed explication of what has to be organized and what is going on in the overall system can be seen from Figure 14.5 concentrating on visual perception.

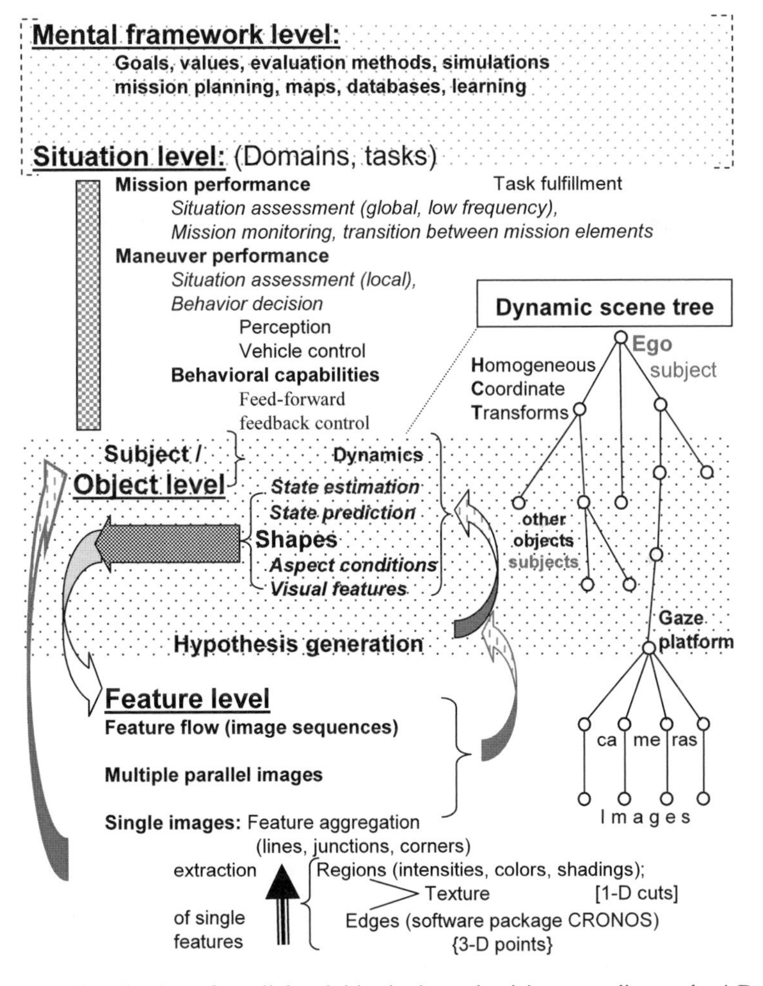

Figure 14.5. Visualization of parallel activities in dynamic vision according to the 4-D approach on four distinct levels: "Feature level" for bottom-up feature extraction and top-down "imagination"; "Object (subject) level" for recognition and extrapolation in time as well as computation in 3-D space; "Situation level" for fusing individual results on objects/subjects in the context of the mission to be performed and goals to be achieved; "Mental framework level" providing the knowledge background for the other levels.

This figure is a somewhat unconventional arrangement of terms, arrows for activity and information flow as well as some elements for knowledge representation like the dynamic scene tree on the right, which is the core element for scene understanding [D.Dickmanns 1997]. The homogeneous coordinate transformations (HCTs) represented by the edges of the graph link image features to internally represented 3-D objects moving over time according to some motion constraints (including the effects of control activity). The recursive estimation process with prediction-error feedback iterates the entries in the HCTs such that the sum of prediction errors squared is minimized. When egomotion (conventionally measured) is involved, this approach realizes motion stereointerpretation in a natural way. Unknown shape parameters can be iterated for each object observed by properly incrementing the number of state variables.

The tree as a whole – besides representing the individual object-to-feature mapping – thus also codes the geometric and shape aspects of the entire situation. Note that since motion constraints for mechanical systems are of second order, the velocity components of the objects observed may also be reconstructed. By determining the sensitivity (first-order derivative or "Jacobian") matrix between features measured and changes in object states, perspective inversion can be bypassed in this approach by a least-squares fit. All these aspects have been discussed in previous chapters down to implementation details; here, they are just recalled to show the interplay among the different components.

Bottom-up detection of single features is the starting point for vision (bottom of Figure 14.5). Combinations of features in a single image, from several images taken in parallel or from image sequences, allow coming up with object hypotheses represented directly in 3-D space and time. Looking at convergence properties and error statistics for each of the n parallel processes that track a single object, each allows perceiving "objects and their motion in 3-D space" (shaded area in center of the figure with object nodes at the end of branches of the scene tree). For each of these objects, the DOB contains the variables of the HCTs that link the object node with the features in the images (= nodes at bottom right). These variables together with the code for computing HCT and for generating 3-D shape from a few parameters allow "*imagination*", the process of generating virtual images from abstract mathematical representations (as in computer graphics). In connection with the models for dynamic motion, the evolution of scenes with several objects/subjects can be predicted, which is the process of forming expectations. Time histories of expectations generate representations of *maneuvers* needed for deeper understanding of motion processes. These *temporally more extended elements of knowledge representation* on a timescale different from "*state reconstruction here and now*" have been missing in many approaches in the past.

Keeping track of the evolution of trajectories of other objects/subjects stored allows recognizing their direction of motion; if parts of stereotypical maneuvers can be identified, such as the onset of a lane change or preparation for a turnoff maneuver, this is considered recognition of an intention of the other subject, and predictions of this maneuver allow computing expectations for deeper understanding of what is happening in the environment. This is part of situation assessment, preceding behavior decision and control output. The results of situation assessment can be stored in the DOB and logged for later (off-line) evaluation.

This part of the system (at the top of Figure 14.5) needs further expansion in the direction of learning, now that the actually needed parts for basic mission performance seem to be in place. Off line analysis of logged maneuvers and mission elements performed should allow adaptation and improvement of quality criteria and of developing adjustments in the subject's behavior (parameter selection and timing) taking time delays observed and other perturbations into account.

Figure 14.6 shows yet another visualization of the same scheme emphasizing the modularity developed with the network of capabilities, realized in object-oriented coding (see also Figures 3.6 and 3.28). In the top row, central decision orders a mission plan to be generated (top left) which is returned as a list of sequential mission elements. CD now activates the complex behavioral capabilities needed to perform the first mission element. Three of them are shown: "Follow road/lane", "Turnoff onto cross road", and "Follow sequence of (GPS) waypoints". Depending on the task of the first mission element, the proper mode is selected, probably with some priority parameters for the implementation level. The arrows emanating from each complex behavioral capability indicate which basic stereotypical capabilities (shown on the broad arrow between BDGA and BDL) have to be available to start the complex behavior. Their availability is checked each time before intended activation, so that actual malfunctions are detected and availability of this behavior is modified (adjusted) or even negated.

"Follow road/lane" thus needs "optimization of viewing behavior" (OVB) for gaze control, "road detection and tracking" (RDT) for road recognition, and "Road/lane running" (RLR) for vehicle guidance (dark solid arrows from top left of shaded area). "Turnoff onto crossroad" would need, beside the ones mentioned before, also "Crossroad detection and tracking" (CRDT). For "Following a sequence of waypoints" a second visual perception capability for "3-D surface recognition" (3-DS) is needed to avoid falling into ditches or other negative obstacles. (Results will be discussed briefly in Section 14.6.6.) A special capability for "Obstacle detection and tracking" (ODT) is shown necessary for going cross-country (fallen trees or sufficiently big rocks, *etc.*). As a new capability for locomotion (vehicle control), "Waypoint navigation" (WPN) is shown.

All these stereotypical capabilities have to rely more or less on "basic skills" (lower row of circles shown). This is again indicated by arrows (not exhaustive). Their availability is checked before the stereotypical capabilities can signal their availability to the next higher level. This is part of the safety concept introduced for the third-generation EMS vision system. The local "behavior decision" units BDGA and BDL (see darkened rectangles left and right) work with the rules for mode transition between behaviors. They also monitor the progress of behavior initiated and have to recognize irregularities from failures or perturbations. (This is another area needing more attention in future more reliable systems for practical applications of this new technology.)

The interested reader may find more details in [Maurer 2000, Gregor et al. 2002, Pellkofer et al. 2001, 2003; Pellkofer 2003; Siedersberger 2004]. The introduction and implementation of networks of capabilities is a rather late development, whose full potential has by far not yet been exploited. We will come back to this point in the Outlook at the end of the book.

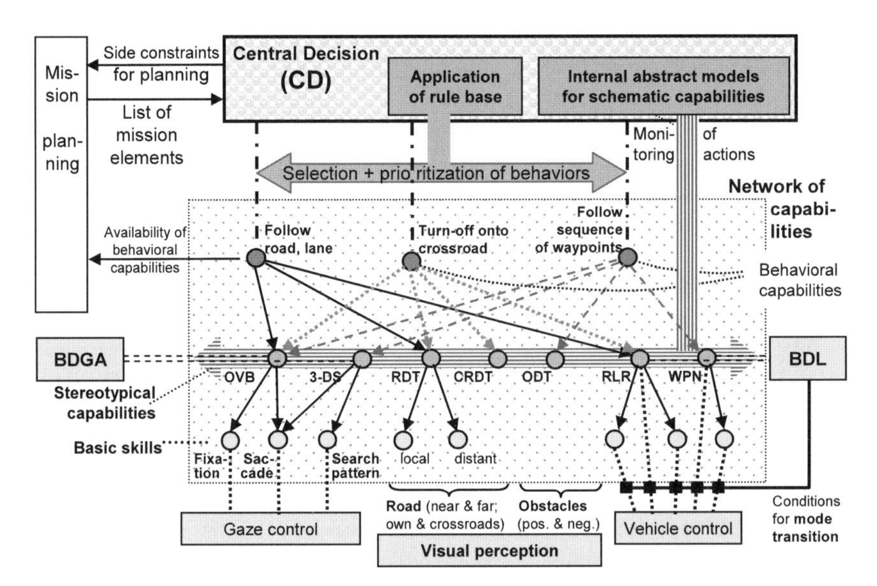

Figure 14.6. Control flow for realizing behaviors according to the approach developed with networks of capabilities for visual perception and motion control in object-oriented programming (see text for explanation) [Gregor *et al.* 2002, Pellkofer 2003, Siedersberger 2004]

14.5 Hardware Realization of Third-generation EMS Vision

Since progress in hardware development for computing and communication is still huge, not much space will be devoted to this point here; the latest hardware used was from the end of last century and is outdated by now. Figure 14.7 gives a summary of the system built from up to four Dual-Pentium® processors with clock rates of less than 1 GHz [Rieder 2000]. Synchronization and data exchange have been performed using an off-the-shelf communication network (Scalable Coherent Interface: SCI®) together with Fast Ethernet® (for initialization). Dynamic knowledge representation (DKR) via scene tree makes the latest estimated states of all processes available to all units (DOB as part of it, see Figure 14.4) via SCI (top row in Figure 14.7).

Three of the four Dual-Pentium PCs were devoted to image feature extraction and visual perception (PC 1 to 3, top left); the fourth one, dubbed "Behavior PC" had two subsystems connected to it as interfaces to measurement and actuator hardware. The "gaze subsystem" for gaze control (GC) receives data from mechanical angle and inertial angular rate measurements of the gaze platform at high frequencies; it commands the actual output to the torque actuators in pan and tilt, which can work at update rates up to 500 Hz (2 ms cycle time for smooth pursuit).

The "vehicle subsystem" for vehicle control (VC, lower right in Figure 14.7) receives all conventional measurement data from the sensors on the vehicle at least at

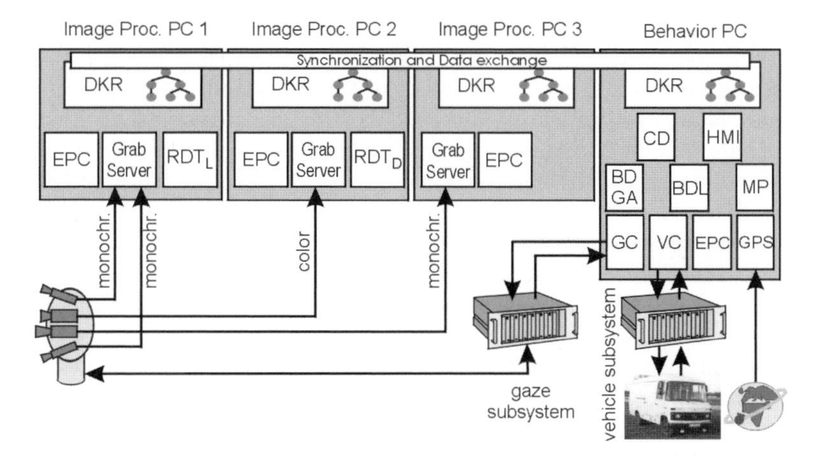

Figure 14.7. Realization of EMS vision on a cluster of four dual-processor PCs plus two subsystems as interfaces to hardware for gaze and vehicle control [Rieder 2000] (see text)

video rate. It also forms the interface to the actuators of the vehicle for steering, throttle position, and braking. The process VC, running on the Behavior PC, implements the basic behavioral skills. The decision processes CD, BDGA, and BDL perform situation assessment based on the data in the DOB, the conventional measurement data, and on data from the Global Positioning System (GPS) arriving once a second for navigation (see corresponding blocks displayed at right in Figure 14.7). MP stands for mission planning (running off–line, usually), and HMI realizes the "human–machine interface" through which the operator can communicate with the system. An embedded demon process (EPC) allows programming and controlling the overall system via fast Ethernet from one external resource.

14.6 Experimental Results of Mission Performance

Before we start discussing results achieved with the third-generation vision system, an early test made with hardware-in-the-loop simulation for preparing these advanced perception processes will be discussed.

14.6.1 Observing a Maneuver of Another Car

The challenge is not just to get a best estimate for the actual state of another vehicle observed, but to find out how much deeper understanding of motion and control processes can be gained by the 4-D approach to dynamic vision. To keep things simple in this first step, the camera observing the other vehicle was stationary with an elevation of 3 m above the ground; it followed the center of the projected moving vehicle without errors (ideal fixation, see Figure 2.17). Vehicle motion and projected shape, according to a shape model as given in Figure 2.15, were simulated on a separate computer, and feature positions were superimposed with noise. The vehicle started straight ahead from rest at the position labeled 0 in Figure 2.17 with an acceleration of 2 m/s² until a speed of 5 m/s (18 km/h) was reached (*dotted gray pulse at the extreme left in Figure 14.8*); this maneuver lasted 2.5 s, which at an estimation rate of 20 times per second yields 50 steps (dimension of the horizontal coordinate in the figure). The solid curves shown in the left graph give the speed and acceleration of the vehicle reconstructed from feature data taking not only the translational degrees of freedom but also the rotational ones as variables to be observed (not shown here). As shown at the top of Figure 2.17, due to changing aspect conditions, some features disappear and others newly appear during this acceleration period. It can be clearly seen that there is a time lag of several cycles until the start of the motion is perceived. Acceleration is reconstructed rather noisy, but "perceived" speed increases almost linearly, although with constant time delay compared to motion generation (this would be the diagonal line from (0, 0) to (50, 5). The instantaneous drop in simulated acceleration at coordinate 50 is, of course,

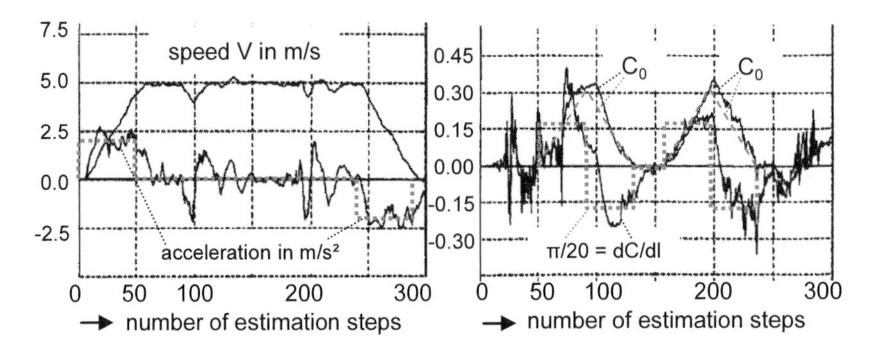

Figure 14.8. Maneuver recognition on an oval track (see Figure 2.17) in HIL-simulation with a real camera in front of a monitor. The image displayed was computer generated with the optical axis from a fixed location 3 m above the ground always fixated on the cg of the car displayed in the center of the image (ideal tracking; see text).

also perceived only with a corresponding time delay; keep in mind that in perceptual reconstruction from vision data, speed is derived from position changes and acceleration from speed changes (buried in the recursive estimation process)!

The vehicle then drives the oval track at constant speed $V = 5$ m/s [0.25 m/(time step)] and with two double pulses in steering rate (lateral control) as shown by the dotted gray polygon in the right-hand part of the figure. The first pulse extends over 40 time steps (50 to 90, = 2 s) with a magnitude of curvature change rate per arc length of $dC/dl = +\pi/20$; it will ideally turn vehicle orientation from 0 to 90° (neglecting the difference between trajectory and vehicle heading changes, see Figure 10.6). The maximal curvature at this point is $C_{max} = 0.25 \cdot 40 \cdot \pi/20 = \pi/10$ m^{-1} (corresponding to a turn radius of a little more than 3 m). The second part of the double pulse then immediately starts decreasing curvature back to zero (achieved at time point 130, see Figure 2.17). There it is seen that during this turn the aspect conditions of the vehicle have changed completely; it was visible from the right side initially, right from the front at point 100, and for the new straight section (130 to 157), it is seen from the left. When looking at the vehicle from the front (around 105) or from the rear (around 190), it can be seen from the left subfigure of 14.8 that longitudinal motion is poorly conditioned for tracking; only the change in size and vertical position of the cg in the image contain information depending on range. Correspondingly, estimated values of speed and longitudinal acceleration show large perturbations resulting from noise in measured feature positions in these phases. Reconstruction of the curvature of the trajectory driven is not too bad, though. The *dashed triangular shapes in gray* are the integrals of the dotted input curve, *i.e.*, they represent the "ideal" curvature at the reference location of the vehicle; the solid lines representing the estimated curvatures from recursive estimation are not too far off in both turns; the second turn runs from time point 157 to 237. Keeping the relation between curvature and steering angle in mind (see Figure 10.6, lower equation), a constant steering rate of the vehicle can easily be concluded, at least for the second turn which started with all initial transients settled.

These results from around 1990 showed rather early that the 4-D approach had its merits both in estimating actual states and in recognizing maneuver elements at the same time but on different timescales.

14.6.2 Mode Transitions Including Harsh Braking

The maneuver tested here is one that is not, usually, part of a regular mission plan, but may occur every now and then during mission performance on high-speed roads. The test vehicle VaMP had received a request to cruise at a set speed of 140 km/h; driving at 40 km/h when registration started (second graph from top), the selected driving mode was "acceleration" (V1, top graph left, Figure 14.9). Due to increasing speed and air drag, the acceleration level decreased with time from about 2 to less than 1 m/s^2 (second graph from bottom). Shortly after 20 s, a vehicle had been detected in the same lane at about 130 m ahead (center graph). Its speed seemed to be rather low and, of course, unreliable due to the usual transients in state estimation after detection.

The system reacted with a rather harsh initial braking impulse at around 23 s (lower two graphs). It was initiated by a mode change on the situation assessment level (*top curve*), which resulted from brake pressure in the feed-forward component (*dash-dotted in lower graph*). The actual pressure buildup shows the usual lag. The resulting deceleration of the vehicle went up to a magnitude of - 8 m/s^2 (second from bottom), reducing speed to about 80 km/h (second from top).

Due to the resulting pitching motion from braking, speed estimation for the vehicle in front showed a strong transient with overshoot (second from top, lower curve). The commanded pressure in the brake system went down to zero for a short time. However, with the transient motion of the subject vehicle vanishing, the difference in the subject's speed and the speed of the vehicle ahead was perceived, and the system decided to regulate this (relatively small) difference by feedback control which also had to realize the intended distance for convoy driving [Bruedigam 1994; Maurer 2000; Siedersberger 2004]. The bottom graph shows the brake pres-

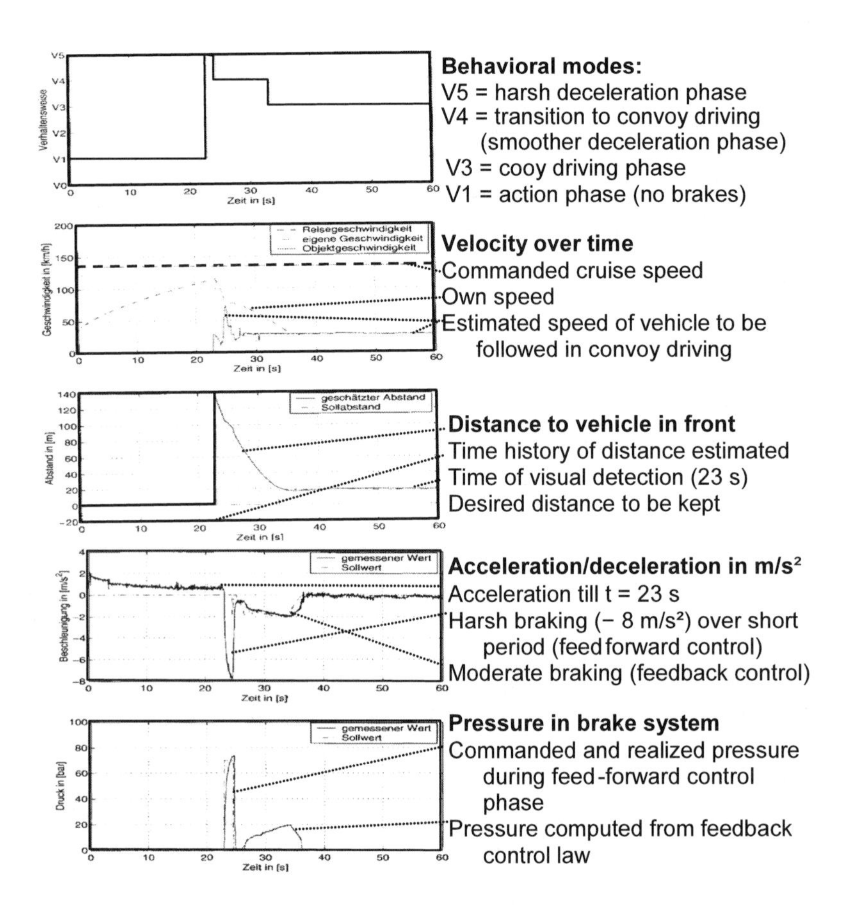

Figure 14.9. Braking maneuvers of test vehicle VaMP (Mercedes 500 SEL) in two phases for smooth transition from v ≈ 120 km/h to convoy driving at V ≈ 35 km/h and about 20 m longitudinal distance with three mode changes (top): (1) from acceleration to harsh braking at time 23 s (second graph from bottom and from top showing deceleration and speed), (2) from harsh to moderate braking at around 24 s, and (3) to convoy driving at about 33 s. The control variable is brake pressure (bottom graph); the controlled state is the distance to the vehicle in front (center graph) [Maurer 2000].

sure as a control output thus determined, and the graph above it shows the commanded (*dash-dotted*) and the realized deceleration (around − 0.2 g).

During this smooth driving phase (mode V4, top graph), the estimated speed of the vehicle ahead becomes stable (lower curve, second graph from top); this speed determines the distance to be kept from the vehicle in front for convoy driving.The transition into this mode V3 (top) occurs at about 33 s. Note that the remaining error in distance for convoy driving is eliminated in this new mode. At around 37 s, the stationary new driving state is reached with braking activities vanishing (acceleration zero for t > 37 s, second graph from bottom).

14.6.3 Multisensor Adaptive Cruise Control

This function is not for fully autonomous driving but for extending and improving an assistance system for distance keeping in convoy driving. These types of systems are on the market under various names for certain premium cars (class name "ACC"). They, usually, rely on radar as the distance sensor. The human driver has to control the vehicle in the lateral degrees of freedom all the time; as an extension to conventional automatic "cruise control" (CC) at constant speed on a free stretch of road, these systems allow braking at soft levels for distance keeping from a vehicle in front.

Though this capability has been demonstrated with the vision system of UniBwM in the framework of the Prometheus project in 1991 (on separate test track, demo in Torino) and since 1993 (in public traffic) already, industry had decided to base the system for market introduction on radar measurements. This approach has all-weather capability, it requires less computing power onboard, and less software development for interpretation. At that time, a single, specially developed vision system for realizing this function reliably would have cost about as much as a premium car. Therefore, the disadvantages of radar: low resolution, small field of view, relatively many false alarms, and problems with multipath propagation, had been accepted initially. To people believing in vision systems, it has always been only a question of time until vision would be added to these systems for more reliable performance. Above all, radar is not capable of recognizing the road, in general. Figure 14.10 shows a concept studied with industry in the late 1990s. Figure 11.26 had schematically shown the advantage of object tracking by a joint radar and vision system. The role of the vision part has already been discussed in Section 11.3.4.1. Here, a survey of the overall system and of system integration will be given. Radar was installed underneath the front license plate (center

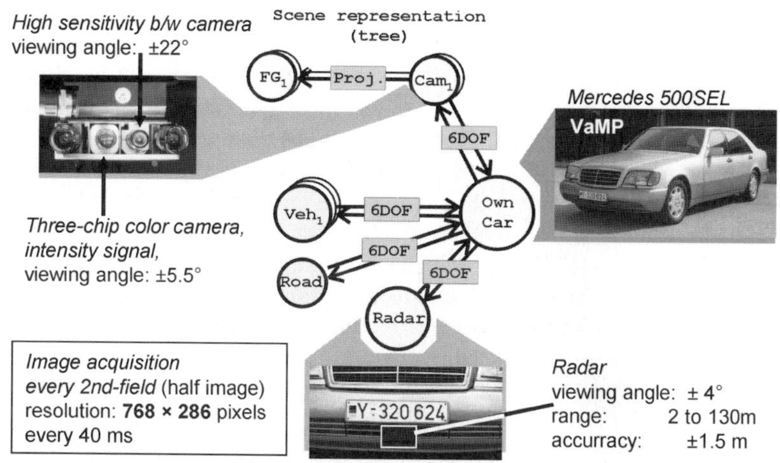

Figure 14.10. Hybrid adaptive cruise control with radar (center bottom) and bifocal vision (top left): System survey with hardware components and condensed scene tree (center, multiple cameras, framegrabbers, and other vehicles)

bottom of Figure 14.10); its range of operation was 2 to 130 m, its viewing angle ± 4°. The position and orientation relative to the car body was represented in the scene tree by three translations and three rotations [six degrees of freedom (DOF)] as for all other objects of relevance (cameras, other vehicles, and the road; see center of Figure 14.10). The two cameras used (top left) had fields of view of 5.5° and 22°; their position was in front of the rearview mirror behind the top center of the windshield. Figure 11.25 shows a typical pair of images analyzed by the 4-D approach.

Vehicle detection was performed by radar; in all regions of special interest, vision looked for features indicative of vehicle candidates. At the same time, lanes were tracked, and both horizontal and vertical curvatures of the road were determined. It turned out that for larger ranges covered by radar, recognition of even small vertical curvatures was important for good tracking results. It has been demonstrated that vision was able to eliminate all candidates based on false alarms from radar and that lateral positions relative to the lanes could be determined precisely. The system automatically switched the reference lane when the driver crossed the lane boundary during lane change. It marked the lower part of the reference vehicle for distance keeping by a red rectangle and of other vehicles tracked by blue ones.

Lane markings recognized were marked by short line elements according to the horizontal curvature model estimated. Vertical curvature was displayed by three yellow bars, the outer two of which showed the perceived horizon, and the center one indicating the actual vertical surface position above or below the planar value. Snapshots (such as Figure 11.25) and figures are hardly able to give a vivid impression of the results achieved. Video films of tests during daylight and night driving are available for demonstration [Siedersberger 2003; Hofmann 2004].

14.6.4 Lane Changes with Preceding Checks

This maneuver is a standard one for driving on multilane high-speed roads. The nominal control time history for applying the skill "lane change" has been discussed in Section 3.4.5.2. Figure 3.27 shows the effect of maneuver time on the evolving trajectory in simulation. An actual lane change with test vehicle VaMP is shown in Figure 14.11 with standard feed-forward and superimposed feedback control for a nominal maneuver time of 8 s. The top left graph allows recognizing the nominal lane change maneuver without a phase of driving straight ahead at the center. Feedback control has been kept running all the time; at the start of the maneuver, the reference for feedback control was modified by adding the values according to the nominal trajectory of the (feed-forward) maneuver to the position of the lane center (which is usually the reference).

It can be seen that the additive corrective commands deform the rectangular pulse-shape considerably (top left). The yaw errors at the beginning and at the end of the maneuver (lower left in Figure 14.11) lead to especially larger increments. The lateral offsets (lower right) from the nominal trajectory never exceed about 25 cm, which is considered good performance for a lane width of 3.75 m and a vehicle width of less than 2 m.

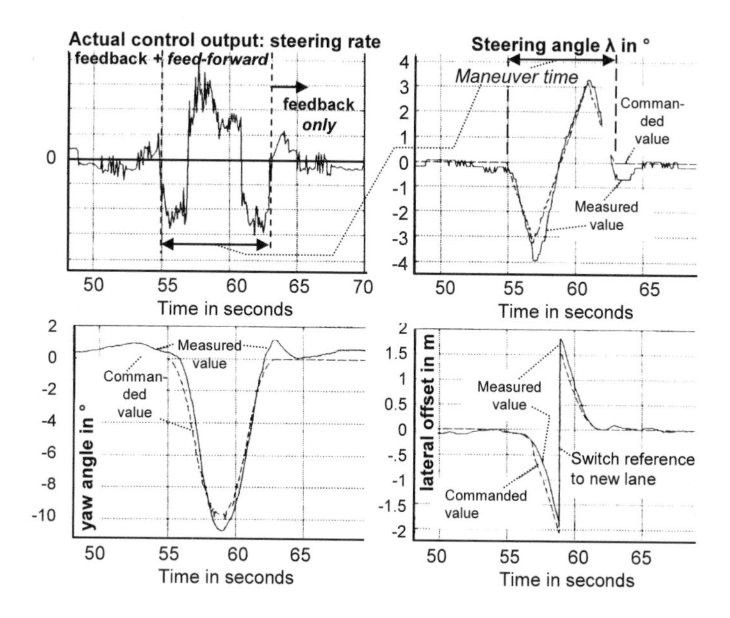

Figure 14.11. Real lane change maneuver with test vehicle VaMP: The actual control output (top left) was the result of both a feed-forward component according to the nominal maneuver and a feedback component trying to drive the difference between actual and nominal trajectory to zero. These differences are displayed in the bottom part: Left the yaw angle and right the lateral offset from the center of the reference lane.

Before such a maneuver can be initiated, it has to be checked by perception that the neighboring lane to be changed to is free in the rear, to the side, and in front. This is part of situation assessment when a lane change is intended. Vehicles with higher speed ΔV_R coming from behind will close the gap d_R in the time span $\Delta T_g = d_R/\Delta V_R$. If this time is larger than the maneuver time for lane change (possibly plus some safety margin), the maneuver is safe from the rear. Checking whether there is another vehicle to the side in the intended lane is a more involved challenge: Either there are special sensors for this purpose, such as laser range finders or special radar, or gaze direction (of the front and the rear cameras) can be changed sufficiently to check the lane briefly by a quick saccade, if required. Another possibility is to keep track of all vehicles leaving the rear field of view and to check whether all these vehicles have reappeared in the front field of view. However, this yields 100% correct results only for a single neighboring lane in the direction of the intended change; otherwise, a vehicle might have changed from the second neighboring lane to the immediate one in the meantime. This approach for situation assessment, minimizing sensor hardware needed, has been successfully tested when driving in the center lane of a three-lane highway in the final demonstration of the Prometheus project on Autoroute 1 (Paris) in 1994 in public traffic [Kujawski 1995].

During the long-distance test drive Munich–Odense in 1995, more than 400 lane changes were performed autonomously after the safety driver triggered the maneuver. Figure 14.12 shows the statistics of a period of about half an hour, displayed

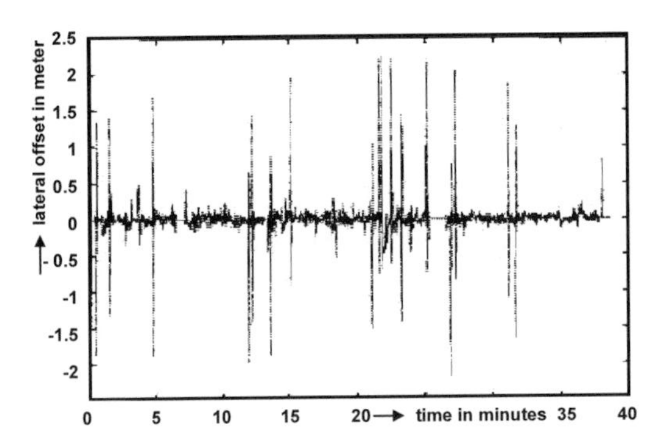

Figure 14.12. Statistic of lane changes performed on the long-distance trip to Odense in November 1995 with test vehicle VaMP. Within about half an hour, 16 lane changes have been performed autonomously after triggering by the safety driver; the reference lane center is switched when the vehicle cg crosses the lane (offset of half the actual lane width)

by the data logged for lateral offsets from the lane center. Lane changes are easily recognized as peaks of magnitude of about half the lane width, since after reaching this boundary, the reference lane is switched, and the sum of the positive and negative peak always yields the actual lane width in the region of the maneuver. As can be counted directly, 16 lane changes were made within about half an hour; those appearing as close pairs represent passing slower vehicles and immediately returning to the old lane.

14.6.5 Turning Off on Network of Minor Unsealed Roads

Based on the results of [Mueller 1995] discussed in Chapter 10, the capability of turning off in the new concept of capability networks has been developed by Luetzeler (2002), Gregor (2002), Pellkofer (2003), and Siedersberger (2004). It has been demonstrated with the test vehicle VaMoRs in various environments. The example discussed here is from a proving ground in the southern part of Germany. The roads are all unsealed and have a gravel surface. Edge detection has been done with CRONOS-software. Figure 14.13 shows three snapshots from the telecamera several frames apart containing a saccade.

Since the turnoff is intended to the left, the crossroad on the left-hand side (left image) and at the crossing (right image) are viewed alternately by saccadic vision. Search regions for edge feature extraction are marked by horizontal and vertical line elements. They cannot be seen in the center image indicating that feature ex-

Figure 14.13. Tele-images during saccadic vision while approaching a crossroad; the center image during a saccade is not evaluated (missing indicated search paths)

traction is suppressed during saccadic motion. [In this case, the saccade was performed rather slowly and lighting conditions were excellent so that almost no motion blur occurred in the image (small shutter times), and feature extraction could well have been done.] The white curve at the left side of the road indicates that the internal model fits reality well.

The sequence of saccades performed during the approach to the crossing can be seen from the sequence of graphs in Figure 14.14 (a) and (b): The saccades are started at time ≈ 91 s; at this time, the crossroad hypothesis has been inserted in the scene tree by mission control expecting it from coarse navigation data (object ID for the crossroad was 2358, subfigure (e). At that time, it had not yet been visually detected. Gaze control computed visibility ranges for the crossroad [see graphs (g) and (h)], in addition to those for the road driven [graphs (i) and (j), lower right]. Since these visibility ranges do not overlap, saccades were started.

Eleven saccades are made within 20s (till time 111). The "saccade bit" (b) signals to the rest of the system that all processes should not use images when it is "1"; so they continue their operation based only on predictions with the dynamic models and the last best estimates of the state variables. Which objects receive attention can be seen from graph [(e) bottom left]: Initially, it is only the road driven; the wide-angle cameras look in the near (local, object ID = 2355) and the telecamera in the far range (distant, ID number 2356). When the object crossroad is inserted into the scene tree (ID number 2358) with unknown parameters width and angle (but with default values to be iterated), determination of their precise values and of the distance to the intersection is the goal of performing saccades.

At around t = 103 s, the distance to the crossroad starts being published in the DOB [graph (f), top right]. During the period of performing saccades (91 – 111), the decision process for gaze control BDGA continuously determines "best viewing ranges" (VR) for all objects of interest [graphs (g) to (j), lower right in Figure 14.14]. Figure 14.14 (g) and (h) indicate, under which pan (platform yaw) angles the crossroad can be seen [(g) for optimal, (h) for still acceptable mapping]. Graph (i) shows the allowable range for gaze direction so that the road being driven can be seen in the far look-ahead range (+2° to −4°), while (j) does the same for the wide-angle cameras (± 40°). During he approach to the intersection the amplitude of the saccades increases from 10 to 60° [Figure 14.14 (a), (g), (h)].

For decision-making in the gaze control process, a quality criterion "information gain" has been defined in [Pellkofer 2003]; the total information gain by a visual mode takes into account the number of objects observed, the individual information gain through each object, and the need of attention for each object. The procedure is too involved to be discussed in detail here; the interested reader is referred to the original work well worth reading (in German, however). The evolution of this criterion "information input" is shown in graphs (c) and (d). Gaze object 0 (road nearby) contributes a value of 0.5 (60 to 90 s) in roadrunning, while gaze object 1 (distant road) contributes only about 0.09 [Figure 14.14 (d)]. When an intersection for turning off is to be detected, the information input of the tele-camera jumps by about a factor of 4, while that of the wide-angle cameras (road nearby) is reduced by ~ 20% (at t = 91 s). When the crossroad is approached closely, the road driven loses significance for larger look-ahead distances and gaze direction for crossroad tracking becomes turned so much that the amplitudes of saccades would

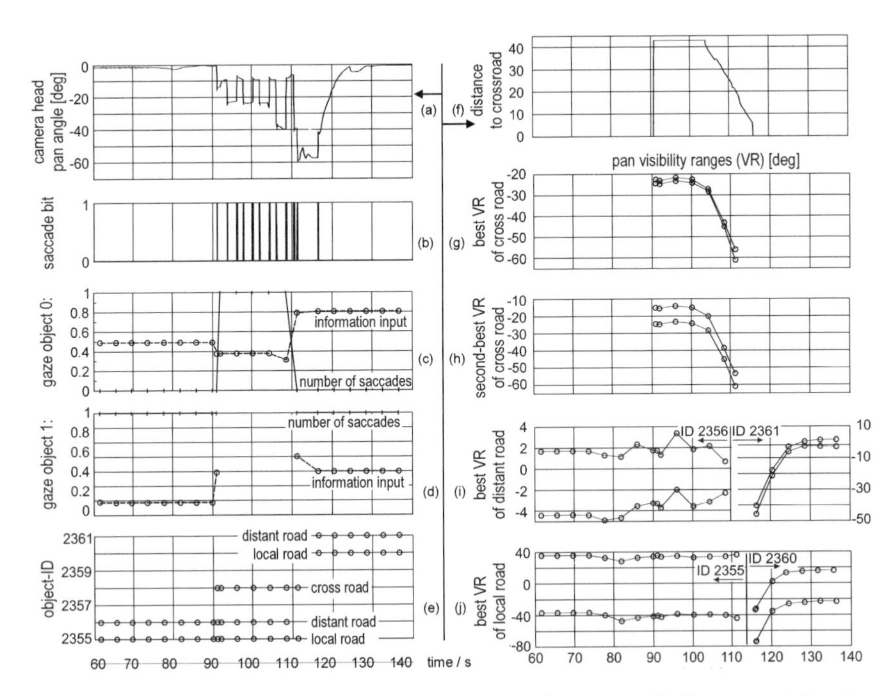

Figure 14.14. Complex viewing behavior for performing a turnoff after recognizing the crossroad including its parameters: width and relative orientation to the road section driven (see text)

have to be very large. At the same time, fewer boundary sections of the road driven in front of the crossing will be visible (because of approaching the crossing) so that the information input for the turnoff maneuver comes predominantly from the crossroad and from the wide-angle cameras in the near range (gaze object 0). At around 113 s, therefore, the scene tree is rearranged, and the former crossroad with ID 2358 becomes two objects for gaze control and attention: ID 2360 is the new local road in the near range, and ID 2361 stands for the distant road perceived by the telecamera, Figure 14.14 (e). This re-arrangement takes some time (graphs lower right), and the best viewing ranges to the former crossroad (now the reference road) make a jump according to the intersection angle. While the vehicle turns into the crossroad, the small field of view of the telecamera forces gaze direction to be close to the new road direction; correspondingly, the pan angle of the cameras relative to the vehicle decreases while staying almost constant relative to the new reference road, *i.e.*, the vehicle turns underneath the platform head [Figure 14.14 (i) and (a)]. On the new road, the information input from the near range is computed as 0.8 [Figure 14.14 (c)] and that from the distant road as 0.4 [Figure 14.14 (d)]. Since the best visibility ranges for the new reference road overlap [Figure 14.14 (i) and (j)], no saccades have to be performed any longer.

Note that these gaze maneuvers are not programmed as a fixed sequence of procedures, but that parameters in the knowledge base for behavioral capabilities as well as the actual state variables and road parameters perceived determine how the maneuver will evolve. The actual performance with test vehicle VaMoRs can be seen from the corresponding video film.

14.6.6 On- and Off-road Demonstration with Complex Mission Elements

While the former sections have shown single, though complex behavioral capabilities to be used as maneuvers or mission elements, in this section, finally, a short mission for demonstration is discussed that requires some of these capabilities. The mission includes some other capabilities in addition, too complex to be detailed here in the framework of driving on networks of roads. The mission was the final demonstration in front of an international audience in 2001 for the projects in which expectation-based, multifocal, saccadic (EMS) vision has been developed over 5 years with a half dozen PhD students involved.

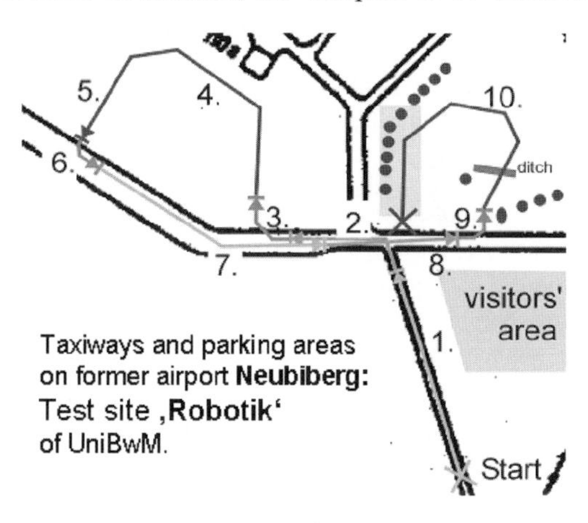

Figure 14.15. Schedule of the mission to be performed in the final demonstration of the project, in which the third-generation visual perception system according to the 4-D approach, EMS vision, has been implemented (see text)

Figure 14.15 shows the mission schedule to be performed on the taxiways and adjacent grass surfaces of the former airport Neubiberg, on which UniBwM is located. The start is from rest with the vehicle casually parked by a human on a single-track road with no lane markings. This means that no special care has been taken in positioning and aligning the vehicle on the road. Part of this road is visible in Figure 14.16 (right, vertical center). The inserted picture has been taken from the position of the ditch in Figure 14.15 (top right); the lower gray stripe in Figure 14.16 is from the road between labels 8 and 9.

In phase 1 (see digit with dot at lower right), the vehicle had to approach the intersection in the standard roadrunning mode. On purpose, no digital model of the environment has been stored in the system; the mission was to be performed relying on information such as given to a human driver. At a certain distance in front of the intersection (specified by an imprecise GPS waypoint), the mission plan ordered taking the next turnoff to the left. The vehicle then had to follow this road across the T-junction (2); the widening of the road after some distance should not interfere with driving behavior. At point 3, a section of cross-country driving, guided by widely spaced GPS waypoints was initiated. The final leg of this route (5) would intersect with a road (not specified by a GPS waypoint!). This road had to be recognized by vision and had to be turned onto to the left through a (drivable)

Figure 14.16. VaMoRs ready for mission demonstration 2001. The vehicle and road sections 1 and 8 (Figure 14.15) can be seen in the inserted picture. Above this picture, the gaze control platform is seen with five cameras mounted; there was a special pair of parallel stereo cameras in the top row for using hard- and software of Sarnoff Corporation in a joint project 'Autonav' between Germany and the USA.

shallow ditch to its side. This perturbed maneuver turned out to be a big challenge for the vehicle.

In the following mission element, the vehicle had to follow this road through the tightening section (near 2) and across the two junctions (one on the left and one on the right). At point 9, the vehicle had to turnoff to the left onto another grass surface on which again a waypoint-guided mission part had to be demonstrated. However, on the nominal path, there was a steep deep ditch as a negative obstacle, which the vehicle was not able to traverse. This ditch had to be detected and bypassed in a proper manner, and the vehicle was to return onto the intended path given by the GPS waypoints of the original plan (10).

Except for bypassing the ditch, the mission was successfully demonstrated in 2001; the ditch was detected and the vehicle stopped correctly in front of it. In 2003, a shortened demo was performed with mission elements (1, 8, 9, and 10) and a sharp right turn from 1 to 8. In the meantime, the volume of the special processor system (Pyramid Vision Technology) for full frame-rate and real-time stereo perception had shrunk from a volume of about 30 liters in 2001 to a plug-in board for a standard PC (board size about 160 × 100 mm). Early ditch detection was achieved, even with taller grass in front of the ditch partially obscuring the small image region of the ditch, by combining the 4-D approach with stereovision. Photometric obstacle detection with our vision system turned out to be advantageous for early detection; keep in mind that even a ditch 1 m wide covers a very small image region from larger distances for the aspect conditions given (relatively low elevation above the ground). When closing in, stereovision delivered the most valuable information. The video "Mission performance" fully covers this abbreviated mission with saccadic perception of the ditch (Figure 14.3) and avoiding it around the right-hand corner, which is view-fixated during the initial part of the maneuver [Pellkofer 2003; Siedersberger 2004, Hofmann 2004]. Later on, while returning onto the trajectory given by given by GPS waypoints, the gaze direction is controlled according to Figure 14.2.

15 Conclusions and Outlook

Developing the sense of vision for (semi-) autonomous systems is considered an animation process driven by the analysis of image sequences. This is of special importance for systems capable of locomotion which have to deal with the real world, including animals, humans, and other *subjects*. These subjects are defined as capable of some kind of perception, decision–making, and performing some actions. Starting from bottom-up feature extraction, tapping knowledge bases in which generic knowledge about 'the world' is available leads to the 'mental' construction of an internal spatiotemporal (4-D) representation of a framework that is intended to duplicate the essential aspects of the world sensed.

This internal (re-)construction is then projected into images with the parameters that the perception and hypothesis generation system have come up with. A model of perspective projection underlies this "imagination" process. With the initial internal model of the world installed, a large part of future visual perception relies on feedback of prediction errors for adapting model parameters so that discrepancies between prediction and image analysis are reduced, at best to zero. Especially in this case, but also for small prediction-errors the process observed is supposed to be understood.

Bottom-up feature analysis is continued in image regions not covered by the tracking processes with prediction-error feedback. There may be a variable number N of these tracking processes running in parallel. The best estimates for the relative (3-D) state and open parameters of the objects/subjects hypothesized for the point in time "now" are written into a "dynamic object database" (DOB) updated at the video rate (the short-term memory of the system). These object descriptions in physical terms require several orders of magnitude less data than the images from which they have been derived. Since the state variables have been defined in the sense of the natural sciences/engineering so that they fully decouple the future evolution of the system from past time history, no image data need be stored for understanding temporal processes. The knowledge elements in the background database contain the temporal aspects from the beginning through dynamic models (differential equation constraints for temporal evolution).

These models make a distinction between state and control variables. State variables cannot change at one time, they have to evolve over time, and thus they are the elements for continuity. This temporal continuity alleviates image sequence understanding as compared to the differencing approach, after having analyzed consecutive single images bottom-up first, favored initially in computer science and AI.

Control variables, on the contrary, are those components in a dynamic system that can be changed at any time; they allow influencing the future development of the system. (However, there may be other system parameters that can be adjusted under special conditions: For example, at rest, engine or suspension system parameters may be tuned; but they are not control variables steadily available for system control.) The control variables thus defined are the central hub for intelligence. *The claim is that all "mental" activities are geared to the challenge of finding the right control decisions.* This is not confined to the actual time or a small temporal window around it. With the knowledge base playing such an important role in (especially visual) perception, expanding and improving the knowledge base should be a side aspect for any control decision. In the extreme, this can be condensed into the formulation that intelligence is the mental framework developed for arriving at the best control decisions in any situation.

Putting control time histories as novel units into the center of natural and technical (not "artificial") intelligence also allows easy access to events in and maneuvers on an extended timescale. Maneuvers are characterized by specific control time histories leading to finite state transitions. Knowledge about them allows decoupling behavior decision from control implementation without losing the advantages possible at both ends. Minimal delay time and direct feedback control based on special sensor data are essential for good control actuation. On the other hand, knowledge about larger entities in space and time (like maneuvers) are essential for good decision-making taking environmental conditions, including possible actions from several subjects, into account. Since these maneuvers have a typical timescale of seconds to minutes, the time delays of several tenths of a second for grasping and understanding complex situations are tolerable on this level. So, the approach developed allows a synthesis between the conceptual worlds of "Cybernetics" [Wiener 1948] and "Artificial Intelligence" of the last quarter of last century.

Figure 15.1 shows the two fields in a caricaturized form as separate entities. Systems dynamics at the bottom is concentrated on control input to actuators, either feed-forward control time histories from previous experience or feedback with direct coupling of control to measured values; there is a large gap to the *artificial intelligence* world on top. In the top part of the figure, arrows have been omitted for immediate reuse in the next figure; filling these in mentally should pose no problem to the reader. The essential part of the gap stems from neglecting temporal processes grasped by differential equations (or transition matrices as their equivalent in discrete time). This had the fundamental difference between control and state variables in the real world be mediated away by computer states, where the difference is absent. Strictly speaking, it is hidden in the control effect matrix (if in use).

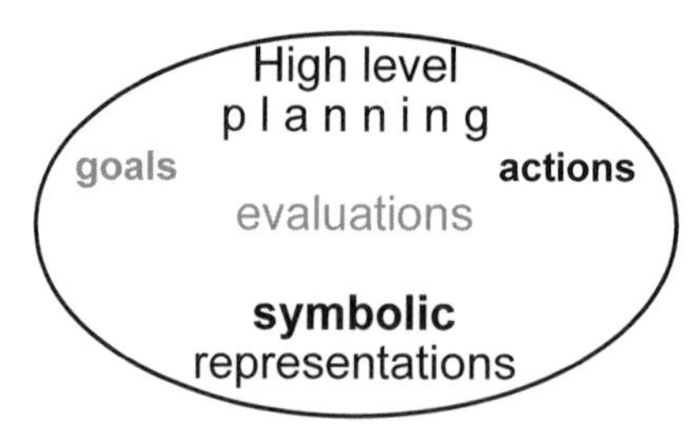

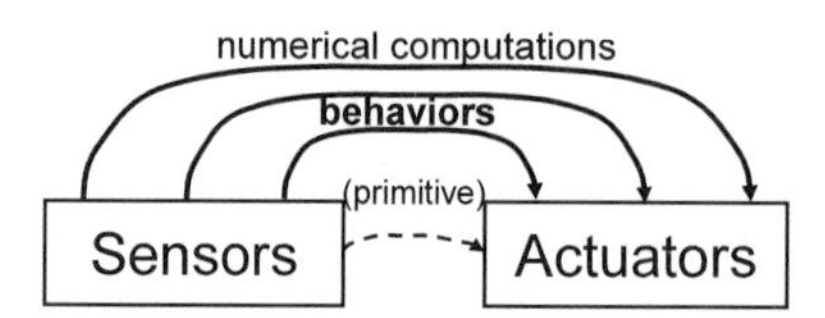

Figure 15.1. Caricature of the separate worlds of system dynamics (bottom) and Artificial Intelligence (top)

Figure 15.2 is intended to show that much of the techniques developed in the two separate fields can be used in the unified approach; some may even need no or very little change. However, an interface in common terminology has to be developed. In the activities described in this book, some of the methods needed for the synthesis of the two fields mentioned have been developed, and their usability has been demonstrated for autonomous guidance of ground vehicles. However, very much remains to be done in the future; fortunately, the constraints encountered in our work due to limited computing power and communication bandwidth are about to vanish, so that prospects for this technology look bright.

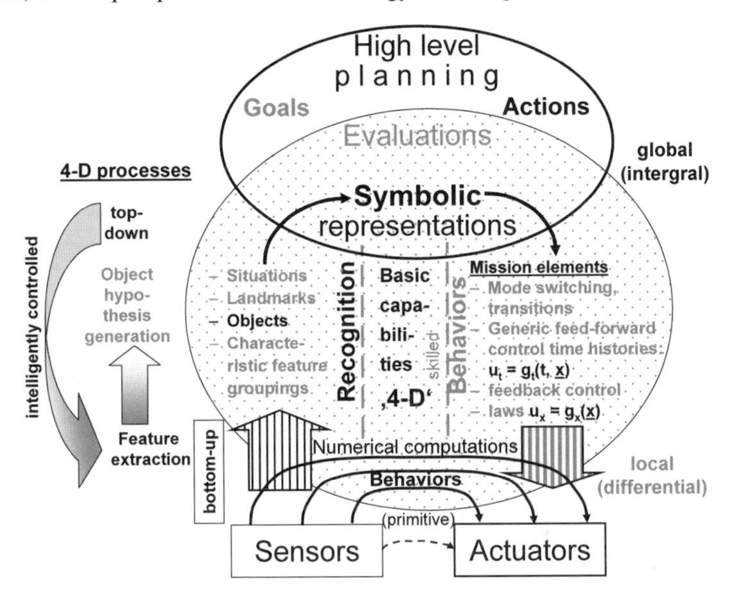

Figure 15.2. The internal 4-D representation of 'the world' (central blob) provides links between the 'systems dynamics' and the AI approach to intelligence in a natural way. The fact that all 'measurement values' derived from vision have no direct physical links to the objects observed (no wires, only light rays) enforces the creation of an 'internal world'.

Taking into account that about 50 million ground vehicles are built every year and that more than 1 million human individuals are killed by ground vehicles every year worldwide, it seems mandatory that these vehicles be provided with a sense of vision allowing them to contribute to reducing the latter number. The ideal goal of zero death-toll seems unreachable (at least in the near future) and is unrealistic for open-minded individuals; however, this should not be taken as an excuse for not developing what can be achieved with these new types of vehicles with a sense of vision, on any sensor basis what-ever.

Providing these vehicles with real capabilities for perceiving and understanding motion processes of several objects and subjects in parallel and under perturbed conditions will put them in a better position to achieve the goal of a minimal accident rate. This includes recognition of intentions through observation of onsets of maneuvering, such as sudden lane changes without signaling by blinking. In this case, a continuous buildup of lateral speed in direction of one's own lane is the critical observation. To achieve this "animation capability", the knowledge base has to include "maneuvers" with stereotypical trajectories and time histories. On the other hand, the system also has to understand what typical standard perturbations due to disturbances are, reacting to it with feedback control. This allows first, making distinctions in visual observations and second, noticing environmental conditions by their effects on other objects/subjects.

Developing all these necessary capabilities is a wide field of activities with work for generations to come. The recent evolution of the capability network in our approach [Siedersberger 2004; Pellkofer 2003] may constitute a starting point for more general developments. Figure 15.3 shows a proposal as an outlook; the part realized is a small fraction on the lower levels confined to ground vehicles. Especially the higher levels with proper coupling down to the engineering levels of automotive technology (or other specific fields) need much more attention.

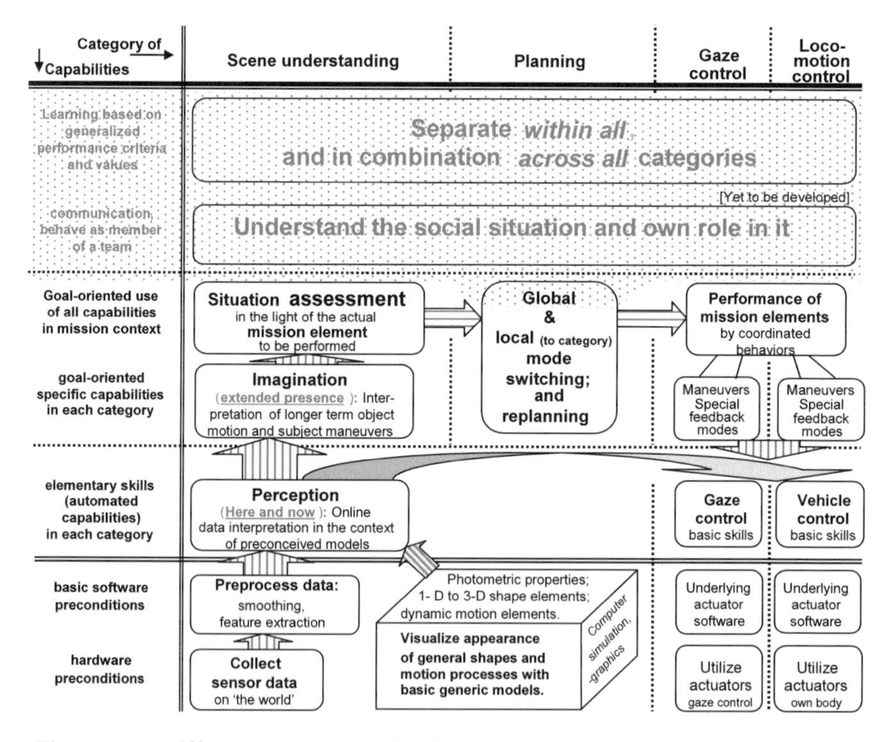

Figure 15.3. Differentiation of capability levels (vertical at left side) and categories of capabilities (horizontal at top): Planning happens at the higher levels only in internal representations. In all other categories, both hardware available (lowest level) and ways of using it by the individual play an important role. The uppermost levels of social interaction and learning need more attention in the future.

Appendix A
Contributions to Ontology for Ground Vehicles

A.1 General Environmental Conditions

A.1.1. Distribution of ground on Earth to drive on (global map)

Continents and Islands on the globe

Geodetic reference system, databases

Specially prepared roadways: road maps

Cross-country driving, types of ground

Geometric description (3-D)

Support qualities for tires and tracks

Ferries linking continents and islands

National Traffic Rules and Regulations

Global navigation system availability

A.1.2. Lighting conditions as a function of time

Natural lighting by sun (and moon)

Sun angle relative to the ground for a given location and time

Moon angle relative to the ground for a given location and time

Headlights of vehicles

Lights for signaling intentions/special conditions

Urban lighting conditions

Special lights at construction sites (incl. flashs)

Blinking blue lights

A.1.3 Weather conditions

Temperatures (Effects on friction of tires)

Winds

Bright sunshine/Fully overcast/Partially cloudy

Rain/Hail/Snow

Fog (visibility ranges)

Combinations of items above

Road surface conditions (weather dependent)

Dry/Wet/Slush/Snow (thin, heavy, deep tracks) /Ice

Leaf cover (dry – wet)/Dirt cover (partial – full)

A.2 Roadways

A.2.1.Freeways, Motorways, Autobahnen etc.

Defining parameters, lane markings

Limited access parameters

Behavioral rules for specific vehicle types

Traffic and navigation signs

Special environmental conditions

A.2.2. Highways (State-), high-speed roads

Defining parameters, lane markings (like above)

A.2.3. Ordinary state roads (two-way traffic) (like above)
A.2.4. Unmarked country roads (sealed)
A.2.5. Unsealed roads
A.2.6. Tracks
A.2.7. Infrastructure along roadways
Line markers on the ground, Parking strip, Arrows,
Pedestrian crossings
Road shoulder, Guide rails
Regular poles (reflecting, ~1 m high) and markers for snow conditions

A.3 Vehicles

(as objects without driver/autonomous system; wheeled vehicles, vehicles with tracks, mixed wheels and tracks)

A.3.1. Wheeled vehicles
Bicycle: Motorbike, Scooter;
Bicycle without a motor: Different sizes for grown-ups and children
Tricycle
Multiple (even) number of wheels
Cars, Vans/microbuses, Pickups/Sports utility vehicles, Trucks, Buses, Recreation vehicles, Tractors, Trailers

A.3.2. Vehicles with tracks
A.3.3. Vehicles with mixed tracks and wheels

A.4 Form, Appearance, and Function of Vehicles

(shown here for cars as one example; similar for all classes of vehicles)

A.4.1. Geometric size and 3-D shape (generic with parameters)
A.4.2. Subpart hierarchy
Lower body, Wheels, Upper body part, Windshields (front and rear)
Doors (side and rear), Motor hood, Lighting groups (front and rear)
Outside mirrors
A.4.3. Variability over time, shape boundaries (aspect conditions)
A.4.4. Photometric appearance (function of aspect and lighting conditions)
Edges and shading, Color, Texture
A.4.5. Functionality (performance with human or autonomous driver)
Factors determining size and shape
Performance parameters (as in test reports of automotive journals; engine power, power train)
Controls available [throttle, brakes, steering (*e.g.*, "Ackermann")]
Tank size and maximum range
Range of capabilities for standard locomotion:
Acceleration from standstill
Moving into lane with flowing traffic
Lane keeping (accuracy)
Observing traffic regulations (max. speed, passing interdiction)

Distance keeping from vehicle ahead
(standard, average values, fluctuations)
Lane changing [range of maneuver times as f(speed)]
Overtaking behavior [safety margins as f(speed)]
Braking behavior (moderate, reasonably early onset)
Proper setting of turn lights before start of maneuver
Turning off onto crossroad
Entering and leaving a circle
Handling of road forks
Observing right of way at intersections
Negotiating "hair-pin" curves (switchbacks)
Proper reaction to static obstacle detected in your lane
Proper reaction to animals detected on or near the driveway
Emergency stops
Parking alongside the road
Parking in bay
U-turns
Safety features (ABS, ESP …)
Self-check capabilities
Tire pressure
Engine performance (a few easy standard tests like "gas pulses")
Brake performance

A.4.6. Visually observable behaviors of others

(driven by a human or autonomously)

Standard behavioral modes (like list of capabilities above)

Unusual behavioral modes

Reckless entrance into your lane from parking position or neighboring lane at much lower speed
Oscillations over entire lane width (even passing lane markings)
Unusually slow speeds with no noticeable external reason
Disregarding traffic regulations [max. speed (average amount), passing interdiction, traffic lights]
Very short distance to vehicle ahead
Hectic lane change behavior, high acceleration levels (very short maneuver times, large vehicle pitch and bank angles, "slalom" driving)
Overtaking behavior (daring, frequent attempts, questionable safety margins, cutting into your lane at short distance)
Braking behavior (sudden and harsh?)
Start of lateral maneuvers before or without proper setting of turn lights.
Speed not adapted to actual environmental conditions (uncertainties and likely fluctuations taken into account)
Disregarding right of way at intersections.
Pedestrians disregarding standard traffic regulations
Bicyclists disregarding standard traffic regulations

Recognizing unusual behavior of other traffic participants due to unexpected or sudden malfunctions (perturbations).
Reaction to animals on the driveway (f(type of animal))
Other vehicles slipping due to local environmental conditions (like ice)

A.4.7. Perceptual capabilities

A.4.8 .Planning and decision making capabilities

A.5 Form, Appearance, and Function of Humans

(Similar structure as above for cars plus modes of locomotion)

A.6 Form, Appearance, and Likely Behavior of Animals

(relevant in road traffic: Four-legged, birds, snakes)

A.7 General Terms for Acting "Subjects" in Traffic

Subjects: Contrary to "objects" (proper), having passive bodies and no capability of self-controlled acting, "subjects" are defined as objects with the capability of sensing and self-decided control actuation. Between sensing and control actuation, there may be rather simple or quite complicated data processing available taking stored data up to large knowledge bases into account. From a vehicle guidance point of view, both human drivers and autonomous perception and control systems are subsumed under this term. It designates a superclass encompassing all living beings and corresponding technical systems (*e.g.*, robots) as members.

These systems can be characterized by their type of equipment and performance levels achieved in different categories. Table 3.1 shows an example for road vehicles.

The capabilities in the shaded last three rows are barely available in today's experimental intelligent road vehicles. Most of the terms are used for humans in common language. The terms "behavior" and "learning" should be defined more precisely since they are used with different meanings in different professional areas (*e.g.*, in biology, psychology, artificial intelligence, engineering).

Behavior (as proposed here) is an all-encompassing class term subsuming any kind and type of 'action over time' by subjects.

Action means using any kind of control variable available to the subject, leading to changes in the state variables of the problem domain.

State variables are the set of variables allowing decoupling future developments of a dynamic system from the past (all the history of the system with respect to body motion is stored in the present state); state variables cannot be changed at one moment. (Note two things: (1) This is quite the opposite of the definition of "state" in computer science; (2) accelerations are in general not (direct) state variables in this systems-dynamics sense since changes in control variables will affect them directly.)

Control variables are the leverage points for influencing the future development of dynamic systems. In general, there are two components of control activation involved in intelligent systems. If a payoff function is to be optimized by a "*maneuver*", previous experience will have shown that certain control time histories perform better than others. It is essential knowledge for good or even optimal control of dynamic systems, to know in which situations to perform what type of maneuver with which set of parameters; usually, the maneuver is defined by certain time histories of (coordinated) control input. The unperturbed trajectory corresponding to this nominal feed-forward control is also known, either stored or computed in parallel by numerical integration of the dynamic model exploiting the given initial conditions and the nominal control input. If perturbations occur, another important knowledge component is knowing how to link additional control inputs to the deviations from the nominal (optimal) trajectory to counteract the perturbations effectively. This has led to the classes of feed-forward and feedback control in systems dynamics and control engineering:

Feed-forward control components $\underline{U}_{ff}$ are derived from a deeper understanding of the process controlled and the maneuver to be performed. They are part of the knowledge base of autonomous dynamic systems (derived from systems engineering and optimal control theory). They are stored in generic form for classes of *'maneuvers'*. Actual application is triggered from an instance for behavior decision and implemented by an embedded processor close to the actuator, taking the parameters recommended and the actual initial and desired final conditions (states) into account.

Feedback control components $\underline{u}_{fb}$ link actual (additional) control output to system state or (easily measurable) output variables to force the trajectory toward the desired one despite perturbations or poor models underlying step 1. The technical field of 'control engineering' has developed a host of methods also for automotive applications. For linear (linearized) systems, linking the control output to the entire set of state variables allows specifying the "eigenmodes" '*at will*' (in the range of validity of the linear models). In output feedback, adding components proportional to the derivative (D) and/or integral (I) of the signal allows improving speed of response (PD) and long-term accuracy (PI, PID).

Combined feed-forward and feedback control: For counteracting at least small perturbations during maneuvers, an additional feedback control component $\underline{u}_{fb}$ may be superimposed on the feed-forward one ($\underline{U}_{ff}$) yielding a robust implementation of maneuvers.

Longitudinal control: In relatively simple, but very often sufficiently precise models of vehicle dynamics, a set of state variables affected by throttle and (homogeneous) braking actions with all wheels forms an (almost) isolated subsystem. It consists of the translational degrees of freedom in the vertical plane containing the plane of symmetry of the vehicle and the rotational motion in pitch, normal to this plane. The effects of gravity on sloping surfaces and the resulting performance limits are included.

Lateral control: Lateral translation (y direction), rotations around the vertical (z) and the longitudinal (x) axes form the lateral degrees of freedom, controlled essentially by the steer angle. Lateral motion of larger amplitude does have an influence also on longitudinal forces and pitching moment.

Maneuvers are stereotypical control output time histories (feed-forward control) known to transform (in the nominal case) the initial system state $\underline{\mathbf{x}}(t_0)$ into a final one $\underline{\mathbf{x}}(t_f)$ in a given time (range) with boundary conditions (limits) on state variables observed. Certain ranges of perturbations during the maneuver can be counteracted by superimposed feedback control.

Maneuvers may be triggered by higher level decisions for implementing strategic '*mission elements*' (*e.g.*, turning off onto a crossroad) or in the context of a behavioral mission element running, due to the actual *situation* encountered (*e.g.*, lane change for passing slower traffic or an evasive maneuver with respect to a static obstacle during 'roadrunning').

Table 3.3 *gives a collection of road vehicle behavioral capabilities realized by feed-forward (left column) and feedback control (right column).*

Mission elements are those parts of an entire mission that can be performed with the same subset of behavioral capabilities and parameters. Note that mission elements are defined by sets of compatible behavioral capabilities of the subject *actually performing* the mission.

Situation is the collection of environmental and all other facts that have an influence on making proper (if possible 'optimal') behavior decisions in the mission context. This also includes the state within a maneuver being performed (percentage of total maneuver performed, actual dynamic loads, *etc.*) and all safety aspects.

General comment:

Dimension: There are only four dimensions in our (mesoscale) physical world: Three space components and time. Rotational rates and velocities are components of the physical state, due to the nature of mechanical motion described by second-order differential equations (Newton's law). These velocity components are additional degrees of freedom (d.o.f.), but not dimensions as claimed in some recent publications. Recursive estimation with physically meaningful models delivers these variables together with the pose variables.

Dimensions from discretization: In search problems it is a habit to call the possible states of a variable the dimension of the search space; this has nothing to do with physical dimensions.

Appendix B
Lateral Dynamics

B.1 Transition Matrix for Fourth-Order Lateral Dynamics

The linear process model for lateral road vehicle guidance derived in Chapters 3 and 7 (see Table 9.1) can be written as a seventh order system in analogue form [Mysliwetz 1990]:

$$\dot{x}(t) = Fx + g'u + v'(t) \qquad \text{(Equation 3.6 with one control variable)} \tag{B.1}$$

$$\begin{pmatrix} \dot{\lambda} \\ \dot{\beta} \\ \dot{\psi}_{rel} \\ \dot{y}_V \\ \dot{C}_{0hm} \\ \dot{C}_{1hm} \\ \dot{C}_{1h} \end{pmatrix} = \left(\begin{array}{cccc|ccc} 0 & 0 & 0 & 0 & 0 & 0 & 0 \\ f_{12} & -1/T_\beta & 0 & 0 & 0 & 0 & 0 \\ V/a & 0 & 0 & 0 & -V & 0 & 0 \\ 0 & V & V & 0 & 0 & 0 & 0 \\ \hline 0 & 0 & 0 & 0 & 0 & V & 0 \\ 0 & 0 & 0 & 0 & 0 & -3 \cdot V/L & 3 \cdot V/L \\ 0 & 0 & 0 & 0 & 0 & 0 & 0 \end{array}\right) \begin{pmatrix} \lambda \\ \beta \\ \psi_{rel} \\ y_V \\ C_{0hm} \\ C_{1hm} \\ C_{1h} \end{pmatrix} + \begin{pmatrix} k_\lambda \\ 0 \\ 0 \\ 0 \\ 0 \\ 0 \\ 0 \end{pmatrix} \cdot u(t) + \begin{pmatrix} 0 \\ 0 \\ 0 \\ 0 \\ 0 \\ 0 \\ n_{C_{1h}} \end{pmatrix}$$

with $f_{12} = 1/(2T_\beta) - V/a$; $T_\beta = V/k_{ltf}$; Equation 3.30.

With T as cycle time for sampling (video frequency), the Laplace transform for the transition matrix is (see, *e.g.*, [Kailath 1980]):

$$A(T) = L^{-1}\{(sI - F)^{-1}. \tag{B.2}$$

For the discrete input gain vector g, one obtains from F and g'

$$g = \int_0^T A(\tau) g' d\tau . \tag{B.3}$$

The noise term also has to be adjusted properly in correspondence with T. The resulting difference equation is B.4. The matrix A and the input gain vector g have the entries given below (note that g has many more entries than g'; this is due to the buildup of state components from constant control input over one cycle!):

$$x(k+1) = A \cdot x(k) + g \cdot u(k) + v(k) . \tag{B.4}$$

$$A = \begin{pmatrix} 1 & 0 & 0 & 0 & 0 & 0 & 0 \\ a_{21} & a_{22} & 0 & 0 & 0 & 0 & 0 \\ a_{31} & 0 & 1 & a_{34} & a_{35} & a_{36} & a_{37} \\ a_{41} & a_{42} & a_{43} & 1 & a_{45} & a_{46} & a_{47} \\ 0 & 0 & 0 & 0 & 1 & a_{56} & a_{57} \\ 0 & 0 & 0 & 0 & 0 & a_{66} & a_{67} \\ 0 & 0 & 0 & 0 & 0 & 0 & 1 \end{pmatrix} \text{ and } g = \begin{pmatrix} g_1 \\ g_2 \\ g_3 \\ g_4 \\ 0 \\ 0 \\ 0 \end{pmatrix} . \tag{a}$$

With the following abbreviations:

$$c_F = V / a; \qquad a_F = -2\mathrm{k}_{\mathrm{ltf}} / V;$$
$$b_F = -\mathrm{k}_{\mathrm{ltf}} / V + c_F; \qquad a_c = -3V / L;$$

the non-vanishing elements a_{ij} of the transition matrix are

$$a_{21} = b_F / a_F \cdot (e^{a_F T} - 1); \qquad a_{22} = e^{a_F T}$$
$$a_{31} = c_F T; \qquad a_{35} = -VT; \qquad a_{36} = V^2 / a_c^2 \cdot (a_c T + 1 - e^{a_c T});$$
$$a_{37} = -V^2 [(1 - e^{a_c T}) / a_c^2 + T / a_c + T / 2];$$
$$a_{41} = (b_F V / a_F) \cdot [(e^{a_F T} - 1) / a_F - T] + c_F V T^2 / 2;$$
$$a_{42} = (e^{a_F T} - 1) V / a_F; \qquad a_{43} = VT; \qquad a_{45} = -V^2 T^2 / 2;$$
$$a_{46} = V^3 [(1 - e^{a_c T}) / a_c^2 + T / a_c + T / 2] / a_c;$$
$$a_{47} = -V^3 [-(1 - e^{a_c T}) / a_c^3 + T / a_c^2 + T / (2 a_c) + T^3 / 6];$$
$$a_{56} = -V (1 - e^{a_c T}) / a_c; \qquad a_{57} = V (a_c T + 1 - e^{a_c T}) / a_c;$$
$$a_{66} = e^{a_c T}; \qquad a_{67} = 1 - e^{a_c T}.$$

The entries in the input gain vector are

$$g_1 = T; \qquad g_2 = b_F / a_F \cdot \{(e^{a_F T} - 1) / a_F - T\}; \qquad g_3 = c_F T^2 / 2;$$
$$g_4 = -b_F V / a_F^2 \cdot [T + a_F T^2 / 2 - (e^{a_F T} - 1) / a_F] - c_F V T^3 / 6.$$

The rows in matrix A are given by the first index; this corresponds to the sequential innovation scheme in square root filtering using row vectors of A. Note that out of the 49 elements of A, 26 are zero (53%); explicit use of this structure can help making vector multiplication very economical: for rows 1 and 7 (first and last), the result has the same value as the multiplicand. For row index 2, multiplication of elements can stop at this row index 2, while for indices 5 and 6, starting multiplication at these indices is sufficient. Efficiently coded, many multiplications may be saved in this inner loop that is always running. Since several multiplications with these vectors occur in recursive computation of the expected error covariance (step 7.2 in Table 6.1), being efficient here really pays off in real-time vision.

B.2 Transfer Functions and Time Responses to an Idealized Doublet in Fifth-order Lateral Dynamics

From Equations 3.38 and 3.45, the analytical solution for the state vector in the 'Laplace s-realm' is obtained. The former equation yields for the 'system matrix'

$$sI - \Phi = \begin{pmatrix} s & 0 & 0 & 0 & 0 \\ -V/(aT_\psi) & s + 1/T_\psi & 0 & 0 & 0 \\ -1/(2T_\beta) & 1 & s + 1/T_\beta & 0 & 0 \\ 0 & -1 & 0 & s & 0 \\ 0 & 0 & -V & -V & s \end{pmatrix}. \qquad \text{(B.5)}$$

By multiplication of Equation 3.45 from the left by $(sI - \Phi)^{-1}$, there follows

$$\underline{\mathrm{x}}_{La}(s) = (sI - \Phi)^{-1} \cdot \underline{\mathrm{b}} \cdot u(s) + (sI - \Phi)^{-1} \cdot \underline{\mathrm{x}}_{La}(0). \qquad \text{(B.6)}$$

The first term defines the five transfer functions of a control input $u(s)$ while the second term gives the response to initial values in the state variables. All these expressions have a common denominator, the characteristic polynomial $D(s)$ of the determinant det $(sI - \Phi)$

$$D = s^3 (s + 1/T_\psi)(s + 1/T_\beta) \,. \tag{B.7}$$

The numerator polynomials of the transfer functions are obtained by the determinants in which the column corresponding to the state variable of interest has been replaced by the coefficient vector $\underline{b}$ for control input. This yields the numerator terms for the heading angle ψ and the lateral position y

$$\begin{aligned} N_\psi &= [V/(a \cdot T_\psi)] \cdot s \cdot (s + 1/T_\beta) \\ N_y &= Va/(2T_\beta) \cdot [s^2 + s/T_\psi + 2V/(a \cdot T_\psi)]. \end{aligned} \tag{B.8}$$

With the doublet input of Equation 3.44, $u_{idd}(s) = A{\cdot}s$, the resulting state variables yaw angle $\psi(s)$ and the lateral acceleration in the y direction $s^2 \cdot y(s)$ are in the Laplace-domain

$$\begin{aligned} \psi(s) &= [N_\psi / D] \cdot A \cdot s \\ &= A \cdot s \; [V/(a\; T_\psi)] \cdot s \; (s \; + 1/T_\beta) / [\, s^3 \cdot (s \; + 1/T_\beta)(s + 1/T_\psi)] \\ &= A \; V/(a \; T_\psi) / [s \; (s \; + 1/T_\psi)] \\ &= A \; V/(a \; T_\psi) \cdot [1/s - 1/(s \; + 1/T_\psi)]. \end{aligned} \tag{B.9}$$

$$\begin{aligned} a_y(s) &= s^2 \cdot y(s) = N_y \cdot A \cdot s^3 / D \\ &= A \; V / T_\beta \cdot s^3 \cdot [s^2 + s/T_\psi + 2V/(a \cdot T_\psi)] / D \\ &= Va/(2T_\beta) \cdot [1(s) + B/(s + 1/T_\psi) - (B \; + \; 1/T_\beta)/(s + 1/T_\beta)], \\ &\qquad \text{with} \quad B \; = \; 2V/[a \cdot (T_\psi / T_\beta - 1)]. \end{aligned} \tag{B.10}$$

The expression $1/s$ in Equation B.9 corresponds in the time domain to a unit step function, and the second term in Equation B.9: $c/(s + 1/T_i)$ to: $c \cdot \exp[-(t/T_i)]$. This yields with Equation 3.36 after back-transformation into the time domain

$$\psi(t)_{\text{doublet}} \; = \; A \cdot k_{ltf} / (a \cdot i_{zB}^2) \cdot [1 - \exp(- t / T_\psi)] \,. \tag{B.11}$$

$1(s)$ in Equation B.10 corresponds to a *'Dirac impulse'* $\delta(0)$ at time 0 with an integral value of 1 (a step function in the integrated variable, the lateral velocity $v_y = \int a_y dt$ experiences a jump from 0 to 1 at t = 0). Introducing this and the relations given in Equation 3.37 into Equation B.10 yields for the time functions

$$a_y(t) = A \cdot k_{ltf} \cdot \left\{ 0.5 \cdot \delta(0) + \frac{V}{a(1 - i_{zB}^2)} \cdot \left[e^{-t/T_\beta} - e^{-t/T_\psi} \right] - \frac{0.5 \cdot k_{ltf}}{V} e^{-t/T_\beta} \right\} \tag{B.12 (a)}$$

$$= \frac{A}{2} \cdot k_{ltf} \cdot \delta(0) \; + e^{-t/T_\beta} \cdot \left\{ \frac{VA \; k_{ltf}}{a(1 - i_{zB}^2)} \cdot \left[1 - e^{-t/T_{\beta\text{mod}}} \right] - \frac{A \; k_{ltf}^2}{2V} \right\}, \tag{b}$$

where $T_{\beta\text{mod}} = T_\beta \cdot i_{zB}^2 / (1 - \, i_{zB}^2)$.

The value $T_{\beta\text{mod}} / T_\beta = i_{zB}^2 / (1 - i_{zB}^2)$ is 2.77 for VaMoRs and 5.67 for VaMP. Figure B.1 shows the principal time histories of the exponential functions in scaled form for VaMoRs. The yaw angle goes from zero to $[A \cdot k_{ltf} /(a \cdot i_{zB}^2)]$ with time constant T_ψ. According to Equation 3.37, T_ψ increases linearly with speed.

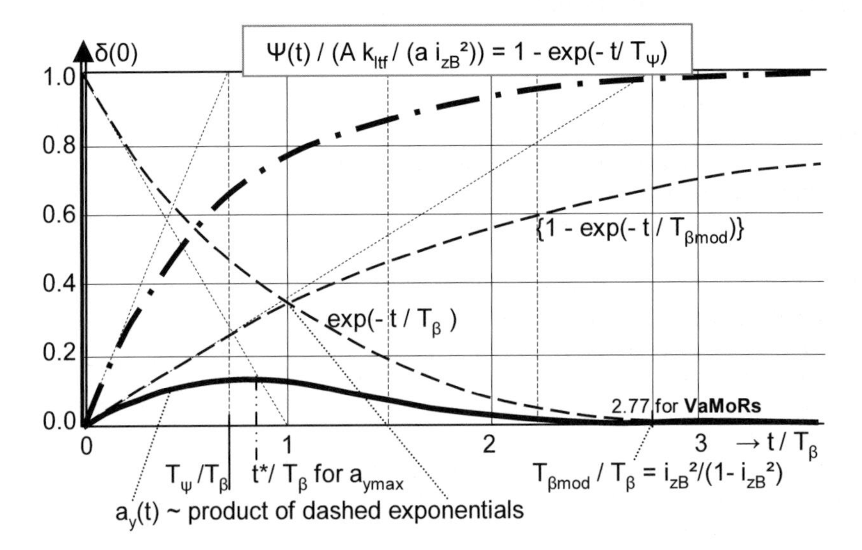

Figure B.1. Scaled dynamic response in yaw angle (*dash-dotted*) and lateral acceleration (*solid curve*) to doublet input in steering rate (see Equations B.11/B.12); the time axis is scaled by T_β

Appendix C
Recursive Least-squares Line Fit

Through a set of measurement points (y_{m1} ... y_{mN}), equidistantly spaced on the abscissa ($x_1 = 0.5$, 1.5 .... $N-0.5$), a straight line shall be fit recursively with interpolated measurement points (y_1 ... y_N), if the standard deviation remains below a threshold value σ_{maxth} and the new measurement point to be added is within a 3σ band around the existing fit. The resulting set of smoothed measurement data will be called a segment or a 1-D blob. The result shall be represented by the average value y_c in the segment, with the origin at the segment center x_c, and the (linear) slope 'a' around this center (see Figure C.1). A deviation from the usual terminology occurs because image evaluation with symmetric masks has its origin right between pixel boundaries; the reference pixel for mask evaluation has been selected at position (0.5, 0.5) of the mask (see Figure 5.19).

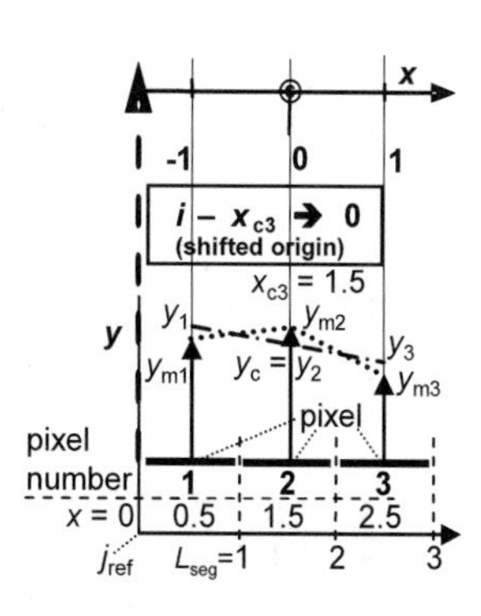

Figure C.1. Nomenclature used with integer basic scale for pixels; origin of centered scale at $N/2$

This definition leads to the fact that x_c is either an integer or lies exactly at the center between two integers ($i - 0.5$). Due to the integer values for the pixels and because a new segment is always started with the *reference coordinate* j_{ref} for $x_0 = 0$, the center of N values is located at

$$x_{c,abs} = j_{ref} + N/2 = j_{ref} + x_c . \tag{C.1}$$

C.1 Basic Approach

Since two measurement points can always be connected by a straight line, interpolation starts at the third point into a new segment. The general form of the interpolating straight line with x_c as the center of all x_i is

$$\begin{aligned} y_i &= y_c + a \cdot \Delta x_i \\ \text{with } \Delta x_i &= x_i - x_c . \end{aligned} \tag{C.2}$$

y_c is always taken at the center of the data set, and 'a' is the slope of the line. This yields for the residues $e_i = y_i - y_{mi}$ (see Figure C.1) the set of equations:

$$\begin{aligned} e_1 &= y_c + (x_1 - x_c) \cdot a - y_{m1}, \\ e_2 &= y_c + (x_2 - x_c) \cdot a - y_{m2}, \\ e_3 &= y_c + (x_3 - x_c) \cdot a - y_{m3}. \end{aligned} \tag{C.3}$$

For easy generalization, this is written in matrix form with the two unknown variables: average value y_c of the interpolating straight line and slope a. The measurement vector y_m has length N, with the original running index i linearly increasing from 1 to N; measurement values of the pixels then are located at x-position [(pixel index i) − 0.5]; the initial value for N is 3 (Figure C.1). The running index for the shifted scale with $x_c = 0$ then goes from $-N/2$ to $+N/2$. With the model of Equation C.2, the errors are

$$[e] = \begin{bmatrix} 1 & x_1 - x_c \\ . & . \\ 1 & x_i - x_c \\ . & . \\ 1 & x_N - x_c \end{bmatrix} \cdot \begin{bmatrix} y_c \\ a \end{bmatrix} - [y_m]_N = A \cdot p - [y_m]_N \,. \tag{C.3a}$$

The sum J of the squared errors can now be written

$$J = \sum_{i=1}^{n} e_i^2 = e^T e = (A \cdot p - y_m)^T (A \cdot p - y_m)\,. \tag{C.4}$$

The same procedure as in Section 5.3.2.1, setting the derivative $\mathrm{d}J/\mathrm{d}p = 0$, leads to the optimal parameters p_{extr} for minimal J:

$$\begin{pmatrix} y_c \\ a \end{pmatrix}_{LS} = p_{extr} = (A^T A)^{-1} A^T y_m \,. \tag{C.5}$$

With $x_i = i - 0.5$ from 0.5 to $N - 0.5$, and with Equation C.3a the product $A^T A$ can be written

$$A^T A = \begin{bmatrix} 1 & .. & 1 & .. & 1 \\ 0.5 - x_c & .. & i - 0.5 - x_c & .. & N - 0.5 - x_c \end{bmatrix} \cdot \begin{bmatrix} 1 & 0.5 - x_c \\ . & . \\ 1 & i - 0.5 - x_c \\ . & . \\ 1 & N - 0.5 - x_c \end{bmatrix} \tag{C.6}$$

$$= \begin{bmatrix} N & a_{12} \\ a_{21} & a_{22} \end{bmatrix}.$$

By always choosing the center of gravity of the abscissa values x_c as reference and by shifting the indices correspondingly (see lower part of Figures C.1 and C.2), there follows

$$a_{21} = a_{12} = \sum_{i=1}^{N} (i - 0.5 - x_c) = 0 \,. \tag{C.7}$$

For each positive index in the shifted coordinates (lower part of figures), there is a corresponding negative one, yielding Equation C.7. This means that $A^T A$ is zero off the main diagonal. For a_{22} in Equation C.6 one obtains

$$a_{22} = \sum_{i=1}^{N} (i - 0.5 - x_c)^2 = \sum_{i=1}^{N} (i^2 - 2 \cdot i \cdot x_c + x_c^2 + 0.25 + x_c - i)\,. \tag{C.8}$$

Introducing the well-known relations

$$\sum_{i=1}^{N} i = N(N+1)/2 \,, \tag{C.9}$$

$$\sum_{i=1}^{N} i^2 = N(N+1)(2N+1)/6\,, \tag{C.10}$$

the following result for a_{22} is obtained with $x_c = N/2$ after several steps

$$a_{22} = N(N^2-1)/12. \tag{C.11}$$

With Equations C.7 and C.11 the inverse of A^TA (Equation C.6) becomes

$$(A^TA)^{-1} = \begin{pmatrix} 1/N & 0 \\ 0 & 1/a_{22} \end{pmatrix}. \tag{C.12}$$

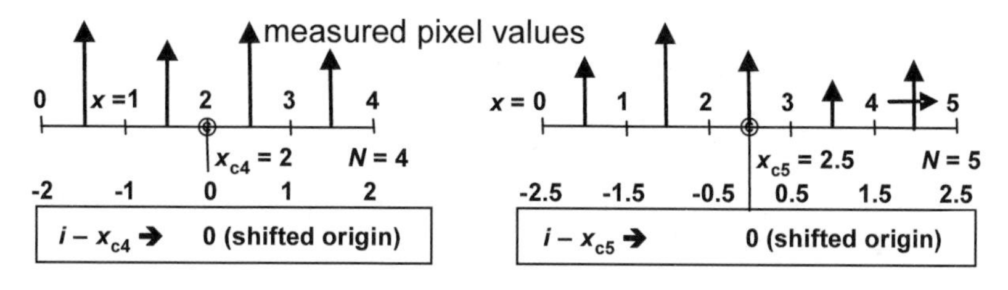

Figure C.2. Centered coordinates for even and odd numbers of equidistant grid points; this choice reduces the numerical workload

To obtain the optimal parameters for a least squares fit according to Equation C.6, the factor A^Ty_m has yet to be determined:

$$A^T y_m = \begin{bmatrix} 1 & .. & 1 & .. & 1 \\ 1-0.5-x_c & .. & i-0.5-x_c & .. & N-0.5-x_c \end{bmatrix} \cdot \begin{bmatrix} y_{m1} \\ . \\ y_{mi} \\ . \\ y_{mn} \end{bmatrix}_N \tag{C.13}$$

$$= \begin{bmatrix} Sym \\ Siym - 0.5 \cdot Sym - x_c \cdot Sym \end{bmatrix},$$

$$\text{with} \quad Sym = \sum_{i=1}^{N} y_{mi} \quad \text{and} \quad Siym = \sum_{i=1}^{N} (i \cdot y_{mi}). \tag{a}$$

Inserting this, the relation $x_c = N/2$, and Equation C.11 into C.5, the following optimal parameters y_c and a are finally obtained:

$$\begin{pmatrix} y_c \\ a \end{pmatrix}_{LS} = \begin{pmatrix} 1/N & 0 \\ 0 & 1/a_{22} \end{pmatrix} \cdot \begin{pmatrix} Sym \\ Siym - \dfrac{N+1}{2} Sym \end{pmatrix} \tag{C.14}$$

or

$$\begin{pmatrix} y_c \\ a \end{pmatrix}_{LS} = \begin{bmatrix} Sym/N \\ \left(Siym - \dfrac{N+1}{2} Sym \right) / a_{22} \end{bmatrix} = \begin{bmatrix} Sym/N \\ \dfrac{6}{N-1} \left(\dfrac{2}{N+1} \cdot \dfrac{Siym}{N} - \dfrac{Sym}{N} \right) \end{bmatrix}. \tag{C.14 (a)}$$

y_c is nothing but the average value of all measurements y_{mi}; to obtain the optimal slope a, the product $i \cdot y_{mi}$ has to be summed, too (Equation C.13a). It is seen that for this least-squares fit with a cg-centered coordinate system ($y(0) = y_c$), where this origin moves in steps by 0.5 with N increasing by 1, just four numbers have to be stored: The number of data points N, the averaged sum of all measured values $y_c = Sym/N$, the averaged sum of all products ($i \cdot y_{mi}$): $Siym/N$, and, of course, the reference for $i = 1$ where the data set started; this yields the optimal parameters: 'y_c = *average value at the segment center*' and 'a = *slope*' for the best interpolating straight line by just a few mathematical operations independent of segment length n.

C.2 Extension of Segment by One Data Point

Let the existing segment have length N; the averaged sums Sym/N and $Siym/N$ (Equation C.13) have been stored. If one new measurement value arrives, its expected magnitude according to the existing model can be computed. The number of measurements increases by 1, and the new segment center x_{ce} shifts to

$$N_e = N + 1; \quad x_{ce} = N_e / 2 = x_c + 0.5. \tag{C.15}$$

The predicted measurement value according to the linear model is

$$y_{mpr} = y_c + a \cdot (N + 1 - x_c) = y_{cpr} + a(N_e - x_{ce}). \tag{C.16}$$

Since the new origin x_{ce} is shifted to the right by 0.5, the expected average value y_{cpr} will be shifted by $a/2$, yielding the rightmost part of Equation C.16. The new measurement value y_{mNe} will be accepted as extension of the segment only if

$$| y_{mNe} - y_{mpr} | \le 3 \cdot \sigma,$$

or

$$(y_{mNe} - y_{mpr})^2 \le 9 \cdot \sigma^2, \tag{C.17}$$

with σ^2 as variance of all previous measurements; otherwise, the segment is concluded with the original value for N.

If the segment is extended, the new parameters for best fit are, according to Equation C.14a

$$y_{ce} = \frac{Sym_e}{N_e} = \left[\left(\frac{Sym}{N} \right) \cdot \frac{N}{N_e} + \frac{y_{mNe}}{N_e} \right];$$

$$a_e = \frac{6}{N_e - 1} \cdot \left(\frac{Siym_e / N_e}{x_{ce}} - y_{ce} \right) \tag{C.18}$$

$$= \frac{6}{N_e - 1} \left\{ \frac{1}{x_{ce}} \cdot \left[\left(\frac{Siym}{N} \right) \cdot \frac{N}{N_e} + y_{mNe} \right] - y_{ce} \right\}.$$

The terms in rounded brackets are the stored (original) values, while the terms in squared brackets are the new values for $N_e = N+1$ to be stored. The definition of the variance is

$$Var(e)_N := \frac{1}{N-1} \sum_{i=1}^{N} e_i^2 = J / N \quad \text{(with Equation C.4)}. \tag{C.19}$$

Inserting the result C.14a for p in J, and exploiting the relations C.1 and C.6 to C.13, one obtains, with $y_m^T y_m = \sum_{i=1}^{N} y_{mi}^2 = Sy2m$ as shorthand notation,

$$Var(e)_N = \frac{Sy2m}{N-1} - f\left(Siym,\ N,\ a,\ x_c,\ y_c\right). \tag{C.19a}$$

To be able to compute the variance recursively, the sum of the squared measurement values *Sy2m* (divided by $N-1$) also has to be stored as an entry into Equation C.19a. The recursive update from N to $N+1$ is given below (on the right side)

$$\frac{Sy2m}{N} = \frac{1}{N}\sum_{i=1}^{N} y_{mi}^2 \quad \text{or} \quad \frac{Sy2m_e}{N_e} = \left[\frac{N}{N_e}\cdot\left(\frac{Sy2m}{N}\right) + \frac{y_{mNe}^2}{N_e}\right]. \tag{C.20}$$

This shows that the new stored value (in square brackets) results from the old one (in rounded brackets) weighted by the factor N/N_e and the squared new measurement value weighted by $1/N_e$. Since the rather complex expressions for the variance are not used in the real-time algorithm, they are not discussed in detail here (see next section).

C.3 Stripe Segmentation with Linear Homogeneity Model

Instead of comparing the magnitude of the prediction error with the 3σ-value of the existing fit, as a criterion for acceptance of the next measurement point, the less computer-intensive criterion in the lower part of Equation C.17 is used:

$$(y_{mpr} - y_{mNe})^2 \le 9\cdot\sigma_N^2 = 9\cdot(Var(e)\,|_N),$$

or even (C.21)

$$(y_{mpr} - y_{mNe})^2 \le \text{variance limit (VarLim)}.$$

VarLim is a fixed threshold parameter of the method; for typical intensity values of video signals (8 to 10 bit, 256 to 1k levels) and the sensitivity of the human eye (~ 60 levels), threshold values $4 \le \text{VarLim} \le 225$ seem reasonable. With Equation C.21, acceptance can be decided without computing the new fit and the new variance. The influence of the new measurement point on the parameters for optimal fit is neglected.

The second method is to compute all new parameters including the new variance and to compare the residue

$$e_{Ne} = y_{cNe} + a_{Ne}\cdot(N_e - x_{ce}) - y_{mNe} \tag{C.22}$$

squared of the new measurement point N_e with the newly determined variance

$$e_{Ne}^2 \le 9\cdot\sigma_{Ne}^2 = 9\cdot(Var(e)\,|_{Ne}). \tag{C.23}$$

When the new value is accepted, store all updated values:

$$N = N_e;\quad \frac{Sym}{N} = \frac{Sym_e}{N_e};\quad \frac{Siym}{N} = \frac{Siym_e}{N_e};\quad \frac{Sy2m}{N} = \frac{Sy2m_e}{N_e};$$
$$y_c = y_{ce};\quad x_c = x_c + 0.5;\quad a = a_e;\quad Var(e)\,|_N = Var(e)\,|_{Ne}. \tag{C.24}$$

Now the next value can be tested with the same procedure.

C.4 Dropping Initial Data Point

This segment reduction at the start may sometimes have beneficial effects on the quality of the linear fit. After several data points have been interpolated by a straight line, the variance of the whole set may be reduced by dropping the first point or a few initial points from the segment. An indication for this situation is given when the first residue on the left side is larger than the standard deviation of the segment. To check this, the interpolated value at location $i = 1$ has to be computed with the parameter set (y_{cN} and a_N) after N data points

$$e_{1N} = y_{cN} + a_N \cdot (1 - x_{cN}) - y_{m1}\,. \tag{C.25}$$

For computational efficiency again the variance is taken as a base for decision: If

$$(e_1)_N^2 \geq Var(e)\,|_N\,, \tag{C.26}$$

dropping the first data point will decrease the variance of the remaining data set. The even simpler check with a fixed threshold

$$(e_1)_N^2 \geq VarLim \tag{C.26a}$$

has proven well suited for efficient real-time computation.

With the stored values of Equation C.24 and always working with locally centered representations, the reduction at the left is directly analogous to the extension at the right-hand side. The problematic point is the sum *Siym* of the products (local index times measurement value) (Equation C.13a and C.18). The new starting point of the segment will become ($x_{ref} + 1$), but the length N_{red} of the segment will be reduced by 1, while the position of its center is reduced by −0.5 for symmetry; i in *Siym* has to be decremented

$$x_{ref} := x_{ref} + 1; \quad N_{red} = N - 1; \quad x_{cred} = x_c - 0.5;$$

$$Siym_{red} = N\sum_{j=2}^{N}(j-1)\cdot y_{mj} = \sum_{i_{red}=1}^{N_{red}} i_{red}\cdot y_{mi_{red}} = \sum_{j=2}^{N} j\cdot y_{mj} - \sum_{j=2}^{N} y_{mj}\,. \tag{C.27}$$

By noting that

$$Siym_N = y_{m1} + \sum_{i=2}^{N} i\cdot y_{mi}; \quad Sym_N = y_{m1} + \sum_{i=2}^{N} y_{mi}\,, \tag{C.28}$$

adding $0 = (y_{m1} - y_{m1})$ to the lower Equation C.27 immediately yields

$$\begin{aligned} Siym_{red} &= y_{m1} + \sum_{j=2}^{N} j\cdot y_{mj} - (y_{m1} + \sum_{j=2}^{N} y_{mj}) = Siym_N - Sym_N; \\ Sym_{red} &= Sym_N - y_{m1}. \end{aligned} \tag{C.29}$$

This leads to the *recursive procedure* that is to be executed as long as the initial residue squared is larger than the threshold ‘variance limit’:

$$\begin{aligned} &N_{red} = N - 1; \qquad x_{ref} := x_{ref} + 1; \qquad x_{cred} = x_c - 0.5; \\ &Siym_{red} = Siym_N - Sym_N; \qquad Sy2m_{red} = Sy2m_N - y_{m1}^2; \\ &Sym_{red} = SymN - y_{m1}; \qquad y_{cred} = Sym_{red}/N_{red}; \\ &a_{red} = \frac{6}{N_{red} - 1}\left(\frac{2}{N_{red} + 1}\cdot\frac{Siym_{red}}{N_{red}} - y_{cred}\right). \end{aligned} \tag{C.30}$$

The maximum number k of points to be dropped will not be large, in general, since otherwise the segment would have been ended during normal extension.

After reducing the segment length at the side of low indices i, increasing the segment at the other end should be tried again. Figure 5.32 has been interpolated without dropping terms at the left side of the segment, from which the line fit was started. The first longer segment from ~ row = 90 to 115 (upper left center in the figure, designated as 'blob 1') could profit from dropping the leftmost data point since a steeper negative value for a_{red} can fit several following data points better. (dotted line, Figure C.3 shows a blown up view of blob 1 of Figure 5.32. The solid line is the least-squares line fit resulting without a check of the residue of the 'first' data point after each update; interpolation is stopped at the solid black dots (lower right) because the variance exceeds the threshold set.

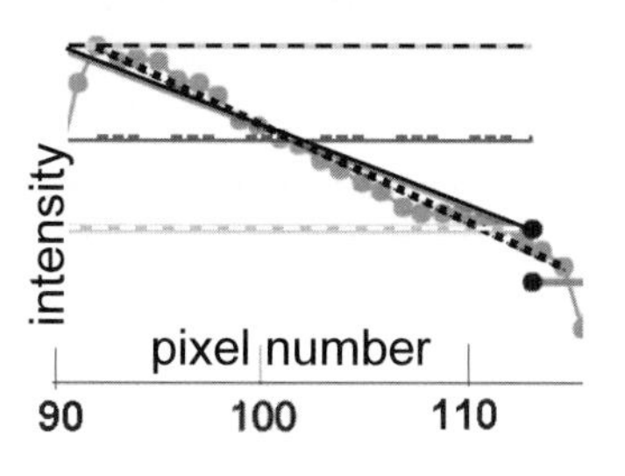

Figure C.3. Large threshold values for starting a line fit (upper left corner) may lead to suboptimal results (see text)

Dropping the 'first' data point (top left) allows a much better fit to the remaining points; it even allows an extension of the segment to larger values L_{seg} (two more data points) The dotted line in Figure C.3 shows an improved fit with reduced total variance.

Figure 5.35 shows several cases of this type of blob data fit, with the method described here, for a number of columns of a video field. In the top left subfigure, the white lines mark the cross sections selected. Some correspondences between object regions in the scene and blob parameters are given. The task of hypothesis generation in vision is to come up with most reasonable object hypotheses given a collection of features (blobs, edges and corners); homogeneous areas with center of gravity, shape as well as shading parameters are a big step forward compared to just edge features with adjacent average gray values available in real-time evaluation about fifteen years ago.

Figure C.4 (next page) shows the flowchart of the segmentation algorithm including size adaptation at both ends for an optimal fit of shaded stripe segments.

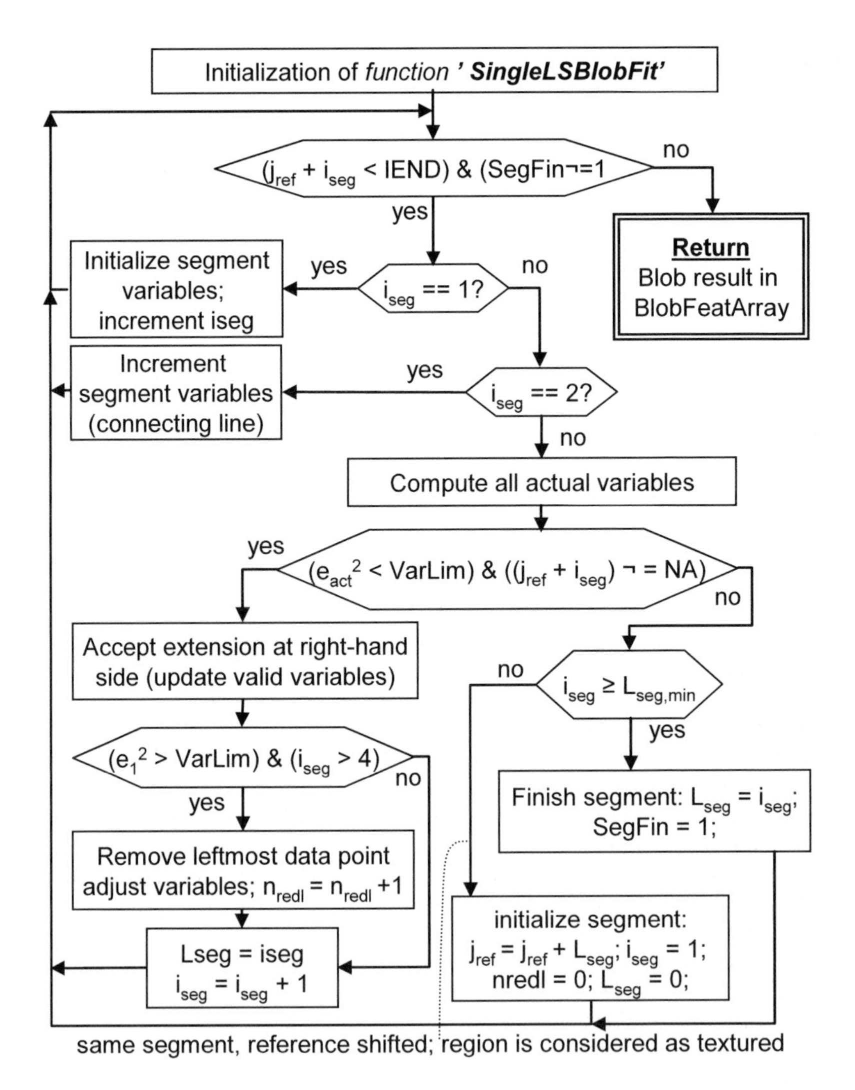

Figure C.4. Flowchart of segmentation method for linearly shaded regions including boundary adaptation at both ends (Matlab®-terminology)

References

Albus J.S., Meystel A.M. (2001): Engineering of Mind. – An Introduction to the Science of Intelligent Systems. Wiley Series on Intelligent Systems

Alexander R. (1984): The gaits of bipedal and quadrupedal animals. Journal of Robotics Research 3(2)

Aloimonos J., Weiss I., Bandyopadhyay A. (1987): Active vision. Proceedings 1st Int. Conf. on Computer Vision: 35–54

Altmannspacher H., Ruhnau E. (eds) (1997): Time, Temporality, Now. Springer–Verlag, Berlin

Arbib M.A., Hanson A.R. (eds) (1987): Vision, Brain, and Cooperative Computing. MIT Press, Cambridge, MA

Arkin R. (1998): Behavior-Based Robotics. MIT Press, Cambridge, MA

Baghdassarian C., Lange H., Sahli, Laurgeau C. (1994): Recognition of arrows in the environment of road markings. Proc. Int. Symp. on Intell. Vehicles'94, Paris: 219–224

Ballard D.H. (1991): Animate vision. Artificial Intelligence 48: 57–86

Ballard D.H., Brown C.M. (1982): Computer Vision. Prentice–Hall, Englewood Cliff, NJ

Bar Shalom Y., Fortmann T.E. (1988): Tracking and Data Association. Mathematics in Science and Engineering, Academic Press

Bar Shalom Y., Li X.R. (1998): Estimation and Tracking: Principles, Techniques, and Software. Danvers, MA; YBS

Behringer R. (1996): Visuelle Erkennung und Interpretation des Fahrspurverlaufes durch Rechnersehen für ein autonomes Straßenfahrzeug. Diss. UniBw Munich, LRT; also: Fortschrittberichte VDI, Reihe 12, Nr. 310

Bergen J.R. (1990) Dynamic Analysis of Multiple Motion. Proc. Israeli Conference on Artificial Intelligence and Computer Vision. Tel Aviv

Bertozzi M., Broggi A., Fascioli A. (2000): Vision-based intelligent vehicles: State of the art and perspectives. Robotics and Autonomous Systems 32: 1–16

Bertozzi M., Broggi A., Fascioli A., Tibaldi A., Chapuis R., Chausse, F. (2004): Pedestrian Localization and tracking System with Kalman Filtering. Proceedings IEEE International Symposium on Intelligent Vehicles, Parma: 584–590

Besl P.J., Jain R.C. (1995): Three dimensional object recognition. ACM Comput. Surveys 17(1): 75–145

Bierman G.J. (1975): Measurement Updating Using the U-D Factorization. Proc. IEEE Control and Decision Conf., Houston. TX: 337–346

Bierman G.J. (1977): Factorization Methods for Discrete Sequential Estimation. Academic Press, New York

Blake A., Zisserman A. (1987): Visual Reconstruction. MIT Press, Cambridge, MA

Blake A., Yuille A. (1992): Active Vision. MIT Press, Cambridge, MA

Blinn J.A. (1977): A Homogeneous Formulation for Lines in 3 Space. Proc. of SIGGRAPH 77. In Computer Graphics 11(3)

Bohrer S., Zielke T., Freiburg V. (1995): An integrated obstacle detection framework for intelligent cruise control on motorways. Proceedings International Symposium on Intelligent Vehicles, Detroit

Braess H.H., Reichart G. (1995a): Prometheus: Vision des 'intelligenten Automobils' auf 'intelligenter Straße'? Versuch einer kritischen Würdigung – Teil 1. ATZ Automobiltechnische Zeitschrift 97, Bd 4: 200–205

Braess H.H., Reichart G. (1995b): Prometheus: Vision des 'intelligenten Automobils' auf 'intelligenter Straße'? Versuch einer kritischen Würdigung – Teil 2. ATZ Automobiltechnische Zeitschrift 97, Bd. 6: 330–343

Brammer K., Siffling G. (1975): Kalman-Bucy-Filter.Deterministische Beobachtung und stochastische Filterung. Oldenbourg Verlag, Muenchen, Wien

Broggi A., Cerri P., Antonello P.C. (2004): Multi-resolution Vehicle Detection using Artificial Vision. Proceedings International Symposium on Intelligent Vehicles, Parma: 310–314

Brooks R.A. (1986): A robust layered control system for a mobile robot. IEEE-J. of Robotics and Automation, 2: 14–23

Brauckmann M.E., Goerick C., Gro J.; Zielke T. (1994): Towards All Around Automatic Visual Obstacle Sensing for Cars. Proceedings International Symposium on Intelligent Vehicles, Paris: 79–84

Bruderlin A., Calvert T.W. (1989): Goal-directed, dynamic animation of human walking. Computer Graphics 23(3), ACM SIGGRAPH, Boston

Brüdigam C. (1994): Intelligente Fahrmanoever sehender autonomer Fahrzeuge in autobahnaehnlicher Umgebung. Diss. UniBw Munich, LRT

Burt P.J., Hong T.H., Rosenfeld H. (1981): Segmentation and estimation of image region properties through cooperative hierarchical computation. IEEE Trans. Systems, Man, and Cybernetics, **11**(12): 802–825

Canny J.F. (1983): A computational approach to edge detection. IEEE-Trans. Pattern Analysis and Machine Intelligence (PAMI) 8 (6): 679–698

Carlson S., Eklundh J.O. (1990): Object detection using model based prediction and motion parallax. In: Proc. European Conference on Computer Vision (ECCV'90), France: 297–306

Davis L., Kushner T.R., Le Moigne J.J., Waxman A.M. (1986): Road Boundary Detection for Autonomous Vehicle Navigation. Optical Engineering, 25(3): 409–414

Debusschere I., Bronckaers E., Claeys C., Kreider G., Van der Spiegel J., Sandini G., Dario P., Fantini F., Bellutti P., Soncini G. (1990): A Retinal CCD Sensor for Fast 2D Shape, Recognition and Tracking. Sensors and Actuators, A21: 456–460

Deriche R., Giraudon G. (1990): Accurate Corner Detection: An Analytical Study. Proc. International Conference on Computer Vision (ICCV), 1990: 66–70

Dickmanns D. (1995): Knowledge based real-time vision. Proc. 2nd IFAC Conf. on Intelligent Autonomous Vehicles, Helsinki

Dickmanns D. (1997): Rahmensystem für visuelle Wahrnehmung veränderlicher Szenen durch Computer. Diss., UniBw Munich, INF. Also: Shaker Verlag, Aachen, 1998

Dickmanns E.D. (1985): 2-D-Object Recognition and Representation using Normalized Curvature Functions. In M.H. Hamza (ed): Proc. IASTED International Symposium on Robotics and Automation, Acta Press: 9–13

Dickmanns E.D., Zapp A. (1986): A Curvature-based Scheme for Improving Road Vehicle Guidance by Computer Vision. In: 'Mobile Robots', SPIE Proc. 727, Cambridge, MA: 161–168

Dickmanns E.D. (1987): 4-D Dynamic Scene Analysis with Integral Spatio-Temporal Models. In: Bolles R, Roth B(eds.): Robotics Research, 4th International Symposium, MIT Press, Cambridge MA

Dickmanns E.D., Zapp A. (1987): Autonomous High Speed Road Vehicle Guidance by Computer Vision. 10th IFAC World Congress, Munich, Preprint 4: 232–237

Dickmanns E.D., Graefe V. (1988): (a) Dynamic monocular machine vision. Journal of Machine Vision and Application, Springer International 1:223-240. (b) Applications of dynamic monocular machine vision. *(ibid)*: 241–261

Dickmanns E.D. (1988): Object Recognition and Real-Time Relative State Estimation Under Egomotion. In: Jain AK(ed) (1988): Real-Time Object Measurement and Classification. Springer-Verlag, Berlin: 41–56

Dickmanns E.D. (1989): Subject-Object Discrimination in 4-D Dynamic Scene Interpretation by Machine Vision. Proc. IEEE-Workshop on Visual Motion, Newport Beach: 298–304

Dickmanns E.D., Christians T. (1989): Relative 3-D state estimation for autonomous visual guidance of road vehicles. In: Kanade, T. *et al.* (eds.): 'Intelligent Autonomous Systems 2', Amsterdam, Dec. 1989, Vol.2 pp. 683-693; also appeared in: Robotics and Autonomous Systems 7 (1991): 113 – 123

Dickmanns E.D., Mysliwetz B., Christians T. (1990): Spatio-Temporal Guidance of Autonomous Vehicles by Computer Vision. IEEE-Trans. on Systems, Man and Cybernetics, 20(6), Special Issue on Unmanned Vehicles and Intelligent Robotic Systems: 1273–1284

Dickmanns E.D., Mysliwetz B. (1992): Recursive 3-D Road and Relative Ego-State Recognition. IEEE-Trans. Pattern Analysis and Machine Intelligence (PAMI) 14(2), Special Issue on 'Interpretation of 3-D Scenes': 199–213

Dickmanns E.D., Behringer R., Dickmanns D., Hildebrandt T., Maurer M., Thomanek F., Schiehlen J. (1994): The Seeing Passenger Car 'VaMoRs-P'. In Masaki I (ed): Proceedings International Symposium on Intelligent Vehicles'94, Paris: 68–73

Dickmanns E.D. (1995): Road vehicle eyes for high precision navigation. In Linkwitz *et al.* (eds) (1995) High Precision Navigation. Dummler Verlag, Bonn: 329–336

Dickmanns E.D., Müller N. (1995): Scene Recognition and Navigation Capabilities for Lane Changes and Turns in Vision-Based Vehicle Guidance. Control Engineering Practice, 2nd IFAC Conf. on Intelligent Autonomous Vehicles'95, Helsinki; also published in Control Engineering Practice (1996), 4(5): 589–599

Dickmanns E.D., Wuensche H.J. (1999): Dynamic Vision for Perception and Control of Motion. In: Jaehne B., Haußenecker H., Geißler P. (eds): Handbook of Computer Vision and Applications, Vol. 3, Academic Press: 569–620

Dickmanns E.D. (2002a): The development of machine vision for road vehicles in the last decade. Proceedings International Symposium on Intelligent Vehicles'02, Versailles

Dickmanns E.D. (2002b): Vision for ground vehicles: History and prospects. International Journal of Vehicle Autonomous Systems, 1(1): 1–44

Dickmanns E.D. (2003): Expectation-based, Multi-focal, Saccadic Vision - (Understanding dynamic scenes observed from a moving platform). In: Olver P.J., Tannenbaum A. (eds) (2003) Mathematical Methods in Computer Vision, Springer-Verlag: 19–35

Duda R., Hart P. (1973): Pattern classification and scene analysis. Wiley, New York

Ehrenfels C.V. (1890): Über Gestaltqualitäten. Vierteljahresschrift für wissenschaftliche Philosophie, Bd 14

Enkelmann W. (1990): Obstacle Detection by Evaluation of Optical Flow Fields from Image Sequences. In: Proc. ECCV 90, France: 134–138

Estable S., Schick J., Stein F., Janssen R., Ott R., Ritter W., Zheng Y.J. (1994): A Real-Time Traffic Sign Recognition System. Proceedings International Symposium on Intelligent Vehicles'94, Paris: 213–218

Fagerer C., Dickmanns E.D., Dickmanns D. (1994): Visual Grasping with Long Delay Time of a Free Floating Object in Orbit. J. Autonomous Robots, 1(1)

Fifth Workshop on Massively Parallel Processing (2005), Denver, CO

Fikes R., Nilsson N. (1971): STRIPS: A New Approach to the Application of Theorem Proving to Problem Solving. Artificial Intelligence, 2: 189–208

Fleischer K., Nagel H.H., Rath T.M. (2002): 3D-Model-based Vision for Innercity Driving Scenes. Proceedings International Symposium on Intelligent Vehicles'02, Versailles

Florack L.M.J., ter Haar Romeny B.M., Koenderink J.J., Viergever M.A. (1992): Scale and the differential structure of images. Image and Vision Computing, 10(6): 376–388

Foley J.D., van Dam A., Feiner S., Hughes J.F. (1990): Computer Graphics – Principles and Practice. Addison-Wesley

Franke U. (1992): Real time 3D-road modeling for autonomous vehicle guidance. In: Johanson, Olson (1992) Selected Papers of 7th Scandinavian Conference on Image Analysis. World Science Publishing Company: 277–284

Franke U., Rabe C., Badino H., Gehrig S.K. (2005): 6D-Vision: Fusion of Stereo and Motion for Robust Environment Perception. Proceedings Deutsche Arbeitsgemeinschaft für Mustererkennung (DAGM), Symposium 2005: 216–223

Freeman H. (1974): Computer processing of line-drawing images. Computing Surveys, 6(1): 57–97

Fritz H. (1996): Model-Based Neural Distance Control for Autonomous Road Vehicles. Proceedings International Symposium on Intelligent Vehicles'96, Tokyo: 29–34

Gauss C.F. (1809): Theoria Motos Corporum Coelestium. Goettingen. Republished in 1857, and by Dover in 1963 by Little, Brown and Co.

Gelb A. (ed) (1974): Applied Optimal Estimation. MIT Press

Giampiero M. (2007): Handbook of Road Vehicle Dynamics. CRC Press (in press)

Graefe V. (1984): Two Multi-Processor Systems for Low Level Real-Time Vision. In: Brady J.M., Gerhard L.A., Davidson H.F. (eds) (1984): Robotics and Artificial Intelligence, Springer–Verlag: 301–307

Graefe V. (1989): Dynamic vision systems for autonomous mobile robots. Proceedings, IEEE/RSJ International Workshop on Intelligent Robots and Systems, Tsukuba: 12–23

Graefe V., Efenberger W. (1996): A Novel Approach for the Detection of Vehicles on Freeways by Real-time Vision. Proceedings International Symposium on Intelligent Vehicles, Tokyo: 363–368

Gregor R., Lützeler M., Pellkofer M., Siedersberger K.H., Dickmanns E.D. (2000): EMS-Vision: A Perceptual System for Autonomous Vehicles. Proceedings International Symposium on Intelligent Vehicles, Dearborn, MI: 52–57

Gregor R., Dickmanns, E.D. (2000): EMS-Vision: Mission Performance on Road Networks. Proceedings International Symposium on Intelligent Vehicles, Dearborn, MI: 140–145

Gregor R., Luetzeler M., Dickmanns E.D. (2001): EMS-Vision: Combining on- and off-road driving. Proc. SPIE-Aero-Sense, Orlando, FL

Gregor R., Lützeler M., Pellkofer M., Siedersberger K.H., Dickmanns E.D. (2001): A Vision System for Autonomous Ground Vehicles with a Wide Range of Maneuvering Capabilities. Proc. ICVS, Vancouver

Gregor R., Lützeler M., Pellkofer M., Siedersberger K.H. and Dickmanns E.D. (2002): EMS-Vision: A Perceptual System for Autonomous Vehicles. IEEE Transactions on Intelligent Transportation Systems, 3(1): 48–59

Gregor R. (2002): Faehigkeiten zur Missionsdurchfuehrung und Landmarkennavigation. Diss. UniBw Munich, LRT

Handbook of Physiology, American Physiological Society:

Brooks VB (ed) (1987): Motor Control, Vol. II, Parts 1 and 2

Darian-Smith I (ed) (1984): Sensory Processes, Vol. III, Parts 1 and 2

Plum F. (ed) (1987): Higher Functions of the Brain, Vol. V, Parts 1 and 2

Hanson A.R., Riseman E. (ed): (1978) Computer Vision Systems. Academic Press, New York

Hanson A.R., Riseman E. (1987): The VISIONS image understanding system – 1986. In: Brown C (ed) Advances in Computer Vision. Erlbaum, Hillsdale, NJ

Haralick R.M., Shapiro L.G. (1993): Computer and Robot Vision. Addison–Wesley

Harel D. (1987): State charts: A Visual Formalism for Complex Systems. Science of Computer Programming, 8: 231–274

Harris C.G., Stephens M. (1988): A combined corner and edge detector. Proc. 4th Alvey Vision Conf.: 147-151

Hillis W.D. (1992) (6th printing): The Connection Machine. MIT Press, Cambridge, MA

Hock C., Behringer R., Thomanek F. (1994): Intelligent Navigation for a Seeing Road Vehicle using Landmark Recognition. In: Close Range Techniques and Machine Vision. ISPBS, Melbourne Australia

Hock C. (1994): Wissensbasierte Fahrzeugfuehrung mit Landmarken fuer autonorne Roboter. Diss., UniBw Munich, LRT

Hofmann U., Dickmanns E.D. (2000): EMS-Vision: An Application to Intelligent Cruise Control for High Speed Roads. Proceedings International Symposium on Intelligent Vehicles, Dearborn, MI: 468–473

Hofmann U., Rieder A., Dickmanns E.D. (2003): Radar and Vision Data Fusion for Hybrid Adaptive Cruise Control on Highways. Journal of Machine Vision and Application, 14(1): 42–49

Hofmann U. (2004): Zur visuellen Umfeldwahrnehmung autonomer Fahrzeuge. Diss., UniBw Munich, LRT

Hogg D.C. (1984): Interpreting images of a known moving object. Ph.D. thesis, University of Sussex, Department of Computer Science

http://iris.usc.edu/Vision-Notes/bibliography/contents.html

Hubel D.H., Wiesel T. (1962): Receptive fields, binocular interaction, and functional architecture in the cat's visual cortex. Journal of Physiology, 160: 106–154

IV'00 (2000): Proceedings of the International Symposium on Intelligent Vehicles. Dearborn, MI, with the following contributions on EMS-Vision: Gregor *et al.* (2000a, b), Hofmann *et al.* (2000), Lützeler *et al.* (2000), Maurer (2000), Pellkofer *et al.* (2000), Siedersberger (2000), [*individual references under these names*]

Jaynes E.T. (2003): Probability Theory, The Logic of Science. Cambridge Univ. Press

Johansson G. (1973): Visual perception of biological motion and a model for its analysis. Perception and Psychophysics 14(2): 201–211

Kailath T. (1980): Linear Systems. Prentice-Hall Inc., Englewood Cliffs, NJ

Kailath T., Sayed A.H., Hassibi B. (2000): Linear estimation. Prentice Hall Inc., Englewood Cliffs, NJ

Kalinke T., Tzomkas C., v. Seelen W. (1998): A Texture-based Object Detection and an Adaptive Model-based Classification. Proceedings International Symposium on Intelligent Vehicles'98, Stuttgart

Kalman R.D. (1960) A new approach to linear filtering and prediction problems. Trans. ASME, Series D, Journal of Basic Engineering: 35–45

Kalman R.D., Bucy R.S. (1961) New results in linear filtering and prediction theory. Trans. ASME, Series D, Journal of Basic Engineering: 95–108.
Kanade T. (ed,) (1987): Three-Dimensional Machine Vision. Kluwer Acad. Publ.
Kenue S. (1989): Lanelok: Detection of lane boundaries and vehicle tracking using image processing techniques: Parts I and II. In: SPIE Proc. Mobile Robots
Kinzel W. (1994a): Pedestrian Recognition by Modeling their Shapes and Movements. In S. Impedovo (ed.) (1994) Progress in Image Analysis and Processing III; Proc. 7th Int. Conf. on Image Analysis and Processing, IAPR, World Scientific, Singapore: 547–554
Kinzel W. (1994b): Präattentive und attentive Bildverarbeitungsschritte zur visuellen Erkennung von Fußgängern. Diss., UniBw Munich, LRT. Also as Fortschrittsberichte VDI Verlag, Reihe 10, Nr. 329
Klass P.J. (1985): DARPA Envisions New Generation of Machine Intelligence. Aviation Week & Space Technology, April: 47–54
Kluge K., Thorpe C. (1988): Explicit models for robot road following. In: Proc. IEEE Conf. on Robotics and Automation
Koch C. (1995): Vision Chips: Implementing Vision Algorithms with Analog VLSI Circuits. IEEE Computer Society Press
Koenderink J.J., van Doorn A.J. (1990): Receptive field families. Biol.Cybern., 63: 291–298
Koller D., Daniilidis K., Nagel H.H. (1993): Model-based object tracking in monocular image sequences of road traffic scenes. Int. J. of Computer Vision, 3(10): 257–281
Kraft H., Frey J., Moeller T., Albrecht M., Grothof M., Schink B., Hess H., Buxbaum B. (2004): 3D-Camera of High 3D-Frame Rate, Depth-Resolution and Background Light Elimination Based on Improved PMD (Photonic Mixer Device) –Technologies. OPTO 2004, AMA Fachverband, Nuremberg, Germany
Kroemer K.H.E. (1988): Ergonomic models of anthropomorphy, human biomechanics, and operator-equipment interfaces. Proc. of a Workshop, Committee on Human Factors, National Academy Press, Washington DC: 114–120
Kuan D., Phipps G., Hsueh A.C. (1986): A real time road following vision system for autonomous vehicles. Proc. SPIE Mobile Robots Conf., 727, Cambridge MA: 152–160
Kuehnle A. (1991): Symmetry-based recognition of vehicle rears. In: Pattern Recognition Letters 12 North-Holland: 249–258
Kuhnert K.D. (1988): Zur Echtzeit-Bildfolgenanalyse mit Vorwissen. Diss. UniBw Munich, LRT
Kujawski D. (1995): Deciding the Behaviour of an Autonomous Road Vehicle in Complex Traffic Situations. 2nd IFAC Conf. on Intelligent Autonomous Vehicles-95, Helsinki
Labayarde R., Aubert D., Tarel P. (2002): Real Time Obstacle Detection in Stereovision on non Flat Road Geometry through 'V-disparity' representation. Proceedings International Symposium on Intelligent Vehicles'02, Versailles
Leonhard J.J., Durrant-White H.F. (1991): Mobile robot localization by tracking geometric beacons. IEEE Transactions on Robotics and Automation 7: 376–382
Loffeld O. (1990): Estimationstheorie. Oldenbourg
Luenberger D.G. (1964): Observing the state of a linear system. IEEE Trans. on Military Electronics 8: 74–80
Luenberger D.G. (1964): Observing the state of a linear system. IEEE Trans. on Military Electronics 8: 290–293.
Luenberger D.G. (1966): Observers for Multivariable Systems. IEEE Trans. Automatic Control, AC-11: 190–197

Lützeler M., Dickmanns E.D. (2000): EMS-Vision: Recognition of Intersections on Unmarked Road Networks. Proceedings International Symposium on Intelligent Vehicles, Dearborn, MI: 302–307

Lützeler M. (2002): Fahrbahnerkennung zum Manoevrieren auf Wegenetzen mit aktivem Sehen. Diss. UniBw Munich, LRT. Also as Fortschrittsberichte VDI Verlag, Reihe 12, Nr. 493

Mandelbaum R., Hansen M., Burt P., Baten S. (1998): Vision for Autonomous Mobility: Image Processing on the VFE-200. In: IEEE International Symposium on ISIC, CIRA and ISAS

Marr D., Nishihara H.K. (1978): Representation and Recognition of the spatial organization of three-dimensional shape. Proceedings of the Royal Society of London, Series B 200: 269–294

Marr D (1982): Vision. W.H. Freeman, New York

Marshall S (1989): Review of shape coding techniques. Image and Vision Computing, 7(4): 281–294

Masaki I. (1992++): yearly 'International Symposium on Intelligent Vehicles', in later years appearing under IEEE – ITSC sponsorship. Proceedings

Maurer M. (2000): Knowledge Representation for Flexible Automation of Land Vehicles. Proc. of the International Symposium on Intelligent Vehicles, Dearborn, MI: 575–580

Maurer M. (2000): Flexible Automatisierung von Strassenfahrzeugen mit Rechnersehen. Diss. UniBw Munich, LRT. Also as Fortschrittsberichte VDI Verlag, Reihe 12, Nr. 443

Maurer M. , Stiller C. (2005): Fahrerassistenzsysteme mit maschineller Wahrnehmung. Springer, Berlin

Maybeck. PS (1979): Stochastic models, estimation and control. Vol. 1, Academic Press, New York

Maybeck P.S. (1990): The Kalman filter: An introduction to concepts. In: Cox, I.J., Wilfong G.T. (eds): Autonomous Robot Vehicles, Springer–Verlag

McCarthy J. (1955): Making Robots Conscious of their Mental State. Computer Science Report, Stanford University, CA

McCarthy J., Minsky M., Rochester N., Shannon C. (1955): A Proposal for the Dartmouth Summer Research Project on Artificial Intelligence, Aug. 31

Meissner H.G. (1982): Steuerung dynamischer Systeme aufgrund bildhafter Informationen. Diss., UniBw Munich, LRT

Meissner H.G., Dickmanns E.D. (1983): Control of an Unstable Plant by Computer Vision. In: Huang T.S. (ed) (1983): Image Sequence Processing and Dynamic Scene Analysis. Springer-Verlag, Berlin: 532–548

Metaxas D.N., Terzopoulos D. (1993): Shape and Nonrigid Motion Estimation Through Physics-Based Synthesis. IEEE Trans. Pattern Analysis and Machine Intelligence 15(6): 580–591

Mezger W. (1975, 3. Auflage): Gesetze des Sehens. Verlag Waldemar Kramer, Frankfurt M. (1. Auflage 1936, 2. Auflage 1953)

Miller G., Galanter E., Pribram K. (1960): Plans and the Structure of Behavior. Holt, Rinehart & Winston, New York

Mitschke M. (1988): Dynamik der Kraftfahrzeuge – Band A: Antrieb und Bremsung. Springer-Verlag, Berlin, Heidelberg, New York, London, Tokio

Mitschke M. (1990): Dynamik der Kraftfahrzeuge - Band C: Fahrverhalten. Springer-Verlag, Berlin, Heidelberg, New York, London, Tokio

Moravec H. (1979): Visual Mapping by a Robot Rover. Proc. IJCAI 1079: 598–600

Moravec H. (1983): The Stanford Cart and the CME Rover. PIEEE(71), 7: 872–884

Mori H., Charkari N.M. (1993): Shadow and rhythm as sign patterns of obstacle detection. In IEEE Int. Symp. on Industrial Electronics, Budapest: 271–277

Moutarlier, P. Chatila R. (1989): Stochastic multisensory data fusion for mobile robot location and environment modeling. In: 5th International Symposium on Robotic Research, Tokyo.

Müller N., Baten S. (1995): Image Processing Based Navigation with an Autonomous Car. International Conference on Intelligent Autonomous Systems (IAS–4), Karlsruhe: 591–598

Müller N. (1996): Autonomes Manoevrieren und Navigieren mit einem sehenden Strassenfahrzeug. Diss., UniBw Munich, LRT. Also as Fortschrittsberichte VDI Verlag, Reihe 12, Nr. 281

Mysliwetz B, Dickmanns E.D. (1986) A Vision System with Active Gaze Control for real-time Interpretation of Well Structured Dynamic Scenes. In: Hertzberger LO (ed) (1986) Proceedings of the First Conference on Intelligent Autonomous Systems (IAS-1), Amsterdam: 477–483

Mysliwetz B. (1990): Parallelrechner–basierte Bildfolgen–Interpretation zur autonomen Fahrzeugsteuerung. Diss., UniBw Munich, LRT

Nevatia R., Binford T. (1977): Description and recognition of curved objects. Artificial Intelligence, 8: 77–98

Newell A., Simon H. (1963): GPS: a program that simulates human thought. In: Feigenbaum E., Feldman J. (eds): Computers and Thought. McGraw-Hill, New York

Nieuwenhuis S., Yeung N. (2005): Neural mechanisms of attention and control: losing our inhibitions? Nature Neuroscience, 8 (12): 1631–1633

Nilsson N.J. (1969): A Mobile Automaton: An Application of Artificial Intelligence. Proceedings International Joint Conference on Artificial Intelligence (IJCAI): 509–521

Nishimura M., Van der Spiegel J. (2003): Biologically Inspired Vision Sensor for the Detection of Higher–Level Image Features. Proc. IEEE Conf. on Electron Devices and Solid-State Circuits: 11–16

Nunes J.C., Guyot S., Delechelle E. (2005): Texture analysis based on local analysis of the Bidimensional Empirical Mode Decomposition. J. Machine Vision and Application, 16: 177–188

Paetzold F., Franke U. (2000): Road recognition in urban environment. Image and vision Computing 18(5): 377–387

Papoulis A. (1962):The Fourier Integral and Its Applications. McGraw-Hill, New York

Pele S., Rom H. (1990): Motion based segmentation. In: Proc. IEEE Int. Conf. Pattern Recognition, Atlantic City: 109–113

Pellkofer M., Dickmanns E.D. (2000): EMS–Vision: Gaze Control in Autonomous Vehicles. Proceedings International Symposium on Intelligent Vehicles'00, Dearborn, MI: 296–301

Pellkofer M., Lützeler M., Dickmanns E.D. (2001): Interaction of Perception and Gaze Control in Autonomous Vehicles. Proc. SPIE: Intelligent Robots and Computer Vision XX, Newton: 1–12

Pellkofer M., Dickmanns E.D. (2002): Behavior Decision in Autonomous Vehicles. Proceedings International Symposium on Intelligent Vehicles'02, Versailles

Pellkofer M. (2003): Verhaltensentscheidung für autonome Fahrzeuge mit Blickrichtungssteuerung. Diss., UniBw Munich, LRT

Pellkofer M., Lützeler M., Dickmanns E.D. (2003): Vertebrate-type perception and gaze control for road vehicles. In: Jarvis R.A., Zelinski A.: Robotics Research. The Tenth International Symposium, Springer–Verlag: 271–288

Pellkofer M., Hofmann U., Dickmanns E.D. (2003): Autonomous cross-country driving using active vision. SPIE Conf. 5267, Intelligent Robots and Computer Vision XXI: Algorithms, Techniques and Active Vision. Photonics East, Providence

PMDTech (2006): See: Kraft *et al.* (2004)

Pöppel E., Chen L., Glünder H., Mitzdorf U., Ruhnau E., Schill K., von Steinbüchel N. (1991): Temporal and spatial constraints for mental modelling. In: Bhatkar, Rege K (eds): Frontiers in knowledge-based computing, Narosa, New Dehli: 57–69

Pöppel E. (1994): Temporal Mechanisms in Perception. International Review of Neurobiology, 37: 185–202

Pöppel E., Schill K. (1995): Time perception: problems of representation and processing. In: Arbib M.A. (ed): The handbook of brain theory and neural networks, MIT Press, Cambridge: 987–990

Pöppel E. (1997): A hierarchical model of temporal perception. Trends in Cognitive Science, Vol.1 (2)

Pomerleau D.A. (1989) ALVINN: An Autonomous Land Vehicle in Neural Network. In: Touretzky D.S. (ed): Advances in Neural Information Processing Systems 1. Morgan Kaufmann,

Pomerleau D.A. (1992): Neural Network Perception for Mobile Robot Guidance. PhD-thesis, CMU, Pittsburgh [CMU-CS-92-115]

Potter J.E. (1964): W Matrix Augmentation. MIT Instrumentation Laboratory Memo SGA 5-64 Cambridge MA

Priese L, Lakmann R, Rehrmann V (1995): Ideogram Identification in a Realtime Traffic Sign Recognition System. Proc. Int. Symp. on Intelligent Vehicles, Detroit: 310–314

RAS-L-1 (1984): Richtlinien fuer die Anlage von Strassen (RAS). Forschungsgesellschaft fuer Strassen- und Verkehrswesen (ed.), Cologne, Germany, edition 1984. [Guide lines for the design of roads]

Rasmussen C. (2002): Combining Laser Range, Color, and Texture Cues for Automated Road Following. Proc. IEEE International Conference on Robotics and Automation, Washington DC

Regensburger U., Graefe V. (1990): Object Classification for Obstacle Avoidance. Proc. of the SPIE Symposium on Advances in Intelligent Systems, Boston: 112–119

Regensburger U. (1993): Zur Erkennung von Hindernissen in der Bahn eines Strassenfahrzeugs durch maschinelles Echtzeitsehen. Diss., UniBw Munich, LRT

Rieder A. (1996): Trinocular Divergent Stereo Vision. Proc. 13th International Conference on Pattern Recognition (ICPR) Vienna: 859–863

Rieder A. (2000): FAHRZEUGE SEHEN – Multisensorielle Fahrzeugerkennung in einem verteilten Rechnersystem fuer autonome Fahrzeuge. Diss. UniBw Munich, LRT

Ritter W. (1997): Automatische Verkehrszeichenerkennung. Koblenzer Schriften zur Informatik, Band 5, Verlag D. Fölbach, Diss., Univ. Koblenz/Landau

Roberts L.G. (1965): Homogeneous matrix representation and manipulation of n-dimensional constructs. MS-1405, Lincoln Laboratory, MIT

Roland A., Shiman P. (2002): Strategic Computing: DARPA and the Quest for Machine Intelligence, 1983–1993. MIT Press

Rosenfeld A., Kak A. (1976): Digital Picture Processing, Academic Press, New York

Ruhnau E. (1994a) The Now – A hidden window to dynamics. In Atmanspacher A, Dalenoort G.J. (eds): Inside versus outside. Endo- and Exo-Concepts of Observation and Knowledge in Physics, Philosophy and Cognitive Science, Springer, Berlin

Ruhnau E. (1994b): The Now – The missing link between matter and mind. In Bitbol M, Ruhnau E. (eds): The Now, Time and Quantum. Gif-sur-Yvette: Edition Frontière

Sack A.T., Kohler A., Linden D.E., Goebel R., Muckli L. (2006): The temporal characteristics of motion processing in hMT/V5+: Combining fMRI and neuronavigated TMS Neuroimage, 29: 1326–1335
Schick J., Dickmanns E.D. (1991): Simultaneous Estimation of 3-D Shape and Motion of Objects by Computer Vision. In Proc. IEEE Workshop on Visual Motion, Princeton, NJ, IEEE Computer Society Press: 256–261
Schick J. (1992): Gleichzeitige Erkennung von Form und Bewegung durch Rechnersehen. Diss., UniBw Munich, LRT
Schiehlen J. (1995): Kameraplattformen fuer aktiv sehende Fahrzeuge. Diss., UniBw Munich, LRT. Also as Fortschrittsberichte VDI Verlag, Reihe 8, Nr. 514
Schmid M., Thomanek F. (1993): Real-time detection and recognition of vehicles for an autonomous guidance and control system. Pattern Recognition and Image Analysis 3(3): 377–380
Schmid M.(1993): 3-D-Erkennung von Fahrzeugen in Echtzeit aus monokularen Bildfolgen. Diss. UniBw Munich, LRT. Also as Fortschrittsberichte VDI Verlag, Reihe 10, Nr. 293
Scudder M., Weems C.C. (1990): An Apply Compiler for the CAAPP. Tech. Rep. UM-CS-1990-060, University of Massachusetts, Amherst
Selfridge O., (1959): Pandemonium: A paradigm for learning. In: The Mechanization of Thought Processes. Her Majesty's Stationary Office, London
Selfridge O., Neisser U., (1960): Pattern Recognition by Machine. Scientific American, 203: 60–68
Shirai Y. (1987): Three Dimensional Computer Vision. Series Symbolic Computation, Springer, Berlin
Siedersberger K.-H. (2000): EMS-Vision: Enhanced Abilities for Locomotion. Proceedings International Symposium on Intelligent Vehicles'00), Dearborn, MI: 146–151
Siedersberger K.H., Pellkofer M., Lützeler M., Dickmanns E.D., Rieder A., Mandelbaum R., Bogoni I., (2001): Combining EMS-Vision and Horopter Stereo for Obstacle Avoidance of Autonomous Vehicles. Proc. ICVS, Vancouver
Siedersberger K.H. (2004): Komponenten zur automatischen Fahrzeugführung in sehenden (semi-) autonomen Fahrzeugen. Diss., UniBw Munich, LRT
Solder U., Graefe V. (1990): Object Detection in Real Time. Proc. of the SPIE, Symp. on Advances in Intelligent Systems, Boston: 104–111
Spillmann W. (1990) Visual Perception. The Neurophysiological Foundations. Academic Press, New York
Spivak M. (1970): A Comprehensive Introduction to Differential Geometry. (Volumes I – V). Publish or Perish, Berkeley, CA
Steels L. (1993): The Biology and Technology of Intelligent Autonomous Agents. NATO-Advanced Study Institute. Ivano, Italy
Talati A., Hirsch J. (2005): Functional specialization within the medial frontal gyrus for perceptual "go/no-go" decisions based on "what", "when", and "where" related information: an fMRI study. Journal of Cognitive Neuroscience, 17(7): 981–993
Talati A., Valero-Cuevas F.J., Hirsch J. (2005): Visual and Tactile Guidance of Dexterous Manipulation: an fMRI Study. Perceptual and Motor Skills, 101: 317–334
Thomanek F., Dickmanns D. (1992): Obstacle Detection, Tracking and State Estimation for Autonomous Road Vehicle Guidance. In: Proc. of the 1992 International Conference on Intelligent Robots and Systems, Raleigh NC, IEEE, SAE: 1399–1407
Thomanek F., Dickmanns E.D., Dickmanns D. (1994): Multiple Object Recognition and Scene Interpretation for Autonomous Road Vehicle Guidance. In: Masaki I. (ed): Proc. of International Symposium on Intelligent Vehicles '94, Paris: 231–236

Thomanek F. (1996): Visuelle Erkennung und Zustandsschätzung von mehreren Straßenfahrzeugen zur autonomen Fahrzeugführung. Diss., UniBw Munich, LRT. Also as Fortschrittsberichte VDI Verlag, Reihe 12, Nr. 272

Thornton C.L., Bierman G.J. (1977): Gram-Schmidt Algorithms for Covariance Propagation. International Journal of Control 25(2): 243–260

Thornton C.L., Bierman G.J. (1980): UDU^T Covariance Factorization for Kalman Filtering. In: Control and Dynamic Systems, Advances in Theory and Application, Vol. 16, Academic Press, New York: 178–248

Thorpe C., Hebert M., Kanade T., Shafer S. (1987): Vision and navigation for the CMU Navlab. Annual Review of Computer Science, Vol. 2

Thorpe C., Kanade T. (1986): Vision and Navigation for the CMU Navlab. In: SPIE Conf. 727 on 'Mobile Robots', Cambridge, MA

Thrun S., Burgard W., Fox D. (2005): Probabilistic Robotics. MIT Press, Cambridge, MA

Tomasi C., Kanade T. (1991): Detection and Tracking of Point Features. CMU, Tech. Rep. CMU-CS-91-132, Pittsburgh, PA

Tsinas L. (1996): Zur Auswertung von Farbinformationbeim maschinellen Erkennen von Verkehrssituationen in Echtzeit. Diss., UniBw Munich, LRT

Tsugawa S., Yatabe T., Hirose T., Matsumoto S. (1979): An Automobile with Artificial Intelligence. Proc. 6th IJCAI, Tokyo: 893-895

Tsugawa S., Sadayuki S. (1994): Vision-based vehicles in Japan: Machine vision systems and driving control systems. IEEE Trans. Industrial Electronics 41(4): 398–405

Turk M.A., Morgenthaler D.G., Grembran K.D., Marra M. (1987): Video road-following for the autonomous land vehicle. Proc. IEEE Int. Conf. Robotics and Automation, Raleigh, NC: 273–280

Ulmer B. (1994): VITA II - Active collision avoidance in real traffic. Proceedings International Symposium on Intelligent Vehicles'94, Paris

von Holt V. (1994): Tracking and classification of overtaking vehicles on Autobahnen. Proceedings International Symposium on Intelligent Vehicles'94, Paris

von Holt V. (2004): Integrale Multisensorielle Fahrumgebungserfassung nach dem 4-D Ansatz. Diss. UniBw Munich, LRT

Wallace R., Stentz A., Thorpe C., Moravec H., Wittaker W., Kanade T. (1985) First Results in Robot Road-Following. Proc. 9th IJCAI: 65–67

Wallace R., Matsusaki K., Goto S., Crisman J., Webb J., Kanade T. (1986): Progress in Robot Road-Following. Proceedings International Conference on Robotics and Automation, San Francisco CA: 1615–1621

Werner S. (1997): Maschinelle Wahrnehmung fuer den bordautonomen automatischen Hubschrauberflug. Diss. UniBw Munich, LRT

Wertheimer M. (1921): Untersuchungen zur Lehre der Gestalt I. Psychol. Forschung, Bd 1

Wiener N. (1948) Cybernetics. Wiley, New York

Winograd T., Flores C.F. (1990): Understanding Computers and Cognition. A New Foundation for Design. Addison-Wesley

Wünsche H.J. (1983): Verbesserte Regelung eines dynamischen Systems durch Auswertung redundanter Sichtinformation unter Berücksichtigung der Einflüsse verschiedener Zustandsschätzer und Abtastzeiten. Report HSBw/LRT/WE 13a/IB/83-2

Wünsche H.J. (1986): Detection and Control of Mobile Robot Motion by Real-Time Computer Vision. In: Marquino N. (ed): Advances in Intelligent Robotics Systems. Proc. SPIE, 727: 100–109

Wünsche H.J. (1987): Bewegungssteuerung durch Rechnersehen. Diss. UniBw Munich, LRT. Also as Fachberichte Messen, Steuern, Regeln Bd. 10, Springer-Verlag, Berlin, 1988

Zapp A. (1988): Automatische Straßenfahrzeugführung durch Rechnersehen, Diss., UniBw Munich, LRT

Zheng Y.J., Ritter W., Janssen R. (1994): An adaptive system for traffic sign recognition. Proc. Int. Symp. on Intelligent Vehicles, Paris

Zielke T., Brauckmann M., von Seelen W. (1993): Intensity and Edge-based Symmetry Detection with an Application to Car Following. CGVIP: Image Understanding 58: 177–190

Index

Printing: Krips bv, Meppel
Binding: Stürtz, Würzburg